Advanced Sciences and Technologies for Security Applications

Indexed by SCOPUS

The series Advanced Sciences and Technologies for Security Applications comprises interdisciplinary research covering the theory, foundations and domain-specific topics pertaining to security. Publications within the series are peer-reviewed monographs and edited works in the areas of:

- biological and chemical threat recognition and detection (e.g., biosensors, aerosols, forensics)
- crisis and disaster management
- terrorism
- cyber security and secure information systems (e.g., encryption, optical and photonic systems)
- traditional and non-traditional security
- energy, food and resource security
- economic security and securitization (including associated infrastructures)
- transnational crime
- human security and health security
- social, political and psychological aspects of security
- recognition and identification (e.g., optical imaging, biometrics, authentication and verification)
- smart surveillance systems
- applications of theoretical frameworks and methodologies (e.g., grounded theory, complexity, network sciences, modelling and simulation)

Together, the high-quality contributions to this series provide a cross-disciplinary overview of forefront research endeavours aiming to make the world a safer place.

The editors encourage prospective authors to correspond with them in advance of submitting a manuscript. Submission of manuscripts should be made to the Editor-in-Chief or one of the Editors.

More information about this series at http://www.springer.com/series/5540

Hamid Jahankhani • Stefan Kendzierskyj
Nishan Chelvachandran • Jaime Ibarra
Editors

Cyber Defence in the Age of AI, Smart Societies and Augmented Humanity

Springer

Editors
Hamid Jahankhani
Northumbria University
London, UK

Stefan Kendzierskyj
Cyfortis, Worcester Park
Surrey, UK

Nishan Chelvachandran
Open Innovation House
Saidot OY
ESPOO, Finland

Jaime Ibarra
Northumbria University
London, UK

ISSN 1613-5113 ISSN 2363-9466 (electronic)
Advanced Sciences and Technologies for Security Applications
ISBN 978-3-030-35745-0 ISBN 978-3-030-35746-7 (eBook)
https://doi.org/10.1007/978-3-030-35746-7

This Springer imprint is published by the registered company Springer Nature Switzerland AG.
The registered company address is: Gewerbestrasse 11, 6330 Cham, Switzerland

Foreword

This timely volume should be required reading for lawyers, academics and other professionals who are attempting to keep pace with the breathless pace of technological change. It should also have strong appeal to those who are concerned about the regulatory and ethical challenges presented by change.

Our private information has never been more threatened by troubling exposure to strangers, often with malign and criminal intent. The armed robbers of yesterday have been replaced by the information criminals of today. They can pick the locks of the private online world as fast as new security is invented. Even private homes and offices can no longer be regarded as safe spaces. Part of the threat comes from state actors, suggested to include Russia, Iran and North Korea but not excluding some Western countries with strong privacy laws. Nevertheless, there are strong arguments made to increase the online power of the State, to protect the private interests of citizens and private business, to guard national security and even to insure the continuity of normal life against asymmetric threats to public utilities.

Concisely but thoroughly, this book, emerging as it does from an outstandingly forward-looking university, looks at the full range of challenges. For example, it features a penetrating chapter in which there is analysis of the threats from AI and deepfake technology to political stability and psychological security: a few years ago, this would have been likened to sci-fi fantasy – today, it is a real-world concern.

Throughout the book, there is discussion of the relationship between threat and response. The UK Parliament, in dealing with counterterrorism legislation, has been troubled deeply by the proportionality between the private and the public. Acknowledging as realistically it does that citizens voluntarily and daily give much of their privacy away to social networking giants, or to online vendors, nevertheless Parliament is astute to its duty to protect citizens from unnecessarily intrusive investigation. Reality, holographs and mixed reality have to be placed in the appropriate regulatory and ethical place in a sector that evolves exponentially faster than the run of our daily lives. Mathematicians have learned how to do things, for example, by algorithmic techniques, without necessarily making a judgement of the effect of their brilliance.

Against the threats briefly discussed above, AI brings huge benefits, which we would wish to harness for the benefit of our human race. The understanding of climate change is an example of the benefits we should rightly harness in the interests of a better future. Once unimaginable medical techniques that enable those who could not to walk, see, hear, speak, translate are all now possible, some already happening.

This volume describes many of those benefits and importantly the moral and ethical implications that arise. Of course, one book cannot be the last word in any debate. However, Professor Hamid Jahankhani has made a major contribution to the discussion, one which he continues through his involvement in the Annual International Conference on Global Security, Safety and Sustainability.

Lord Carlile of Berriew CBE QC LLD, London, UK Alex Carlile
September 2019

Contents

Part I
Cyber Defence & Critical National Infrastructure (CNI)

Critical National Infrastructure, C4ISR and Cyber Weapons in the Digital Age

Stefan Kendzierskyj and Hamid Jahankhani

Abstract Cyber-attacks have become more sophisticated over recent years with the different configuration types and various industry sectors have suffered from a range of these attack vectors resulting in some devastating outcomes. These have manifested in the shape of ransomware, malware, manipulation methods, phishing and spear-phishing. Bearing the brunt of many of these attacks has fallen in the area of critical national infrastructure (CNI) and for a variety of reasons from the sensitive data that can be accumulated through to knowing an impact in an area of CNI will have potential devasting effects or leave instability/uncertainty to become a risk. Whilst data breaches are serious incidents, in most organisations, there is a growing concern regarding attacks that are designed to have a more destructive effect, such as the Ukraine cyber-attack in 2015 that resulted in a shutdown of the power grid. Or perhaps the sophisticated attack a year later in Kiev that although causing a brief power blackout, had the manifestations of more concerns in how that attack was built and delivered. Or the WannaCry ransomware attack in 2017 that caused widespread chaos with healthcare institutions unable to carry out any tasks since access to data/systems was unavailable. These critical national infrastructure (CNI) attacks into sectors such as healthcare, energy, etc., cause data breaches/disruptions and are also able to leverage vulnerabilities in the industrial processes, especially where legacy infrastructure contains ICS and SCADA systems. Perhaps the state sponsored cyber-attacks cause the most concern as they tend to be at the more sophisticated level of the spectrum and maximize on amount of potential harm that is delivered. Hence why Command, Communications, Computers, Intelligence, Surveillance and Reconnaissance systems (C4ISR) should have higher degrees of interoperability and be integrated and responsive in the

S. Kendzierskyj (✉)
Cyfortis, Worcester Park, Surrey, UK
e-mail: Stefan@cyfortis.co.uk

H. Jahankhani
Northumbria University, London, UK
e-mail: Hamid.jahankhani@northumbria.ac.uk

H. Jahankhani et al. (eds.), *Cyber Defence in the Age of AI, Smart Societies and Augmented Humanity*, Advanced Sciences and Technologies for Security Applications, https://doi.org/10.1007/978-3-030-35746-7_1

current and accelerating climate of digital warfare. It is clear that cyber command has become as an important priority as the initiatives considered over land, sea and air and considerable investment is going into ensuring its development and advancement.

Keywords CNI · Defence · C3I · C4ISR · Cyberattacks · ICS · SCADA

1 Introduction

The Government and military command structure assess world events in order to threat model and analyse the potential imminent and future threats that pose to the country's critical national infrastructure (CNI). This naturally is an essential task in order to understand the threat landscape and what type of mitigations or proactive responses need to be deployed; which can be anything from informative messaging to a preventative cyber strike against the party posing the threat. This latter measure has become an effective tool and considered as strong an impact as a conventional missile strike. CNI is a much desired attack vector for state sponsored or hostile parties, cyber criminals and terrorists that can have a gamut of reasons to cause disruption, financial chaos, weakening of a state, espionage for economic, military, political and commercial benefit. The traditional threat is difficult enough with managing aspirational domestic and international terrorist attacks, which are fortunately limited in volume. But given the onset of the digital age and fast moving technology landscape of the Internet of Things (IoT), with many devices connected to the internet, has meant the cyber threat has increased exponentially with so many possible points of entry. Cyber criminals, whilst trying to further their prosperity, can also cause damage to the national structure, credibility, intellectual property and financial loss on a large scale. Or worse impacts with intentions that state sponsored actors usually have.

1.1 Critical National Infrastructure

Critical National Infrastructure (CNI) can be the physical or virtual systems, assets, facilities, processes and networks that are of extreme vital importance that is necessary for a country to operate. In the UK, the Centre of Protection of National Infrastructure (CPNI) lead on physical and personal security and work in partnership with the National Cyber Security Centre (NCSC) who ensure the resilience of the UK's critical national infrastructure NCSC [31]. A report by NCSC [30] outlines the national cyber security strategy from 2016 to 2021; its vision is by 2021 that the 'UK is secure and resilient to cyber threats, prosperous and confident in the digital world', NCSC [30].

In the UK there are 13 national infrastructures sectors (whilst the US has 16 sectors) as follows:

- Chemicals
- Civil Nuclear
- Communications
- Defence
- Emergency Services
- Energy
- Finance
- Food
- Government
- Health
- Space
- Transport
- Water

According to the Public Summary on Sector Security report by the Cabinet Office, [4] the UK government has a core objective in reducing CNI's vulnerability to threats, improving resilience by the ability to recover from disruption. This is built on four themes that provide infrastructure security and resilience:

Resistance: Protection of infrastructure from physical damage and also includes reducing vulnerability through physical, personnel and cyber security measures.
Reliability: Infrastructure protection under adverse conditions and so mitigate against a damaging event.
Redundancy: Ensuring backup installations, systems, processes and spare capacity.
Response & Recovery: The ability for effective, rapid response and recovery from disruptive events (Table 1).

1.2 Command, Control, Communications, Intelligence (C3I and C4ISR)

A key component to managing battlespace through operational dominance and information superiority is ensuring Command, Control, Communications and Intelligence (C3I) is interoperable, integrated and latest technologies/strategies are deployed. This interaction and dependency on each other of C3 and Intelligence, Surveillance and Reconnaissance through computers has given rise to the term C4ISR (with Computers as the common denominator). Interoperability across military organisations is becoming a key theme and to solve this requires deeper networking and integration of C4ISR. Importance is given to turning data into knowledge and understanding when to turn that knowledge into action.

Table 1 CNI roles and responsibilities: Cabinet Office, [4]

Roles and responsibilities	
Infrastructure owners & operators	Day-to-day responsibilities for operations. Risk assessments taken on assets and decision making on maintenance, training and methods to progress any organisational improvements and assets for its security and resilience.
Regulators	Lead Government Departments need support from regulators to ensure legislation/regulation are followed. Operations can be intervened to ensure security, standards or accountability is there and if necessary request they are met before continuing business as usual.
Local authorities & emergency services	The Civil Contingencies Act 2004 states local authorities and emergency services carry our risk assessments to help identity likelihood and impact of emergencies.
Government agencies	A number of agencies provides infrastructure advice on risks and mitigations such as the Centre for Protection of National Infrastructure for advice to businesses. Or National Cyber Security Centre (NCSC) whose mission is to reduce the cyber security threat to UK.
Lead government departments	Responsible for sector-level security and resilience policies.

Challenges are known and a report by Booz Allen Hamilton [2], explains a mix of:

- **Operational;** where systems are standalone or proprietary/stovepiped
- **Technology;** persistent and damaging cyber-attack threats or lack of cybersecurity
- **Process;** acquisition issues with complex and complicated mechanisms
- **Budgetary;** funding and budget constraints.

Within the Booz Allen Hamilton [2], report, a survey undertaken sees 52% believe interoperability as an issue that cannot be resolved without full integration and networking of C4ISR.

Considering state sponsored cyber-attacks and risk imposed by countries that are significantly more advanced, poses the question to ensure cyber capabilities within military command structure are given the correct amount of priority. As recent example shows, Iran recognising and creating a new military command and control unit to withstand a US cyber-attack, Doffman [8], particularly as the US had launched an offensive cyber strike on Iran to disable computer systems used to control rocket and missile launch capability in the previous month of this announcement. Cyber strikes of this nature mean the delivery mechanism is not always needed to be a conventional missile strike. Penetration of the Iranian Command and Control centre sends a message to the rest of the world regarding cyber strengths and willingness to deploy them as a defensive attack.

2 Cyber Operations as a Domain of Warfare

Cyber warfare has quite recently been recognised as a domain of warfare, though the formal classification and designation varies across different countries and international bodies. The US Department of Defence's Joint Publication (JP) 3–12 Cyberspace Operations omitted the term cyber war entirely in favour of 'cyberspace operations' [17]. This new term is generally defined as "the employment of cyberspace capabilities where the primary purpose is to achieve objectives in or through cyberspace". This definition is intentionally ambiguous and covers operations where the sole conflict occurs in cyberspace and those undertaken to support traditional military operations [17]. NATO in the 2016 Warsaw summit recognised cyberspace as a domain of operations in which NATO must defend itself as effectively as it does in air, on land and at sea [29]. US military doctrine includes space as another domain of warfare though this inclusion seems to vary depending upon the author.

More recently over the last 10 years the domain of cyber operations has encompassed attacks on CNI on an increasing scale across many countries. The objectives vary as to the reasons and sometimes as a form of defence. Take Stuxnet, a highly sophisticated computer worm and where its mission was to infect CNI in a way that manifests itself physically. Mostly accepted now (but not completely validated) as a US/Israeli led attack on Iran's nuclear program for physical disruption; Fruhlinger [11]. The Iranians, in retaliation, mobilized their cyber resources and targeted 46 US financial organisations (e.g. New York Stock Exchange, JP Morgan, etc.) and sustained attacks over 5 months, Halpern [16].

The Russians view cyber operations very differently than its Western counterparts. Russian military theorists generally do not use the terms cyber or cyber warfare. Instead, they conceptualize cyber operations within the broader framework of information warfare, a holistic concept that includes computer network operations, electronic warfare, psychological operations, and information operations [7]. Russians, like the Chinese, tend to use the word information, conceptualizing cyber operations within the broader rubric of information warfare. In other words, cyber is regarded as a mechanism for enabling the state to dominate the information landscape, which is regarded as a warfare domain in its own right [7]. But the results directed are still the same whatever terminology is used within countries. Given that there are many vulnerabilities to exploit and the know-how and technical sophistication is there to deliver these attacks then suggestions on how to improve CNI security are all the more relevant. Russia has multiple hacker teams motivated by their group instructions. Groups known as Energetic Bear, Dragonfly, Koala, Iron Liberty, uses a number of attacks such as watering-hole where Trojans are planted and websites infected to phishing emails targeting vendors of industrial control systems (ICS). Sandworm is another Russian hacker group that deployed the Ukraine cyber utility attacks in 2015 and another attack in 2016 but is not just specific to electricity grid attacks and has been waging a number of attacks through all Ukrainian sectors such as deleting/encrypting terabytes of data held by government agencies, Greenberg [13–15]. ESET has advised that the NotPetya ransomware attack that damaged thousands of networks in Ukraine and globally has a close resemblance to Sandworm's trademarks of fake ransomware that give no true options to decrypt affected files. Another Russian cybercriminal group, Silence APT, targets financial institutions in more than 30 countries, Khandelwal [22]. The group has evolved into a sophisticated, advance persistent threat (APT) group threatening banks worldwide. Again, as with others, spear-phishing is the most common entry point and once into the organisation would deploy its techniques, tactics and procedures to further install its malware and other.

North Korea has used cyber-attacks as an innovative method to fund their weapons programme according to a United Nations report, Nichols [32]. This has been estimated at $2 billion as a fund to help programs for its weapons of mass destruction, mainly achieved through cyber-attacks on cryptocurrency theft and cyberspace has been used to launder the stolen money with direction and leadership given by the Reconnaissance General Bureau (North Korean military agency). With sanctions bans imposed by the Security Council, its purpose to stifle funding for the weapons programs, has likely helped fuel North Korea's objective to carry out these type of purpose driven cyber-attacks, which is in some ways a form of cyberwarfare.

APT41, a prolific Chinese cyber group, carries out state-sponsored activity in espionage and combined with financial motivations, Fraser et al. [10]. APT41 uses non-public malware usually kept aside for espionage activities and particularly targeting healthcare, telecommunications, and technology organisations. For its financially motivated operations APT41 is able to transfer across different operating environments of Windows and Linux to access in video gaming production environments. From here source code is taken and digital certificates so that malware can be endorsed. Malicious codes are inserted into legitimate files and naturally distributed (leveraging over 46 different malware families and tools). Typically, as in other industries, spear-phishing emails are the initial entry method before deploying more tactics, techniques and procedures.

2.1 CNI Threat Factors

Cyber-attacks on critical infrastructure are increasing in velocity and have advanced threat actors CERT-UK [5]. The threat to CNI is very real as there has been a number of successful attacks over recent years. The computer systems underpinning CNI that, for example, hold sensitive data in healthcare facilities or Industrial Control Systems (ICS) that operate, control and monitor physical structures such as nuclear power plants, electricity grids, railway networks, may have a number of vulnerabilities that cyber-attackers are drawn to; Logsdon [23]. These networks, in order to control and monitor efficiently, are connected to the Internet and so the connection to cyber physical attacks are established here. Not only is there possible major disruption, loss/compromise of services and potential threat to life but also has an impact on national security and why the overarching role of CPNI and NCSC plays an important part. The Cambridge Centre for Risk Studies estimate that if the UK power supply were to suffer a major cyber-attack then losses to UK GDP over 5 years could be as much as £442 billion, Morbin [24].

Interestingly research by the Pew Research Center, Poushter and Fetterolf [35], undertook a survey that shows multiple countries believe government data, critical national infrastructure and national elections will be targeted for future attacks (27,612 respondents to the survey). One research statistic regarding the likelihood of a cyber-attack on national security information or typically the type held in CNI, shows as likely with a median of 74% across 26 countries. The US surveyed, equated to eight-in-ten believing cyber-attacks to national infrastructure will be damaged (83%), national security information will be accessed (82%) or elections will be tampered (78%).

Globally all are aware of the huge digital transformation undertaken over the last decades and entering the Industry 4.0 age. Whilst an amazing technology convergence has happened and clearly has benefits both to organisations and users it has also taken an interconnected approach where new technologies merge with legacy systems. The more interconnectivity and the more potential vulnerabilities.

Typically, within CNI there is this mix and convergence of old and new technologies just due to the sheer size and growth over the years. The structure and set up of CNI needs deeper analysis as to why it is under attack and level of ease of access to attackers. As mentioned a lot of the infrastructure contains Industrial Control Systems (ICS) and Supervisory Control and Data Acquisition (SCADA) and in many cases are legacy and older technology. Previously SCADA was a system that was not particularly well known, and its mode of preservation was security through obscurity. But over the recent years knowledge of SCADA has become widespread and used in the CNI sectors, such as railway networks, traffic, airline, etc., Logsdon [23], and attacks more openly shared, and vulnerabilities and off-the-shelf exploits more commonly shared.

Barwick [1], comments that the rise reported in Dell 2015 Security Annual Threat Report, shows the global attacks rise from 163,228 in January 2013 to 675,186 in January 2014 and data driven from analysis by Kaspersky Lab shows the number of vulnerable products in ICS as per Table 2 and the energy sector having the most vulnerable number. That would lead to what was mentioned earlier, that ICS and SCADA type networks have many vulnerabilities to exploit and how this change in cyber operations has shifted focus to CNI. It's clear the reasons of attacks escalating and CNI being targeted, is due to their potential vulnerabilities and the role of importance in controlling industrial processes in the physical world.

Globally, Fig. 1 shows the top affected countries (UK is reportedly at a low level of 14.5%).

2.2 *Types of Attack*

Most of CNI is privately owned, and so the risk and how to mitigate is in the hands of the operators but is crucial they still partner and work with CPNI, NCSC and

Table 2 Number of vulnerable products used in CNI sectors. Kaspersky Lab ICS CERT [21]

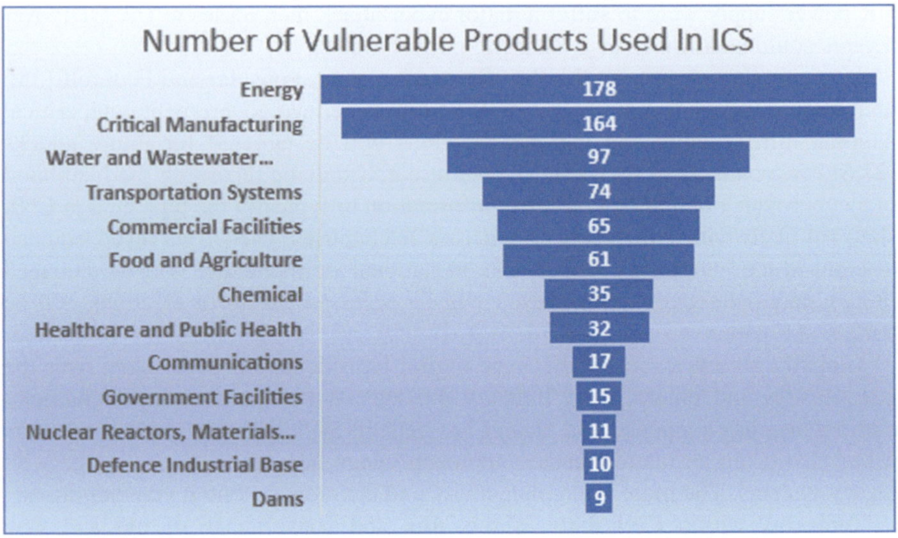

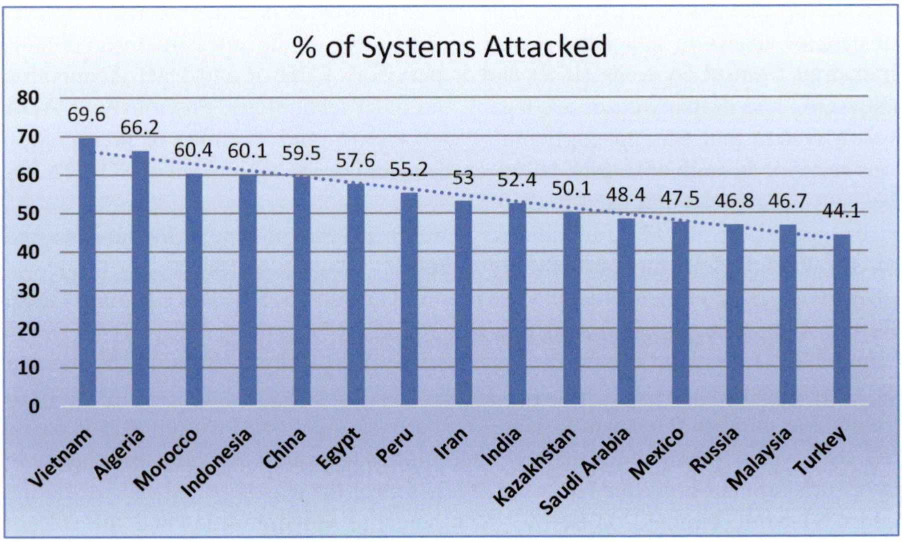

Fig. 1 Top 15 Countries by Percentage of ICS Computers. Kaspersky Lab ICS CERT [21]

regulators such as the Independent Commissioners Office (ICO). There may be a range of reasons that cyber-attackers are attracted to CNI from general criminal behaviour, espionage, physical disruption, etc. Depending on the cyber-attacker's background they may vary from individuals or activists with limited knowledge or resources (or hire malware-as-a-service) to state sponsored attackers that will have access to abundance of resources and sophisticated toolsets. There are commonly known types of methods used for cyber-attacks such as ransomware, Distributed Denial of Service (DDoS), phishing, spear-phishing and web-based attacks. Many attacks start from the users clicking on phishing or spear-fishing attacks and from these attackers can gain (and this may be over a period of time) permissions and passwords to complete a sophisticated attack at a later date. Most of cloud computing acts as a Trusted Third Party (TTP) and this has the disadvantage of having a single point of failure. As in CNI there will be mission critical data being held and poses a threat to the organisation and individual when the data is breached. In the case of CNI, breached data could end up very quickly being sold on the deep web and at a high value and transfer.

Generally, there has been a significant increase from ransomware and data breach incidents. Where this stands out the most in CNI is in the healthcare sector where more than half of all cyber-attacks took place here rising from 34% in 2016 to 58% in 2017 Jay [20]. Some ransomware attacks were widespread, causing large-scale damage such as the malware attack that started on 12 May 2017 across the National Health Service (NHS), infecting 230,000 computers and transmitting across 150 countries Strategic Comments [36]. Interestingly, a report commissioned by the National Audit Office shows an assessment of 88 out of 236 National Health trusts found that none passed cyber-security standards, and this was the status prior to the attack NAO [27]. Many healthcare institutions are running on systems that require investment and have vulnerabilities but equally have systems that can't be disrupted as this can have detrimental impact on life (e.g. surgical operations cancelled).

2.3 CNI Attacks Through Vulnerable Systems: ICS/SCADA

Cyber-attacks against industrial targets have doubled over the last 6 months according to research undertaken, Osborne[33] with over 50% occurred in the manufacturing sector. Whilst the objectives of cyber-attacks does exist to have covert surveillance and data theft as top priorities, there is also a growing trend of destructive behaviour in the types of malicious code that can lock business systems, delete critical data or cause fear and uncertainty when the objective is not to extort money or data but provide malicious impacts.

Data and processes secured through ICS and SCADA systems in CNI infrastructure have their valuable assets/data susceptible to cyber-attack. The shock in 2010 of how Stuxnet first targeted and destroyed uranium enrichment centrifuges in Iran, Fruhlinger [11], has questioned the focus on where and when this malware may happen again in another shape or form; see Fig. 2 for timeline examples from Stuxnet onwards of destructive malwares.

According to Greenberg, [13–15] that rare piece of malware has reappeared in the Middle East. Triton (also known as Trisis) is built to damage/tamper ICS, targeting equipment in the oil and gas sector, nuclear and manufacturing plants (Schneider Electric Trionex products). The vulnerability is simple, since the safety industrial

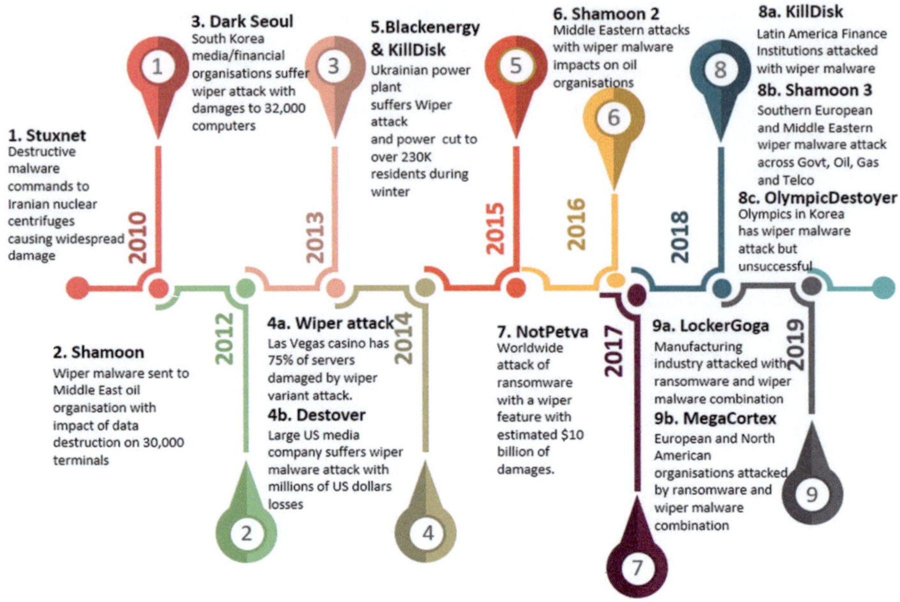

Fig. 2 Destructive malware attacks; examples over timeline

systems are designed to be more autonomous from other equipment as so to monitor any arising and dangerous changes to conditions. Alerts can then be generated and ensure correct shutdowns are initiated to provide the preventative measures against accident, human errors or worse as sabotage. Attackers usually obtain the foothold by conventional methods such as phishing and once in the core distributed control systems, attackers could then launch the Triton malware to whatever desired impact. The danger in Triton's code was its capability to disable Triconex safety measures and inject new commands into its operations. Even if the commands are not accepted by the Triconex equipment it can effectively crash the system. These would be the failsafes that are there to shut down equipment, so disabling the ability to stop the impact from its course set. For critical national infrastructure this is a particularly worrying situation as the attacker has a much more heightened and effective attack with a range of impacts to be selected. All can appear to be normal and working as the malware masks the true picture and depends on what process the ICS is monitoring as to kill people, machinery to be destroyed, etc.

The question of who was behind this attack and malware isn't easy to identify, as the object is to covertly disrupt. Iran has been mentioned, Greenberg [13–15], due to its history of Middle East attacks. For example, an Iranian malware known as Shamoon destroyed thousands of computers at Saudi Aramco in 2012 and thought to be a retaliatory move for Stuxnet revenge against the West. After 2 years Shamoon has reappeared again in a new wave of attacks in the Middle East with up to 400 Saipem servers affected and a few other organisations also hit the same week.

This latest variant is twice as destructive as it contains a wiper to delete files from the infected computers before the Shamoon malware wipes the master boot record Symanntec [37].

The damages to an enterprise organisation, if a destructive cyber-attack is successful is estimated as an average of over 12,000 workstations would be damaged plus 512 h to rectify or start some recovery process. Also, according to the Ponemon Institute the average cost of a data breach being estimated at USD 3.92 million, Osborne [33].

3 CNI Case Studies in the Energy and Healthcare Sector

According to Muncaster [25, 26], around 90% of CNI providers advised that their operational technology equipment had been damaged by cyber-attacks over the last 2 year period.

As explained in previous sections that with a lot of vulnerable equipment across CNI infrastructure makes a large and varied target to deploy against and with a range of impacts as to what end outcome is delivered. The following studies are some CNI examples of type of cyber-attacks with impacts that can have huge and worrying factors.

3.1 Ukraine Power Grid – CNI Attacks Through ICS/SCADA Systems

Ukraine suffered two power grid attacks across as many years and although more coverage was given to the first attack it was the second attack that demonstrates more concerns over malware used, its intentions and ability to be redeployed with swappable or plug-in components that pose a wider geographical risk than just regionally limited to Ukraine. The latter attack was also automated instead of manually gaining access to the Ukrainian utilities networks and manually switch off sub-stations as was achieved in the first attack.

During the first attack a high-profile cyber-attack on Ukraine's power grid in 2015 was initiated through phishing attacks, where the recipient clicked through a link/attachment and allowed attackers access into the network (see Fig. 3 below). The attack was manifested 6 months before the outage and meant that the attack analysis of the infrastructure, security credentials were compromised and set in place for activation when required. Here, the attackers demonstrated a variety of capabilities, spear phishing, variants of the BlackEnergy 3 and KillDisk malware and control of MS Word documents, that gave an anchor into the main IT operations of the electricity organisations. This anchoring provided the stronghold position to harvest credentials and access the ICS network. The attack crippled 27 power

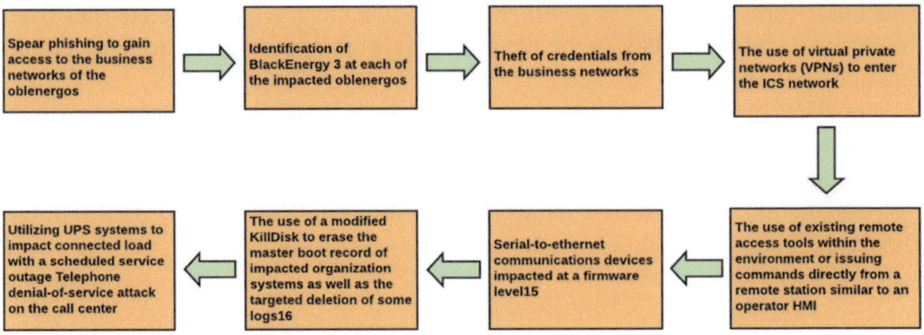

Fig. 3 Diagram Flow of Ukraine 2015 cyber-attack

distribution operation centres across the country. To heighten the attack effect, the call centre that supported the energy network was also inundated with telephone call traffic to create interference. The attack affected 225,000 customers and although power was manually restored some hours later, the business was running reduced operations for some months Parliamentary Office of Science and Technology [34]. This capability to perform long-term reconnaissance undetected, and to successfully execute a complex and combined attack across many sites does provoke to believe origins are of a state sponsored attack.

Further Ukraine attacks in 2016 created a loss of energy consumption in Kiev for one night and were said to be more sophisticated and complex than 2015 and were part of a series of 6500 cyber-attacks over the period of a few months and blame attributed to Russian security Higgins [18]. The outage was not a long incident, approximately an hour, which initially seems not a major incident. But of more concern is the complexity and indications that Ukraine was used in a dry run type reason. It appears that Ukraine is being used as a testing ground for Russian hackers to test out exploits, vulnerabilities and observe devastation effects and objectives were to destabilize the economy and political situation. Spear-phishing (targeting of specific victims using social engineering techniques) was used in an initial attack to gather passwords for specific workstations and custom malware script was written (not the work of amateurs and why it suggests sophisticated state sponsored techniques). It's been attributed to being one of the most evolved malwares developed for power grid attacks and named as Industroyer or Crash Override (as alternatively known) and researchers have stated as only the second ever purpose built malicious code destined to disrupt physical systems (Stuxnet being the first), Greenberg [13–15].

The example of the sophisticated method and malware shows the concerns over ability to automate mass power outages with adaptable swappable or plug-in components that allow modification to deploy to other electricity utility companies. Having that swappable component as part of the design means a redeployment to Europe/US or elsewhere can be easily achieved by adapting the protocols. Interestingly as well as being creative in terms of its adaptability it could also be destructive by destroying the infected files in order for requirement to cover any tracks or signature of origin.

The automated method of being programmed to communicate directly with the power grid means there are fewer humans managing the attack; that in itself meant less preparation, and Crash Override could achieve blackouts at a much a faster pace. Also, the language used in terms of obscure protocols and commands the controls use to handle power is also another interesting point of analysis. It could be programmed to detonate at a pre-set time. Greenberg [13–15], also mentions analysis by ESET uncovers a further disturbing feature that could cause physical damage to power equipment. The exploit is found in Siemens equipment part (Siprotec digital relay) where the device opens circuit breakers if it detects power levels that are dangerously high. The malware could disable this feature or take a more destructive form if used in combination with overloading the charge on grid components and prevent the kill-switch from being operational and cause expected overheating, damage or as ESET debates that a well-crafted attack on multiple points in a power grid and cause what is known as a cascading outage. This is where the power overload cascades as an outage from one grid to another.

The type of Ukraine attack is a direct attack on CNI and was not just a local concern to that region but a potential wake-up call to power grids across the globe. It's not fully proven who was behind the attacks but mostly thought to be a group of Russian government hackers, waging attacks on Ukraine since 2014, called Sandworm; Greenberg [13–15]. State sponsored cyber-attacks can have devasting effects as the sophistication level is more refined and waves of attacks, as in the Ukraine example, can be harboured on defence, finance and port authority systems before this second Ukrainian outage occurred and shows the delicate nature of CNI. As well as devasting attacks that shut down or majorly disrupt day to day basic operations, there is also the risk of cyber-attacks gaining access to core mission critical data. At times attacks of this sophistication are used to mask these activities of gaining access to core data.

3.2 National Health Service Infrastructure Study

Patient and healthcare records represent the most valuable data to cyber attackers both in their saleable value and type of data they represent. Medical records are among the most complete set of records available and so are in demand for a variety of reasons. It is said to be ten times the value of credit card details sold on the deep web as for obvious reasons you can cancel a credit card, but data points of identity are not amendable. According to a breach incidents report by Gemalto [12], published in March 2017 covering the first half of 2017, healthcare was the hardest hit sector in terms of the number of data breaches, with 228 recorded data breaches, and up to 26 million records stolen. Many patients and individuals are not aware if their data has been compromised in any way because healthcare data breaches are in the millions of beached records. The following information in Fig. 4 extracted

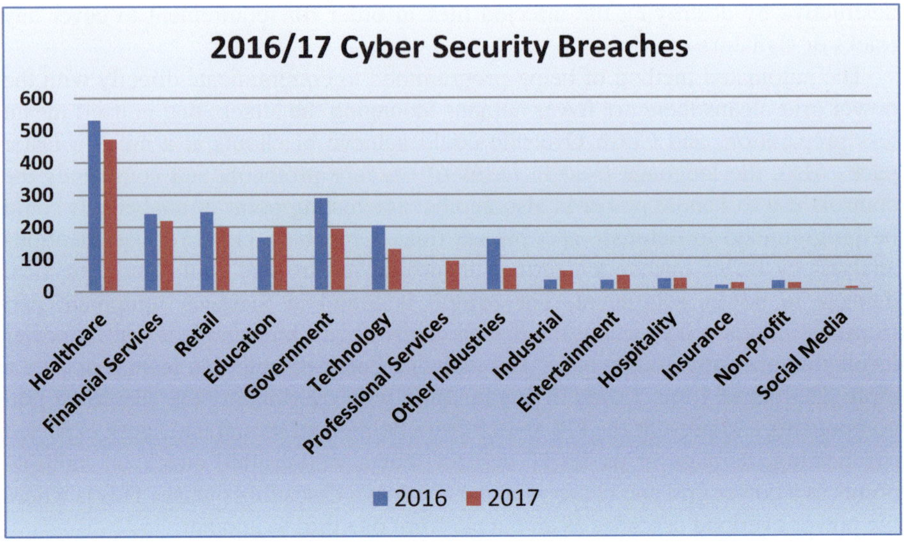

Fig. 4 Number of cyber breaches 2017 comparison to 2016. Breachlevelindex [3]

from the Breachlevelindex [3], depicts a strong trend of cyber-attacks in healthcare services.

Further analysis of the data and it can be seen that there is a direct correlation to whom is doing the attack and what is the impact; and looks to be malicious outsider attacks and the obtaining of identity theft as the most common combination. Other typical high issue losses occur through accidental error and malicious insider attacks.

This relates with the Information Commissioner's Office (ICO) who confirms that human error being at the forefront of data disclosure [19], which includes theft, incorrect posting of data, failure to redact data has been pointed out as being some key reasons. The WannaCry ransomware incident in 2017 caused huge disruption and impact and sits as one of the four types of attack that could disrupt the ability to provide care; these are Ransomware, Data Breaches, DDoS Attacks and Insider Treat). It is widely recognised that over one million NHS patients' private data was stolen during the ransomware attack of 2017. This is a recurring theme and many reasons associated to this which all need application and methodology. These are the information governance, training and education, up to date patching of systems, technology intrusion and prevention systems and so on. There is also the possibility to apply blockchain to healthcare; some pilots and successful live deployments are already underway over the last few years and some described earlier in this chapter. This would take care of the problems associated with interoperability and data sharing in a secure way. Plus, the benefits of immutability, smart contracts and a better way to protect privacy of patient data. With so many off the shelf tools available and malware-as-a-service and ransomware-as-a-service it means it is not only the data residing in central systems that is under threat but also the data surrounding IoT and IoMT.

Besides ransomware there are other tools that have been responsible for healthcare attacks. These include Aircrack, WarDriving and Kismet for wireless devices, Snort for stealthy port scans, NMAP for network mappings and operating system scans, including other tools such as Brutus, Cane and Able, Jack the Ripper and L0ptcrack for password cracking. Also, when looking at robotics and medical devices there are some concerns. It seems password is not required for logging in and access. For Robotic Process Automation (RPA) task bots are vulnerable as well to unauthorised access as the RPA task bot may not be password protected. For malicious reasons, harm can be delivered to the healthcare provider as files may be deleted or amended. Rights and permissions are also a problem area if misused and data stored locally for manipulation as plain text by the task bot could be accessed by those who should not have access to the data. Personal Health Information (PHI) will be recorded and stored on the Electronic Health Record (EHR). There should be fundamental steps taken to protect and secure the data in a range of measures and this means that not only the data needs to be protected from access directly, but updates to that data must also be protected in transit. Healthcare, and in particular the National Health Service, are part of critical national infrastructure and a well-known fact that it is under heavy cyber-attacks.

4 Cyber Weapons and Evolution in the Digital Age

With conventional conflicts fought on land, sea and air, cyber space is quickly becoming a fourth conflict area and one that can bring to bear as devastating results. The digital age has brought about a more dynamic, integrated and coordinated effort by both government and military agencies and also more autonomous groups with specialist skill-sets but sponsored by states. Announced by the National Security Agency (NSA) to establish a new defence cyber-security division to focus on the foreign cyber threats to the US, Cimpanu [6], shows the priority given to cyber strategy.

Part of cyber defence is effective cyber weapons or a cyber arsenal. Usually shrouded in secrecy the US had taken the decision to be more overt with its arsenal over recent years. According to Halpern [16], the time during Obama's Administration knowledge of US cyber weapons were strictly kept covert and classified. Fast forward to the Trump Administration and the Department of Defense (DoD) has not only acknowledged existence of such arsenals but an open declaration to using them in any form of preventative action and as a defence mechanism. In June 2019, there were some attacks on oil tankers in the Gulf of Oman that the US believes Iran were behind and subsequently a Global Hawk drone was shot down by an Iranian surface-to-air missile and as a result the US launched a cyber-attack against Iranian maritime operations. What was interesting here is the purposeful 'leak' regarding the US deploying such cyber weapons.

Cyber Command (such as the USCYBERCOM) can deploy very specific strategies and arrange how a cyber weapon should be executed; that is if it should target a wide network or standalone computer. They are becoming an integral and important part of the military and offensive teams' objectives are to hunt for vulnerabilities in another country's infrastructure and perhaps remain undetected whilst obtaining information. Hence an offensive cyber reaction mentioned earlier regarding the oil tanker attacks may have been the result of being familiar with targets and layout as time spent in advance, Halpern [16]. Its clear cyber command and offensive cyber weapons have grown in priority with the Pentagon elevating Cyber Command to equal status as nine other combat commands. One of the advantages of offensive cyber weapons is the fact that it can reduce escalation since the attack can be very specific, cause no loss of life and limit damage to infrastructure to serve more as a warning rather than retaliatory strike.

By 2023, a new NATO military command centre is planning to be fully operational to defend and deter computer hackers, Emmott [9]. This is an action to help defend against the hundreds of hacking attempts on NATO which happen monthly. Some of these attacks are executed by North Korea, China and Russia and are particularly sophisticated. NATO has already recognised cyberspace as another front to defend as well as the traditional land, sea and air. Interestingly, with the positioning of the planned NATO cyber command centre has also the possibility to evoke Article 5, a collective defence clause and cornerstone of NATO principles that now takes cyber defence as part of the ideology, NATO [28]. It effectively means that the highest ranking General is able to make fast decisions on cyber strike weapons, very similar as in the same decision making manner as conventional weapons and as the NATO collective principle.

5 Conclusions

It is clear that CNI has to be safeguarded with highest priority since it is a main attack vector due to its positioning within a country's infrastructure. There should be much more joined up thinking in cyber defence approaches both to defend against regular cyber-attacks either criminally motivated or destructive down to the level of state sponsored or highly organised and funded cyber hacking groups. The cyber command strategy and directives under C4ISR where cyber space can be considered as high a priority as land, sea and air warfare helps elevate the approach to sophisticated attacks and improve a more strategic defence layer to those type attacks. It also has become clear that a cyber strike is a very potent weapon that has certainly had more importance placed amongst normal defence strategy as the digital age and evolution of technology is taking an increasingly faster progression. Perhaps also worth noting is the mix of legacy older technology, such as the example of Industrial Control Systems, with the more advanced Internet of Things and fog layer computing has produced an additional risk and more vulnerabilities for attackers to take advantage of.

References

1. Barwick H (2015) Attacks against SCADA systems soar. Available at: https://www.computerworld.com.au/article/572668/attacks-against-scada-systems-soar/. Accessed 18 July 2019
2. Booz Allen Hamilton (2016) C4ISR survey summary report. Available at: https://www.boozallen.com/d/insight/publication/challenges-facing-c4isr-integration-for-the-military.html. Accessed 18 July 2019
3. Breachlevelindex (2018) Data breach database. [Online]. Available at: https://breachlevelindex.com/data-breach-database#. Accessed 18 July 2019
4. Cabinet Office (2018) Public summary of sector security and resilience plans. Available at: https://assets.publishing.service.gov.uk/government/uploads/system/uploads/attachment_data/file/786206/20190215_PublicSummaryOfSectorSecurityAndResiliencePlans2018.pdf. Accessed 19 July 2019
5. CERT-UK (2016) Annual report 2015/16. https://www.ncsc.gov.uk/content/files/protected_files/report_files/CERT-UK-Annual-Report-2015-16.pdf. Accessed 18 July 2019
6. Cimpanu C (2019) NSA to establish a defense-minded division named the cybersecurity directorate. ZDNet. Available at: https://www.zdnet.com/article/nsa-to-establish-a-defense-minded-division-named-the-cybersecurity-directorate/. Accessed 18 July 2019
7. Connell M (2017) Russia's approach to cyber warfare. Center for Naval Analyses, Arlington. Retrieved from http://www.dtic.mil/docs/citations/AD1032208. Accessed 6 Oct 2018
8. Doffman F (2019) Iran launches new military command and control unit to withstand U.S. cyberattack. Forbes. Available at: https://www.forbes.com/sites/zakdoffman/2019/07/07/iran-launches-new-military-command-and-control-unit-to-withstand-u-s-cyberattack/. Accessed 18 July 2019
9. Emmott R (2018) NATO cyber command to be fully operational in 2023. Reuters. Available at: https://www.reuters.com/article/us-nato-cyber/nato-cyber-command-to-be-fully-operational-in-2023-idUSKCN1MQ1Z9. Accessed 30 July 2019
10. Fraser et al (2019) APT41: a dual espionage and cyber crime operation. Fireeye. [Online]. Available at: https://www.fireeye.com/blog/threat-research/2019/08/apt41-dual-espionage-and-cyber-crime-operation.html. 10 Aug 2019
11. Fruhlinger J (2017) What is Stuxnet, who created it and how does it work? Available at: https://www.csoonline.com/article/3218104/malware/what-is-stuxnet-who-created-it-and-how-does-it-work.html. Accessed 30 July 2019
12. Gemalto (2017) Trust in a connected world. [online]. Available at: https://www.gemalto.com/investors-site/Documents/2018/Annual-report-2017.pdf. Accessed 21 July 2019
13. Greenberg A (2017a) Unprecedented malware targets industrial safety systems in the middle east. Wired. Available at: https://www.wired.com/story/triton-malware-targets-industrial-safety-systems-in-the-middle-east/. Accessed 21 July 2019
14. Greenberg A (2017b) 'Crash override' the malware that took down a power grid. Wired. Available at: https://www.wired.com/story/crash-override-malware/. Accessed 21 July 2019
15. Greenberg A (2017c) Your guide to Russia's infrastructure hacking teams. Available at: https://www.wired.com/story/russian-hacking-teams-infrastructure/. Accessed 18 July 2019
16. Halpern S (2019) How cyber weapons are changing the landscape of modern warfare. The New Yorker. Available at: https://www.newyorker.com/tech/annals-of-technology/how-cyber-weapons-are-changing-the-landscape-of-modern-warfare. Accessed 18 July 2019
17. Hermann Jr RM (2017) Cyber war in a small war environment. Doctoral dissertation, Utica College, April. Retrieved from https://pqdtopen.proquest.com/doc/1892789852.html?FMT=ABS&pubnum=10271200. Accessed 18 July 2019
18. Higgins K (2017) Latest Ukraine blackout tied to 2015 cyberattackers. Dark Reading. Available at: https://www.darkreading.com/threat-intelligence/latest-ukraine-blackout-tied-to-2015-cyberattackers/d/d-id/1327863. Accessed 18 July 2019

19. ICO (2018) What action we've taken in Q4, what you've reported to us and what you can do to stay secure. [Online]. Available at: https://ico.org.uk/media/action-weve-taken/reports/2014675/data-security-trends-pdf.pdf. Accessed 18 July 2019
20. Jay J (2018) Healthcare sector suffered more than half of all cyber-attacks in 2017. Available at: https://www.scmagazineuk.com/healthcare-sector-suffered-more-than-half-of-all-cyber-attacks-in-2017/article/763532/?utm_source=hs_email&utm_medium=email&utm_content=62703473&_hsenc=p2ANqtz%2D%2Drrl9qlLIbs5RTh5U6NBARNQlIEVWdyAsQwfiNL80sIcAw9MmgCC8exXjGCVox_WaTosWWVdCYDeELjiApOO4g0Wn7w&_hsmi=62703473#new_tab. Accessed 18 July 2019
21. Kaspersky Lab ICS CERT (2018) Threat landscape for industrial automation systems in H2 2017. Available at: https://securelist.com/threat-landscape-for-industrial-automation-systems-in-h2-2017/85053/. Accessed 18 July 2019
22. Khandelwal S (2019) Russian hacking group targeting banks worldwide wit evolving tactics. The Hacker News. [Online]. Available at: https://thehackernews.com/2019/08/silence-apt-russian-hackers.html?m=1. Accessed 21 Aug 2019
23. Logsdon M (2016) Why companies using SCADA systems need to wake up to the increased threat of cyber-attacks. Available at: https://www.scmagazineuk.com/why-companies-using-scada-systems-need-wake-increased-threat-cyber-attacks/article/1477598. Accessed 18 July 2019
24. Morbin T (2016) £442 billion potential loss in UK power sector cyber-attack. Available at: https://www.scmagazineuk.com/442-billion-potential-loss-uk-power-sector-cyber-attack/article/1477490. Accessed 18 July 2019
25. Muncaster P (2019a) Nine in ten CNI providers damaged by cyber-attacks. Available at: https://www.infosecurity-magazine.com/news/nine-10-cni-providers-hit-damaging-1/. Accessed 18 July 2019
26. Muncaster P (2019b) Nine in ten CNI providers damaged by cyber attacks. InfoSecurity. Available at: https://www.infosecurity-magazine.com/news/nine-10-cni-providers-hit-damaging-1/. Accessed 1 Aug 2019
27. NAO (2017) Investigation: WannaCry cyber-attack and the NHS. Available at: https://www.nao.org.uk/wp-content/uploads/2017/10/Investigation-WannaCry-cyber-attack-and-the-NHS.pdf. Accessed 18 July 2019
28. NATO (2018) Collective defence – Article 5. North Atlantic Treaty Organisation. Available at: https://www.nato.int/cps/en/natohq/topics_110496.htm. Accessed 18 July 2019
29. NATO (2019) NATO cyber defence. Available at: https://www.nato.int/nato_static_fl2014/assets/pdf/pdf_2019_02/20190208_1902-factsheet-cyber-defence-en.pdf
30. NCSC (2016) National cyber security strategy 2016 to 2021. NCSC. Available at: https://www.gov.uk/government/publications/national-cyber-security-strategy-2016-to-2021. Accessed 18 July 2019
31. NCSC (2018) We work for the government and critical national infrastructure. Available at: https://www.ncsc.gov.uk/information/we-work-government-and-critical-national-infrastructure. Accessed 18 July 2019
32. Nichols M (2019) North Korea took $2 billion in cyberattacks to fund weapons program: U.N. report (Reuters). Available at: https://uk.reuters.com/article/us-northkorea-cyber-un/north-korea-took-2-billion-in-cyberattacks-to-fund-weapons-program-u-n-report-idUKKCN1UV1ZX. Accessed 13 Aug 2019
33. Osborne C (2019) Cyberattacks against industrial targets have doubled over the last 6 months. ZDnet. Available at: https://www-zdnet-com.cdn.ampproject.org/c/s/www.zdnet.com/google-amp/article/cyberattacks-against-industrial-targets-double-over-the-last-6-months/. Accessed by 12 Aug 2019
34. Parliamentary Office of Science and Technology (2017) Cyber security of UK infrastructure. Available at: http://researchbriefings.files.parliament.uk/documents/POST-PN-0554/POST-PN-0554.pdf. Accessed 18 July 2019

35. Poushter J, Fettrolf J (2019) International publics brace for cyberattacks on elections, infrastructure, national security. Pew Research Center. Available at: https://www.pewresearch.org/global/2019/01/09/international-publics-brace-for-cyberattacks-on-elections-infrastructure-national-security/. Accessed 18 July 2019
36. Strategic Comments (2017) The WannaCry Ransomware attack 23(4):vii–ix, Taylor and Francis. Available at: https://doi.org/10.1080/13567888.2017.1335101. Accessed 18 July 2019
37. Symanntec (2018) Shamoon: destructive threat re-emerges with new sting in its tail. Available at: https://www.symantec.com/blogs/threat-intelligence/shamoon-destructive-threat-re-emerges-new-sting-its-tail. Accessed 18 July 2019

Cyberwarfare – Associated Technologies and Countermeasures

Nishan Chelvachandran, Stefan Kendzierskyj, Yelda Shah, and Hamid Jahankhani

Abstract With the development of automated and AI technology permeating into all sectors of public, private and industry life, the interconnectivity of once remote, siloed and air gapped systems is on the increase. Whilst this affords productive, streamlined and efficient ways of working, monitoring and maximise the effectivity of these systems, it is the connectivity, that can create a critical vulnerability. This vulnerability, is the source of exploitative measures that we refer to in the context of cyberwarfare. Where state and or adversarial threat actors can, utilising mechanisms on the internet, infiltrate, manipulate and attack these systems, to great and potentially devasting effect. It is paramount that the appropriate measures are taken to minimise the risk of these threats and vulnerabilities, through the review and security of internal systems, but also understanding where the vulnerabilities in the systems could lie, and to what effect they would cause should they be exploited. It is also important to understand not only the capabilities of how to respond should such an attack take place, but also the proportionality and legal of such responses.

Keywords Tallin · Cyberwar · Cyberops · SCADA · ICS · Intelligence · Cyberthreat · Attribution · Countermeasures

---

N. Chelvachandran (✉)
Open Innovation House, Saidot OY, ESPOO, Finland
e-mail: nishan@cyberreu.co.uk

S. Kendzierskyj
Cyfortis, Worcester Park, Surrey, UK
e-mail: stefan@cyfortis.co.uk

Y. Shah · H. Jahankhani
Northumbria University, London, UK
e-mail: Hamid.jahankhani@northumbria.ac.uk

H. Jahankhani et al. (eds.), *Cyber Defence in the Age of AI, Smart Societies and Augmented Humanity*, Advanced Sciences and Technologies for Security Applications, https://doi.org/10.1007/978-3-030-35746-7_2

1 Cyberwar – A New Weapon of Mass Effect

Cyber warfare involves the actions by a nation sate or international organisation to attack and attempt to damage another nation's computer or information networks. Cyber war targets systems that are critical to maintaining a nation's way of life, in part, to cause widespread panic and uncertainty. These systems can include financial systems, energy power grids or plants, healthcare, water, communications systems, transport systems and food and agricultural systems. As Critical National Infrastructure, C4ISR systems, Supervisory Control and Data Acquisition (SCADA) and Industrial Control Systems (ICS) become both more interconnected, and augmented through automation and AI powered decision making processes, cyber security has become one of the main concerns in securing and defending these critical systems. The intrinsic and high impact that these systems play within a Nation's critical national infrastructure, and in the coordination and implementation of public and private sector services, means that these mechanisms are a prime target for adversarial actors, both in the unknown or rogue threat actor context, or in conventional warfare constructs. Cyberwarfare is and further becoming as devastating if not more so, that both conventional weapons and those of mass destruction, whilst, seeming conducted remote, with no risk to life for the threat actor during the action taken. In the example of a SCADA system, for example in a water treatment facility, a remote operator can take control of the system and issue commands to open a value, setting or changing temperature points, stopping pumps, or spoofing sensor readings to affect chemical adjustments for the treatment of the water. Such an attack could prove difficult to initially detect and have a massive impact on a population area of a targeted country or state. A real-world example of such an attack on critical national infrastructure was an attack in December 2015, whereby the Ukraine was affected by a massive power outage after the national electrical grid suffered an attack on one of its SCADA systems. This caused about 230,000 people to not have power for several houses. There have also been reports of attacks on key infrastructure systems of a small dam in New York State and the Wolf Creek Nuclear Operating Corporation. According to a report from the UK's General Communications Headquarters (GCHQ), and the National Cyber Security Centre (NCSC), there are also concerns about suspicious attacks that have occurred on the UK Energy Sectors.

It is observed that much of the critical national infrastructure in western countries, is not solely operated, maintained and managed by public sector agencies, but that the private sector are also key stakeholders in the infrastructure, its operation and maintenance. As such, it is important that the private sector also plays a significant role with public sector agencies to ensure these securities of such infrastructure, and that the risk from the threat landscape is minimised.

Considerations should also be made towards the wider security and governance constructs. Cybersecurity in itself does not solely revolve around network security or the technological layer of these systems, but also the information and governance frameworks that are in place. The security and management of data and information, to prevent breaches and leaks are just as critical. In most cybersecurity practice, humans are often seen to be the weakest link in information security.

SCADA systems play a large role in critical national infrastructure; however, attention should also be paid to the Internet of Things, or IoT. IoT, while widespread in consumer markets in various products, smart speakers, appliances, toys, it is their sensor functionality that places a key role in infrastructure systems as we see today. For example, in healthcare critical infrastructure, IoT devices and sensors are used in hospitals and clinics, pharmaceutical and testing labs, as well as human interactive devices, such as medical devices and implantable devices. IoT attacks can be classified into three types of attack vectors:

1. Computational capabilities
2. Listening Capabilities
3. Broadcasting Capabilities

With computational capabilities, data modification or impersonation attacks can be launched by an adversary. With listening capabilities, a threat actor can listen and perform eavesdropping or track critical healthcare data within the healthcare infrastructure. And on broadcasting capabilities, a threat actor can replay the critical data from healthcare infrastructure.

Another area of IoT rich implementation is in transportation, but with the development of autonomous transportation, and smart traffic and transportation management. IoT devices have been the backbone for the growth and development of this sector, with technologies utilising both various systems, and data collection and use in a wider network in order to enable the infrastructure. Such technologies include:

- Machine Vision
- In-Vehicle Devices
- Global Positioning System (GPS)
- Acoustic Sensors
- Radar
- Light Detection and ranging
- In-vehicle sensors
- Electronic devices
- Roads/Structures on which the vehicles drive
- Odometrical sensors
- Maps

Each of these strands presents an avenue and potential threat vector for adversaries to target to exploit or influence transportation infrastructure. The manipulation of traffic management systems to cause gridlock and congestion in a major city, whilst superficially could be seen as an inconvenience, can quickly turn into a matter of critical national security, which coupled with attacks on other infrastructure systems, such as electrical or gas grids.

IoT devices are also affected by a number of major security issues:

- Flash network traffic: Sudden High number of end-user devices and new things (IoT).
- Security of radio interfaces: Radio interface encryption keys sent over insecure channels.
- User plane integrity: No cryptographic integrity protection for the user data plane.
- Mandated security in the network: Service-driven constraints on the security architecture leading to the optional use of security measures.
- Roaming security: User-security parameters are not updated with roaming from one operator network to another, leading to security compromises with roaming.
- Denial of Service (DoS) attacks on the infrastructure: Visible nature of network control elements, and unencrypted control channels.
- Signalling storms: Distributed control systems requiring coordination, e.g. Non-Access Stratum (NAS) layer of Third Generation Partnership Project (3GPP) protocols.
- DoS attacks on end-user devices: No security measures for operating systems, applications, and configuration data on user devices

2 5G as an Enabler

In order for the widespread IoT integrated in infrastructure and wider systems, the devices will not only need to interact with each other and with a base station to response to signals, it requires a faster and stable internet connection, which enables higher data rates for the purpose of information transfers. The development of the 5G protocol, enabling higher data transfer speeds with lower latency means that as 5G begins to roll out across the telecoms infrastructure, there will in turn be a boom in the enabling and utilisation of IoT and automation across sectors and industry. However, exchanging data in terms of transmitting and receiving information via a local network could be feasible but the more complex the amount of information gets, the more data rates it requires. Millimetre wave communication technology is one of the core parts of 5G networks and is expected to offer wireless data transfer by settling for a higher bandwidth. However, the drawback to this technological concept is that the transmission distance of this particular wave is known to be limited into 100 m in the atmosphere. And so, while the line of sight transmission distance degrades quickly compared with previous generations of mobile telecoms protocols, a proposed workaround or solution would be the use of a densification strategy, in that areas utilising 5G technology would have increased radio head, node and relays, to increase the density of the covered area, and so minimise degradation in the signal. However, this requires a great deal of improvement and investment into the mobile telephony infrastructure.

With the implementation of new telephony infrastructure, one of the significant features of 5G to consider, is data handling and storing solutions. The telecoms manufacturer, Huawei points out that *"security"* as such, remains an indispensable factor for business continuity. Furthermore, Huawei suggests the consideration of applying privacy and security properties from former generations of mobile network to the upcoming mobile network (5G) so that business continuity can be provided.

By mitigating the impact of security breaches and understanding the influence that risk factors have, business continuity can be subject to audit through consistent safeguarding. With the 5G network adding function and enhancement to the reliability and availability of faster wireless service to applications, appliances and other 5G driven technologies, the security issue gains importance and further highlight to 5G. 3GPP's newest Release 15, introduces the development for additional space for massive connections between devices but also to deliver faster services with reduced latency. Under section 7.3 of 3GPP's technical specification in Release 15, it is stated that this newest release builds on the LTE features for Machine-Type Communications (MTC) introduced in Release 13 and Release 14 by adding support for new use cases and general improvements with respect to latency, power consumption, spectral efficiency, and access control.

Although, 5G will be capable to cover high numbers of devices, machines and other appliances, the amount of data retrieved and processed will increase enormously.

That is when the confidentiality of vulnerable information may get violated.

Another vulnerability to the 5G network as well as wider critical infrastructure, is the interconnection and dependence on the infrastructure as a whole to be operational as a whole. A crucial point for communications is the power grid. **Power supply** depicts a crucial point when assessing risks, the 5G network has on users and the security structure of a nation. The collapse or disruption of a wired power supply systems might a huge cascade effect on wider systems within the network chain, such as data handling and electrical systems.

Wireless communication systems have been prone to exploitation of security vulnerabilities from their inception. However, with the proliferation of 4G and soon to be 5G networks, the proliferation of smart devices into the mobile domain, with multimedia traffic, and new services with vast quantities of data, have greatly diversified and given huge complexity to the corresponding threat landscape. The dynamic threats that will come from the proliferation of 5G, will also affect the interconnected systems that will rely on 5G. It is crucial to address these issues both in these existing systems, and potential new systems that will be realised with the use of 5G.

A privacy by design approach should be adopting with 5G and it's use, where privacy is considered from the beginning, with features built-in in its implementation. Better mechanisms for accountability, data minimisation, transparency and openness should also be drafted.

3 Cyberthreat Intelligence

In order to appropriately predict, respond and mitigate or minimise the growing threat from Cyberwar, resources and work must be conducted to explore and review the threat intelligence of would-be victim systems. Cyber threat intelligence refers to the intelligence collected before a threat actor attacks a victim system. Organisations and agencies use cyber threat intelligence to mitigate risks relating to cyber-attacks, such as zero-day exploits, internal and external threat actors or Advanced Persistent Threats (APTs). This approach allows for a proactive stance to cybersecurity to be taken and utilise possible countermeasures in advance of an attack. There are multiple sources by which such intelligence can be gathered, such as Open Source Intelligence (OSINT), Social Media Intelligence (SOCMINT), human intelligence (HUMINT), technical intelligence and intelligence from the dark web.

The UK's NCSC classifies Cyberthreat Intelligence into 4 categories:

1. Strategic Cyber Threat Intelligence – This utilises data, high level information and a timely warning of cyber threats, consumed at board or other senior level decision making level. Strategic cyber threat intelligence formulates an overall picture of both the intention and capabilities of threat actors and their impact at a high level.
2. Tactical Cyber Threat Intelligence – This involves data that is obtained from real-time monitoring of systems. This data refers to real time systems events linked to an adversary's actions against a system. This intelligence is used by cyber defensive roles to ensure that their response and investigation systems are prepared for possible tactics used by adversaries.
3. Operation Cyber Threat Intelligence – This involves the fata that gives details about a specific incoming attack. This can include malware, campaigns or cyber weapon tools. The insights gained from this intelligence guides and supports responses to specific incidents, as well as aiding in the assessment of determining future attacks and the organisational ability to do so.
4. Technical Cyber Threat Intelligence – This refers to the data that is consumed through technical means. For example, a suspected malicious IP address of a threat action. This intelligence has a short lifespan, as a threat actor can change their IP change. However, as part of the larger cyber intelligence landscape, technical intelligence helps defensive operators take preventative action, as well as in the investigative reviews post attack.

4 Attribution

The importance of attribution splits the global cybersecurity communities, both in private sectors, influencers and the operating community. In political and traditional legal and warfare frameworks, the attribution of a threat, potential attack or attack sustained by a victim state by an adversarial nation state is a factor a precursor to a pre-emptive or retaliatory attack against the adversary. Intelligence and other factors also play into responses, however, an adversary or must be known in order to direct a response.

In the cyber landscape however, this is not as straight forward as in a traditional or conventional scenario. The availability and capabilities of cyber weaponry or mechanisms that can be weaponised are not solely available to military or defensive agencies or organisations. Solo operators or lone wolves can be as effective in an attack or response as a nation state operator. By virtue of privacy and other such technological apparatus available, attribution is not as simple as once thought. VPN technology enables a user to encrypt their IP address and traffic through a data tunnel and "appear" to operate from another geographic location. The Dark Web and TOR network also allows users to mirror and redirect their traffic through multiple nodes across the world, making tracing very difficult. Of course, a cyber attack in itself may not necessarily be performed by a real time user. Malware, viruses and botnets can infect and spread across many machines undetected, before opening their payload and becoming operational, instigating coordinated attacked. Attribution in this sense becomes more difficult, as many of the tools, code and software used in the commission and formulation of these "cyber weapons", are available "off the shelf", and available for purchase by a user who may not necessarily have or require the full technical knowledge and capabilities to code such scripts.

There is also the recurring issue of legislation and jurisdiction in this instance, in that, the traffic pertaining to the adversarial attack may have been routed through a nation to give the appearance that they are the perpetrators. Or, if the operator is identified as not being affiliated or acting on behalf of a nation state, then when and where the operator was during the time that the attack was perpetrated needs to be taken into account. The complexity of which legislation applies and in which context, is something that has been discussed in the two published Tallinn Manuals, for Cyberwar and Cyberoperations. These manuals discuss and compile legislation from multiple states, relating to maritime law, international law, legislation relating to surveillance and data protection, as well as security and rules of engagement. Authors of the manual also argued that as cyber warfare increasingly becomes part of and a mechanism utilised in wider conflict, that peacekeeping organisations such as the UN will need to have a cyber defence or peacekeeping capability. However, research into cyber warfare has shown that there is still no answer to what actually constitutes an armed attack in cyber space, and what the ethical boundaries of cyberwarfare are. As such, there currently would be great difficulty for the UN Security Council to agree upon how to enforce peace in a cyber conflict.

A multilateral approach can be used to detect attacks, and potentially perform attribution, as even large intelligence organisations have limited technological and human budgetary capabilities to effectively achieve global coverage.

The existence of the 5-eyes intelligence forum, consisting of the UK, US, Canada, Australia and New Zealand, is one of example of this. This expanded to the 9 eyes forum after the attacks on the US on September 11th, to a wider cooperation including Denmark, France, the Netherlands and Norway. It has continued to expand, and now is referred to as 14-eyes intelligence cooperation, including Belgium, Italy, Spain, Sweden and Germany.

5 Countermeasures

Under international law, a state is entitled to take countermeasures for breaches of international law against it, that are attributable to another state. Countermeasures are acts by an injured state against another state that would ordinarily be unlawful but are legally justified as responses to the offending state's unlawful activity. The use of countermeasures is subject to strict conditions. The purpose is to encourage the offending state to stop its unlawful activity, rather than to punish. The countermeasures must also be proportionate. And they must not use force.

As discussed in the Tallinn Manual, there is no reason why cyber operations may not in principle be used as a countermeasure in response to a breach of international law. There is nothing in their nature to make an exception for them.

In order to consider countermeasures to cyber-attack, a comprehensive approach to cybersecurity should be taken. The holistic review of internal systems and processes in place, to identify vulnerabilities and making appropriate amendments and adjustments to harden vulnerable systems, is the best way of countering the threat of a cyber-attack. By the virtue of any cyberattack, is the exploitation of vulnerabilities in a system. At a very minimum, internal practices must be conducted to create a minimal base level of security. These can include:

- Continuous Risk Assessment
- IT Environmental Health
- Authentication
- Internal Commitment and responsibility
- Data Retention
- Access to information
- Preventative, detective and Corrective security controls

To further mitigate and harden such internal vulnerabilities, multiple global intelligence agencies agree that the following should also be addressed:

5.1 Education

Employees of an organisation must be aware of the kinds of attacks that can occur and what they should do about them. This includes learning proper operating procedures, the key attack targets, and the classic attack methods. Some studies have shown education to be more effective than any other countermeasure for protecting information systems since knowledge of information-systems security is not a requirement for most jobs.

5.2 Legal Responses

In most western allied states, laws prohibit eavesdropping on communications and damage to computers. But most attackers do not worry about getting caught, since it is hard to track them down and laws are hard to apply. Laws can however be effective against repeat offenders within a given legal jurisdiction, like spies selling secrets.

5.3 Patches

It is important to fix flaws or bugs in software as soon as they are discovered, since attacks are typically launched within days of the discovery of major flaws. Manufacturers provide "patches", "security updates", or "service packs" to fix flaws, in the form of modified software that you must go to their Web site to download. Software that has been sold for a significant period of time generally requires fewer future patches because programmers have had more time to find and fix its flaws.

5.4 Backups

Since many attacks destroy data or programs, making copies of digital information is essential to recovery from attack. Backups need to be done for any critical information and need to be stored some distance from the systems they track so no common disaster affecting both locations is likely. Optical-disk storage is preferable for backups because it cannot be as easily damaged as magnetic media can be. A backup can be an entire duplicate computer system when it is important to maintain continuous operation.

5.5 Access Controls

Automated access controls are important for cyberspace. Access controls for computers are generally managed by passwords that must be supplied to log on and use resources. Controls can be set for individuals or for groups of people, and they can apply separately to reading, writing, or execution of resources, or to the ability to extend those privileges to other users. Access controls for networks are enforced by "firewalls", dedicated computers on a local-area network that restrict traffic to and from the network according to simple rules on such features as origin and communications protocol. Unfortunately, access controls are vulnerable to many attacks mentioned above, and will not generally protect against attacks by insiders like staff.

5.6 Encryption

Encryption hides data in some form that cannot easily be read; you then supply a character-string "key" to decode it when you need it Any attempts to modify encrypted data will result in undecipherability, so you can tell if encrypted messages or programs have been modified. Strong and virtually unbreakable methods of encryption have been developed recently with public-key cryptography. Encryption methods can also be used for authentication or to provide digital signatures on documents to prove who wrote them and when. Encryption has been touted as a solution to many security problems, however, it is not a panacea. If an attacker gains system-administrator privileges, they may be able to get keys or disable encryption methods without a user's knowledge.

5.7 Intrusion Detection and Computer Forensics

Logging records the events on a computer system or network. This can generate enormous amounts of data, so intrusion-detection systems can be set up to check and record just the events that might indicate an attack, alerting system administrators when matters become serious. IDSs can be located on individual or on networks. They are important defensive tools against a broad range of known attacks including Trojan horses. Most look for or bit patterns of known attacks, but a few look for or statistically suspicious behaviour and thus can detect some new kinds of attacks. IDSs are useful but are not perfect since attackers try hard to disguise their attacks.

For new or complex attacks, computer forensics capabilities are needed, utilising methods for inspecting computer storage after an attack to determine how the attack was accomplished and what damage it did. Forensics includes a wide variety of techniques and requires an intelligent investigator to use considerable judgment. Thus, it requires time and can only be done after the attacker is gone.

5.8 Honeypots

Honeypots and honeynets (networks of honeypots) provide richer log information about cyber-attacks. These are systems with no legitimate purpose other than to receive attackers, so everyone using them other than their system administrator is inherently suspicious. Honeypots need not explicitly invite attackers once they are on the Internet, attackers can find them with automated tools. However, they can be dangerous if attackers use them as springboards to attack other sites. For this reason, reverse firewalls of various kinds must keep the attack from spreading. But an attacker may infer the existence of the honeypot from the restrictions of the reverse firewall, so a honeypot cannot remain effective forever.

5.9 Intrusion Prevention Systems

Most of the methods discussed so far just react to attacks. The alternative is an active network defence, which in its simpler forms is called an intrusion-prevention system. This includes simple things like turning off the Internet connection or logging out a user when they become sufficiently suspicious as judged by an intrusion-detection system. It can also include forms of limiting damage such as denying the user certain resources, downgrading their priority, or delaying them.

5.10 Back Tracing

Back tracing is a form of active network defence that tries to find where an external attack is coming from so as to stop it more easily. Back tracing is virtually impossible with serious attackers, who take care to come in via a long sequence of sites through many countries and jurisdictions; it is hard to get the cooperation of all those jurisdictions.

5.11 Counterattacking

A more irresponsible form of active network defence is trying to counterattack whatever machine is attacking you. However, this won't work against insiders. Since most serious attacks use intermediate machines to attack yours, such a response will often only hurt a site or computer that is an innocent bystander. Even if it works and you do hurt the attacker, attacks could easily escalate with resultant collateral damage.

5.12 Deception

Deliberate deception has also been proposed for active network defence. Systems could lie, cheat, and mislead attackers to prevent them from achieving their goals. Deception is particularly useful for time-critical military-style attacks such as those by cyber-terrorists or information-warfare experts, when just delaying an attack a while could buy time to find a more permanent defence. Deception has been used in honeypots to keep the attacker interested. Fake files can be put on a honeypot to make it look more like a normal machine, and fake sites can be programmed to respond like real network nodes. Deception is equally useful against insider and outsider attacks.

In regard to external countermeasures, as they involve an act, in that there are in response to an act committed unlawfully against the victim state, there are strict conditions that need to be met. The countermeasure may not be conducted until the injured state has notified the state responsible that it intends to take countermeasures and gives the responsible state an opportunity to desist in its unlawful conduct. However, in the context of a cyber response, the notification requirement is subject to a condition of feasibility, as the advanced notification of an impending cyber countermeasure could allow the responsible state to foil such a response.

The countermeasures should also be proportionate to the injury to which they respond. They have to be "commensurate with the injury suffered, taking into account the gravity of the internationally wrongful act and the right question". There may be provisions in treaties that detail the taking of specific responses in the even of breaches, if so, the injured state much resort to them before taking any countermeasures.

Operationally, countermeasures that are utilised also do not need necessarily need to be "in-kind" nor directed at the entity that authored the internationally wrongful act. A state victim of an attack may response with cyber measures that the sovereignty of the responsible state, for example. This is reflected in maritime law, where a state that has been targeted by another state's unlawful cyber operations would be entitled to close its territorial sea to vessels of the responsible state, that are transiting in innocent passage. Another cyber response to an attack could be to direct the response at private corporations within the responsible state, so long as the response and operations are proportionate to the originating attack and comply with the requirements of countermeasures.

6 Conclusion

As critical national infrastructure utilises emerging technologies in order to maximise efficiency and effectivity, it also opens up these critical networks and systems to non-conventional cyber-attacks. The threat landscape is as complex as the system itself, with vulnerabilities open for exploitation by both independent operators, rogue state operatives, state backed threat actors and state mechanisms themselves. Such is the benefit of utilising cyberoperations mechanisms in terms of speed, impact and minimising human impact by means of the deliverer of the attack vector, that cyber capabilities are now a ratified capability in many nations globally. It is imperative that a greater understanding of both the implications of an sustained attack, and the results of perpetrating a cyber attack filtered into the legislative and governance mechanisms of nation states, and that agreed conventions and law defining peace and wartime mechanisms is adhered to, and understood in the context of a cyber war.

Bibliography

1. 3GPP (2017) SA3-security. The Third Generation Partnership Project (3GPP)
2. Agiwal M, Roy A, Saxena N Next generation 5G wireless networks: a comprehensive survey. IEEE Commun Surv Tutor 18(3):1617–1655. thirdquarter 2016
3. Akghar B, Yates SJ (2011) Strategic intelligence management for combating crime and terrorism. In: Akhgar B., Yates S. (eds) Intelligence Management. Advanced Information and Knowledge Processing. Springer, London
4. Alliance N (2015) NGMN 5G white paper. Next generation mobile networks, White paper
5. Cook A, Smith R, Maglaras L, Janicke H (2016) Measuring the risk of cyber attack in industrial control systems. BCS eWiC
6. Cook A, Nicholson A, Janicke H, Maglaras L, Smith R Attribution of cyber attacks on industrial control systems. EAI Endors Trans Ind Netw Intell Syst 3(7):151158
7. Cyber attack led to bristol airport blank screens. https://www.bbc.com/news/uk-england-bristol-45539841
8. Energy sector on alert for cyber attacks on UK power network. https://www.ft.com/content/d2b2aaec-4252-11e8-93cf-67ac3a6482fd. Accessed on 5 Feb 2018
9. Ericsson GN (2010) Cyber security and power system communication—essential parts of a smart grid infrastructure. IEEE Trans Power Deliv 25(3):1501–1507
10. Evans M, He Y, Maglaras L, Janicke H (2018) Heart-is: a novel technique for evaluating human error-related information security incidents. Comput Secur
11. Ferrag MA, Maglaras LA, Janicke H, Jiang J, Shu L (2018) A systematic review of data protection and privacy preservation schemes for smart grid communications. Sustain Cities Soc 38:806–835
12. Ferrag MA, Maglaras L, Argyriou A, Kosmanos D, Janicke H (2018) Security for 4g and 5g cellular networks: a survey of existing authentication and privacy-preserving schemes. J Netw Comput Appl 101:55–82
13. Freudiger J, Manshaei MH, Hubaux J-P, Parkes DC (2009) On noncooperative location privacy: a game-theoretic analysis. In: Proceedings of the 16th ACM conference on computer and communications security, ser. CCS '09. ACM, New York, pp 324–337
14. Fujita H, Gaeta A, Loia V, Orciuoli F (2018) Resilience analysis of critical infrastructures: a cognitive approach based on granular computing. IEEE Trans Cybern:1–14
15. Geraci G, Dhillon HS, Andrews JG, Yuan J, Collings IB (2014) Physical layer security in downlink multi-antenna cellular networks. IEEE Trans Commun 62(6):2006–2021
16. Gope P, Hwang T (2016) Bsn-care: a secure iot-based modern healthcare system using body sensor network. IEEE Sensors J 16(5):1368–1376
17. Huawei (2016) 5G security: forward thinking. Huawei, Technical .report
18. Knapp ED, Langill JT (2014) Industrial Network Security: Securing critical infrastructure networks for smart grid, SCADA, and other industrial control systems. Syngress, Waltham
19. Kulkarni P, Khanai R, Bindagi G (2016) Security frameworks for mobile cloud computing: a survey. In: 2016 international conference on electrical, electronics, and optimization techniques (ICEEOT), pp 2507–2511
20. Maglaras LA, Jiang J (2014) Intrusion detection in SCADA systems using machine learning techniques. In: Science and information conference (SAI), IEEE, pp 626–631
21. Maglaras LA, Kim K-H, Janicke H, Ferrag MA, Rallis S, Fragkou P, Maglaras A, Cruz TJ (2018) Cyber security of critical infrastructures. ICT Express 4(1):42–45
22. Nicholson A, Watson T, Norris P, Duffy A, Isbell R (2012) A taxonomy of technical attribution techniques for cyber attacks. In: European conference on information warfare and security, p 188
23. ONF (2013) SDN security considerations in the data center. Open Networking Foundation
24. Panayiotou CG, Ellinas G, Kyriakides E, Polycarpou MM (2016) Critical information infrastructures Security. Springer, Berlin/Heidelberg

25. Petit J, Shladover SE (2015) Potential cyberattacks on automated vehicles. IEEE Trans Intell Transp Syst 16(2):546–556
26. Pipyros K, Thraskias C, Mitrou L, Gritzalis D, Apostolopoulos T (2018) A new strategy for improving cyber-attacks evaluation in the context of Tallinn manual. Comput Secur 74:371–383
27. Polla ML, Martinelli F, Sgandurra D A survey on security for mobile devices. IEEE Commun Surv Tutor 15(1):446–471. First 2013
28. Ralston PAS, Graham JH, Hieb JL (2007) Cyber security risk assessment for SCADA and DCS networks. ISA Trans 46(4):583–594
29. Robinson M, Jones K, Janicke H (2015) Cyber warfare: issues and challenges. Comput Secur 49:70–94
30. Robinson M, Jones K, Janicke H, Maglaras L (2018) An introduction to cyber peacekeeping. J Netw Comput Appl 114:70–87
31. Robinson M, Jones K, Janicke H, Maglaras L (2018) Developing cyber peacekeeping: observation, monitoring and reporting. Gover Inform Q 36(2):276–293
32. Rye dam attack. https://www.newsweek.com/cyber-attack-rye-dam-iran-441940
33. Saalbach K (2017) Attribution von cyber-attacken – methoden und praxis
34. Schmitt MN (2013) Tallinn manual on the international law applicable to cyber warfare. Cambridge University Press, Cambridge
35. Stellios I, Kotzanikolaou P, Psarakis M, Alcaraz C, Lopez J (2018) A survey of iot-enabled cyberattacks: assessing attack paths to critical infrastructures and services. IEEE Commu Surv Tutor 20(4):3453–3495
36. Ten C-W, Manimaran G, Liu CC (2010) Cybersecurity for critical infrastructures: attack and defense modeling. IEEE Trans Syst Man Cybern Part A Syst Hum 40(4):853–865
37. Ukraine cyber attack energy. https://www.wired.com/story/crash-override-malware/
38. Vikas SS, Pawan K, Gurudatt AK, Shyam G (2014) Mobile cloud computing: security threats. In: 2014 international conference on electronics and communication systems (ICECS), pp 1–4
39. Wolf creek nuclear plant hit cyberattack. https://www.theenergytimes.com/cybersecurity/wolf-creek-nuclear-plant-hit-cyberattack
40. Zonouz SA, Rogers KM, Berthier R, Bobba R, Sanders WH, Overbye TJ (2012) Scpse: Security-oriented cyber-physical state estimation for power grid critical infrastructures. IEEE Trans Smart Grid 3(4):1790–1799

The Malicious Use of AI-Based Deepfake Technology as the New Threat to Psychological Security and Political Stability

Konstantin A. Pantserev

Abstract Contemporary psychological warfare has a number of instruments, including deepfakes, in which the human image is synthesized, based on AI algorithms. At first deepfakes appeared for entertainment. Special software based on artificial intelligence offers the opportunity to create clones that look, speak and act just like their templates. However, today the potential for deepfakes to be used maliciously is growing, whereby one creates a clone of a well-known figure and manipulates his or her words. This chapter analyses a wide range of examples of deepfakes in the modern world, as well as the Internet-services that generate them. It will also consider the possibility of using artificial intelligence to prevent their spread, as they constitute a serious threat to psychological security.

Keywords Psychological warfare · Deepfakes · Artificial intelligence · Deep learning · Fakes · Neural networks · Psychological security · Disinformation

1 Introduction

A new epoch is emerging in the history of world conflicts: an age of psychological warfare. Of course, psychological warfare as a phenomenon is not new, but today the development of information technologies (ICT) has rapidly increased its role in securing world dominance and geopolitical leadership, especially when even small military conflicts can now develop into global nuclear confrontation. Therefore, the contemporary global information and communication space has become the key battlefield and it is possible to say that we are now living in the era of global psychological warfare.

K. A. Pantserev (✉)
Saint-Petersburg State University, Universitetskaya nab. 7–9, Saint Petersburg, Russia
e-mail: k.pantserev@spbu.ru

H. Jahankhani et al. (eds.), *Cyber Defence in the Age of AI, Smart Societies and Augmented Humanity*, Advanced Sciences and Technologies for Security Applications, https://doi.org/10.1007/978-3-030-35746-7_3

The key feature distinguishing such warfare from more conventional forms is the absence of any rules. Thus, whereas conventional military conflict should be conducted according to international conventions elaborated by the United Nations (and relatedly a number of international bodies exist with missions to supervise military conflict), psychological warfare can be conducted by any means. Consequently, mass disinformation and the production of fake news has become the key instrument of modern psychological warfare, following the principle of "if you can't convince somebody, misinform him or her".

AI-based deepfake technology is just one recent technological innovation that initially appeared as a form of entertainment but very quickly became a dual-use technology. This means that when used maliciously, deepfakes can threaten the psychological security of any state and international stability more broadly. In the following chapter the author will attempt to clarify the malicious use of deepfake technology.

2 From Fake News to Deepfakes

Contemporary information technologies are modernizing every day and offer considerable possibilities for users both in the field of business and entertainment. Thus, in order to provide some fun, there have appeared special web applications that produce fake news.

For example, the application *Fake TV News Maker*, which is positioned as the funniest generator of fake television news, offers the possibility to create TV news by using the interface of well-known news programmes. Using this application, one can create, change and distribute any news he or she wants. There is also the potential to replace famous speakers of TV news programmes who work for the largest information agencies and broadcasting companies, such as the BBC, ABC and CBS. *Fake TV News Maker* provides three categories for creativity: news, rumours and sport. Such news are carefully designed to appear like professional journalistic materials (what, when and where something happened), supplemented with photos that create the maximum realistic effect and thereby prevent fake and real news from being distinguished. The application also offers users the opportunity to share such "breaking news" with friends via e-mail, Bluetooth, Dropbox, Google Drive, Facebook, Flipboard, Google+, Hangouts, Linkedln, Picasa, SMS, Telegram, Twitter, WhatsApp and other social media and chats.

Another fake news generator, the *Fake Newspaper Maker*, provides opportunities for those who dream of seeing his or her news in newspapers with a large circulation around the world.

One application, *Fake News & Charts for iPad*, offers the possibility for those who are tired of fake news being distributed by unfair media and politicians to create his or her own "real" news.

Of course, one can find on the web a number of other fake news generators, such as *#FakeNews*, *Journalist CreativeBot*, *Fake News Creator*, *Fake Breaking News Maker*, *Make Fake News* and *Fake News Editor*.

For those who choose to prank, there is an application called *Text Now* that helps in the creation of fake text messages for "funny communication". This application offers the opportunity to create fake chats between any famous or even fictitious persons. There are also a number of other applications of such kind, like *Fake Message*, *Fake Text Message* and *Fake Call and Fake SMS*. And finally there is the *Fake News Generator*, the application which directly offers in one of its advertisements the chance to make fake news stories in order to fool friends and even the masses.

The appearance of all of these applications demonstrates that there is a demand for such production on the web. However, we suggest looking into the opposite site of all these jokes because they undermine the credibility and value of information distributed on the web and force us to seek an answer to the question regarding how we can trust such information: "How do we know that the computer is behaving as we expect it to or that an e-mail from our colleague is actually from that colleague?" [25].

Thus it becomes evident that the uncontrolled and spontaneous distribution of fake news represents a real and very serious threat to the psychological security of any country, and is capable of provoking psychological terrorist attacks that may cause fatal consequences.

Recent novelties in the field of artificial intelligence have also brought to life a new phenomenon, called deepfakes, which again initially appeared as a form of fun but today represent a serious challenge to psychological security.

3 Deepfakes: The Entity of the Technology

Deepfake, a portmanteau of "deep learning" and "fake", is a method of synthesizing a human image based on AI algorithms. Deepfake technology is based on two innovations in the field of machine learning: neural networks and generative adversarial networks, whose mission is to make deepfakes extremely realistic [6]. Neural networks represent a key element of machine learning technology.

> These are brain-inspired networks of interconnected layers of algorithms, called neurons, that feed data into each other, and which can be trained to carry out specific tasks by modifying the importance attributed to input data as it passes between the layers. During training of these neural networks, the weights attached to different inputs will continue to be varied until the output from the neural network is very close to what is desired, at which point the network will have 'learned' how to carry out a particular task [15].

Neural networks use the principle of the human brain's functioning: the more the human brain is exposed to information, the more accurately it can repeat it. Accordingly, the greater the number of examples downloaded to the neural network, the more carefully and correctly it can produce new examples [17].

In terms of deepfakes, the more video and audio data they download into the neural network, the more accurate the new audio and video will be, to the extent that it becomes impossible to determine whether this or that speech of this or that person is real or fake.

But, as Dack [7] argues, "neural networks are only half of the equation. Without generative adversarial networks, deep fakes would not be as realistic as they are". Generative adversarial networks were invented by Ian Goodfellow. He was formerly a Google researcher at the *Google Brain*, a special research team established in the early 2010s. This research team's mission was to make a breakthrough in the field of deep learning artificial intelligence. He is currently employed by *Apple*, another well-known corporation that conducts a research in the field of artificial intelligence. Goodfellow et al. [10] developed the idea of combining two neural networks together in order to make them compete with each other and hence improve the final product.

The principle of the work of generative adversarial networks is as follows. The first neural network, which is called the "generator", produces a new fake video or audio by copping the data set that has been downloaded. Next the original data set and the deepfake created by the first neural network are downloaded to the second neural network, which is called the "discriminator". Its mission is to distinguish a fake video from a true one [9]. If the discriminator is able to determine the fake video or audio, the generator tries to learn how the discriminator understood which video was fake and subsequently makes appropriate corrections. With each new iteration it becomes more and more difficult to distinguish a deepfake.

This presents the greatest problem. At present, when a deepfake is discovered, the appropriate correction is made and it will prove more difficult to discover the deepfake next time. Each detection of the deepfake improves it. Of course, researchers are working hard in order to improve methods of detecting deepfakes. For example, they pay attention to the frequency of the flicker of the image, natural micro-changes of the color of the face, or the irregularity of head or body movements, and so on. But all of those methods that help in identifying a deepfake today will fail to discover it in the future. Eventually, 1 day, there will appear a super-realistic fake video or audio that will be impossible to distinguish from a real one. One can use this for fun, just as a joke to fool one's friends, or (and much more seriously), to maliciously fool the masses by fake speeches of famous and influential persons such as politicians.

> Thanks to advances in artificial intelligence (AI) and computer-generated imagery (CGI) technology, over the coming decade it will become trivial to produce fake media of public figures and ordinary people saying and doing whatever hoaxers can dream of—something that will have immense and worrying implications for society [21].

The danger is that anybody can make any politician say whatever he or she wants, and then publishes this fake speech on YouTube or Facebook, on a fake website of the well-known mass media or on a fake social media profile of this or that politician. It will subsequently be shared by millions of people on social media. The fake video or audio can very quickly spread on the web and cause unexpected consequences, such as by ending the political career of this or that person or even affecting the complexity of international relations between countries, possibly resulting in war.

> Progress in AI will enable new varieties of attacks. These attacks may use AI systems to complete certain tasks more successfully than any human could, or take advantage of vulnerabilities that AI systems have but humans do not [2].

Furthermore, some experts claim that "deep fakes don't need to be undetectable or even convincing to be believed and do damage. It is possible that the greatest threat posed by deep fakes lies not in the fake content itself, but in the mere possibility of their existence" [24]. This is because people believe in what they want to believe and they do not necessarily care whether a video is fake or real. Thus the main purpose is to feed people with the information they want to hear from the mouth of this or that politician. And this represents the greatest danger posed by deepfakes.

4 The Malicious Use of Deepfake Technology: From Fun to Psychological Warfare

As has been shown above, deepfakes can threaten personal, public and even national psychological security. It represents a very significant challenge of the contemporary digital age, as highlighted by many experts who are busy in the field of psychological and cyber security. For example, the company McAfee announced in March 2019 that it is today impossible to detect a change to a face using the naked eye. Thus, Steve Grobman, chief technology officer of McAfee, and Celeste Fralick, the company's chief data scientist, predicted in their keynote speech at the RSA Conference on cyber security issues in San Francisco that the use of new technologies by hackers is just a question of time.

Grobman and Fralick claimed in their speech that there is a special area in the field of cyber security called adversarial machine learning. Experts in this field are studying cyber-attacks to the machine learning classifiers. Grobman and Fralick argued that the method of replacing images represents a serious threat and can be used for the distortion of the work of the image classifier. As an example, they demonstrated one approach of how to deceive people using artificial intelligence: producing a real photo and carefully changing a small part of it. Through such minimal changes, a photo of penguins can consequently be interpreted by artificial intelligence as a frying pan. It is evident that false operation on a more serious scale can cause catastrophic consequences. Grobman stressed that deepfake technology represents a weapon that can be used for different purposes. It is impossible to prevent the malicious use of this technology, yet this is necessary to establishing a line of defense.

In order to prove that this threat is real, Grobman and Fralick presented a video in which Fralick's words were coming out from Grobman's face, even though Grobman never said them. Fralick concluded that this example "just shows one way that AI and machine learning can be used to create massive chaos. It makes me think of all sorts of other ways in the social engineering realm that AI could be used by attackers, things like social engineering and phishing, where adversaries can now create automated targeted content" [27]. Every day such technology is improving.

Consequently, deepfake videos have become one of major challenges to the national security of any country, as they make people say or do things that they have never said or done. It is also necessary to point out that this technology is only 2 years old, and yet it has already made significant progress, with both the expert community and politicians starting to speak of the threat represented by deepfakes.

We'd like to remind that initially deepfake technology appeared as a fun in the "pornography industry where it referred to the process of inserting celebrities' faces into pornographic scenes" [7].

Such fake video clips first appeared in December 2017, when a user with the nickname *Deepfakes* published a pornographic video involving the famous actress Gal Gadot on the social media platform Reddit. Yet this video was a fake. In fact: he simply put the face of the Hollywood star onto the body of a pornographic actress with the aid of artificial intelligence. He "used TensorFlow, image search engines, social media websites and public video footage to insert someone else's face (it was a Gal Gadots' face—K.P) onto preexisting videos frame by frame" [13]. Since this video clip we have seen the widespread distribution of similar pornographic video clips involving other celebrities on the web. Users find an appropriate pornographic actress who meets all of their criteria and with the aid of the neural network change her face to that of the famous person of their choosing. The episode with Gal Gadot was just the beginning. Emma Watson, a British actress and model, was another victim of such fake pornographic films. The rapid growth of such videos has been clear. Thus, one can find on the web similar fake pornographic clips involving other famous women, whether Chloë Moretz, Jessica Alba, Scarlett Johansson or Maisie Williams.

Furthermore, fake videos in which a woman's face is changed into that of a man have also appeared, such as one of the actress Amy Adams being altered to that of the actor Nicolas Cage (to see this video follow the link: https://youtu.be/RdH7JoZZC2M). This "work" is actively used as an example of the possibilities of this technology, because it proves that it is possible to change a woman's face to a man's face. Nicolas Cages' face has subsequently appeared in many other domains, including movies such as *The Dark Knight Rises* and *Man of Steel* in which he never played. He has even replaced Sean Connery in the famous film *Dr. No*, Stephen Dillane in *Game of Thrones* and Harrison Ford as Indiana Jones in *Raiders of the Lost Ark*. It is difficult to determine why Nicolas Cage is so popular in such fake videos. Maybe people perceive his face type as the most suited for replacement. Perhaps the reason lies in the popularity of the actor. Indeed, Nicolas Cage "has become a meme of sorts in recent years in some corners of the internet. The deliberate incongruity of putting his image in unlikely places offers obvious opportunities for subversive humor" [22].

The considerable popularity and frightening realism of such fake video clips has forced Reddit, Twitter and even Pornohub to stop the distribution of video clips created using AI-based deepfake technology. For example, the account of Deepfakes has been banned from Reddit. But the technology has already been successfully tested and today nobody can stop the continued distribution of fake videos that threaten any famous person, who must understand that 1 day he or she may be manipulated to appear in compromising videos. "Creating these deepfakes isn't difficult or expensive in light of the proliferation of A.I. software and the easy access to photos on social media sites like Facebook" [23].

Thus, the key feature of the contemporary digital age is the ease by which information technologies can be used. Indeed, the user does not need to know how neural networks work; he or she simply requires basic knowledge about computers and the Internet, find an appropriate application on the web, download it and have some fun.

Given the fact that AI technologies have become more and more accessible to ordinary users, coupled with easy access to photos and videos of former partners, colleagues and other people on Facebook and other social media, we are observing growing demand in programme tools that offer the possibility of creating fake videos.

One of these is *FakeApp*, a desktop application that can change faces in videos. It has been elaborated by the user *Deepfakes* on *Reddit* and is now distributed for free on the Internet. With the aid of neural networks, this application analyses the original faces of peoples and then tries to convert them into those of celebrities selected by the user. *FakeApp* is based on TensorFlow, Google's open-source platform for developing AI algorithms and the open-source library *Keras* [1]. It has a rather simple interface and offers detailed instructions on how to install and use the application. This simplification of complicated technology gives any user the opportunity to create fake videos and subsequently distribute them on the web via social media. The only thing he or she needs is an appropriate number of photos of this or that person so the application can learn and produce a highly realistic fake video clip. Here lies a great danger. As Chesney and Citron [6] argue:

> Imagine a video depicting the Israeli prime minister in private conversation with a colleague, seemingly revealing a plan to carry out a series of political assassinations in Tehran. Or an audio clip of Iranian officials planning a covert operation to kill Sunni leaders in a particular province of Iraq. Or a video showing an American general in Afghanistan burning a Koran. In a world already primed for violence, such recordings would have a powerful potential for incitement. Now imagine that these recordings could be faked using tools available to almost anyone with a laptop and access to the Internet—and that the resulting fakes are so convincing that they are impossible to distinguish from the real thing. Advances in digital technologies could soon make this nightmare a reality. Thanks to the rise of "deepfakes"—highly realistic and difficult-to-detect digital manipulations of audio or video—it is becoming easier than ever to portray someone saying or doing something he or she never said or did.

Thus, the expert community marks out both a positive and negative side of this technological novelty. According to Francis Tseng, co-publisher of *New Inquiry*, an online magazine of cultural and literary criticism, on the positive side "deepfake technologies can bring to life new forms of art and even be used to create new genres of media ... And there can appear a whole culture of bootleg films produced in this way" [22].

But the use of deepfakes also has a negative side.

> As the technology improves, it will likely be used in more dangerous and antisocial ways. For example, it has the potential to turbo-charge fake news. When paired with technology that can synthesize real people's voices, apps such as FakeApp could make it extremely difficult for ordinary people to distinguish what's real from what's fake. And such technology could well be used to harass and blackmail people, putting them—virtually—in compromising situations… (and – K.P.)… lift cyberbullying to a whole new level [22].

In April 2018, a filmmaker named Jordan Peele sought to prove that this threat is a reality by using *FakeApp* to make a video in which former United States President Barak Obama insults the incumbent President Donald Trump. The voice of Obama is imitated by Peele himself. Of course, the production of this video required considerable time as well as specialists qualified in using special effects. In addition to *FakeApp*, *Adobe After Effects* was used for the editing of the video and its dynamic images, the development of the composition, animation and different effects. In total, it took about 56 h of automated processing of the video stream under the control of the specialist in special effects.

> Such technology should completely unnerve everyone of every political stripe, religion, sports affiliation, philosophical school and Jane Austin Book Club. Which is to say everyone everywhere. Because the capability works both ways, it's just as easy for your rivals to ruthlessly attack you with it as you them. It allows you to be "outed" by your enemies, for example, even if you have nothing to be outed for. It can now be done because it can. Any of us can be viciously slandered publicly by virtual "people" we don't know from Adam and who don't know us. Trash-talked and abused on the internet, TV and radio in every depraved manner imaginable, and unimaginable [26].

The rapid growth of such videos can be easily observed. Thus, a YouTube blogger with the nickname *Ctrl Shift Face*, who is familiar with deepfake technology, has introduced Jim Kerry in the role of Jack Terrence in Stanley Kubrick's film *The Shining*. The blogger manipulated the movie so that comic actor Kerry, in the role of Terrence (originally performed by Jack Nicolson) would punch a hole in the door with an axe and stick his head through it. This video clip has been published as open access on YouTube and is accessible at https://www.youtube.com/watch?time_continue=35&v=Dx59bskG8dc.

In discussing *Ctrl Shift Face*, it is necessary to mention that this YouTube blogger has rich experience in the production of deepfakes. Indeed, he has already changed Schwarzenegger to Stallone in *Terminator 2* (video clip accessible at https://www.youtube.com/watch?v=AQvCmQFScMA). But according to the evaluation of experts, his scene from *The Shining* is much more realistic. Kerry's face almost avoids "sliding" from the head of Nicholson, even when he makes sudden movements or covers his face with his hands. And what is next? YouTube followers of *Ctrl Shift Face* often request that he make a reverse replacement with Nicolson performing as Ace Ventura in the comedy film *Ace Ventura: Pet Detective*. The appearance of such video clips proves that deepfakes are not limited to erotic films. Therefore, if you want to see your favorite actor in roles originally performed by another person, the only thing you need is access to the appropriate AI algorithms.

Deepfakes also can be used to restore poor-quality videos. Indeed, a team of YouTube filmmakers from the channel the "Corridor" decided to modernize some of the worst scenes in the history of the film industry using special effects. For instance, they changed the scene in *The Mummy Returns* in which the King of Scorpions played by Dwayne Douglas Johnson first appears, the special effects of which have been greatly criticized. They thus used deepfake technology and the face of Dwayne Douglas Johnson to improve the scene. The final product as well as the filmmakers' explanations are available at https://youtu.be/KH1V6CHO1Jk.

A final example is that of American YouTubers *Sam* and *Niko*, who sought to create an amusing video with large numbers of views by producing a real film involving a fake Keanu Reeves, titled *Keanu Reeves Stops a Robbery*. It can be accessed at https://www.youtube.com/watch?time_continue=11&v=3dBiNGufIJw and has more than 1.4 million views. The story of this film is simple. Two men meet Keanu Reeves in the supermarket. One of them wants the actor to sign an autograph, but when they begin to talk, a robber enters the supermarket and tries to take money from the cashier. Keanu Reeves then attempts to stop the robbery without anybody suffering. He suggests that the robber take all of the money in his pockets, and on hearing the police arrive, he asks the robber to take his car and escape. But it is too late. The robber returns, followed by a cop. He kills the policeman and Keanu kills the robber. Keanu then gives the autograph to the man who had requested it, and leaves the supermarket. End of film. The value of this film is that it was not merely a scene from an existing movie: some YouTube bloggers decided to make a new film and invited Reuben Christopher Langdon, an American motion-capture and voice actor, to perform as Keanu Reeves. With the aid of deepfake technology, they altered Langdon's face to that of Reeves. The result turned out to be super-realistic.

All of the video clips described above were created in a short period of time in 2019. The extreme popularity of such videos enables us to conclude that deepfake technologies, introduced at the end of 2017, are rapidly spreading on the web and can serve different purposes. They can be used for fun by talented YouTube bloggers; they also can be used in the film industry when it is necessary to edit an old film, to improve a weak scene or to finish the film following the death of an actor. The film industry already has experience of using computer reconstructions of deceased actors. One of the most well-known examples comes from the film *Rogue One: A Star Wars Story*, in which British actor Peter Cushing appears in the role of Grand Moff Tarkin more than 20 years after his death in 1994! His image was reconstructed with the aid of a common gateway interface overlaid on a real actor, Guy Henry, who ensured a motion capture. The continued modernization of this technology will render this complicated technological process much easier.

However, such technologies can also be used maliciously, for example by compromising a famous person by making it appear that they are participating in intimate episodes or by spreading panic among the masses using speeches of different well-known figures. For example, in April 2018 there appeared a fake video involving Mark Zuckerberg, the Facebook founder, describing how he is going to close down Facebook and make its entire services toll.

This final point offers the opportunity to conclude that "[i]n our present age of misinformation, society will soon have to deal with deepfakes that can threaten national security. Consider a deepfake of President Trump announcing impending nuclear missile attack on North Korea" [14, p. 102].

Let's take another example. The Flemish Socialist Party created a fake video in which President Trump, during one of his speeches at the White House, calls for his country's exit from the Paris Climate Agreement and calls upon Belgium to follow the American example and withdraw as well (to see this video follow the link: https://www.facebook.com/Vlaamse.socialisten/videos/10155618434657151).

Towards the end of the video, Trump says, "We all know climate change is fake, just like this video". Of course, when evaluating this video one can see that Trump does not look as real and natural as he should, but this represents an example of how anybody can make a politician or any other person say anything he or she wants using deepfake technology. The rationale of the Flemish Socialist Party was to simply "start a public debate" and "draw attention to the necessity to act on climate change". At the conclusion of the video, it "calls on signing a petition that urges investing in renewable energies, electronic cars and public transport. The petition also calls on closing the Doel nuclear plant in Flanders". In order to create this video the "Flemish Socialist Party even used services of professional video studio checked with them that this is a legitimate procedure" [30]. This last statement demonstrates how today there is no law or any other power to prevent the further distribution of such fake videos. And what comes next?

According to Simon Chandler [4], a freelance technology journalist:

> Deepfakes are only the first step in a chain of technological developments that will have one distinct end: the creation of AI clones that look, speak, and act just like their templates. Using neural networks and deep learning programs, these clones will first exist in video and in virtual worlds. Whether you're knowingly involved or not, they'll provide exacting reproductions of your facial expressions, accent, speech mannerisms, body language, gestures, and movement, going beyond the simple transplanting of faces to offer comprehensive, multidimensional imitations.

This danger even forced the American TV company CNN to speak about the threat posed by deepfakes. The episode is entitled: "Rise of the "Deep Fake": Doctored Digital Videos Posing Threat to 2020, Security" (to see this episode please follow the link: https://edition.cnn.com/videos/tv/2019/01/29/lead-jake-tapper-dnt-deep-fakes-politics.cnn). On this episode is provided a fragment of the report to Congress made by Dan Coats, Director of National Intelligence, in which he offers a very real warning about media manipulation posing a major threat to US security. According to Coats, the country remains a superpower, but it has real competitors, with new technologies offering them opportunities to narrow the gap very significantly. One example is the famous fake video of Barak Obama, produced by Jordan Peele and described above. This video is frequently used to illustrate the possibilities of AI-based deepfake technology. Further details have been provided through an interview with Jeff M. Smith, a researcher from the National Center for Media Forensic of the University of Colorado at Denver, where he explains the basic principles of the work in deepfake technology.

Eileen Donahoe, a former US ambassador to the United Nations Human Rights Council, has also described the dangers of deepfakes in an interview with CNBC. She argued that deepfakes should be seen as the next generation of disinformation.

> In a year or two, as the algorithms continue improving, it's unclear whether the average person will even be able to discern authentic videos from fakes. At that point, even video evidence becomes questionable, and perhaps even unbelievable. In a world that already can't agree on simple facts, the future looks pretty terrifying [1].

Based on the facts stated above, one can conclude that today the malicious use of ICT is growing in unprecedented ways. We are already suffering from information garbage because the vast majority of information being disseminated on the web is produced by ordinary people who often use pseudonyms, rendering their identification very challenging. Now we come to the question of how can we trust information from anonymous or semi-anonymous sources? How can we distinguish true information from fake? Deepfakes render identification much more difficult or even almost impossible.

> Deep fakes represent a turning point in information warfare. They will increase the reach of fake news and decrease our connection to a shared understanding of facts. If people cannot trust what they see and hear with their own eyes and ears online, then they will choose what they want to believe [7].

Therefore, it is crucial to think about this threat very seriously and to elaborate real working instruments on how we can counteract the further distribution of deepfakes.

5 Countering the Deepfakes: From Theory to Practice

In discussing the necessity of tackling the distribution of deepfakes, one can argue that it is the key task of the state to protect its citizenry from toxic content. The state possesses different instruments with various levels of severity:

> ... online, by shutting down political websites or portals; offline, by arresting journalists, bloggers, activists, and citizens; by proxy, through controlling Internet service providers, forcing companies to shut down specific websites or denying access to disagreeable content; and, in the most extreme cases, shutting down access to entire online and mobile networks [16].

But the question is whether all of the measures mentioned above would be able to achieve positive results. Probably not. Once a website is closed down by the authorities, a number of new ones with almost the same content will appear. Furthermore, the arrests of popular bloggers, activists and leaders of oppositional movements can trigger mass protests. Therefore, more accurate and workable instruments need to be developed. In our opinion, solving the problem lies in two dimensions: a technological one and a legislative one.

As Gregory [12] argues, "in the category of technical solutions, many platforms, researches and startups are exploring using AI to detect and eliminate deepfakes.

There are also new innovations in video forensics that aim to improve our ability to track the authenticity and provenance of images and videos, such as *ProofMode* and *TruePic*, which aim to help journalists and individuals validate and self-authenticate media".

Today, computer science specialists are working hard to create appropriate algorithms capable of detecting deepfakes. Such algorithms could then be used by all major social media platforms, including Facebook, Twitter and YouTube, which are supposed to check all uploaded videos for deepfakes before they are made visible and accessible to other users.

But the problem is how to ensure deepfake detection with 100% probability. According to John Villasenor [29], "deepfake detection techniques will never be perfect. As a result, in the deepfakes arms race, even the best detection methods will often lag behind the most advanced creation methods". Consequently, elaboration of an effective and workable deepfake detection method constitutes a complicated task. Today there are a number of technical specialists working on this technological puzzle, proposing algorithms aimed at detecting deepfakes. Scholarly interest in this subject is also growing.

Indeed, Li Yuezun and Lyu Siwei from the Computer Science Department of the University at Albany, State University of New York have attempted to elaborate a method that would be able effectively distinguish AI-generated fake videos from real videos. In 2018 they proposed a method "based on detection of eye blinking in the videos, which is a physiological signal that is not well presented in the synthesized fake videos" [18]. However, this method cannot guarantee the detection of deepfakes completely because it is impossible to do so only based on the eye blinking. Therefore, Li and Lyu continued their research and in 2019 proposed an updated version of their method.

> Method is based on the observations that current DeepFake algorithm can only generate images of limited resolutions, which need to be further warped to match the original faces in the source video. Such transforms leave distinctive artifacts in the resulting DeepFake videos, and we show that they can be effectively captured by convolutional neural networks (CNNs). Compared to previous methods which use a large amount of real and DeepFake generated images to train CNN classifier, our method does not need DeepFake generated images as negative training examples since we target the artifacts in affine face warping as the distinctive feature to distinguish real and fake images [19].

David Guera and Edward Delp from Purdue University have proposed the use of neural networks to detect discrepancies between multiple frames in a video sequence that often occur as a result of face replacement. Their method "uses a convolutional neural network (CNN) to extract frame-level features. These features are then used to train a recurrent neural network (RNN) that learns to classify if a video has been subject to manipulation or not" [13].

A research team from the University of California, Department of Electrical and Computer Engineering, Naval Air Warfare Center Weapons Division, California, and Mayachitra Inc., Santa Barbara, has also developed a method for the detection and localization of fake images using resampling features and deep learning. In fact, they proposed two methods:

> In the first method, the Radon transform of resampling features are computed on overlapping image patches. Deep learning classifiers and a Gaussian conditional random field model are then used to create a heatmap. Tampered regions are located using a Random Walker segmentation method. In the second method, resampling features computed on overlapping image patches are passed through a Long short-term memory (LSTM) based network for classification and localization [3].

Both of those methods are aimed on the detection of digital manipulations such as scaling, rotation and splicing which are normally used in fake videos.

Social media platforms such as Twitter and Facebook are also involved in the elaboration of effective algorithms for the detection of deepfakes. For example, Antonia Woodford, the Facebook product manager, announced in September 2018 that Facebook has created a machine-learning model that would be able to detect "potentially bogus photos or video, and then sends these to its fact-checkers for review. Third-party fact-checking partners can use visual verification techniques, including reverse image searching and image metadata analysis to review the content" [28]. The deepfake content will be detected it is removed from the platform.

> Facebook intends to use its collection of reviewer ratings of photos and videos to improve the accuracy of its machine-learning model in detecting misinformation in these media formats. It's defined three types of misinformation in photos and video, including: manipulated or fabricated content; content that's presented out of context; and false claims in text or audio. Facebook offers a high-level overview of the difficulties identifying false information in image and video content compared to text, and some of the techniques it's using to overcome them. But overall the impression is that Facebook isn't close to having an automated system for detecting misinformation in video and photos at scale [28].

The latter statement by Tung demonstrates that Facebook remains far from elaborating a real workable algorithm for the detection of deepfakes. Furthermore, these methods and algorithms are mostly checking content that has been already published on social media. Therefore, before the deepfake is identified and banned, it may have been viewed by millions of people. This is why it is extremely important to think about how to identify deepfakes before they are made available to Internet users. This is the task of moderators of social media platforms, who must use appropriate AI-based algorithms in order to prevent the further distribution of deepfakes on the web. However, the problem is that today deepfake technology "is evolving so rapidly that as quickly as we can find ways to counter it, its creators can adapt it to make it more convincing" [8].

The final implementation of all of these algorithms aimed at detecting and blocking deepfakes can additionally face serious legislative problems because this field remains almost fully unregulated by lawmakers. This is why it is extremely important to think about the modernization of all national systems of laws in the field of ICT. The case of deepfakes proves that today information technologies are developing much more quickly than the national legislature, and large gaps exist in law in this sphere because almost every day new technological solutions that need to be regulated by law appear. The EU's measures, for example, "are still designed to target the disinformation of yesterday rather than that of tomorrow" [20].

This perspective is confirmed by Chesney and Citron [5], who state that today there is "no current criminal law or civil liability regime bans the creation or distribution of deep fakes. A threshold question is whether such a law would be normatively appealing and, if so, constitutionally permissible".

This is because this issue has a large range of legislative conundrums. "[T]here could be a lot of interesting [intellectual property] cases if amateur filmmakers start synthesizing films using the likenesses of celebrities and start profiting of that". [22].

Hence the first legislative problem detected is the need to stop the illegal distribution of images of persons on the web (we would like to remind the reader here that in order to create a deepfake, a large number of images of a person need to have been uploaded). This is a very difficult task because every day users upload millions of private images to different social media platforms not thinking that 1 day somebody will use them maliciously.

The next legislative puzzle lies in the nature of those synthesizing films. If one uses the face of this or that person in the creation of the fake film without his or her agreement, is it a crime? The national legislature of any country must find an answer to this question. Finally, a legislative solution on how to protect the private life of any person is required in the digital age.

But when elaborating anti-deepfake laws, it is necessary to consider the fact that sometimes deepfakes can have a positive impact. In other words, it is very important to try to find a solution regarding "how to distinguish malicious deepfakes from other usages for satire, entertainment and creativity, how to distinguish levels of computational manipulation" [11]. This is why it is first necessary to define the malicious use of deepfake technology. Only then it will become possible to obligate social media platforms to monitor, verify and remove toxic content.

However, any state adopting anti-deepfake laws can face a number of serious problems. From the first site it is evident that when countering fake videos, it is necessary to develop criminal laws regarding terrorist propaganda, inciting ethnic hatred, distributing knowingly fake information, using images of people without their permission, or invading their privacy. On the other hand, every person has civil instruments available when protecting his or her interests. Thus, he or she can sue for slander or for being portrayed in a false light. He or she can also sue for the use of his or her image without permission, seeking to prove that somebody else is benefiting from it. But when filing a lawsuit or elaborating anti-deepfake laws, it is necessary to consider how best to harmonize counteraction to deepfakes with freedom of speech and expression, which are considered fundamental human rights and are strictly protected by international conventions and the constitutions of the vast majority of countries. In the United States, for example, freedom of speech is protected by the First Amendment to the Constitution. Moreover, somebody can try to prove that deepfakes represent new forms of art and self-expression.

This would appear to constitute a serious obstacle to the adoption of anti-deepfake law. Yet at the same time, without such a law it seems unlikely that social media platforms will start to implement algorithms targeted at identifying and blocking fake video clips. Therefore, the problem of the mass distribution of deepfakes will remain highly significant for the foreseeable future.

6 Conclusion

To conclude it is necessary to highlight how today "we are on the verge of having neural networks that can create photo-realistic images or replicate someone's voice in a pitch-perfect fashion" [15]. This technology is known as deepfakes and represents a method of synthesising human images based on AI algorithms. The technology has existed for only 2 years but it has already achieved significant results. Now it produces super-realistic fake videos that are almost impossible to identify by the naked eye.

Initially the technology appeared as a form of fun in the pornographic industry, when somebody changed the face of an obscure porn actress to that of a celebrity. Thereafter a large number of similar videos have appeared on the web. *FakeApp*, a special desktop application that simplifies the process of creating fake videos, has also been developed, offering opportunities to anybody with basic computer skills to create deepfakes.

Today the use of AI-based deepfake technology is not limited to the pornographic industry and almost everybody can create deepfakes for entertainment, for business or for malicious use. Therefore, it is possible to recognize both the positive and malicious use of deepfakes.

Undoubtedly deepfake technology presents a wide range of possibilities. As Chesney and Citrion [5] argue, deepfakes can be used in education and offer the opportunity "to manufacture videos of historical figures speaking directly to students, giving an otherwise unappealing lecture a new lease on life. The technology opens the door to relatively cheap and accessible production of video content that alters existing films or shows, particularly on the audio track, to illustrate a pedagogical point". Deepfake technology can also be used in the film industry as it is now possible to use images of actors who have died to make new films or improve scenes of low quality. Finally, deepfakes open up a wide range of opportunities in the development of interactive television, as a user can change the actors involved. Thus, one must view deepfakes not only as a new technology but as a new form of art and self-expression.

However, deepfake technology can also be used maliciously and poses considerable danger both to personal and national security. It is necessary to remember that initially deepfakes were used in the pornographic industry to replace a porn actress' face with that of a celebrity, a "joke" that compromises the latter by her "participation" in the film.

In discussing the malicious use of deepfakes, it is necessary to recognize three levels of use: for individuals and organizations, for society and for the entire nation. According to Chesney and Citrion [5], deepfakes offer great opportunities for plotters to exploit and sabotage others in order to obtain financial or other benefits (this is true of the first level of malicious use of deepfakes, targeted at individuals and organizations). For example, "blackmailers might use deep fakes to extract something of value from people, even those who might normally have little or nothing to fear in this regard, who quite reasonably doubt their ability to debunk the fakes persuasively, or who fear in any event that any debunking would fail to reach far and fast enough to prevent or undo the initial damage. In that case, victims might be forced to provide money, business secrets, or nude images or videos (a practice known as sextortion) to prevent the release of the deep fakes" [5].

The second level of malicious use is a threat to society. Fake videos offer major opportunities for plotters and terrorists to make politicians and other officials say and do things that they have never said or done.

> Fake videos could place them in meetings with spies or criminals, launching public outrage, criminal investigations, or both. Soldiers could be shown murdering innocent civilians in a war zone, precipitating waves of violence and even strategic harms to a war effort. A deep fake might falsely depict a white police officer shooting an unarmed black man while shouting racial epithets. A fake audio clip might "reveal" criminal behavior by a candidate on the eve of an election [5].

All of the examples mentioned above can cause deep political crises, end political careers, or even worse disturb relations between countries and thereby undermine international stability. This, the third level of malicious use, represents the greatest threat to national security.

Based on the above, it is possible to conclude that this technology bears a wide range of malicious use. But given that it can also be used for constructive purposes, one should think not how to completely stop the distribution of deepfakes, but rather to stop the distribution of toxic content.

In the opinion of this author, solving this conundrum will only be possible by combining technological and legislative methods. At the legislative level, it is necessary to elaborate a legal understanding of the malicious use of deepfakes and who (for example, service providers or social media platforms) should be responsible for detecting and blocking toxic content. At the same time, a workable AI-based algorithm aimed at quickly identifying and blocking deepfakes created for malicious purposes should be developed.

Given that this technology has only existed for 2 years, we do not expect a quick solution to this problem, although some algorithms aimed at identifying deepfakes have already been proposed and major social media platforms such as Facebook are conducting studies to try to block this content as soon as it is identified. Nevertheless, no country has yet adopted an anti-deepfake law. This is a considerable problem, because legal frameworks of deepfakes with clearly defined possibilities and cases of the legal use of such technology should be developed. Moreover, the main conundrum to be solved by lawmakers is ensuring a balance between forbidding the free distribution of deepfakes and protecting freedom of speech and self-expression, fundamental human rights protected by both international and national law. As has already been highlighted in this paper, deepfakes can be deemed a new form of art and self-expression. Thus, a simple banning of any fake video would grossly violate freedom of speech and self-expression.

Furthermore, the problem is that destructive elements that use deepfakes maliciously may also refer to the principal of the freedom of speech. This is why it is so crucial to solve this legal conundrum and distinguish the use of deepfakes for malicious purposes from the legal use of such technology as a new form of art. Considering the fact that the technology to quickly identify deepfakes continues to lag behind in its development relative to that of their production, fake videos are likely to continue spreading widely on the web.

Until this occurs, people will continue to face the serious challenge of navigating around the information space and distinguishing true from fake information: even video scenes that look very realistic could in fact be fake. Thus, we come to the necessity of improving the basic communicative culture and information literacy of ordinary people.

This is a complicated process that requires systematic preventive work at all educational levels, from school to university. Scholars with different areas of expertise (for instance, psychology, social sciences and computer sciences) should be involved in this process. Together they should elaborate a curriculum for preventive work. Simultaneously, it seems extremely important that: (1) mass media informational and analytical publications emerge in which experts explain examples of the malicious use of information technologies and how one should navigate in the growing informational flow, distinguishing true information from false; (2) the organization of psychological master classes and seminars at schools, universities and other educational centers in order to improve people's communicative culture and information literacy; (3) conducting preventive conversations with parents, who should explain to their children from an early age the potential dangers posed by the Internet. Furthermore, in order to organize systematic work on this issue, it is extremely important that every state elaborate and adopt the "National Concept on the Organization of Preventive Work with Population", including the core principles of work aimed at improving people's communicative culture and information literacy. Only then will it become possible, if not to stop the further distribution of deepfakes for malicious purposes, but at least to neutralize their negative impacts.

Acknowledgements The author acknowledges Saint-Petersburg State University for research grant 26520757.

References

1. Browne R (2018) Anti-election meddling group makes A.I.-powered Trump impersonator to warn about 'deepfakes'. https://www.cnbc.com/2018/12/07/deepfake-ai-trump-impersonator-highlights-election-fake-news-threat.html. Accessed 19 July 2019
2. Brundage M et al (2018) The malicious use of artificial intelligence: forecasting, prevention, and mitigation. https://arxiv.org/ftp/arxiv/papers/1802/1802.07228.pdf. Accessed 5 July 2019
3. Bunk J et al (2017) Detection and localization of image forgeries using resampling features and deep learning. https://arxiv.org/pdf/1707.00433.pdf. Accessed 30 July 2019
4. Chandler S (2018) Deepfakes 2.0: the terrifying future of AI and fake news. https://www.dailydot.com/debug/deepfakes-ai-clones-fake-news. Accessed 4 July 2019
5. Chesney R, Citron D (2018) Deep fakes: a looming challenge for privacy, democracy, and national security. https://papers.ssrn.com/sol3/papers.cfm?abstract_id=3213954. Accessed 26 July 2019

6. Chesney R, Citron D (2019) Deepfakes and the new disinformation war: the coming age of post truth geopolitics. Foreign Affairs, January/February. https://www.foreignaffairs.com/articles/world/2018-12-11/deepfakes-and-new-disinformation-war. Accessed 26 July 2019
7. Dack S (2019) Deep fakes, fake news, and what comes next, 20 March. https://jsis.washington.edu/news/deep-fakes-fake-news-and-what-comes-next. Accessed 12 July 2019
8. Fillion RM (2019) Fighting the reality of deepfakes. https://www.niemanlab.org/2018/12/fighting-the-reality-of-deepfakes. Accessed 30 July 2019
9. Fletcher J (2018) Deepfakes, artificial intelligence, and some kind of dystopia: the new faces of online post fact performance. Theatr J 70(4):455–471
10. Goodfellow IJ et al (2014) Generative adversarial networks, 10 June. https://arxiv.org/pdf/1406.2661.pdf. Accessed 12 July 2019
11. Gregory S (2018) Deepfakes and synthetic media: survey of solutions against malicious usages. https://blog.witness.org/2018/07/deepfakes-and-solutions. Accessed 23 July 2019
12. Gregory S (2019) "Deepfakes" are here, now what? https://internethealthreport.org/2019/deepfakes-are-here-now-what. Accessed 29 July 2019
13. Guera D, Delp E (2018) Deepfake video detection using recurrent neural networks. https://engineering.purdue.edu/~dgueraco/content/deepfake.pdf. Accessed 4 July 2019
14. Harris D (2018) Deepfakes: false pornography is here and the law cannot protect you. Duke Law Technol Rev 17(1):99–128
15. Heath N (2018) What is AI? Everything you need to know about artificial intelligence: an executive guide to artificial intelligence, from machine learning and general AI to neural networks. https://www.zdnet.com/article/what-is-ai-everything-you-need-to-know-about-artificial-intelligence. Accessed 27 July 2019
16. Hussain M et al (2013) State power 2.0: authoritarian entrenchment and political engagement worldwide. Ashgate, Surrey
17. Karras T et al (2018) Progressive growing of GANs for improved quality, stability, and variation, 26 February. https://arxiv.org/pdf/1710.10196.pdf. Accessed 12 July 2019
18. Li Y, Lyu S (2018) In ictu oculi: exposing AI generated fake face videos by detecting eye blinking. https://arxiv.org/pdf/1806.02877.pdf. Accessed 26 July 2019
19. Li Y, Lyu S (2019) Exposing deepfake videos by detecting face warping artifacts. https://arxiv.org/pdf/1811.00656.pdf. Accessed 23 July 2019
20. Meserole C, Polyakova A (2018) The West is ill-prepared for the wave of "deep fakes" that artificial intelligence could unleash. https://www.brookings.edu/blog/order-from-chaos/2018/05/25/the-west-is-ill-prepared-for-the-wave-of-deep-fakes-that-artificial-intelligence-could-unleash. Accessed 30 July 2019
21. Price R (2017) AI and CGI will transform information warfare, boost hoaxes, and escalate revenge porn. https://www.businessinsider.com/cgi-ai-fake-video-audio-news-hoaxes-information-warfare-revenge-porn-2017-8. Accessed 20 July 2019
22. Price R (2018) People are using creepy, cutting-edge AI technology to splice Nic Cage into every movie they can of. https://www.businessinsider.com/nicolas-cage-inserted-movies-fakeapp-ai-technology-2018-1?r=UK. Accessed 20 July 2019
23. Roberts J-J (2019) Fake porn videos are terrorizing women. Do we need a law to stop them? https://fortune.com/2019/01/15/deepfakes-law. Accessed 23 July 2019
24. Schwartz O (2018) You thought fake news was bad? Deep fakes are where truth goes to die. The Guardian, 12 November. https://www.theguardian.com/technology/2018/nov/12/deep-fakes-fake-news-truth. Accessed 26 July 2019
25. Singer PW, Friedman A (2014) Cybersecurity and cyberwar: what everyone needs to know. Oxford University Press, Oxford
26. Snedeker R (2018) 'Deep fake': Obama didn't say what he just said. YIKES! 8 May. https://www.patheos.com/blogs/godzooks/2018/05/deep-fake-obama-video-yikes. Accessed 15 July 2019

27. Takahashi D (2019) McAfee shows how deepfakes can circumvent cybersecurity. https://venturebeat.com/2019/03/05/mcafee-shows-how-deep-fakes-can-circumvent-cybersecurity. Accessed 25 July 2019
28. Tung L (2018) Facebook's fact-checkers train AI to detect "deep fake" videos. https://www.zdnet.com/article/facebooks-fact-checkers-train-ai-to-detect-deep-fake-videos. Accessed 25 July 2019
29. Villasenor J (2019) Artificial intelligence, deepfakes, and the uncertain future of truth. https://www.brookings.edu/blog/techtank/2019/02/14/artificial-intelligence-deepfakes-and-the-uncertain-future-of-truth. Accessed 30 July 2019
30. Von der Burchard H (2018) Belgian socialist party circulates 'deep fake' Donald Trump video. https://www.politico.eu/article/spa-donald-trump-belgium-paris-climate-agreement-belgian-socialist-party-circulates-deep-fake-trump-videoE. Accessed 19 July 2019

Considerations for the Governance of AI and Government Legislative Frameworks

Nishan Chelvachandran, Sonja Trifuljesko, Karolina Drobotowicz, Stefan Kendzierskyj, Hamid Jahankhani, and Yelda Shah

Abstract The speed and proliferation of AI and algorithmic technology has far outpaced that of the development of the legislative frameworks to which to govern them, to ensure their appropriate, safe and permissive use. It is not suggested that the development of these technologies and integrations are thwarted or inhibited, but more that there is a holistic review and understanding of the complex integrations between the moral, ethical, technological and legal concepts that their use brings. Multiple approaches must be made, utilising top down legislative mechanisms, bottom up consumer and citizen led engagement approaches, and cross sector and industry led standardisation and frame working. Such a cyclical process would ensure that the continual development and evolution of the appropriate instruments keep in pace with technological development. And with such synergic pace, will bring allow such considerations to be made at the design phase technological solutions, rather than taking a reactionary and sometimes unknown approach. A full understanding of new and emerging technologies is needed, how they interact and are interconnected, as well as their vulnerabilities, and causal effects, both direct and indirect, of the use of algorithmic and automated technology.

N. Chelvachandran (✉)
Open Innovation House, Saidot OY, ESPOO, Finland
e-mail: nishan@cyberreu.co.uk

S. Trifuljesko
University of Helsinki, Helsinki, Finland

K. Drobotowicz
Aalto University/KTH Royal Institute of Technology, Espoo, Finland/Stockholm, Sweden, UK

S. Kendzierskyj
Cyfortis, Worcester Park, Surrey, UK
e-mail: stefan@cyfortis.co.uk

H. Jahankhani · Y. Shah
Northumbria University, London, UK
e-mail: Hamid.jahankhani@northumbria.ac.uk

H. Jahankhani et al. (eds.), *Cyber Defence in the Age of AI, Smart Societies and Augmented Humanity*, Advanced Sciences and Technologies for Security Applications, https://doi.org/10.1007/978-3-030-35746-7_4

Keywords AI · Artificial Intelligence · Governance · Legislative · Smart devices · Human rights · Data · 5G · Ethics

1 Governance

Governments, Advisory and Governance bodies use traditional mechanisms by which governance and legislative frameworks are drafted and formalised utilising a "top-down" methodology, whereby consultancy can be conducted by panels of high level experts, and lawmakers, elected officials and advisors, in order to craft an the appropriate instruments to govern. Such a methodology, whilst providing a founding pillar to governance and legislative practice in many western democracies, is struggling to keep pace with the advancement of augmented and automated technologies, especially relating to the realm of Artificial Intelligence. Availability of continually improving and increased compute power is increasing at an exponential rate. Both consumers, private enterprise and public sector bodies are either exploring or utilising such technologies. The rapid approach to innovation that is commonly known as "move fast and break things", is clearly one of the drivers of the pace of technological progress, however, when coupled with the slower top down methodology of crafting governance processes, we see a legislative "lag", a chasm forming between where technology and use cases reside, and where the closest appropriate legislative frameworks lie [3]. As such technologies are have been used in "high impact areas", those which have direct influence and impact on the lives of users and citizens, when the "breaking of things" occurs, the lack of and need for appropriate legislation becomes clear. The use of AI is permeating into many and all areas of consumer and public constructs, where a decision-making process is in place, or a human decision tree utilising multiple data sets, can be augmented algorithmically. Despite the potential for enhancing efficiency and decision-making, it has also raised a great deal of concern, and some controversy in its application.

Artificial Intelligence in its simplest form, it an algorithmic mechanism utilised to perform tasks and make decisions, which are seemingly "intelligent". AI itself consists of multiple sub forms, examples of which are heavily utilised by government and industry alike. Machine Learning, for example, utilises datasets to make predictions, based on that data. NLP or natural language processing is a mechanism by which a natural human language is understood and processed. One of the more widespread and commonly known AI tools is computer vision, or image recognition, whereby algorithms process, identify and categorise images based on their perceived content. All of these mechanisms make the predictions and decisions based on learned logic from the datasets on which they have been trained. Concerns have been raised, both from theoretical discussion and actual real-world instances, where bias and discrimination are potentially being encoded into automated decisions.

There is no right or wrong answer to this dilemma that faces society in the Age of AI, however, there are steps in the right direction. Neither a top down nor bottom up solution in singularity will effectively synergise governance and technology. Instead, we must see an evolution of the management and devops cycles, a cyclical and continual ecosystem involving high level, consumer, citizen, private and public parties, involved in the drafting and formation of governance and legislative frameworks. It will be through enhanced cyclical processes involving multiple stakeholders throughout the process. AI in its use and by its nature can and will become pervasive, and so there is a need to establish such legislation and governance, and ensure that it is reflective of societal values, harbouring trust and the positive and effective use of AI and algorithmic systems.

2 Smart Devices

Arguable all the evidence indicates the need for an urgent intervention for a universal governance solution to address the security threats and privacy issues that have been discussed and mitigate any further future attacks on smart devices including, IoT and IoT landscape. Subsequently; the risk that IoT poses both to the consumer and the economy at large has also caught the attention of governments both nationally and internationally.

The UK government [7] announced that the security of the consumer IoT is considered a very serious issue as they recognised that many IoT devices sold to consumers are devoid of basic cyber security provision. They acknowledge that the current status quo is not sustainable, particularly in situations where the level of risks from compromised devices is often not isolated to just the undermining of a single user's privacy and personal safety but has been known to escalate to wider economy through DDOS (Distributed Denial of Service) attacks such as Mirai Botnet in 2016.

Considering the above-mentioned concerns, the UK government published a Code of practice in 2018 for IoT supplemented by guidance information to support industry to implement good security practices for consumer IoT. Despite these efforts and initiatives from the government to equip industries with the required tools in addressing these issues, there has been no significant improvement rather the consumer IoT market remain inundated with devices with zero level of security.

Now the government is taking a different approach by seeking to shift the responsibility away from the consumer for the security their own device and pushing the expectation to the device manufacturer to ensure that strong cyber security is built into these products from inception. The government recognise the importance of security to the consumer owing to a survey conducted recently. The population sample was 6482 consumer and the result showed that the third most important issues to consumers when they are purchasing an IoT device is security above privacy and design. In addition, those who did not rank security as a top-four consideration, 72% were of the notion that security is built into these devices as a default. Clearly there is a dearth of transparency between what consumer expects to buy and what they are buying. Hence the UK government attempt to restore the transparency by ensuring that device manufacturer openly share information about the cyber security of a device with consumers thereby empowering the consumer with information in deciding whether to purchase the product.

The case for a unified security and privacy governance and standardisation of IoT is eminent as more products and appliances that were traditionally offline can now be made into intelligent smart devices that add value to our every-day lives. The IoT paradigm introduces an environment where consumers now entrust an increasing amount of their personal data to online devices and services hence the urgent need for cyber security of these product. The government have taken a keen interest in combatting the security and privacy challenges of IoT as the wider economy now faces growing threats of large-scale cyber-attacks that hackers unleash by exploiting vulnerabilities of consumer IoT devices. In addition to the Code of Practice for Consumer IoT Security the Government is now working in collaboration with the Department for Digital, Culture, Media and Sport (DCMS) as well as the National Cyber Security Centre (NCSC) to introduce new mandatory industry requirements in which consumer smart devices are designed with basic level security. The DCMS released a consultation document as part of the proposal in May 2019. The proposal outlined in the consultation document offers insights to improve consumer's online security and privacy often compromised using insecure devices.

3 Legislation

To address the legislative gap relating to AI, we must first explore the realms in which AI is now operating, and where it is being proposed to be utilised, and by whom. We must identify what data is being used, the originator, application and outputs. This also brings into question the issue of jurisdiction.

The application of AI in private and public sectors varies and much across geographical regions as the legislation, political stance, population and cultural norms and acceptability of such technology. Aside from the challenges of jurisdiction, many of the legislative challenges that face cyber security today and, in the future, can be found in the issues of legislating and governing AI.

Further research and definitions are required into the categorisation of AI itself, which is turn allows for appropriate legislation to be considered, or new legal instruments to be drafted. What must also be considered is the implications and consequences of the use of AI in specific sectors. This is especially true of high impact sectors, human-affective sectors, such as Defence, Healthcare and the Judiciary.

To address this issue of both the "known unknowns", and "unknown unknowns" of which legal instruments govern such AI technologies and use therefore, some jurisdictions in the US, have opted to ban the use of such algorithmic mechanism, until such time as their use can be appropriately and effectively regulated. This can be seen with the Facial Recognition software ban, passed by the city of San Francisco, banning the use of Facial Recognition Software by public entities. Similar banes are under consideration in other jurisdictions in the United States, such as in the Commonwealth of Massachusetts.

In Illinois, the use of AI in the recruitment and hiring process, by means of interview bots has been restricted. The popular use of AI in this process was to use interview bots to evaluate personal characteristics of the interviewee, such as facial expressions, body language, tone of voice and vocabulary used. The software would then provide a report with feedback on the employability of the candidate. The law passed by the Governor of Illinois, known as the Artificial Intelligence Video Interview Act, is a disclose and informed consent rule, that requires employers to notify applicants for positions based in the state of Illinois, of any plans to have their video interviews analysed electronically. The employers would also need to explain to the job applicants how the AI analysis technology works and what characteristics will be used to evaluate them. The applicants consent to the use of the technology also needs to be obtained.

This is not to say that work is not currently underway to implement new legislation aimed directly at this issue. The US Federal Algorithmic Accountability Act was introduced in the US Congress in April 2019, seeking to enhance federal oversight of AI and Data Privacy. This act would allow the regulation of AI and automated decision systems that makes a decision or augments human decision making which impacts consumers. In this instance, organisations would be required to audit for bias and discrimination in their algorithmic systems and take appropriate correct action to resolve any issues that have been identified. The bill proposes that the responsibility for such oversight will be with the Federal Trade Commission. Whilst it is unclear if this act will be enacted into law, it is illustrative of the recognition that lawmakers in the US and indeed, globally are and will need to start to give to address this legislative inadequacy. Irrespective of this act, the State of California has passed the California Consumer Privacy Act, which can be seen to be like that of the European Union General Data Protection Regulation. As illustrated by these examples in the US, the evolution and "upgrade" of nominal data protection legislation to encompass AI seems to be the logical next step. Artificial Intelligence, by virtue, is an advanced data analytics mechanism, with data utilisation at the core of its functionality.

Earlier in 2019, The Office of the President of the United States issued an Executive Order on Maintaining American Leadership in Artificial Intelligence, with the White House launching AI.gov; a platform designed for government agencies to share AI Initiatives. The Office of Management and Budget in the US is expected to be issuing draft guidelines for the AI Sector this year.

The European Union has various groups and strategies, discussing and reviewing, more holistically, the realms in which AI is infiltrating. High Level Expert panels, from Academia and Industry are formulating draft frameworks and proposed better practices in which to governed and advise on ensuring that appropriate thought of consequence is given in the design and implementation of AI.

4 Frameworks

Another approach in the governance of AI is through the use of appropriate frameworks and standardisation. Legislation and structure and define the confines in which AI and its component mechanisms should operate within a legal jurisdiction.

However, governance frameworks are required to define the better practice, and set out greater breath of considerations and details which may not be defined by law.

Ethical considerations and implications are playing a large part in the draft formulation of such frameworks.

IEEE is one of the world's largest non-profit standardisation organisations, that are currently working on the drafting and formalising of a series of standards known as the P7000x; standards for the ethical use of what IEEE define as algorithmic and automated systems. These standards are drafted and compiled by workgroups of volunteers; subject matter experts, engineers, developers and business leads. They have also devised a framework for the Ethically aligned design of Automated and Algorithmic Systems. This is to address the problems and issues that are now coming to the fore, in a holistic manner, rather than treat such governance as a reactive "bolt on" solution.

Standardisation as a mechanism can align and both industry and the public sector, however, the appetite to subscribe to such standardisation and the application to such frameworks needs to exist. This falls back on the requirement of legislation to also act as a driver for standardising best shared practice.

Artificial Intelligence, as is and will be utilised, is the hybridisation of Humanity and technology, through social, cultural and ethical practice and consideration, to the technological compute processes involved in the mechanisms of deep learning, neural networks and machine learning [1].

And so, in the drafting of such frameworks, social values as well as techno-scientific dogma must be explored and agreed upon. However, this poses a complex challenge to the standardisation and framework process, and ethical practices and considerations in themselves, much like law, subject to jurisdiction or geographical variations and factors. The practices and societal objectives agreed by the majority in Western cultures, greatly differ from that of the Asian East, or Sub-Saharan African Continent. Priorities should be given to agreed inalienable fundamental rights, as a baseline for these considerations to drafting framework. As with the legislative diversity of thought, similar holistic considerations must be made with the standardisation and governance frameworks.

The Ethically Aligned Design Framework from IEEE defines agreed ethical and values-based design, development and implementation of AI systems, guided by the following principles:

- Human Rights
  AI shall be created and operated to respect, promote, and protect internationally recognised human rights
- Well Being
  AI creators shall adopt increased human well-being as a primary success criterion for development
- Data Agency
  AI Creators shall empower individuals with the ability to access and securely share their data, to maintain people's capacity to have control over their identity.
- Effectiveness

AI creators and operators shall provide evidence of the effectiveness and fitness for purpose of AI

- Transparency
  The basis of a particular AI decision should always be discoverable
- Accountability
  AI shall be created and operated to provide an unambiguous rationale for all decisions made.
- Awareness of Misuse
  AI creators shall guard against all potential misuses and risks of AI in operation
- Competence
  AI Creators shall specify, and operators shall adhere to the knowledge and skill required for safe and effective operation.

Each of these principles set the defining tone in which their framework for the ethically aligned designed of AI is set.

As well as ethical foundations, areas of impact should also be considered.

The further exploration into these topic areas reveals that whilst there is a great challenge to seek the appropriate alignment, and standardisation to build the frameworks for the governance and legislation related to AI, at this is not a problem or challenge that can solved in silo, and despite different conceptual or fundamental ways of implementation and practice, that collaborative efforts must be made.

Much of the work currently underway in this regard, is finding affiliation with the UN strategies around the Sustainable Development Goals, or SDGS [12]. Together with NGOs, the complex diverse challenges that face with world today, are can comparatively draw with those in AI, and in the wider context of Cyber Defence.

Human Rights are the backbone for all legislation throughout the democracies of society, and as such, is fundamental in the legislation and governance of AI, in that its use and implementation should be reflective of those Human rights. For these rights to be respected, lawmakers and legislative frameworks need to have these rights explicitly translated into their corresponding instruments. The recommendations given by IEEE are also that AI and related algorithmic systems should always be subordinate to human judgment and control. They also recommend that AI technology itself, should not be granted the rights and privileges equal to human rights.

5 Data Rights and Accountability

With AI utilising large amounts of data from multiple, traditionally siloed sources to garner their predictions and decisions made. And while data protection and fair usage legislation does exist, it is clear that this already needs to go a step further and give specificity when concerning data use by AI. People should have the right to access, benefit from and share their own data and the insights it provides.

Mechanisms are therefore required to help create and curate the conditions and terms regarding the access to their identifying and personal data.

It is also clear that AI technologies are simulating the attributes of human beings in terms of partial autonomy, and the ability to perform specific tasks. And so, there are broader legal questions that must be explored to ascertain how accountability and the allocation of liability can be apportioned when such systems cause harm.

It is for Government and industry stakeholders to identify the types of decisions, and functions that should not be delegated to an AI technology. By keeping the human in the loop in these sensitive, high impact areas, can ensure that rules and standards can govern these uses effectively, in the absence of new, specific frameworks and legislation. The outputs, products and manifestations created by algorithmic systems should also be protected and governed by national and international law.

Another area of consideration is trust, and effectivity. To realise the positive, and potential benefits of the implementation of AI, it's adoption and deployment must be responsible. AI will not be trusted unless it can be shown to be effective in its use. Harm caused by AI or as a result of a decision made by an algorithmic system can undermine its value and in turn, prevent further use and adoption. This is already being seen in the instances of Live Facial Recognition usage for Policing purposes, both in the UK and US, whereby, comprehensive mechanisms and processes were not in place to accommodate for any potential mistakes or inaccurate outputs from the algorithm used. Misidentification, or the biased identification of minority groups as suspected criminals, based on minimal and non-diverse training sets, and have led to the subsequent suspension of their use, as well as a public mistrust in both the technology and the authorities that use it. As part of the governance process, the practitioners and operators of the technology, and the "Humans in the loop" as part of the process, should have a greater understand on both the predictions and outputs of the algorithms, but all the possible failings and sampling errors that can occur, and how these should be taken into consideration when interpreting results.

Whilst the operations of AI systems need to be transparent to a wide range of users and stakeholders, there should also be a different level of transparency between stakeholders, dependant on the use and implementation of the AI system.

At face value, this may fall into the category of explainability of the AI system, however, the tasks performed by AI are often far more complex than those performed by previous technologies. A simple explanation to the decision made my not be available where the task is non-deterministic. This is very much the case with systems that interact with the physical world, such as those involved in medical diagnoses, or autonomous vehicles. The complexity of AI technology, and the non-intuitive way it operates, by design, will also make it difficult for users to understands the actions of it. Given that there is a degree of inexplicability to the function of AI, it is all the more important that transparency plays a key role in the governance of the technology. This is also vitally important in the context of auditing and investigation, dependant on the circumstance, when decisions, or the processes leading to and proceeding the decision-making process need to be scrutinised.

With the official introduction of 5G in 2020, ethical aspects need a predefined review with regards to user safety and public privacy. Furthermore, regulative agreements between government, network providers and public users are needed and shape and manage the overall degree of safety and security.

IEEE's [8] globally developed standards and use cases covers areas that are being monitored within 5G, for instance enabling smart cities and the Internet-of-Things, interoperability of technology as well as autonomous driving, which are connected to the internet. IEEE [8, p. 1] also addresses potential issues, such as

- [...] "Convergence of fixed, mobile, and broadcast services
- Multi-tenancy models
- Sustainability, scalability, security, and privacy management
- Spectrum
- Software enablement for Software-defined Networking (SDN), Network Function Virtualization (NFV), Mobile Edge, Fog Computing, and Virtualization" [...], which are already widely discussed throughout the dissertation.

In a report of GSMA [6, p. 4] "intelligent connectivity" is what is known as the potentially rising combination of 5G, IoT, smart landscapes and Artificial Intelligence (AI). Particularly, the ethics behind the junction of 5G and AI is of interest.

Seeburn [11] highlights the positive and rising features of 5G as fast, reliable and providing a proficient quality of service, which itself shifts technology through a transformation process in a sense that the handling of internet seems to be changing.

On the other hand, Seeburn [11] acknowledges the importance of finding an efficient solution for enclosing AI and 5G together. He goes further by recognizing that AI is intended to operate systems and machines with comparatively human intelligence while being reliable faster because systems executing tasks and analysing data are trained to eventually perform autonomously whilst acting cost-efficient. Merging speed, dependability and human-like intelligence levels, while factoring the technical aspect, rises both safety and ethical concerns [11].

However, the Internet of Things as well as AI are exposed to significant penetration attacks. Especially with the migration from current 4G/LTE- network to the 5G, the threat impact and its probability increases. In addition to that, GSMA [6] states in their new report that the mobile telecommunications industry (MTI) experienced IP-based threats in earlier years while the threat number will increase eventually the more heterogenic services the MTI carry in the long run.

The general consensus in the philosophical area defines ethics as a collective system of moral principles, which systematically distinguishes and analyses an individuals or societal decision-making process as well its impact on individuals or a society. However, nowadays ethics is an applicable tool on various societal fields, such as business and technology. The relationship between technology and ethics is interesting to witness, when considering a bidirectional impact. While a formally developed "Code of professional ethics" showed results within technological, scientific or engineering organisations shaping a project implementation until its final outcome, there are evidences of directional impacts on technology based on consumers choice, legal scripture and public technological and scientific participations [9].

In a research conducted by Chandran and Lobo [2] different methodologies were applied to examine the various approaches within corporate environments. In order for 5G to be implemented and linked with AI and IoT, individuals responsible for striving to secure privacy and data safety must adhere to IT-principles and ethics. Chandran and Lobo [2] comparatively distinguish between three essential concepts, the "Ethics Approach", the "Compliance Approach" and the "Value-based Approach". While corporate decision-making processes should be influenced and applied by principles of ethics to recognize the importance of ethical issues is inherent to the ethics approach [2]. The compliance-based approach within business environments relies on legislative actions for regulation, which then strive to integrate many legal scriptures that are compliant with the core of their businesses, such as [...] "technical legislation" [...] [2, p. 2]. Therefore, implementing 5G while establishing a link with core services running on the new network, must ensure intense data privacy and proper data compliance with international legislation, technical legislation, such as the code of ethics, code of conduct, as well as the general data privacy regulation (GDPR).

However, the European Union's General Data Protection Regulation setting a regulatory move towards data protection and the use of AI. Article 5 of the GDPR [10] refers to the limitation of organisation's data collection and processing to a minimum of what is required. This will eventually, limit corporations from acquiring more data without previously analysing it. That is when the lack of control begins because root of these security issues needed to be treated in earlier stages. Therefore, GDPR [[10], Art. 14.2 g; Art. 22], also points out that the provision of transparency and fair use of data must be given to the data subject especially when organization are able to utilize personal data of a data subject for automated decision procedures only.

For obvious reasons, article 22 demonstrates the regulation for the massive use of artificial intelligence within Europe. On an individual level this article sets a fundamental component for data safety as well as ethical protection. However, on the contrary, article 22 seems to decrease business growth and its continuity, which ultimately results in economic drawbacks because further researches on AI within the European parameter cannot fully evolve due to the legislation.

The technical specification of ETSI [4, p. 13] on the other hand, seems to be aligned with the GDPR in terms of deletion of consumers data because section 4.11; Provision 4.11-1 points out that [...] "Devices and services should be configured such that personal data can easily be removed from them when there is a transfer of ownership, when the consumer wishes to delete it, when the consumer wishes to remove a service from the device and/or when the consumer wishes to dispose of the device"[...]. The GDPR [10] equivalent is formalised in Article 17 – Right to erasure ('Right to be forgotten').

6 Moral and Ethical Considerations

The moral and ethical Considerations of the digitalization of societies and what can be referred to as the augmentation of Humanity, is only now being considered in the technology and academic communities. Now the prolificy of digitalization in society is spreading further from evolution of organizational functions and corporate processes, to tangible changes in societal norms, considerations must be made, both moral and ethical, and factored into governance and legislative frameworks [13]. These guidelines must be holistic in consideration and overarching by design, so the broad spectrum of considerations can be included, reflective of the areas of society and the socio-economic and demographic integrations we are seeing now and, in the future, to come. It is only through the inclusive and diverse considerations, through consultation, design and implementation, can we ensure that the digital evolution, or revolution, is reflective of the societal and cultural tenets which are to be digitalized.

The possibilities and scope for societal digitalization and the augmentation of humanity is, with the advancement and prolificy of technology, is becoming unbounded. Augmented Humanity is seen as the bridge of technology and human-computer interaction. Whilst augmented reality (AR) enables a digital overlay for a user's real world lives, through camera integration and displays, augmented humanity is the direct integration of technology into a user's body, or cyberorganic integration. This can already be seen through the implementation and utilization of implantables and smart devices, ranging from RFID chips implanted into the hand of a user for easy access and ID verification for use and access to vehicles, restricted work premises access or payment, to medical devices such as insulin pumps or pacemakers, with "smart" capabilities, to allow physicians to monitor patients remotely, and more acutely manage conditions that would otherwise have required patients to repeatedly travel into hospitals and clinics. On face value, the advantages and benefits of such technological improvement and advancement is clear, however, on a deeper study, it is clear that many other factors, such as security implications, moral, ethical, and even legal use of such integrated technology may pose a grave threat, that, if left unaccounted for, could lead to the great *problems* for societies and nations at large.

7 Misuse

Perhaps the most identifiable attribute in the context of Cyberdefence is the misuse of an AI system or technology. Legislative and governance frameworks provide a mechanism, by which proactive and reactive measures shape the appropriate, legal use and implementation of AI. However, this can also be effective if there is an awareness of hose the technology can be misused.

As with all new technologies, there is a greater risk of both accidental and deliberate misuse. The impact of hacking, data misuse, exploitation of vulnerable users and system manipulation is far greater when involving AI. AI systems have already been reportedly hacked, for instance, with autonomous vehicles [5]. The Tay AI Chatbot from Microsoft was also manipulated when it mimicked deliberately offensive users online.

The understanding and consideration of such misuse needs to be key in the creation and implementation of AI, and well as respective legal and governance instruments. Creators, users, and lawmakers need to ensure that there is a paradigm shift in culture and knowledge around AI and AI systems. It is only through the holistic cyclical review, testing and validation of these AI technologies, against equally evolving and iterated frameworks and governance models can we ensure that the foundations for more comprehensive legislation is in place to help protect and minimise the risk, both to users and the wider populous, but also the critical systems and national infrastructure that AI is and soon will be intrinsic in maintaining.

8 Conclusion

Legislation or governance frameworks in singularity are not and cannot be a panacea solution to appropriately implementing and designing AI. Traditional mechanisms to which these instruments are designed and ratified are not keeping up with the pace of the technological development and implementation of AI technologies. In the absence of appropriate legislation and governance frameworks, mistakes will, can and have been made. With the infiltration and pervasion of AI into critical and high impact systems, it is paramount that a holistic approach is taken by multi-level stakeholders, to ensure that a cyclical and informed approach is taken, and that the formulate legislation and frameworks are as overarching as the overarching AI and automated technologies to which they apply.

References

1. BSI (2016) BS8611: robots and robotic devices. Guide to the ethical design and application of robots and robotic systems
2. Chandran S, Lobo A (2016) Ethics and compliance in corporations: value based approach. In: 2016 IEEE International Symposium on Ethics in Engineering, Science and Technology (ETHICS), 1(1), 1–4
3. Dillmann R (2010) KA 1.10 benchmarks for robotics research. http://www.cas.kth.se/euron/eurondeliverables/ka1–10-benchmarking.pdf
4. ETSI (2019). "ETSI TS 103 645 V.1.1.1 (2019-02)- Technical Specification: Cyber; Cyber Security for Consumer Internet of Things". Available at: https://www.etsi.org/deliver/etsi_ts/103600_103699/103645/01.01.01_60/ts_103645v010101p.pdf
5. Greenberg A (2016) Hackers fool Tesla S's autopilot to hide and spoof obstacles. Wired
6. GSMA (2019) Intelligent connectivity: how the combination of 5G, AI, big data and IoT is set to change everything. Available at: https://www.gsma.com/IC/wp-content/uploads/2019/02/22209-Intelligent-connectivity-report.pdf

7. HMG UK House of Commons (2016) Decision making tansparency – report of the UK House of Commons Science and Technology Committee on robotics and artificial intelligence
8. IEEE (2019) Ethically alligned design
9. Layton D, Shakib J (2014) Interaction between ethics and technology. In: 2014 IEEE international symposium on ethics in science, technology and engineering, 1(1), 1–5
10. OJ L 119 (Official Journal of the European Union) (2016). Available at: https://eur-lex.europa.eu/legal-content/EN/TXT/PDF/?uri=CELEX:32016R0679
11. Seeburn, K. (2019). "5G and AI: A Potentially Potent Combination". [online] Available at : http://www.isaca.org/Knowledge-Center/Blog/Lists/Posts/Post.aspx?ID=1146#Comments
12. UN (2018) Sustainable development goal (SDG) indicators
13. Winfield A, Jirotka M (2017) The case for an ethical black box. Artificial intelligence

Part II
Augmented Humanity & Digital Society

Augmented Humanity: Data, Privacy and Security

Liam Naughton and Herbert Daly

Abstract Wearable devices have already changed the way in which humans communicate with the digital world. Advances in so called "in-body" devices may further revolutionize the way in which humans learn, play and work. However, new technology brings with it new risks and vulnerabilities. Augmented Human technologies have the potential to help human actors and organizations make better decisions. The data produced must be secured, collated and processed. Unless the integrity of the data is assured these decisions cannot be relied upon. There are also issues related to the privacy of data generated by augmented humans. Sharing and accessing data across multiple jurisdictions presents challenges around consistent application of regulatory frameworks especially regarding data ownership and security.

Keywords Augmented humanity · Privacy · Cyborg · Medical augmentations · Big data · ZigBee · Bluetooth · WIFI · Cyber security · Security

1 Introduction

The term *Augmented Humanity* (AH) is generally credited to former Google CEO Eric Schmidt who used the term in his keynote speech at the IFA (Internationle FunkAusstellung) conference in Berlin in 2010 [12]. Schmidt's discussion on how recent technological developments were nearing the realm of science fiction inspired him to coin the term. There have been many developments since then and it is perhaps difficult to settle on a definition of what AH really is and where the boundaries lie between AH and augmented reality and virtual reality. Certainly the lines are beginning to become blurred and we are entering a time where augmented

L. Naughton (✉) · H. Daly
School of Mathematics & Computer Science, Wolverhampton Cyber Research Institute (WCRI), University of Wolverhampton, Wolverhampton, UK
e-mail: l.naughton@wlv.ac.uk; herbert.daly@wlv.ac.uk

H. Jahankhani et al. (eds.), *Cyber Defence in the Age of AI, Smart Societies and Augmented Humanity*, Advanced Sciences and Technologies for Security Applications, https://doi.org/10.1007/978-3-030-35746-7_5

reality and physical reality are indistinguishable. Others talk about "man and machine integrated systems" and the idea that AH can enhance human biological capabilities in order to survive and surpass cognitive and physical abilities to achieve the next evolutionary stage. In the business world, one might describe AH as the practise of using artificial intelligence (AI) to gain competitive advantage or indeed one may frame AH as the answer to how AI and human intelligence (HI) can add value to each other. Recently, Augmented Humanity has been defined as "what happens when humans work in harmony with technology and machine intelligence to expand and enrich life, helping us to experience more and in deeper ways, to make better decisions and to fulfill our potential as humans" [15].

For the twenty-first century AH involves augmenting humans with devices which can collect data from the individual and from the individuals environment and transmit this data to an external device or service. Depending on the device this data may be intensely private to the individual but it may also be data which the individual wishes to share with external actors such as medical professionals. The data may include aspects which are personal or private to other actors in the users environment e.g. photographs. In any case there are privacy and security issues which must be addressed around AH data.

2 Augmented Humanity: Cyborgs

At this point it is helpful to discuss the fundamental role of data in the study and creation of Augmented Humanity as distinct from earlier ideas such as the Cybernetic Organism or Cyborg. The field of Cybernetics was proposed by Weiner [37], succinctly describing it as "the scientific study of control and communication in the animal and the machine". This wide ranging movement gave birth to many significant concepts in the study of the relationship between people and technology and is documented by Principia Cybernetica Web [24]. A Cyborg may be described as "an organism which is part animal and part machine" or focusing on utility "an organism with a machine built into it with consequent modification of function" [24]. Automatic and adaptive behaviour is achieved through autopoiesis or feedback response mechanisms.

Clarke [8] describes in details some differing kinds of Cyborg construction or intervention and distinguishes between the Prosthetic and the Orthotic. The prosthetic "provides the human body with previously missing functionality or overcomes defective functionality" while the orthotic "supplements or extends a humans capabilities" [8]. These definitions cover a broad range of uses the of technology for example they include users of artificial limbs (Prosthetic) and users of binoculars (Orthotic) as Cyborgs. Helpfully Clarke [8] also classifies Endo, Exo and External in each case where these are within the body, joined to the extremities of the body, or unconnected to the body respectively.

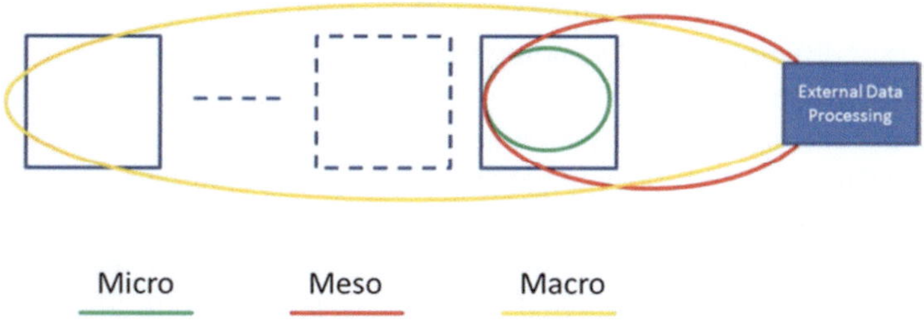

Fig. 1 Feedback response

Classification	Description
Endo-Prosthetic	Supporting capabilities integral to the body e.g. cardio pacemaker
Exo-Prosthetic	Supporting capabilities at the body extremities e.g. artificial limb
External- Prosthetic	Supporting capabilities external to the body e.g. walking stick
Endo-Orthotic	Extending capabilities integral to the body e.g. [36]
Exo-Orthotic	Extending capabilities at the body extremities e.g. smart contact lenses
External-Orthotic	Extension of capabilities external to the body e.g. Infrared goggles

A Cyborg then is, organic but at least in part a constructed artifact using technology to support or extend its natural capabilities through feedback response mechanisms (autopoiesis). In the simplest cases the mechanisms may be purely mechanical (springs responding to pressure) or purely organic (a decision made by the user based on observation). For our discussion in this chapter the Augmented Human is a distinctly data centric form of Cyborg with advanced capabilities for processing and sharing data.

An Augmented Human may display any of features classified above and various examples are discussed. Of the six classifications Endo-Orthotic examples are currently the hardest to find. Endo-Orthotic interventions challenge medical ethics, by using technology internally to enhance performance of an otherwise healthy person is considered questionable by many. Warwick [36] details an experimental intervention where an electrode array was interfaced surgically with the nervous systems of a consenting participant. The subject was then able to use the implant to collect data and operate devices remotely (Fig. 1).

External-Prosthetic and External-Orthotic are examples of using technology for help or support. For the Augmented Human however these devices may be used to communicate directly with Exo or Endo devices and so enhance their overall performance.

While some artifacts are wholly abstract, created from data, and others are wholly material; an Augmented Human has both an abstract and a material aspect. The "adaptive behaviour" which Wiener [37] focuses on is achieved through the "control and communication" of abstract data within the subject and in connection with its material being.

Moreover the Augmented Human (AH) is potentially capable of achieving autopoiesis at different levels including external layers of processing. The external processing may be Prosthetic or Orthotic depending on the analysis or services provided. For example external systems may be used to monitor and respond to long term health data collected about an individual. They may in turn integrate data about a group of users providing collective insights. They may also be used to provide services or extensions to existing capabilities e.g. remote payments. However we may view these as different levels of feedback response Micro, Meso and Macro enabling longer term more strategic behaviour, or service provision, with respect to the environment.

Feedback level	Description
Micro-Autopoiesis	Feedback response processed internally supporting reaction to localized data
Meso-Autopoiesis	Feedback response processed externally supporting complex processing of aggregate data for an individual
Macro-Autopoiesis	Feedback response processed externally supporting complex processing of longer term events aggregated across groups

Wu [38] describes a "Cyborg Intelligence" approach using sensory information fusion and machine learning techniques. Their hierarchical conceptual framework (See Fig. 2) describes an inequivalent three layer model for information processing and interaction in both machines and organisms.

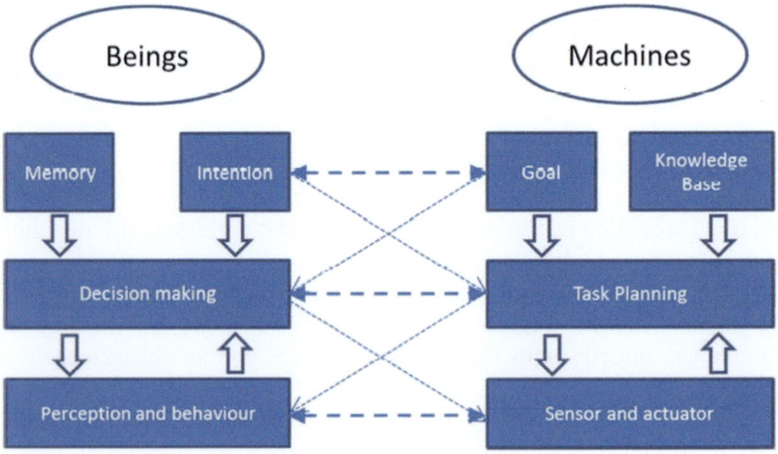

Fig. 2 Hierarchical conceptual framework [38]

3 Developments in Augmented Humanity

Many of the ideas around AH have their origins in the mid-twentieth century when scientists first started to think about the ways in which humans could be "enhanced". In his seminal work on cybernetics William Ross Ashby [2] wrote about amplifying human intelligence. Ashby suggested that intellectual power may be equivalent to "power of appropriate selection" and he argued that since power of selection could be amplified using artificial means then so too could intellectual power.

At about the same time Licklider [18] envisaged a world where man and computers would be "coupled together very tightly" and he predicted that "the resulting partnership will think as no human brain has ever thought and process data in a way not approached by the information handling machines known today". Licklider also recognized the role that the computer would play in what he referred to as "man-computer symbiosis". Together with colleagues at the Defense Advanced Research Agency (DARPA) Licklider laid the foundations for a future where computers could work with humans rather than as tools merely for computation. Licklider's colleague Douglas Englebert, [10], spoke about "taking a systematic approach to improving the intellectual effectiveness of the individual human being". He produced a conceptual framework for augmenting human intellect and went on to found the Augmentation Research Center (ARC). ARC developed new tools for information processing and played a major role in the development of the personal computer. A visionary aspect of the work of both Licklider and Engelbert was to see computers as devices which could help humans process data more efficiently. Engelbert described "a way of life in an integrated domain where hunches, cut-and-try, intangibles, and the human feel for a situation usefully co-exist with powerful concepts, streamlined terminology and notation, sophisticated methods, and high-powered electronic aids". Since the pioneering work of Licklider, Engelbart and others huge advances in human augmentation have been made.

3.1 Medical Augmentations

Much of the stimulus for developing augmentation solutions for humans has come about as a result of efforts to correct deficiencies or injuries. One of the most important devices developed is the cochlear implant. The modern cochlear implant was developed independently in the late seventies by two research teams based in Australia and Austria. The development of the cochlear implant marked the first time that a human sense had substantially been restored using a medical augmentation. A cochlear implant is a surgically implanted neuroprosthetic device which provides a sense of sound to individuals suffering hearing loss. The implant works by sending electric signals directly to the auditory nerve. The implant usually has an internal and an external component. The external component uses a microphone to pick up sound from the local environment. It then filters the sound to prioritize speech. The processed signal is then transmitted to the internal receiver which converts the received signal into impulses which stimulate the cochlear nerve causing it to send signals to the brain. In 2013 the developers of the cochlear implant were honored with the Lasker-DeBakey Clinical Medical research Award [17]. The cochlear implant is just one example of an augmentation which has been developed as a corrective approach to a condition. The technology behind the cochlear implant has stimulated research in other areas including eye prosthetics. One such example is the Retinal Implant project at MIT [39]. This innovation involves placing an array of electrodes behind the retina. The array receives images from a camera and then stimulates the retinal ganglion cells.

Another prominent area for medical human augmentations involves the development of prosthetic limbs. Physical augmentations for amputees now see bone-anchored prosthetics where a titanium prosthesis is directly grafted to the human skeleton eliminating the need for a socket interface [31]. Such biomechanical systems have revolutionized the treatment of amputees. Even more recently we have seen neural augmentations such as the NeuroLife system [4] which uses a brain implant and an electrode sleeve to give paralysis patients back control of their limbs. The general aim of such research has been to develop augmentations that can be controlled by the brain while also providing sensory feedback.

A prolific area of research into human augmentation since the early twentieth century involved the development of cardiac pacemakers. The first implantable pacemaker was developed in the 1950s and the decades since have seen a host of improvements and advances on this technology. The current state of the art for such technology includes the implantable cardioverter defibrillator (ICD), a device implanted in the body which can manipulate the heart rate and even perform emergency defibrillation. The ICD is a life changing augmentation for individuals at risk for sudden cardiac death.

3.2 Augmentation in the Twenty-First Century

Licklider and Engelbart recognized that human intelligence enhancement would enable humans to process information more efficiently. It is only in the very recent past that innovations have been developed which truly subscribe to this vision. We are now seeing a plethora of augmentations which involve on and in body devices which communicate with the individual as well as with outside controllers. For example, Spotify have recently developed a sensor which monitors the users heart rate and then uses an algorithm to choose music to suit the mood of the user. Meanwhile, Google Verily Lenses contain tiny integrated circuits, sensors and wireless communication capabilities for self contained wireless sensing on the surface of the eye. “The team has been working to engineer novel solutions to the technical challenges of significant miniaturization for autonomous sensing systems and dramatic reduction of power consumption to permit tiny batteries” [35]. Samsung have recently (July 2019) secured a patent to develop an augmented reality contact lens [22]. A key feature of this research is that the lens is designed to communicate with an external device e.g. a smartphone. The device "may include an antenna through which information may be transmitted to or from an external device, a capacitor configured to supply power to the display unit and a portion of the peripheral device, a control unit configured to control operations of the display unit and the peripheral device, a motion sensor configured to detect movement of the smart contact lens, and a thin-film camera" [22]. There are already a wide range of real-time in-ear translation systems readily available. Some models such as Google's Pixelbuds communicate through a smart phone while others such as the WT2 Plus Earbuds work via a cloud based translation service. The WT2 Plus system involves a pair of buds where everything the first user speaks is communicated to the cloud by the users earbud. The translated language is then communicated from the cloud directly to the second user's earbud using the chosen language of the second user.

Proteus, a digital medicine company, makes smart pills embedded with a sensor which can be tracked by a credit card sized patch worn on the patients stomach. "The sensor can either be stamped into a pill or included alongside a traditional medication and then encased in a translucent shell that breaks down when a patient swallows it. Then, patients attach a credit card-sized adhesive sensor anywhere on their stomach. The sensor tracks when the pill is ingested" [5]. The technology can also be used to see how active patients are by tracking their movements.

A defining characteristic of all of the twenty-first century AH devices described above is that they involve on or in body sensors transmitting data to external services. Whenever data is being transmitted, particularly personal data, there are a wide range of privacy and security considerations that must be taken into consideration.

4 Data from AH

All of the devices described in Sect. 3.2 produce data in one form or another. Some of the data is processed locally e.g. with a smartphone while more of the data is transmitted to the cloud for processing and subsequent analysis with feedback communicated back through the same channels in many cases. To date much of the data from AH devices has concentrated on lifestyle and health benefits. In 2016 it was reported that one in six consumers in the United States were currently using wearable technology, including smart-watches or fitness bands with the number of wearable fitness devices alone predicted to grow to over 100 million by 2019 [20]. Such devices have the potential to allow users direct access to personal analytics that can "contribute to their health, facilitate preventive care, and aid in the management of ongoing illness" [23]. The graphic in Fig. 3 is taken from [23] and it illustrates just some of the ways in which data is communicated via wearable consumer devices.

"Heart rate can be measured with an oximeter built into a ring, muscle activity with an electromyographic sensor embedded into clothing, stress with an electodermal sensor incorporated into a wristband, and physical activity or sleep patterns

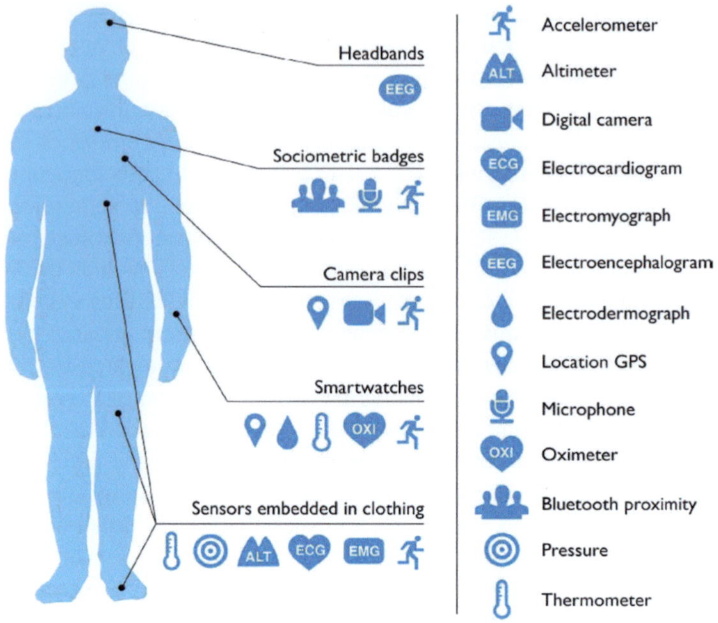

Fig. 3 Data from consumer wearables [23]

via an accelerometer in a watch. In addition, a female's most fertile period can be identified with detailed body temperature tracking, while levels of mental attention can be monitored with a small number of non-gelled electroencephalogram (EEG) electrodes. Levels of social interaction (also known to affect general well-being) can be monitored using proximity detections to others with Bluetooth or Wi-Fi-enabled devices. Consumer wearables can deliver personalized, immediate, and goal-oriented feedback based on specific tracking data obtained via sensors and provide long lasting functionality without requiring continual recharging. Their small form factor makes them easier to wear continuously." [23].

The data gathered from the various sensors and AH devices employed generally needs to be processed in a different ways. Some data processing may occur in the physical environment of the user e.g. via a smartphone or smart-watch, while further processing may take place after the data has been communicated to the cloud. A variety of options are then available for deep analysis and processing. This data will be useful for medical research, for customizing the behaviour of AH devices to match the users individual traits and characteristics e.g. Spotify choosing music based on the user's heart rate. the potential for such data is almost limitless.

"For the data transmission portion of the process, any variety of standard communication protocols may be used including Wi-Fi, Bluetooth, ANT, ZigBee, USB, and 2G, 3G, and 4G. An important recent innovation is Bluetooth low-energy (BTLE) which allows mobile devices to send data more efficiently with much greater battery efficiency than traditional Bluetooth, essentially enabling the regular ongoing if not continuous transmission of relevant data" [29].

The increasing ease of capturing, storing and manipulating data has given rise to a variety of technologies for sharing datasets and visualization tools. In the past, the cost and expertise required for working with large-scale datasets and visualizations generally limited access to institutional professionals, but cost decreases and tool improvements have made data collection and manipulation more available to the individual. One of the most interesting areas for individuals to measure is the self. An underlying assumption for many self-trackers is that data is an objective resource that can bring visibility, information and action to a situation quickly, and psychologically there may be an element of empowerment and control. Quantified self-tracking is being applied to a variety of life areas including time management, travel and social communications. Quantified self-tracking is the regular collection of any data that can be measured about the self such as biological, physical, behavioral or environmental information. Additional aspects may include the graphical display of the data and a feedback loop of introspection and self-experimentation. Health aspects that are not obviously quantitative such as mood can be recorded with qualitative words that can be stored as text or in a tag cloud, mapped to a quantitative scale, or ranked relative to other measures such as yesterdays' rating. Many health self-trackers are recording measurements daily or even more frequently (blood pressure for example) [28]. Furthermore, a number of scientific and popular publications describe methods and techniques for using consumer wearables as "self-hacking devices – to improve sleep, manage stress, or increase productivity [29]".

4.1 Data-Information-Knowledge

The field of Information Science and Knowledge Management has long debated the relationship between Data, Information, Knowledge and Wisdom. Of these wisdom is the most difficult to define and so is often excluded from the discussion. It is clear however that for Augmented Humanity applications to achieve their potential the management of data must support higher levels of insight and services.

Zins [40] explores expert views in the interrelationship between the concepts of Data, Information and Knowledge sometimes described as the "D-I-K model". In the context of the Augmented Human the distinction between the three relates directly to what may be achieved by collecting and transforming sensor data. It is sometimes suggested that the most significant relationship is in the ratio of the value between them as exemplified by Ackoff [1] "An ounce of information is worth a pound of data. An ounce of knowledge is worth a pound of information". The transformation between these categories is the effect of processing.

These maxims suggest essentially that a significant amount of data leads to information, and a significant amount of information leads to some knowledge. Though this view is somewhat superficial it may support the basic understanding of how data produced by Augmented Humans (AH) should be processed and managed. Firstly, key to the development of insight is multi-layered processing and refinement.

Although the terms are sometimes used interchangeably, even informally differences are apparent. A single data point, which some but not all of Zins [40] experts call a datum, allows for little or no inference; it is essentially a symbol. A data stream produced by a sensor can be analysed for its inherent properties, however without a contextual interpretation, such as whether it refers to heart rate, body temperature or blood pressure it is impossible to infer, or learn any useful information about the condition of the source. AH data requires appropriate interpretation frameworks to become information and appropriate analysis to become useful knowledge.

Schmarzo [25] describes the analytical approach for working with large data sets and strategies for identifying useful insights. Though there may be scope to apply such techniques for off-line reporting, the interactive nature for processing data produced by AH applications limits the scope of its effectiveness.

5 Security Issues with AH

The lessons of the recent past suggest that there are many security and privacy issues surrounding AH devices. One area which has gained much attention following the Facebook-Cambridge Analytica (see [6]) scandal is the ability of devices and their vendors to harvest vast quantities of personalized data without the consent of the user. In the Facebook-Cambridge Analytica case the data was used for political advertising purposes in an effort to manipulate the outcomes of elections. Users personal data was collected via their personalized Facebook accounts and became a target for external agencies. The legal gray areas around ownership of such data across jurisdictions also contributes to the problem. In the case of AH devices users are faced with the immediate problem of data ownership. After the data has been collected and communicated to the cloud it often becomes the property of the device manufacturer. While the user may own the device they may not have ownership of the data it records. Depending on the terms and conditions the user has agreed to, however unwittingly that may be, manufacturers may have permission to sell a users a data to third parties. This data may include highly personal data such as gender, weight, and GPS data.

On the other hand, it may be desirable for the user to share the data collected by one or more AH devices particularly if the devices form part of a personalized healthcare monitoring plan. There are many scenarios where it will be beneficial to the user for their data to be included in research studies and for this reason a security and privacy framework is necessary. It is not enough to simply anonymize user data since there are highly sophisticated algorithms available which are capable of identifying a user based on their digital behaviour. In fact the Facebook-Cambridge Analytica scandal described a scenario where a profile of each of two million users was created with "hundreds of data points per person" [14]. Combining similar algorithms with the data generated from AH devices will produce a digital fingerprint of each user which can be used to identify them. Perhaps even more alarmingly "Research on "digital traces" from other sources (e.g., social media) demonstrates that these can be alarmingly accurate when it comes to predicting personality and risk-taking behaviors, two very individual and personal traits" [23].

Another area of concern is the potential for direct attacks against individual AH devices. In 2008 modern implantable defibrillators were shown to be vulnerable to unauthorized communication, potentially harmful device reprogramming, and unauthorized data extraction [19]. The response from the Heart Rhythm Society (US) noted that the devices "were not designed to withstand a terrorist attack (see [13]). Advances in security for medical AH devices have been made since then however the threat of a cuber attack against an individual device cannot be discounted. There are a variety of reasons why a hacker might choose to attack an individual device besides the compromising of user data including damaging the reputation of the manufacturer or financial gain. It is also possible that device security may be compromised accidentally. Malware designed to attack a cloud based system may render an AH device useless either temporarily or permanently. As well as transmitting user data to an external service many AH devices may require occasional software updates. This is another vector which could be capitalized upon by an external hacker. AH devices may also be vulnerable to Denial of Service (Dos) attacks where the devices are flooded with so much communication that they are unable to receive essential communications. AH devices are by their nature low power and that renders them susceptible to attacks which seek to drain the batteries of such devices by repeatedly awakening them from 'sleep mode'. Depending on the device this could have catastrophic consequences. An attack on a non-essential system such as the cochlear implant may be deemed low risk due to the nature of the augmentation it provides whereas devices including pacemakers which have the ability to sustain life require additional safeguards.

5.1 AH Data and the Regulatory Landscape

From the previous discussion then, clearly, effective storing and processing of data is key to reaping the wider benefits of Augmented Humanity applications. Though it is possible as early systems have demonstrated to process data locally to its collection point this inherently has limits. For example for applications which process visual data; captured images, video, alternative visualization (e.g. infra-red, radioactivity) localized storage and processing will always be a limitation in terms of capacity and cause of systems failure. In certain kinds of systems, upgrades and repairs may be trivial, however those with intrinsic features may only be upgraded rarely or perhaps not at all. Many of the advanced use cases for AH, particularly those related to leisure and commerce, require the exchange of value, or some form of tokenisation. Typically these must be processed, or at least recognized beyond a single unit in order to have validity. Where the aim of augmentation is to overcome environmental issues the data collected is typically of greatest value in understanding the problem as a whole, co-ordinating the collaboration of groups or possibly collecting evidence of a defined quality required for later action. There is also, of course the potential for AH projects to evolve beyond their original brief. Data, once collected has the potential to be used or enriched beyond original conception. As more data sources become publicly available, for example through open linked data initiatives, it may be possible to create services, or studies, based on multiple data sources enriched or combined. (pollution and health?) (Stress at work?) (Disaster and emergency?) Moreover the use of artificial intelligence and machine learning to integrate autonomous systems in these applications presents issues about how data is integrated practically into the decision making process. Additionally a particular feature of the AH data is the potential for issues around privacy. These concerns will shape not only how data could be used, but also how it must not be used or shared in the development of services.

5.2 General Data Protection Regulation

The General Data Protection Regulation [9] is a regulation in European Union (EU) law on data protection for all individual citizens of the (EU) and the European Economic Area (EEA). It also addresses the transfer of personal data outside the EU and EEA areas. GDPR contains provisions and requirements relating to the processing of the personal data of individuals. Generally, if a user wants to use an AH device they have little choice but to agree to the manufacturers terms and conditions and this often gives consent for the manufacturer to collect and process the users personal data. The most relevant clause is GDPR Article 25, Data Protection by Design and by Default. Data protection by design (often 'privacy by design') is a concept which is generally attributed to Cavoukian [7]. The principle asserts that "we build privacy directly into the design and operation, not only of technology, but also of operational systems, work processes, management structures, physical spaces and networked infrastructure" [7]. Previously the adoption of data protection by design and default has been voluntary and a matter of good practice. GDPR makes it necessary for each data controller to consider "having regard to the state of the art and the cost of implementation and the nature, scope, context and purposes of processing as well as the risk of varying likelihood and severity for the rights and freedoms of natural persons, the controller and the processor shall implement appropriate technical and organizational measures to ensure a level of security appropriate to the risk". The elements of GDPR and, particularly, data protection by design and default provides guidance on the amount of data that AH devices can collect and process. However, in such a rapidly changing technological area it is difficult see how the GDPR can be considered 'future proof'. Nevertheless, it has raised consumer awareness to the extent that users of AH devices are likely to be more aware and cautious about the consequences of consenting to the terms, conditions and privacy policy of manufacturers.

5.3 U.S. Regulatory Frameworks

The regulatory framework in the U.S. is less clear-cut than in the EU. There are various federal acts which may be relevant including the Food, Drug and Cosmetic Act (FD&C) [33] which covers certain medical devices, the Health Insurance Portability and Accountability Act (HIPAA) [32] which gives users rights over some of their information, and the Federal Trade Commission Act (FTC) [34] which is perhaps the most important act with regard to AH devices. The Food and Drug Administration (FDA) does not classify wearable technology as a medical device within the FD&C act, rather it considers them to be low risk general wellness products which are not regulated by the FDA. Apart from specific clearly defined medical devices the FD&C act may not apply to AH devices in general. The HIPAA may provide some limited protection for data collected via AH devices but, as with the FD&C act, there must be a clear medical function to the design of the device or the collection and processing of the data it generates. The FTC act prohibits companies from engaging in deceptive or unfair acts or practices, including failing to comply with an organizations own privacy policy. The act is enforced by the FTC commission which can bring legal action against organizations that have violated consumer privacy rights and/or failed to maintain the security of sensitive data. Nevertheless, the liability, if any, of AH device manufacturers is far from clear under the above acts.

5.4 Cross Jurisdiction Privacy

Clarke [8] discusses at length the issues of Cyborg rights which are, by extension, rights of the augmented human. Privacy as a right however is not significantly explored and in the case of the augmented human this is a unique and significant vulnerability. Smallwood [27] discusses this issues and frameworks around information governance and more broadly Personally Identifiable Information (PII). This would apply in particular to operational data associated with the augmented human such as codes and identifiers related to the exchange of information and may at some points in the data transfer cycle be dealt with using encryption. The key issues presented by encryption are that it is typically a resource intensive process and software may require updating in the event of a security breech.

Ownership of data about oneself is the subject of discussion across jurisdictions. In the case of the AH although they may be protected by some legislation, the question of consent in the automatic generation of data is critical to establishing rights both of ownership and of privacy. For example, if an augmented human produces data that may be used to diagnose a medical condition, a number of questions emerge; Is their data anonymous or does it identify them uniquely? Do they have the right to access, copy or delete the data? Do they have the right to consent or opt out of different kinds of data analysis that may be applied?

As people travel globally and processing infrastructure may be needed locally to provide appropriate services to augmented humans, designers of these must consider the compliance issues for the legal requirements with respect to data privacy in different regions of the world. The table below illustrates only some of the legislation that may need to be taken into account (Table 1).

Table 1 Global legal frameworks relating to PII for AH data adapted from Smallwood [27]

Nation	Region	Privacy legislation
Morocco	Africa	Data Protection Act
South Africa	Africa	Economic Communications and Transactions Act 2002
Australia	Asia/Pacific	Privacy Act 1988
Hong Kong	Asia/Pacific	Personal Data Ordinance
Japan	Asia/Pacific	Personal Information Protection Act 1988
Philippines	Asia/Pacific	Data Privacy Act 2011
European Union	Europe	European Union Data Protection Directive of 1998 and EU Privacy Law 2002 (2002/58/EC)
France	Europe	Data Protection Act 1978 (Revised 2004)
Germany	Europe	Federal Data Protection Act 2001
Ireland	Europe	Data Protection Act 2003
United Kingdom	Europe	UK Data Protection Act 1998
Canada	North America/Central	Privacy Act 1983 and Personal Information Protection and Electronic Data Act (PIPEDA) 2000
Canada	North America/Central	Privacy Act 1983 and Personal Information Protection and Electronic Data Act (PIPEDA) 2000
Mexico	North America/Central	Federal Law for the Protection of Personal Data Possessed by Private Persons
United States of America	North America/Central	Privacy Protection Act 1980 and Video Privacy Protection Act 1988
Argentina	South America	Personal Data Protection Act 2000 (Habeas Data)
Brazil	South America	Article 5 of the 1988 Constitution
Chile	South America	Act on the Protection of Personal Data 1998
Colombia	South America	Law 1266 of 2008 and Law 1273 of 2009

6 Blockchain and Augmented Humanity

6.1 Blockchain Overview

Blockchain was invented in 2008 by a person (or several persons) using the name Satoshi Nakamoto [21] as the public transaction ledger of the cryptocurrency bitcoin. A blockchain is a constantly growing list of transactions which are collected into batches called blocks. Each block contains a cryptographic hash of the previous block. Blockchain is an immutable, distributed ledger that can record transactions between parties in an efficient and verifiable way. The immutable nature of blockchain means that a record cannot be altered retrospectively without altering all subsequent blocks. A transaction is added to the blockchain only after it has been validated via a consensus mechanism (e.g. mining). Blockchain technologies can be categorized into two main types: Public blockchains and Private blockchain. The differences between the two types will now be outlined.

6.1.1 Access & Permissions

Public blockchains are generally permissionless. Anyone can read or write data to the blockchain and their is no predetermined criteria to take part. All transactions are visible to all other participants. Examples of public blockchains include Bitcoin [3] and Ethereum [11]. Private blockchains only allow certain authorized entities to participate. They have the ability to grant specific rights and restrictions to participants on the network. They are considered to be more centralized than public blockchains since only a small group of participants control the network. An example of such a private blockchain is Hyperledger [30].

6.1.2 Consensus

Transaction on public and private blockchains are verified using a consensus based system but there are different ways in which consensus can be reached. With public blockchains consensus mechanisms are often based on an incentive scheme which rewards participants for making some contribution to the network. One of the most widely used consensus schemes, which is used with several cryptocurrencies, is 'proof-of-work'. In such a scenario 'miners' contribute their computing power to solve cryptographic problems to verify transactions and they are rewarded in the form of a token. Proof-of-work is often computationally intensive and expensive in terms of time. This often leads to slow transaction speeds and high electricity costs. Other consensus schemes such as 'proof-of-stake', 'proof-of-capacity' and 'proof-of-elapsed-time' can also be used. In a private blockchain consensus is often reached via 'selective endorsement'. In this framework only a defined group of entities can verify transactions. An example of such a consensus framework is 'proof-of-authority'.

6.2 Blockchain for AH Data Security and Privacy

Blockchain technology can be used in any system which involves a database and as such it can provide tools to handle data issues around authentication, security and privatization. The immutable nature of the technology means that once a transaction has been made it cannot be altered or tampered with and this provides security. Modern blockchain technologies now come equipped with robust authentication and identification systems that permit different levels of access to the blockchain. For example, one type of access might allow users to write data to the ledger whereas another type of access might allow other users to act as monitors while still a third type of access might allow a user to administer the transfer of value (coins, tokens etc.) between users. An advantage to such a system is that it keeps different types of users and their access to the data compartmentalized. Such a system of access permissions could be produced for AH data which allowed e.g. users to write data to the blockchain via their AH devices while still allowing e.g. medical research professionals to access the data for research purposes. In the light of recent scandals, such as the Facebook-Cambridge Analytica scandal, consumers are becoming more conscious about the security of their personal data. Trust has disappeared from the process so users need a platform where data can be shared in a secure and tamper-proof way.

6.2.1 Validation

AH data stored in a blockchain is encrypted so that unauthorized data modification is a difficult task. The use of cryptographic signatures (hash functions) means that users can verify that a file has not been tampered with by merely inspecting the signature rather than analysing the entire file. Signatures can be cross checked with others across all of the blocks to verify their integrity. If an unwanted agent (e.g. hacker) changes a record then the signature will be invalid. In this way blockchain facilitates reliable data validation (Fig. 4).

Fig. 4 Processing

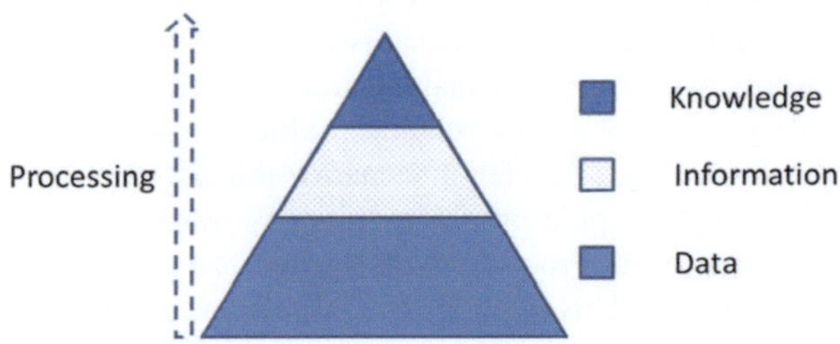

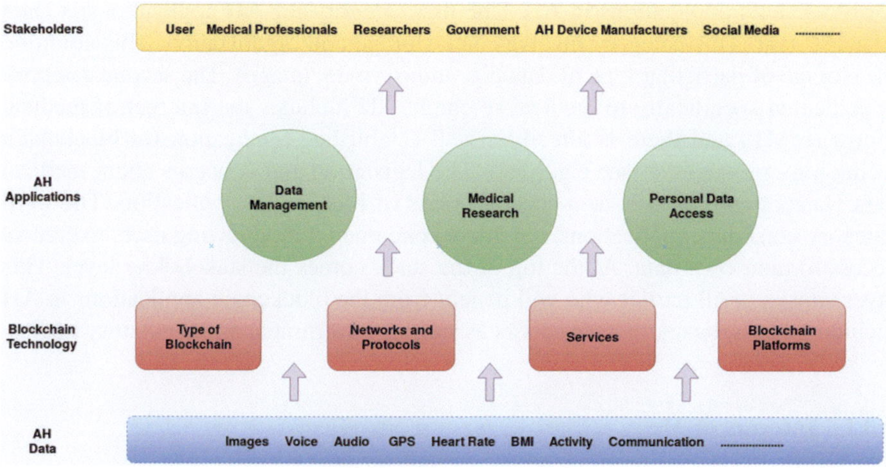

Fig. 5 Workflow of blockchain AH applications

6.2.2 Decentralized Verification

The decentralized nature of blockchain means that it does not rely on any single point of control. Every device on the network has access to a complete copy of the data. The lack of a single gatekeeper makes the system more secure. Consensus is reached across the network in a democratic way to validate transactions and record data. This ensures that the data stored is accurate and trustworthy. For example, the integrity of research carried out using AH data can be trusted since all nodes have access to the data, a difficulty with much current research in fields such as medicine.

6.3 Applications

Figure 5 illustrates a blockchain based workflow for AH applications. The first layer of the diagram consists of the AH data collected by the various AH devices. Given the diverse nature of the devices available this includes GPS data collected by trackers, voice recorded data captured by ear buds, audio data recorded from the user environment, medical data such as heart rate, level of activity as well as communication data transmitted to and from the users' various AH devices. Blockchain technology sits on top of the raw data. In this workflow this layer is divided into four components although there may be others. Each of the platforms has different characteristics and enables users to instigate and manage transactions. Once the chosen technology has been implemented the next step involves integrating this technology with the wider system. In Fig. 5 we have outlined three broad areas for blockchain applications in AH. The first category for AH applications is Data Management. This category involves data storage and organization. This could be the storage of particular type of data e.g. audio, voice, images. The second category is dedicated specifically to medical research. AH includes the Internet of medical things (IoMT) and there is already a well established application for blockchain technology in this area (see e.g. [26]). The legislation and concerns about medical data warrant treating this as a separate class of blockchain application. The third category considers applications which are concerned with allowing users to control access to their own data. At the top of the stack comes the stakeholder layer. This layer consist of all parties who will benefit from the blockchain applications in AH including users, business, researchers as well as government and regulatory bodies.

6.3.1 Internet of Me

In [16] the authors provide a comprehensive review of the use of blockchain technology in healthcare and also suggest directions for future research. One focus of the article (Sect. 5) is the Internet of Medical Things (IoMT). 'With IoMT "healthcare equipment such as heart monitors, body scanners and wearable devices can gather, process and share data over the internet in real time. For example, with the advancement of AI, healthcare providers, using the IoMT paradigm, can capture an image, identify malignant parts or even suspicious cells, and share such knowledge with those who have the right to access the information" [16]. IoMT is certainly a subset of Internet of things (IoT), but some aspects of AH can also be considered as a subset of IoT. One can characterize these aspects of AH which are a subset of IoT as 'Internet of Me" (IoMe). For example devices such as cochlear implant and implanted defibrillators can be considered as both situated within IoMT and IoMe. In similar fashion to [16] one can illustrate IoMe within blockchain via the diagram in Fig. 6.

The user is the source of all data in IoMe. The next level consists of the IoMe AH devices which are generally either attached to or implanted in the user. They typically generate a large volume of data. The devices may be directly connected to the Internet or they may be in e.g. bluetooth communication with a local device e.g. smartphone which is itself connected to the internet. The data from this stage is stored in e.g. cloud storage. At the next stage "AI will help blockchain to create intelligent virtual agents, which in turn can create new ledgers automatically. In case of sensitive medical data, where security is the first priority, decentralized AI system could help block chain to reach highest security" [16]. The final stage involves the end users and this can include health professionals, marketing analysts, employers or government organizations.

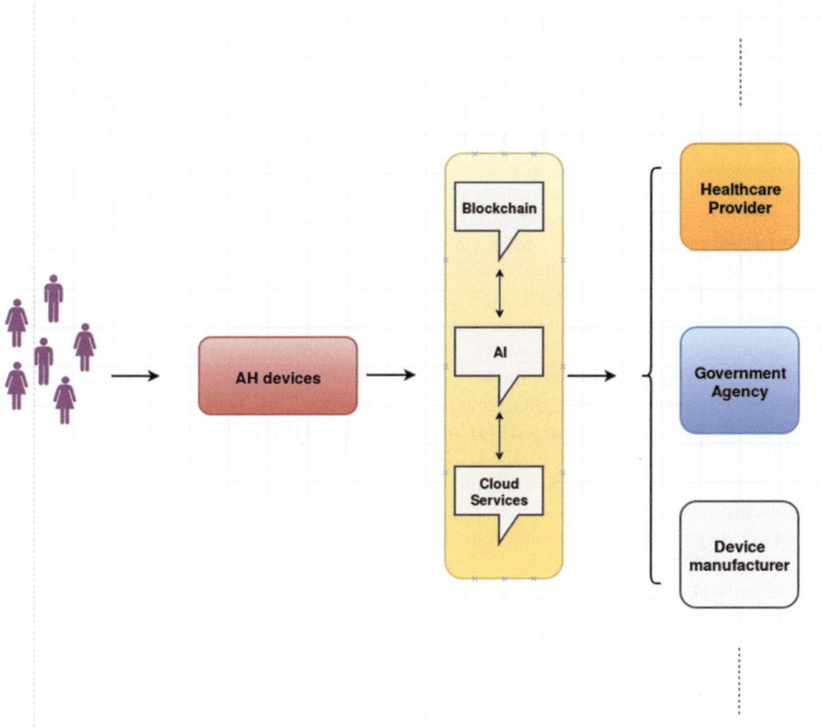

Fig. 6 Data flow in AH

7 Concluding Remarks

There are many concerns around the security and privacy of data generated by AH devices. The rapidly developing nature of the field of AH means that there is also a lack of a clear overarching framework for the regulation of the data landscape concerning AH. Questions around ownership and protection of data still require clear answers. How best to combine the various regulatory frameworks that are available in different jurisdictions is a question that still requires much consideration. Fortunately, there is a body of research beginning in the mid twentieth century with the work of Ashby [2] and others which laid the foundation for the next sixty years of development. The more recent work of Wu [38] and Clarke [8] on Cyborg rights and intelligence provide a starting point for considering how best to approach issues around acceptable treatment of AH data from the user point of view. There are also opportunities, particularly in terms of using AH data for research. Blockchain is one technology which can play a major role. Similar to its applications in IoT and IoMT the applications of blockchain in IoMe will be many. More evaluation of blockchain applications in IoT and IoMT is needed before the potential in IoMe can be fully realised. More work is needed in this area to fully understand the implications and applications of AH data.

References

1. Ackoff RL (1999) Ackoff's best: his classic writings on management. Wiley, New York
2. Ashby WR (1956) An introduction to cybernetics. Wiley, New York. https://www.biodiversitylibrary.org/bibliography/5851
3. bitcoin.org (2019) Bitcoin. https://bitcoin.org/en/. Accessed 18 Aug 2019
4. Bouton CE et al (2016) Restoring cortical control of functional movement in a human with quadriplegia. Nature. https://doi.org/10.1038/nature17435
5. Business Insider (2019) A silicon valley company just launched 'smart' cancer pills that track you with tiny sensors stamped into your medications
6. Cadwalladr C, Graham-Harrison E (2018) Revealed: 50 million Facebook profiles harvested for Cambridge analytica in major data breach. The Guardian
7. Cavoukian A (2010) Privacy by design: the definitive workshop. A foreword by Ann Cavoukian, Ph.D. Identity in the Inf Soc 3(2):247–251
8. Clarke R (2010) Cyborg rights. Proc Int Symp Technol Soc 30:9–22
9. Council of European Union (2014) Council regulation (EU) no 269/2014. http://eur-lex.europa.eu/legal-content/EN/TXT/?qid=1416170084502&uri=CELEX:32014R0269
10. Engelbart D (1962) Augmenting human intellect: a conceptual framework. AFOSR. Stanford Research Institute, Menlo Park
11. ethereum.org (2019) Ethereum. https://www.ethereum.org/. Accessed 15 Aug 2019
12. Gannes L (2010) Eric schmidt: Welcome to "age of augmented humanity"
13. Halperin D, Heydt-Benjamin TS, Ransford B, Clark SS, Defend B, Morgan W, Fu K, Kohno T, Maisel WH (2008) Pacemakers and implantable cardiac defibrillators: software radio attacks and zero-power defenses. In: 2008 IEEE symposium on security and privacy (SP 2008), pp 129–142
14. Hern A (2018) Cambridge analytica: how did it turn clicks into votes? The Guardian
15. Isobar (2019) Augmented humanity, Isobar trends report
16. Khezr S, Moniruzzaman M, Yassine A, Benlamri R (2019) Blockchain technology in healthcare: a comprehensive review and directions for future research. Appl Sci 9:1736
17. Lasker Foundation (2013) 2013 lasker debakey clinical medical research award
18. Licklider JCR (1960) Man-computer symbiosis. IRE Trans Hum Factors Electron 1:4–11
19. Maisel WH, Kohno T (2010) Improving the security and privacy of implantable medical devices. New Engl J Med 362(13):1164–1166. PMID: 20357279
20. Moar J (2018) Juniper research smart wearable devices. Fitness, healthcare, entertainment and enterprise 2013–2018
21. Nakamoto S (2009) Bitcoin: a peer-to-peer electronic cash system. Cryptography Mailing list at https://metzdowd.com
22. Patently Mobile (2019) Samsung wins patent for augmented reality contact lenses
23. Piwek L, Ellis DA, Andrews S, Joinson A (2016) The rise of consumer health wearables: promises and barriers. PLOS Med 13(2):1–9
24. Principia Cybernetica Project (2019) Welcome to principia cybernetica web. http://pespmc1.vub.ac.be/. Accessed: 05 Sep 2019
25. Schmarzo B (2013) Big data: understanding how data powers big business. Wiley, Indianapolis
26. Seliem M, Elgazzar K (2019) BIoMT: blockchain for the internet of medical things
27. Smallwood R (2014) Information governance: concepts, strategies, and best practices. Wiley CIO. Wiley, Hoboken
28. Swan M (2009) Emerging patient-driven health care models: an examination of health social networks, consumer personalized medicine and quantified self-tracking. Int J Environ Res Public Health 6(2):492–525
29. Swan M (2012) Sensor mania! the internet of things, wearable computing, objective metrics, and the quantified self 2.0. J Sens Actuator Netw 1(3):217–253
30. The Linux Foundation (2019) Hyperledger. https://www.hyperledger.org/. Accessed: 02 Sep 2019

31. Thesleff, Alexander, Brånemark R, Håkansson B, Ortiz-Catalan M (2018) Biomechanical characterisation of bone-anchored implant systems for amputation limb prostheses: a systematic review. Ann Biomed Eng. https://distill.pub/2017/aia
32. United States (2004) The health insurance portability and accountability act (hipaa). Washington, D.C.: U.S. Department of Labor, Employee Benefits Security Administration
33. US Congress (1934) United states code: federal food, drug, and cosmetic act, 21 U.S.C. Retrieved from the Library of Congress,
[Periodical]. https://www.loc.gov/item/uscode1934-006021009/
34. U.S. Government (2018) Federal trade commission act
35. Verily (2019) Smart lens program
36. Warwick K, Gasson M, Hutt B, Goodhew I, Kyberd P, Schulzrinne H, Wu X (2004) Thought communication and control: a first step using radiotelegraphy. IEEE Proc Commun 151:185–189
37. Wiener N (1948) Cybernetics; or control and communication in the animal and the machine. Wiley, New York
38. Wu Z, Zhou Y, Shi Z, Zhang C, Li G, Zheng X, Zheng N, Pan G (2016) Cyborg intelligence: recent progress and future directions. IEEE Intell Syst 31(6):44–50
39. Wyatt Jr J (2011) The retinal implant project. Research Laboratory of Electronics, MIT. http://www.rle.mit.edu/media/pr151/19.pdf
40. Zins C (2007) Conceptual approaches for defining data, information, and knowledge. JASIST 58:479–493

Consumer Awareness on Security and Privacy Threat of Medical Devices

Anthonia Sagay and Hamid Jahankhani

Abstract The Internet of Things (IoT) are being enthusiastically adopted by consumers. By the year 2020 the sum of 31 billon IoT devices will be deployed globally. Subsequent as the IoT device landscape is expanding at such speed, so does the threat landscape and vulnerabilities it introduces increases. Thus, making IoT devices easily prone to attacks or to be used to for launching attacks at large economical scale and society is seeing a growth in the scale and frequencies of these attacks. The large scale of attacks and frequency have caught global attention and causing governments to take the security and privacy threats of IoT very seriously and the UK government amongst others are now turning these concerns into actionable measures by considering ways of protecting consumers against the vulnerabilities and threats of IoT. It is part of these actionable measures that the NCSC (National Cyber Security Centre) recently published in a report about the new laws being proposed by the government to strengthen IoT devices. This chapter will look at the IoT security threats and privacy issues, it will explore whether the growing concern of the government to protect consumer has a foundation by investigating consumers awareness and attitude towards IoT security threats and privacy issues and propose a framework to facilitate the introduction of the new initiative of the government to bring in laws to govern IoT products thereby shifting the responsibility of the security threats to the manufacturers and away from the consumer.

Keywords IoT · IoMT · Privacy · Abuse · Cyber attack

A. Sagay · H. Jahankhani (✉)
Northumbria University, London, UK
e-mail: Hamid.jahankhani@northumbria.ac.uk

H. Jahankhani et al. (eds.), *Cyber Defence in the Age of AI, Smart Societies and Augmented Humanity*, Advanced Sciences and Technologies for Security Applications, https://doi.org/10.1007/978-3-030-35746-7_6

1 Literature Review

The Internet of Thing (IoT) is a technological development phenomenon which is enhancing more and more ubiquitous connectivity around the world. It has succinctly eliminated the barriers in product design capabilities by allowing everyday basic devices to be internet-enable thereby adding significant value that were not previously possible or available to these devices. Consumer devices such as webcams, thermostats, watches, TV and many more now have internet capabilities and functionalities. According to J. Hou, L. QU and W. Shi (2019) IoT has enlarged the communication capabilities of Information Communication Technologies (ICTs) from "Any Time" and "Any Place" to "Any Thing".

IoT has introduced a connectivity paradigm of Machine-to-Machine (M2M), Machine-to-Man and Man-to-Man with identification management and control processes. The breath of IoT landscape is so vast and a typical architecture connects from front-end devices to back-end frameworks running in the cloud. Generally, this architectural landscape will include a significant number of smart devices, sensors that sense information from different environment and share them with cloud services for further processing (M. Aly, F. Khomh and M. Haoues et al. 2019).

The key role of the smart sensors in IoT is to assemble, measure and evaluate data and this function is what makes IoT so attractive and powerful as the measured data can be utilised to meet the requirements of any industry. Empirical research demonstrated that IoT offers the healthcare industry a great opportunity in refining operational adequacy, developing and enhancing patient care as well as promoting innovation. The healthcare industry bolstered as IoT made it possible to enable everyday device to provide intelligent data wherever and whenever simply by attaching these devices to the patient; information can be gleaned unhindered by any network or services. By building sensors into these simple things which are then embedded/worn in or on the body the healthcare industry can gather enormous amount of data about patient's health status. Having access to data in this manner opens new possibilities for the healthcare industry and radicalised treatment offered to patients and reduces the care of cost with enhanced results. However, all of these introduces multi-level complexity as more vulnerabilities are introduce creating severe security challenges across the IoMT landscape.

1.1 Security Threats and Privacy Issues of IoMT

The Norwegian research organisation SINTEF reported that in the past 2 years, 90% of the world's data has been produced at a speed exceeding 205,000 gigabytes per second and this was approximated to the equivalent of 150 million books. The data collection of IoT spans the healthcare, retail, transport, manufacturing and many more industries for which IoT can provide smart services by extracting valuable information from diverse collection of data at the IoT end-point devices, which has significant impact on social production and people's life. Base on the important role that data plays in IoT, it can be inferred that discussing IoT without considering data is incomplete. The healthcare IoT market is expected to reach $117 billion by 2020 according to market research. This rapid growth of IoMT has raised considerable concerns around disclosure of personal privacy information particularly around sensitive medical data.

According to an article published by the British Medical Journal in July 2017 the Healthcare sector is more susceptible to cyber-attacks than other sectors owing to the inherent weakness in its security position. The article stated that it is one of the most targeted sectors globally. Amongst the 223 organisations that participated in the survey 81% were from the medical sector and over 110 million patients in the US had their data compromised in 2015 alone. Furthermore, only 50% of the providers were confident that they could defend themselves against cyber-attack and record shows a 300% increase in attacks in the past 3 years. The health sector is an attractive target for two reasons: it offers a rich source of valuable data and it is an easy target. Data is at the core of IoT so much so that researcher are inferring that just as monitoring blood in the human body provides valuable insights into people's health, observing data in an IoT environment could provide significant insight into the security of IoT. Evidently the healthcare sector is a storehouse of valuable data and according to the British Medical Journal another primary reason it is targeted is for financial reward and benefits owing to the nature of the data that can be gleaned. The sum of 80 million records were stolen from Anthem, a US health insurance company and the monetary value of this data on the dark web was estimated to range in billions of dollars. Unlike credit card data that can easily be reset, an individual's medical record could contain sufficient information for a perpetrator to open a bank account, obtain loans or acquire a passport basically fully cloning the victim's identity.

1.2 Recent Cyber-Attacks on Health Sector and the IoMT Threat Landscape

In May 2015 according to the BMJ (British Medical Journal) there was a global cyber-attack unleashed in form of the WannaCry Ransomware; although this attack was not specifically targeted at the healthcare sector and affected around 200,000 systems in more than 150 countries according to the reports. An estimate of about 50 hospitals in the UK were directly hit by the WannaCry Ransomware attack whilst many more in anticipation shut down computer systems causing considerable disruption, impacting the delivery of care, jeopardising patient safety and potentially eroding trust.

In 2016 the Hollywood Presbyterian Medical Centre was compromised due a ransomware attack causing it to shut down its network for 10 days resulting in staff not having access to medical records or being able to use medical equipment until the ransom was paid. The cost of this attack was estimated at $17,000. Another ransomware incident also reported in 2016 was an attack on an English hospital and the impact meant all operations were mandatorily cancelled and patients were transferred to other facilities for 2 days. Freedom of Information request in the UK reported that between the years 2015–16 around 50% of NHS trusts were affected by ransomware in the preceding year. The Australian Red Cross Blood Service reported a breach in 2016 which resulted in the publication of 1.28 million records with large amount of sensitive data, including donor's at-risk sexual behaviour on a public website.

According to a recent article published by Fortinet cited by Adefala (2018) as the healthcare sector technology (IoMT) grows, so does the cybersecurity attack surface. Frost and Sullivan forecast that by 2021 IoMT will reach a growth of $72.02 billion with over 30 billion connected medical devices in the healthcare ecosystem [7]. IoT has transcend the medical sector by introducing numerous IoMT-based platforms, applications and services that enabled remote health monitoring, fitness programs, chronic diseases and elderly care. Guan et al. states that IoMT offers unconventional solutions to the challenges of traditional medical system such as lack of doctors, health resources and research data. In addition, the rapid development has enhanced traditional medical systems in diverse areas, such as disease diagnosis and analysis. Furthermore, the health data gathered in IoMT enables researched to diagnose and predict diseases.

The attractiveness of IoMT combined with its terminal devices is causing exponential growth in the data collected. With the endless possibility that IoMT is offering the healthcare sector, it growing popularity is understandable. Notwithstanding this rapid and excessive growth is a major contributing factor in the expansion of the attack surface making it extremely difficult to address using traditional devices and strategies. Hence the urgent need for cybersecurity to protect the confidentiality, integrity and availability of valuable healthcare data.

The futuristic trend of IoT has not only successfully revolutionised the healthcare sector but has also enable a complete merger of the cyber world to the physical world to create what researchers are calling the cyber-physical world. Due to the cyber-physical nature of IoT, there is a need to consider the security of IoT from a unified perspective by considering both safety and security. Wolf and Serpanos [10] introduced the concept of considering the cyber-physical characteristics of IoT in view of a unified security model from the perspective of safety and security. Unquestionably, IoMT can be classified as a safety-critical cyber-physical system because it comprehensively considers both the reliability and safety of conventional medical devices, as well the dynamics and generic nature and the scalability capabilities of traditional IoT. IoMT devices are designed to constantly interact with the physical world.

Therefore, it can be inferred that safety and security should be considered as a critical challenge for IoMT especially given the severe consequences of an attack and the extensive attack surface. The physical devices in the IoMT infrastructure are embedded with sensors to form a connected ecosystem which is then tagged around the patient to capture, measure and identify key data; stratify risks; make decisions and initiate the necessary action plan. These sensors and controller embody the communication bridge between the cyber and physical world which ironically are major contributors to the vast security threats and privacy issues facing IoMT landscape. These sensors and controllers utilise applications available on phones or web therefore, from a security perspective these devices are susceptible to attacks and exploit in the same manner as a traditional endpoint device such as desktop computer.

Thus, once an attacker can identify a vulnerability, the damages could range from taking total control of the system, accessing and altering the data, flooding and overwhelming the system; the possibilities of malicious attacks are endless. Nevertheless, there are some significant differences that must be considered for IoMT security over traditional technology. Firstly, the accelerated adoption of IoMT has been identified by existing research to pose great security threat and privacy issue owing to the absence of proper security guidance, a landscape of uncertain liability, new standards and emerging polices and regulations.

Typical example of the landscape uncertainty was identified in a recent article on the Metro published in July 2019 regarding the ownership of digital footprint in the event of the death of an individual. The article reported that according to Survey conducted last year by YouGov only 7% of participant consented to keep their social media account active upon their demise, although another study by Oxford Internet Institute (OII) approximated that by the year 2100 the number of dead people whose account will still be active on Facebook will be 4.9 billion. There is now a debate around the ownership of data upon the death of an individual and the question of who should own the data. Should it be Facebook or the deceased family and friends.

Craig Badrick reported in his article publish in January 2019 that the FDA (Food and Drugs Administration) estimated that for over 1000 IoT devices in use 164 are subject to attacks. Subsequently as the hospitals introduce more and more applications for IoMT they risk the likelihood of introducing devices that may put their operations and patient's life in jeopardy. Arguable manufacturers must be made accountable as currently majority of the IoMT devices are not specifically optimised for hospital security network. Regulators such as FDA and other industry standards are falling behind the times and only 17% of medical manufacturers have been reported to be taking steps towards preventing attacks. Given the severity of the nature of the risk and what it is at stake it is shocking to report that security attributes in IoMT devices at unreliable at best. It is common place to find an IoMT with unencrypted communications, weak or non-existent password protection, or setup that make it more problematic or impossible to patch the device for improved security.

The IoMT threats are grave as it impacts both individual patients as well as the entire hospital system. 2017 recorded the recall of 465,000 pacemakers owing to a report that they have been hacked putting patients' lives at risk. In addition, another case reported that about 95% of healthcare institutions have at some point been targeted. Another incident reported in 2015 by the Health and

Human Service Office of Civil Rights that 112 million health records had been breached or compromised in that year. The compounding evidence calls for urgent standardisation across the IoMT ecosystem infrastructure and an intervention with clearly defined accountability for both the manufacturers of devices and those using the device.

1.3 Characteristic of IoMT Comparable to IoT and Cybersecurity Requirements

The IoT landscape is very broad and as such the consultation and the consultation-stage impact assessment set out to define consumer IoT products and included in the examples produce was wearable health trackers which falls under the IoMT category and is the focus of this study. Hence the need to consider the characteristics of IoMT comparable to IoT in other to ascertain if a single framework can be applied across board as best practice with the flexibility for manufacturers and organisation to make adjustments suitable to their environment or perhaps a distinct framework may be required for IoMT.

The interconnection of IoMT are not limited to personal medical devices but it extends from devices to healthcare providers such as hospitals, medical researchers or private companies. Furthermore, existing research identifies that personal smart product are generally wearables it is therefore understandable that the DCMS consultation document has identified and defined IoMT under this category. Gatouillat et al. suggest that owing to the strict ethical concerns of the medical community, biomedical devices must adhere to the following three requirements:

1. Reliability – the expectation here is that the functional goals of the system must always be reliable and should not be susceptible to abrupt failures under normal operating conditions. Fundamentally, the potential diagnostic nature of IoMT-based systems puts reliability at the core of every system component to ensure the correctness and validity of information collected.
2. Safety – this implies that a safe system ought not to cause harm to its operating environment therefore IoMT particularly in the context of medical actuators concrete evidence should be available to ascertain that the system will not cause harm to its user.
3. Security – Medical systems ought to be unyielding against external threats and attacks particularly owing to the sensitive and personal nature of the information they accumulate.

According to Alsubaei et al. [1, 2] the healthcare industry has the highest number of IoT devices; ranking at about a third of all IoT devices and this number is expected to increase by 2025 which will make healthcare the largest sector dominating the IoT device market with an estimated percentage of around 40% of the total global worth of IoT technology ($6.2 trillion). The uptake of IoT in the healthcare sector is unprecedented and currently the number of organisations in the healthcare sector that have adopted IoT technologies is approximated at 60% and this is expected to increase to 87% by 2019.

Based on the extraordinary growth, evolution and dominant of IoMT technologies the evidence suggests that there is an urgent need to address the security threat and privacy issues in this sector especially when the consequences range from severe impact on patient's wellbeing, damaging outcome on medical data privacy, brand reputation, business continuity and financial stability. Furthermore, the array of complexity as identified by Jalali and Kaiser [7] including the dearth of consensus amongst internal stakeholders on security requirements, the disparate technology environment couple with the complexity of multiple channel of IT technology acquisition, internal politics complicated by the intricacies of functions contained within the organisation and additional regulatory pressure. Thus, this study will focus on proposing a framework for IoMT devices that support the implementation of the new laws being proposed by the government.

1.4 Fundamental Objectives of the DCMS Proposed New IoT Security Law

DCMS commission Harris Interactive to conduct a consumer IoT (Internet of Things) security labelling survey in March 2019. The survey identified that consumers have a complacent attitude toward seeking out security information about their smart devices. According to the survey 72% of the respondents naively assume that security features are built into these devices as a default. To alleviate and manage the unprecedented consumer assumption which poses a risk not just to the consumer but the wider economy at large the DCMS are mandating IoT device manufactures to introduces labels that clearly highlights and outlines the security features of a device to help consumers to be better informed about the security attributes of smart devices when making purchase. Subsequently these labels aim to reinforce consumers confident by emphasising that devices meet security standards and provide information on the minimum period for manufacturer security updates. The images below represent the draft design for proposed label at this initial consultation stage. Although there are still concerns and other issues surrounding this design, but these discussions are outside the remit of this study (Fig. 1).

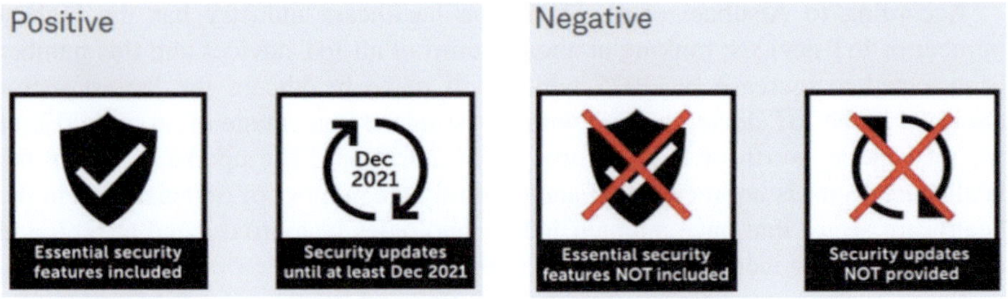

Fig. 1 Draft design of the DCMS IoT security label

1.5 Best Practice and Framework Consideration for Cyber Security Comparable to the New IoT Law

Robert Meyer an expert in assessing the relationships between frameworks proposed in his presentation (November 2016) the core values of Information Technology frameworks. According to Meyer IT frameworks can offer the following:

- Better value creation through effective and innovative use of enterprise IT
- Increased business user satisfaction with IT engagement and services
- Increased compliance with relevant laws, regulations and polices
- Improved relationship between business needs and IT objectives
- Increased financial return from the governance over enterprise IT by obtaining the greatest value from investments in technology
- Connection to, and where relevant alignment with other major frameworks and standards in the marketplace

Information Technology has been described as ubiquitous and critical for business matters within companies, between interconnected companies and/or private individuals for cloud computing solutions, Internet of things, connected and mobile devices and many more internet usages. Due to this indispensable nature of IT risk management has become prevailing [3]. Essentially risk management activities within all domains ought to be under control either for dedicated risk management purposes or for a broader perspective in management systems. Hence the domain focus of this study IoT/IoMT must adhere to risk management controls.

Significant amount of the discussion of this chapter has been centred on the security and privacy risk of IoT/IoMT and according to Brenner [3] there is unquestionable ties between information security and risk management therefore the proposed framework to facilitate the implementation of the new IoT security law will be anchored on best practice standard(s).

SO/ICE 27001 provides a set of guidelines and requirements for developing and implement an information security system and it has controls built in that ties it closely to risk management. In addition, this standard takes an agnostic approach for any specific technology offering the organisation the opportunity the best and most practical controls for their organisation [3]. Pulling this back into the context of this study one of the key requirements of the proposed new IoT security law is that manufacturers still have the flexibility of innovation whilst implementing the appropriate security solutions on the devices.

This standard has been selected specifically based on its relevance to the discussion of this study; also, having established that one of the advantages of standard and framework is the flexibility to align and synergise guidelines and controls to form a robust solution. Furthermore, the selected standard is agnostic in its approach but comprehensive in providing controls around risks and security management.

ISO Standards are common place in today's business world as acceptable standards for benchmarking and identifying organisations who follow best practice. Typically, frameworks are designed to be adopted and tailored to an organisation needs in terms of policies, procedures, industry and/or services rendered. Much of the discussions in this chapter thus far have focused on the security and privacy risk of IoT and as such the framework proposed will consider various aspects of Risk and Information Security Management thus the following ISO standards will be considered for inspiration; ISO/IEC 27001 – Information Technology Security Techniques and Information Security Management Systems Requirements; ISO 13485 Medical Devices – Quality Management Systems and Requirements for Regulatory Purposes; ISO/IEC 30161 Internet of Things – Requirements of IoT data Exchange Platform for Various IoT Services and BIS 31000 International Risk Management Guidelines [9]. NIST IR 8228 will also be considered because it covers Cybersecurity and Privacy Risk for IoT and offers some useful hierarchical structures for identifying/grouping of attack vector surface as well as a good construct of how to mitigate the risks. Furthermore, the scope of this study is around IoT Security and Privacy threats that affects consumers and the guidelines proposed in the NIST IR 8228 is relevant to the objective of this study. Reference will be made to other types of risks that should be considered alongside the Security and Privacy threat focus of this study including safety, reliability and resilience owing to the nature of the knock-on effects that isolation of one risk can have on other risks.

2 Users Profiling and Smart Device Usage

Harris Interactive conducted a study on behalf of the DCMS which reported that 72% of the participants believed that security features were built into smart devices by default whereas the government are on a mission to protect consumers after having identified that majority of the smart devices in the public market domain do not even meet basis security requirements. Evidently there are discrepancies in these two camps.

This study intends to examine consumer awareness and attitude towards security threats and privacy issues of IoT in other to highlight the urgent need for the education of the consumer. In addition, the study will investigate the ongoing efforts of the government to protect the consumer against the security and privacy risks of IoT. To conduct this investigation a quantitative, descriptive questionnaire survey was used as a primary data source and literature reviews as well as government consultation and legislative documents as secondary data source [8]. The questions were designed using an online survey tool as it offered simple but professional looking user-friendly design, provided different medium to distribute survey to participants and collates all the responses in a central location. Most important of all it allowed respondents to remain anonymous and saved resources in terms of time and money.

The audience targeted to participate in the questionnaire survey were between ages 16–65 that owned some form of smart device. Participants were randomly selected as the basis of the study is to identify the topic from the perspective of the general public. A total of 256 participant received a web link to the questionnaire via social media platforms such as LinkedIn and Facebook. Friends and family were also approached to participate in the experiment and were encouraged to share the link to others in their social network to diversify and ensure that the generalist criteria of the required responses are adhered to.

Quantitative research method was selected for this study because it focuses primarily on numerical data and interprets this information using statistic under a reductionist, logical and strictly objective paradigm. Traditionally social science research often utilises existing completed studies in form of literatures that relates to or addresses the hypothesis [8] cited Spyros Konstantopoulos). Thus, narratives gleaned from existing research contributed significantly in designing the survey questions to establish status quo on consumer's awareness and attitude towards security threats and privacy issues around IoT. Questions were constructed to examine the following four areas:

- Consumer awareness/knowledge of the concept of digital footprints
- Consumer smart device security awareness
- Consumer awareness of the potential damage that a security or privacy breach can cause
- Consumer priority preference of smart device benefits versus security and privacy concerns

The survey comprised of 32 questions and was disseminated to 256 participants of which 133 responses were received.

As previously mentioned, secondary data in form of literature review contributed significantly to the findings of this study. The secondary data reinforces the concerns in the dearth of awareness of consumers on the topic of security threats and privacy issues of IoT. Evidence of this was derived from an existing research conducted by Harris Interactive in February 2019 which revealed that consumers have a complacent attitude towards seeking out security information about their smart device and 72% of respondents innocently assumed that security features are built in to smart devices by default. Another secondary data which also strengthened the argument of the study was taken from a survey conducted by Internet Society in May 2019 which reported that consumers have serious concerns about the security of their smart devices but are not knowledgeable about how to adapt and adjust device settings in a manner that might deter these fears.

Ethical consideration has become critical particularly on the recent entrance of GDPR regulation. Therefore, in a bid to ethically align this study the following considerations were considered. Scope of study was clearly defined, and all participants were provided with explicit explanation of the purpose to which the data is being collected and how it will be used prior to the collection of the data. Furthermore, participants were informed that all data collected will be used specifically and solely for the purpose of this research after which the data will be destroyed once the findings of the research are concluded.

The questionnaire survey was conducted in July 2019 and contained 33 questions. Two hundred and fifty six people received the weblink to complete the questionnaire. Respondent age group were between 16–65. About 99% of the total participants indicated that they owned smart devices from popular brands. The brands listed were Apple (58% of respondents), Samsung (43% of respondents), Microsoft (11% of respondents), Huawei (9% of respondents), Fitbits (10% of respondents) and other less popular brands (15% of respondents). Sixty-two percent of respondents confirmed that they actively engaging with the smart features on their smart devices, whilst 38% report that they do not actively engage with the smart features on their smart device. Eighty-eight percent of respondent confirmed that they engage with the smart features of their smart device via their mobile phone, 4% reported they use their tablet and 7% engage via their laptop. Respondent reasons for not engaging with smart features on smart device were (a). Too complicated (13%), (b). cannot be bothered (37%), (c). see no benefits (12%), (d). concerned about security (13%) and (e). concerned about their information (23%).

2.1 Understanding the Consumer Awareness

This study is conducted to understand the consumer awareness on the implications of their interactions and usage of their smart devices. To that end respondent were asked the importance of the benefits they derive from the information receive from their smart device. Fifty-two percent responded that the benefits are important, 33% agreed that the benefits are somewhat important, 12% reported that benefits are not important. The survey results revealed that 62% of respondent understand the term "digital footprint" whilst 38% do not understand the term. Furthermore, 79% of respondents stated that they are aware of the possibilities of leaving a digital footprint trail whilst 21% reported are not aware that their interactions with the smart features on their smart devices leaves behind digital footprints. When asked about the type of digital footprints, social digital footprint had the highest level of awareness amongst respondents at 80% and financial digital footprint was second at 70%, medical digital footprint came third at 39%, then economic digital footprint 24% reported they were aware of this, 22% confirmed they were aware of environmental digital footprint and 20% stated they are aware of biological digital footprint. These findings are an indication that there is a degree of awareness amongst consumer about the different types of data and information trail that are left behind as a result of their interactions and engagement with their diverse smart devices, but this is not enough evidence to ascertain if consumers understand the implications of what this translate to, neither does it reveal the consumer reasoning regarding safety awareness when engaging with their smart device. To put this in context if we look at cigarette pack the message "Smoking Kills" is clearly inscribed on the package and there is enough information as well as awareness on the dangers of cigarettes. The responses gleaned from the findings of this study indicates that more in terms of educating consumers about security and privacy considerations when interacting and engaging with their smart devices is required.

2.2 Consumer Security Awareness

The study other objectives included ascertaining consumer security awareness and attitude whilst interacting with the smart features of their smart devices and to this end consumers were asked if they were aware that most smart device have default password. The findings were as follows: 54% of respondent stated that they are aware that their smart device has a default password and 46% reported that they were not aware that their smart device has a default password. Respondents were asked about their awareness regarding the need to change the default password on their smart device regularly and the findings reported that 68% of respondent were aware of the need to change the default password on their smart device regularly, whilst 32% stated that they were not aware of the need to change the default password on their smart device on a regular basis. Another consideration was to ascertain if respondent know how to go about changing the default password on their smart device and the findings reported that 63% of respondent know how to change the default password on their smart device, whilst the remaining 37% do not know how to go about changing the default password on their smart device. On the final aspect of the security awareness and attitude, respondent were asked if they are likely to read security instructions if it were to be included in their smart device when they purchased it and the findings showed that 42% of respondents are likely to read the security instructions, 24% are indifferent so they are neither likely or unlikely to read security instructions, 32% are unlikely to read security instructions included in their smart device upon purchase. These findings reinforce the findings from the previous section that consumer need to be educated on the seriousness of security threats and privacy issues regarding their smart devices. The evidence clearly indicate that consumers lack awareness and have a complacent attitude around safety and security when interacting with the smart features on their smart device.

2.3 Benefits Versus Security and Privacy Concerns

Another objective of the of the study was to understand consumers attitude towards the benefits derived from their smart device versus their concern over security and privacy breach as a result of their interaction with their smart device. Respondent were asked a series of questions and the survey results revealed that 82% of respondents are aware that a security and privacy breach of their smart device can impact others across their network and 18% of respondent were unaware of this. When asked about attitude about their digital footprint being captured in a remote location as a result of their interaction with smart device, the result showed that 24% agreed that they would be concerned about this, 19% were indifferent, 57% of respondents disagree that this would concern them. Respondents were asked about the importance of their smart device to their every-day life and results revealed that 68% agree that their smart device was critical, and they cannot do without it, 24% were different about the importance as they neither agreed nor disagreed and 8% disagreed that their smart device was critical to their life. When asked about their attitude toward the use of their data by smart device manufacturer, 70% of respondent agree that they will make an effort to understand how their data is used by smart device manufacturer, 22% were indifferent about how smart device manufacturers use their data and 8% of respondent disagree that they would be interested in how smart device manufacturers used their data. Further probe about the use of their data reveal that 70% of respondents agree they would make an effort to get clarification on the use of their data if they do not fully understand something, 20% of respondent were indifferent and 6% of respondent disagree that they would make an effort to get clarification on the use of their data if they do not fully understand something. When asked about the importance safeguarding their digital footprint over the benefits derived from smart device, 68% of respondents agree that safeguarding of the digital footprint is more important than the benefits derived from their smart device, 25% of respondents were indifferent about the safeguarding of the digital footprint being more important than the benefits derived from their smart device and 4% disagree that safeguarding of the digital footprint is more important than the benefits derived from their smart device. When asked about their understanding of the consequence of a security or privacy breach, 81% of respondent reported that they fully understand that a security or privacy breach could lead to minor or colossal fatalities, 12% were indifferent about their understanding of the consequences of a breach and 7% reported that they do not fully understand the consequences of a breach. 44% of respondent revealed that prior to taking part in this study they did not consider security or privacy as a concern when acquiring smart device, 19% revealed they were indifferent about their consideration of security and privacy when acquiring smart device prior to this study and 34% stated that they do consideration of security and privacy when acquiring smart device prior to this study. When asked about their attitude about security and privacy going forward after participation in this study 83% of respondents stated that security and privacy will certainly be a consideration henceforth when acquiring smart devices,

14% were indifferent about what they whether they will consider security and privacy when acquiring smart devices after having taken part in the study and 3% of respondent disagree to consider security and privacy as part of selection criteria for acquiring smart devices even after taking part in the study [5].

The evidence and findings of this study shows that consumers do appear to have a degree of concern about security threat and privacy issues around IoT smart devices, but the finding also revealed conflicting attitude between consumer security and privacy concerns and the benefits derived from their IoT smart device. Furthermore, the finding of this study aligns with findings from studies conducted by Internet Society and Harris Interactive. Both studies identified that consumers do have genuine concerns about security and privacy when it comes to IoT smart devices and the Harris Interactive study clearly identified consumer complacent attitude to seek out knowledge for themselves and educate themselves to improve their basic awareness on what security and privacy consideration should be considered when acquiring IoT smart devices. The Internet Society also reported in their study consumers are concerned about security and privacy, but they lack the know-how on how to adapt and adjust their device settings to alleviate these fears and concerns. Based on these findings it can be inferred that there is a need for government intervention and the consequently the government are already gearing up to address this issue as a matter of urgency as previously identified in this study.

3 IoMT Cybersecurity Framework Design

Alter (2003), Bunge (1985) and Simon (1996) suggest that information systems designed to support organisations are complicated, artificial and purposeful. The common composition of this design includes people, structures, technologies and work systems. For this study the Hevner et al. [6] Information Systems Research Framework diagram below will serve as a guide to design the proposed IoMT Security and Privacy framework (Fig. 2).

The illustration above is presented by scholars as a conceptual framework designed to aid the understanding, execution and evaluation of IS research merging behavioural-science and design-science paradigm. The framework serves as a tool used to position and compare these paradigms. The Environment according to Simon (1996) describes the problem space where the phenomena of interest resides. Silver et al. (1995) suggest that the environment is made up of people, (business) organisation plus their existing or planned technologies. Therein lies the definition of the goals, tasks, problems and opportunities that the business need from the perspective of the people within the organisation. These perceptions are influenced by the roles, capabilities and characteristics of the people within the organisation. Furthermore, business needs are identified by assessing and evaluating the context of organisational strategies, structure, culture and existing business processes. All the above-mentioned are then positioned comparatively to current technology infrastructure, applications, communication architectures and development capabilities.

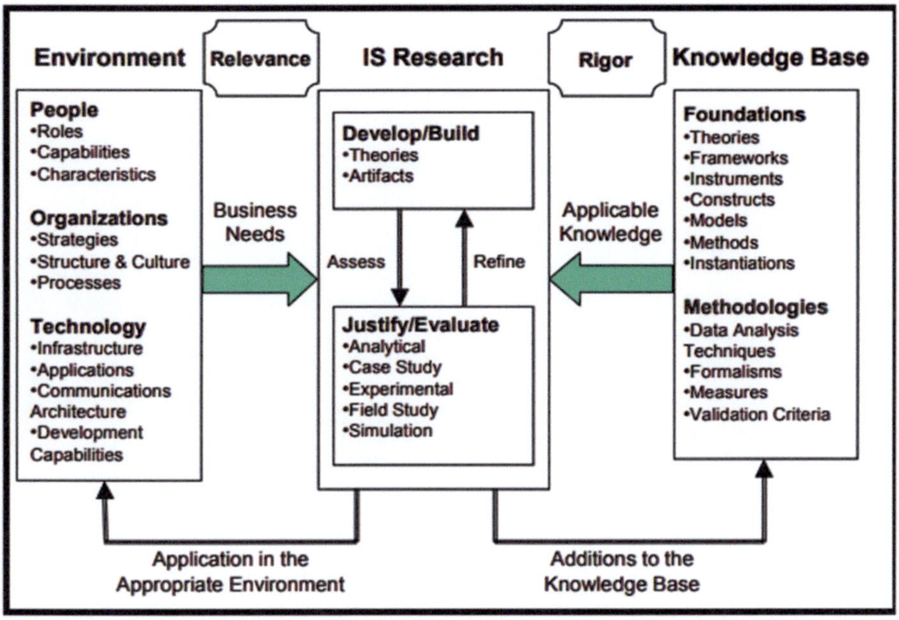

Fig. 2 Information systems research framework. (Source: Ref. [6])

The summation of all these essentials contributes to defining the business need or "problem" from the researcher perspective [6]. Thus, framing research activities to deal with the business needs gives credibility and relevance to the research.

The bedrock of IS (Information System) framework design according to behavioural and design science is a combination of people, structure, work system (processes) and technology as illustrated in the IS conceptual framework diagram in Fig. 3. Thus, IS conceptual framework will be used as the building block for the proposed IoMT framework for this study. The IoMT Security and Privacy Framework is designed to introduce structure to the key areas that have been identified to represent vulnerability bottlenecks within the healthcare sector.

3.1 Management Information Systems

The IoMT Security and Privacy framework provide a holistic view to support all cross functional and inter-organisational business processes and this will be supported by robust Management Information system that will outline succinct business outcomes including adequate measures and controls. These management systems will include the organisations acceptable risks tolerance and prescriptive actions for managing risks at different levels. The information systems inculcate the governance of information security management across all the management

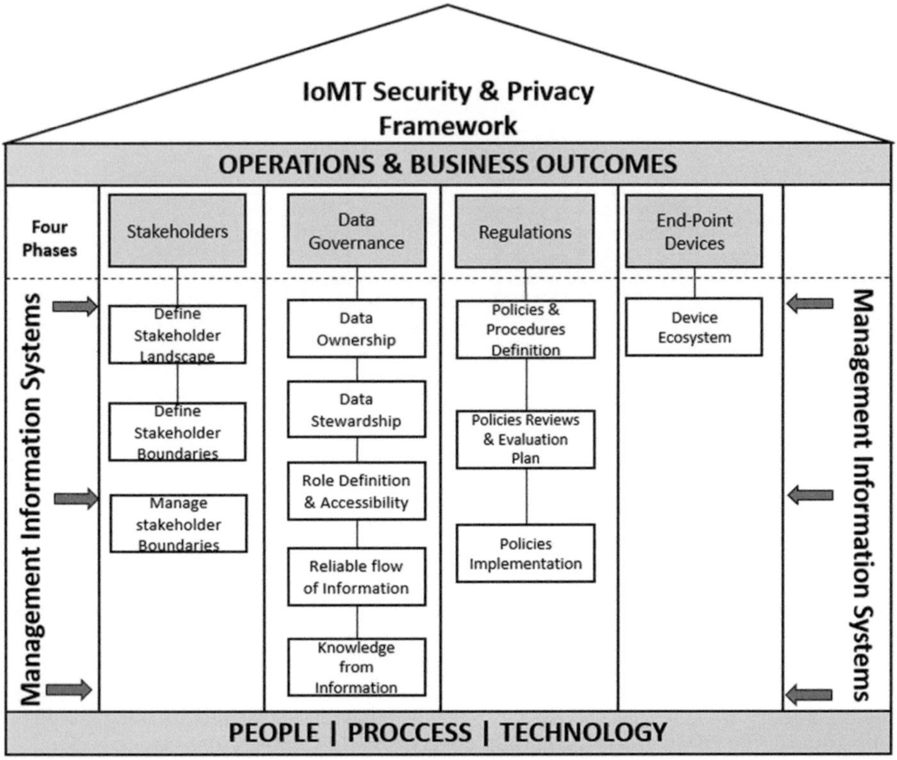

Fig. 3 IoMT security and privacy framework. (Source: Sagay A)

systems and define policies. It will also cover risk management adopting the agnostic characteristic of ISO27001. The ISO27001 standard dictates that security policies must be clearly define and documented procedures must be in place for assessments and treatment of risk. The management Information system represents a holistic perspective and as such the overarching security governance must be all encompassing therefore the Information security Governance Framework will be considered as a good fit to reinforce and ensure a robust security and privacy environment across functions and business activities (Fig. 4).

3.2 Stakeholder

Stakeholders are the people that have a keen interest and/or affected by activities within the organisation or more specifically the healthcare sector. Different stakeholders have different needs and requirement and as such it is important to define the different types of stakeholders by mapping out the entire high-level

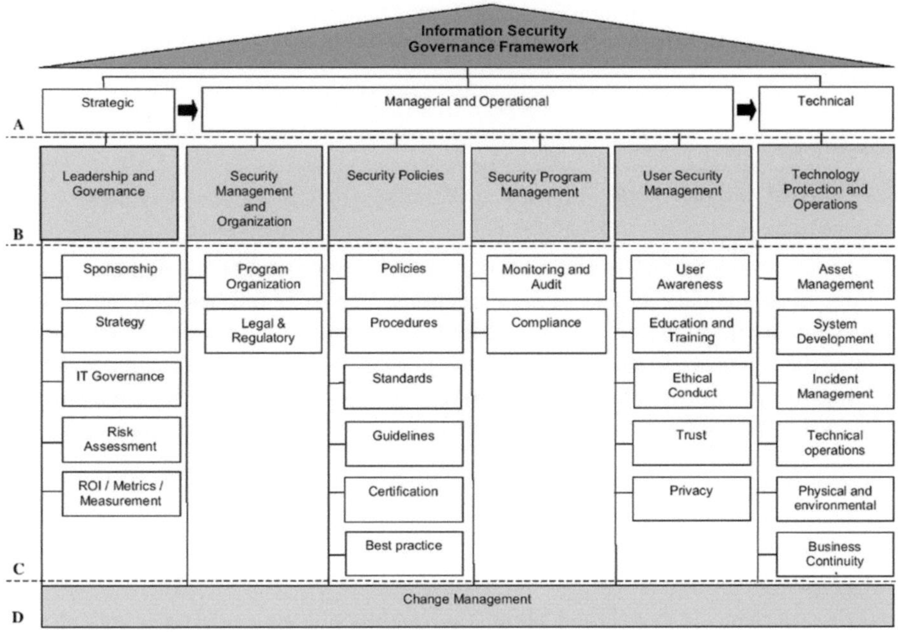

Fig. 4 Information security governance framework. (Source Veiga, A. D. and Eloff, J. H., 2007)

view of all stakeholder landscape that will be affected by the diverse activities and functions across the organisations. The above-mentioned will be handled by the Define Stakeholder Landscape step. Another aspect that will need to be considered is the definition of the different roles and level of involvement for the different stakeholders this will ensure expectations are properly managed and a good governance process across the entire stakeholder management process. This later part will be handled by the Define Stakeholder Boundaries step and will include detailed stakeholder matrix. The dynamic of business activities can cause the role and level of stakeholder involvement to change and as such there needs to be adequate process and controls to manage these changes. This will be handled by the Manage Stakeholder Boundaries step.

3.3 *Data Governance*

Data is the key commodity of the healthcare sector and represents the focal point of target of cyber-criminal activities and must be protected at all cost. Good data governance guarantees secure accessibility to top quality data that allows integrated data-driven decision making resulting in measurable outcomes [4]. Five key principle have been identified for the successful implementation of a robust data governance solution and they include: Data Ownership, Data Stewardship, Role Definition and Accessibility, Reliable Flow of Information and Knowledge from information.

3.4 Data Ownership

Data ownership is primarily about accountability, responsibility and conduct around the organisations data. It set out the guidelines, standards and best practice of data management within the organisation. The underlying focus is ensuring behavioural control measures are in place outlining the correct definition, production, organisation and use of information. Given the IoMT data are stored in the cloud the policies will need to include controls and measures to manage data stored in the cloud.

3.5 Data Stewardship

Data stewardship is concerned about the quality of data and is centred around industry standard Data quality framework. The framework is an iterative process and supports collaborative working which promotes transparency and helps to achieve the benefits of good quality data. Data quality is an essential requirement for making data informed decisions (Fig. 5).

3.6 Role Definition and Accessibility

Privacy, compliance and security are defined under role definition and accessibility. The healthcare sector operates an inherent risk environment owing to the sensitivity of the data hence why data governance is integral to the industry. Ensuring that adequate risk management strategies and embedding risk awareness culture within operational activities is paramount [11]. Furthermore, alignment with other business functions such as record retention compliance requirement will result in a successful and robust data governance.

3.7 Reliable Flow of Information

Good data governance needs to have a solid Information Architecture and Integration that will promote and support the standardisation of common data definitions and ensure these definitions are made available across different platform resulting in good and well-informed decision making. The benefits of having common data is

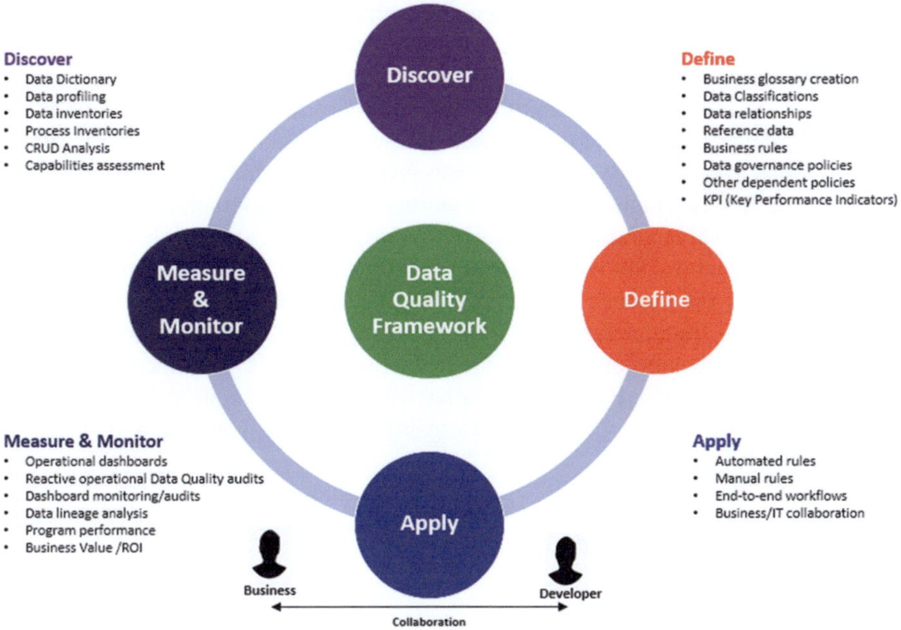

Fig. 5 Data quality framework. (Source: Sullexis Consulting)

that it can be utilised in multiple locations to define current and future capabilities within the organisation, design a durable architectural ecosystem and encourage organisation wide data integration.

3.8 Knowledge from Information

Organisation are heavily reliant on their body of data knowledge especially in the era of big data where data represents competitive advantage and as such reporting and analytics of organisation business data is critical for informed decision making. Data is at the core of all healthcare section activities and the entire IoMT ecosystems extrude data and as such a good measure of quality control will need to be put in place.

3.9 Regulations

Legislative and compliance requirements help organisation to promote and incorporate best practice across functions and business activities. The healthcare sector is heavily regulated, but majority of its legislation focus on patients care and licensing requirement for medical personnel. The risk landscape is constantly changing and there is an urgent need for a culture change within the healthcare sector because cyber security responsibilities can no longer be considered as a problem for the IT department. NIST DES (Data Encryption Standards) Standards offers guidance and best practice relevant to the primary commodity of the healthcare sector. Specifically emphasising the importance of cryptographically protecting sensitive and/or valuable data against disclosure or undetected modification during transmission or whilst it's in storage. A good regulatory framework provides well defined Policies and Procedures and must be embedded within the core activities of the organisation. Regular Reviews and Evaluation of Policies and Procedures will result in a culture change and remove the danger of treating Cyber Security risk as a one-off independent activity and a good Plan for the Implementation of Policies will reinforce and send a message across the organisation of its priority and importance.

3.10 End-Point Devices

The threat landscape of IoMT is vast and growing rapidly especially the end-point devices. The discovery or implementation of any solution to a problem requires an in-depth understanding of the complexity and challenges of the problem environment in other words IoMT Security cannot be planned for, monitored, managed or controlled if the complexity and challenges are not identified and fully understood. The FDA (Food and Drug Administration) defines IoMT end-point devices as "Instruments, apparatus, implement, machines, contrivance implant, in vitro reagent, or other similar or related article, including a component part or accessory intended for use in the diagnosis of disease or other conditions or in the cure, mitigation, treatment, or prevention of disease". Thus, a discovery and identification of data communication and transmission between end-point devices and other component within the IoMT infrastructure can offer valuable insight for a robust security solution.

3.11 Device Ecosystem

According to the Global System for Mobile Communication Association (GSMA) endpoints are physical computing devices responsible for performing motoring activities such as detecting, and it operates as part of an internet-connected product or services including wearable devices. Typically, endpoint device will also connect to hospital networks as well as other medical devices. The end-point communication ecosystem provides transparency by creating visibility potential data entry and exit points for greater control and traceability. Furthermore, this transparency will provide insights for tailored security consideration as one sight cannot fit all given the complexity and disparate nature of the requirement of the healthcare sector.

3.12 People, Process and Technology

Achieving the benefit of good and effective governance cannot be a one-time exercise or activity but rather a continuous cyclical and iterative process that is executed by people and overseen by robust and well-defined technology solutions. The Healthcare sector is complex in its diversity and as such adopting a one size fits all security solution presents a challenge. The model is built on three principles:

- Ontological approach gives autonomy and singularity to its object and still allows the object qualities to exist independently.
- Centred around stakeholders by considering the disparate security requirements and responsibilities of stakeholders within the healthcare sector. The diverse roles mean that different stakeholder's (Patients, Medical Professionals and System Administrators) need will require a different type of interaction with the solution.
- Scenario-based concept considers the heterogeneity of the IoMT device landscape which will also require solutions to be considered according to the business security requirements.

4 Conclusion

This study has proposed an IoMT Security and Privacy Framework based on the key concept of design science paradigm of people, processes and technology in addition to adopting as well as adapting existing best practice standards that are in alignment with the objectives of the framework. Discussions also included the attractiveness of high-level security and privacy breaches of healthcare sector for criminal due to the financial gain and the patient centric nature of the industry means its lagging behind in cybersecurity expertise therefore making it an easily accessible target. In addition, it discussed the unique security and privacy challenges of IoMT particularly homing in on the complexity of the diverse stakeholder security and responsibility requirements and the challenges of the heterogeneity of end-point devices making the idea of a single solution of one-size fits all not advantageous for this environment. Having discussed and considered all of these things the IoMT Security and Privacy Framework was then created with emphasis on data governance because it represents the most valuable commodity within the healthcare sector, stakeholders as they are responsible for executing activities within the operational environment and unique processes tailored and designed to meet the diverse security and privacy requirement as dictated by the environment as well as the stakeholder's responsibilities and requirement. This approach takes a holistic view of the organisation strategies and management information systems as it provides visibility across cross functional and business integrated activities. Furthermore, the IoMT Framework represents a good fit for the proposed new law as manufacturer will have the benefits of innovative design for products whilst still ensuring devices have the appropriate security and privacy requirements. Conceptually the IoMT Security and Privacy Framework was built on the inherent research principle which suggests that framing research activities to deal with the business needs gives credibility and relevance to the research.

References

1. Alsubaei F, Abuhussein A, Shiva S (2017) Security and privacy in the internet of medical things: taxonomy and risk assessment. In: 2017 IEEE 42nd conference on Local Computer Networks Workshops (LCN Workshops). IEEE, pp 112–120
2. Alsubaei F, Abuhussein A, Shiva S (2019) Ontology-based security recommendation for the internet of medical things. IEEE Access 7:48948–48960
3. Brenner J (2007) ISO 27001: risk management and compliance. Risk Manage 54(1):24
4. Data Governance Program. Available at: https://cio.ubc.ca/data-governance/data-governance-program. Accessed on: 24 July 2019
5. Farahat IS, Tolba AS, Elhoseny M, Eladrosy W (2018) A secure real-time internet of medical smart things (IOMST). Comput Electr Eng 72:455–467
6. Hevner AR, March ST, Park J, Ram S (2004, March) Design science in information systems research. MIS Q 28(1):75–105. https://doi.org/10.2307/25148625. Management Information Systems Research Center, University of Minnesota. https://www.jstor.org/stable/25148625
7. Jalali MS, Kaiser JP (2018) Cybersecurity in hospitals: a systematic, organizational perspective. J Med Internet Res 20(5):e10059
8. Osborne JW (ed) (2008) Best practices in quantitative methods. Sage, Los Angeles/London
9. Sethi P, Sarangi SR (2017) Internet of things: architectures, protocols, and applications. J Electr Comput Eng 2017:9324035
10. Wolf M, Serpanos D (2017) Safety and security in cyber-physical systems and internet-of-things systems. Proc IEEE 106(1):9–20
11. Wrestling the data quality bull: using informatic IDQ so upstream business. Available at: http://sullexis.com/blog/wrestling-the-data-quality-bull-using-informatica-idq-so-upstream-business-users-can-grab-data-quality-by-the-horns-and-wrestle-it-to-submission/. Accessed on: 24 July 2019

Biohacking Capabilities and Threat/Attack Vectors

Jaime Ibarra, Hamid Jahankhani, and Jake Beavers

Abstract The Internet of Things is a cutting-edge technology that organisations are adopting them in order to increase their business productivity and speed the operations. It has been involved for homes, companies, industries and now it is present in healthcare. However, due to lack of standardisation and accelerated competition, providers are deploying devices focused on innovation without having the proper balance between security, performance and ease of use. This is leading to new attacking vectors easing attackers to penetrate systems with confidence and without the need to be an expert in hacking thanks to the variety of open source tools available on the Internet e.g. Kali Linux, Github. The increased number of cyber attacks through IoT devices has complicated the performance of forensic investigators, reaching to Chains of Custody (CoC) easy to challenge by defenders and the rejection of investigation cases. Healthcare organisations has become the most attractive targets for cyber crime due to the variety and value of information allocated on Electronic Health Records (EHR).

This chapter aim to highlight the Biohacking capabilities and presents a Digital Forensic Investigation Process Model (DFIPM) addressing IoMT devices and assuring data privacy during the process.

Keywords Biohacking · Attack vector · IoMT · Healthcare data · GDPR · Digital forensics · Cloud computing

J. Ibarra · H. Jahankhani (✉)
Northumbria University, London, UK
e-mail: Hamid.jahankhani@northumbria.ac.uk

J. Beavers
Sheffield Hallam University, Sheffield, UK

H. Jahankhani et al. (eds.), *Cyber Defence in the Age of AI, Smart Societies and Augmented Humanity*, Advanced Sciences and Technologies for Security Applications, https://doi.org/10.1007/978-3-030-35746-7_7

1 Introduction

Cyber Security vulnerabilities have the potential to exist in any computer, it is easily forgotten that everything ranging from our smart phones to an MRI scanner are basically computers. If a malicious attack is performed on a server it can bring down a website, on a pacemaker this has the potential to kill. The FDA (US Food and Drug Administration) recently recalled half a million pacemakers, due to a security vulnerability within the devices that could have been fatal [31]. Implanted Medical devices in particular have been steadily on the rise since the conception of medical technology, however in turn so has the dependency on such devices by patients. Pacemakers, insulin pumps and even neural implants are commonplace in everyday life [10, 11]. There are an estimated 25,000 people every year in the UK that have a pacemaker fitted [30], this does not even include those outside of the UK or those who have other medical implants fitted. This figure is set to rise further with the ease of access to advanced medicine in the UK, as well as the longer lives that humans are experiencing due to the advances in modern medicine [9].

Many of these varying implantable devices, and other types of medical equipment, have been proven time and again to have had flaws in their security. With little sign of meaningful changes to correct this concerning issue, it has become a hot topic in the news and in media. In recent years there has been an increasing amount of attention towards medical device hacking [26], though no meaningful changes have been made to existing laws and legislation. The issue is a disconnect between the medical manufacturing industry and the field of Cyber Security, at first glance you could almost assume that these devices are being developed with only basic security principles in mind.

Malicious attackers are enhancing their tactics, techniques and procedures (TTPs) in order to cause security breaches within organisations leading to data theft, manipulation or blackmailing for instance. An article from Forbes (2019) claims that Electronic Health Records (EHRs) can be worth $1000 (£778) for hackers and therefore the steady increase of cyber-attacks towards the medical sector. One of the most relevant breaches affecting medical processes was the WannaCry ransomware attack over England National Health Service (NHS) [20] that caused a total of 19,000 appointments cancelled and £92 million in investment to remediate and recover from the incident. In addition, an article presented by DiGiacomo [11] presents that in January of 2018 there were reported approximately 115 cyber-attacks, which the one with highest damage rate was over Health South-East RHF, a healthcare organisation that manages hospitals in Norway with a possibility that over 2.9 million users are potentially affected by the breach [5].

Various governing bodies have discussed the idea that the internet should be a human right, providing all of humanity with information and tools that can be as helpful as they are dangerous. It has been proven on numerous occasions that a whole range of medical equipment can be hijacked by a third party, ranging from X-Ray systems, CT Scanners and even Blood Refrigeration Units [36]. Yet despite this knowledge, there has been little advancement towards even the regulation of security within such devices, thus attacks that were used in 2008 may still be viable in 2018. There are governing bodies who regulate the manufacturers of medical devices, however, there appears to be an oversight when it comes to the regulations to enforce adequate security.

2 Value of Healthcare Data for Organised Crime

The article from Morgan [25] points out how data breaches on healthcare are increasing steadily, reaching to a number of 20,836,531 records leaked. Attackers are showing their high interest in this type of information within healthcare services as shown on the previous section, and furthermore, the selling or health records in the black market are rising sharply.

Healthcare has become part of CNI because of the sensitivity of data held by these organisations. Furthermore, the fact that IoT has been involved in this sector enhancing services and easing patient's life style connecting more devices to the internet implies more risks associated in terms of cybersecurity.

The research from Ibarra et al. [13] claims that EHRs offer a significant wealth of information, attracting hackers to exploit and steal. It contains information such as:

- Demographic information.
- Full names, same as shown personal IDs, driver licenses, passports.
- Address history.
- Work history.
- Names, ages, contact details from relatives, which can belong to parents, siblings, life partners or any representative the health provider contacts this person in case the patient faces an emergency.
- Financial information, including bank details, credit/debit cards.
- National Insurance Number (Social Security Number outside the UK).
- Medical history, which contains sensitive information. It includes details of previous medical appointments along with details from doctors, nurses. Moreover, it likely has critical information such as allergy details, surgeries the patient was submitted, results from medical diagnosis such as xrays, electromagnetic resonance. The appointments listed include diagnosis, prescriptions, treatments and dates for the next medical control organised in a chronological manner.

EHRs contains precise details of the victim's life. Once a health provider was subject to a security breach compromising patient records, customers who got involved within the breach can likely get exposed to extortive blackmails for a lifetime. Furthermore, if EHRs contain additional information such as cancer diagnoses, STDs, psychological conditions established (i.e., asperger syndrome, autism, depression, alcoholic), the victim can be exposed to public embarrassment or political assassination depending on the goals of the attackers.

The research from Terry [33] claims that the development of electronic patient health record (E-PHR) systems, the usage of personal health technologies and the Internet of things (IoT), caused policy-makers to highlight a big concern regarding the massive increase of IoT devices used by consumers, whilst data is created and processed every second therefore, the increase of cyber threats and attacking vectors. In addition, he also points out a great challenge to protect healthcare data in the future. This is due to the lack of training and preparation for precision medicine, and usage of robotics for sensitive procedures like surgeries for instance. Therefore the need for the deployment of reliable frameworks, methodologies and standardisation of technologies that could allow organisations to protect their digital assets and respond effectively against any security breach attempt. In addition, the process model proposed for forensic investigation would support businesses to learn from their previous mistakes in order to harden the security posture. Nowadays cyber security is a vital component for businesses to continue competing within the market and it is paramount to adopt the last updated technologies along with training and awareness methodologies before, during and after an incident. This could support investigators to execute the expected top level reports in order to track the origin and author of the unauthorised activities.

3 GDPR in Healthcare

It is necessary to understand how the implemented GDPR has taken effect across IoT networks in order to determine the deliverables, and ascertain with the main stakeholders within the usage of IoT-based medical devices. This will be performed by critically analysing the research from O' Connor [29], which proposed an approach of the "Privacy by Design" principle for IoT environments. This research highlights the importance of an electronic consent (eConsent) in order to proceed with data management, being proactive and not reactive for instance. This regulation points out with emphasis the importance of assuring data privacy and the protection of owners when their personal data has being compromised. In forensic investigation, the assurance of transparency and privacy are vital because businesses must keep producing during an incident, otherwise it would imply significant financial losses, customer dissatisfaction and therefore, their reliability would decrease. In addition, the GDPR sets up to £17 m in fines if the local authority considers that the organisation was considered incapable of executing the necessary steps to protect personal information.

The research from Shu and Jahankhani (2017) claims the impact of the GDPR on the Information Governance Toolkit pointing out on healthcare. Mentioning the impact of cloud computing for allocation of huge amounts of data with focus on six benefited assets as shown in Fig. 1. Governance is one of the main components for effective cyber security regardless of the speciality area. Effective communication and control allows to achieve regulatory compliance and policy enforcement assurance. During a forensic investigation, it is possible that policies

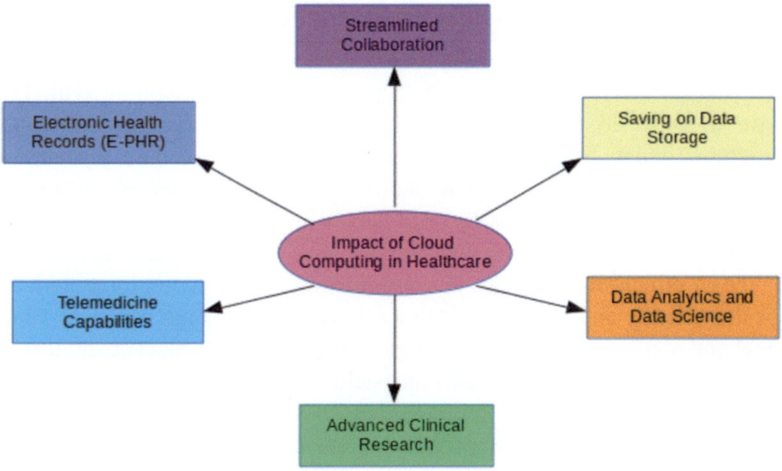

Fig. 1 Impact of cloud computing in healthcare

and procedures would get changed as result of the lessons learned during a cyber incident along with modifications on the technical infrastructure and configuration management procedures.

4 The Role of Forensic Investigation in IoMT

Forensic investigation is an essential part of incident response in cyber security following the NIST Cyber Security Framework. It shows details of the performance of the intrusion, actions and methods that attackers performed, along with assets compromised (systems and/or data). Furthermore, it allows organisations to learn from those mistakes in order to mitigate risks and if applicable the modification or implementation of new cyber security strategies either technical, organisational or legal because compliance is a feature in this field.

Healthcare providers must get adapted to the last updated guidelines, frameworks and standards because the are holding sensitive data that could cause unmeasurable damage if it gets leaked. In addition, the adoption of IoT in medical environments expanded the risks within the industry and because IoT is not standardised yet, can cause extensive trouble for investigators to collect evidence and present comprehensive reports either for courts and the compromised organisation in order to mitigate risks.

However, the threat landscape is subject to modifications and it would depend whether the Internet of Things gets standardised or not during the next years. Otherwise, it will difficult the job from forensic investigators leading to cases rejected or lost due to lack of relevant evidence supporting the investigation.

4.1 Challenges of Digital Forensic Investigation in IoMT

Digital Forensic (DF) investigation is a process that works along with Incident Response in order to extract information from a particular device, system or infrastructure, which is submitted to analysis, preservation and presentation of digital evidence that can be used to identify activities related to security/policy violation or crime. Nevertheless, there is not a standardised model that can provide an overview of the entire investigation. In fact, some of these came from the experience of ethical hackers, system administrators and law enforcement entities without the solidity and consistency that involves every stage of the investigation (technical and non-technical). An investigator might present relevant and incriminating evidence in a comprehensive and consistent manner targeted to legal authorities, otherwise the case may be lost or discarded during the investigation process [22, 23, 24]. Considering that most of devices are unlikely to show or contain the necessary consent from users [29], the limitations that Internet of Things (IoT) present in terms of hardware and software, the complexity in its architecture, no standardisation present, along with the recently enforced European Global Data Protection Regulation (GDPR), the requirement to define a comprehensive and holistic forensic investigation model that ensures data privacy and compliance maintaining most discretion during an investigation in order to protect people during and after a security breach.

Khan et al. [18] claim that forensic investigation in the Internet of Things demands solutions from researchers, security and IoT experts, along with cloud computing providers to secure the infrastructure during a security incident. Nowadays, it is a fact that one of the main targets for malicious attackers are EHRs from patients and therefore, investigators muse assure privacy to data owners during the investigation process. This is because stolen records can lead to severe damage including terrorist-based attacks attempting against the person's life.

The information showed below following Fig. 2, shows the challenges that forensic investigation presents in medical IoT along with details of each component mentioned in the mind map. Considering that IoT works on a similar way as cloud it has been divided into three stages that require investigation. Firstly the device from users, secondly the network where the information is being transmitted and finally the cloud servers. It is important to recall that all digital evidence extracted and sent to courts must be reliable, authentic, complete, believable and admissible in order to present it showing the overall of the investigation.

In addition the evidence analysed must contribute to the incrimination of the malicious actor involved with the unauthorized action performed. Details of every component of forensic investigation in IoMT along with their own challenges are shown below.

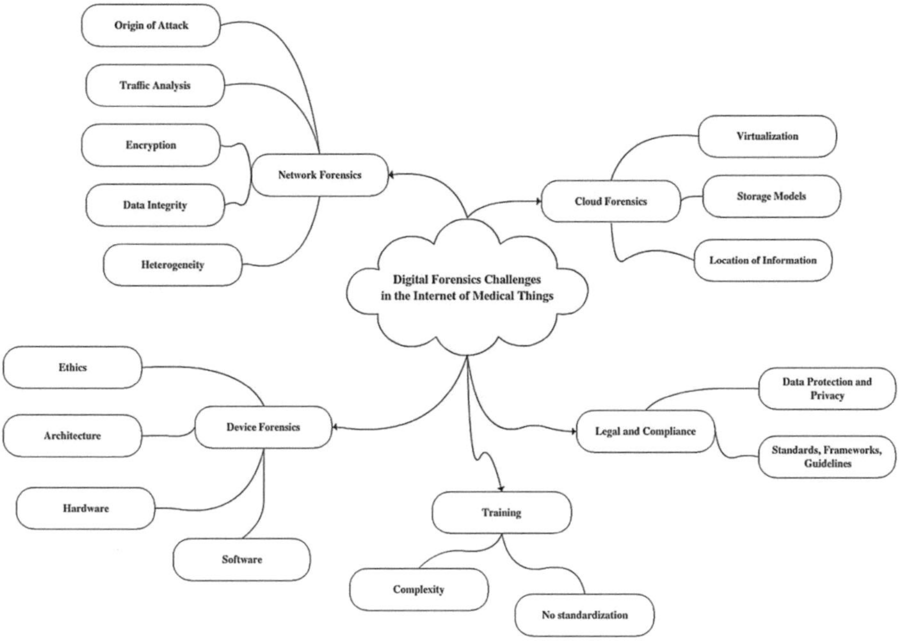

Fig. 2 Digital forensics challenges in IoMT

5 Attack Vectors

All medical implants are required to operate on the MICS band [35], this range is 402 MHz to 405 MHz. There have been many successful hacking attempts on implants by hijacking the RF module [12], this is the most commonly used communication method for implants, this however is due to change and be updated to Bluetooth technology. There is a serious risk for medical equipment within third parties companies and institutions such as the NHS, there were 93 cyber-attacks taken place in healthcare organizations from 2013 to 2016 [2].

The paper "Hacking NHS Pacemakers: A Feasibility Study" [1] demonstrates a blackbox test on NHS implants, in this case pacemakers were chosen. Based on the results the most common attack vectors can be defined as:

- Denial of Service (DoS)
- Replay Attacks
- Code Injection Attacks

5.1 *Denial of Service (DoS)*

DoS is a type of cyber-attack, the intended aim of which is to take the targeted source offline [7]. The methodology behind this attack is to overload the target by overpowering its resources, this is achieved by sending a multitude of spam data signals at the same time. This attack cannot work if the intended target has enough resources available to cope with the extra workload, in these instances more devices are required to perform the attack and succeed. DoS attacks can be combined with a code injection attack, the idea behind this is to execute spam code whilst flooding the connection to intensify the effect.

The primary defence methods for this type of attack are as follows:

- Disabling the wireless functions of the target to stop all communications
- Increase the resources available to the target so it can cope with the extra load
- Limit communication to only specific pre-authorised devices

In RF terms, the equivalent of a DoS attack is signal jamming. This is achieved by broadcasting on the same frequency but at a higher power than the target, effectively this is spamming the airwaves in the same way that a DoS attack spams wireless communications. This results is the device being unable to cope with the high levels of interference and in theory, may cause erratic behaviour in the unit such as performing at a slower rate or even powering off entirely [16].

There are few ways to protect an RF device against signal jamming, the most efficient way is to attempt to mask the transmission so the attacker does not know which frequency to jam. Code Division Multiplexing (CDM) is an alternative method of combating signal jamming in UHF systems (Thakur n.d). CDM works by spreading the spectrum of the signal into multiple channels, then each channel is encoded with its own unique code. Only the receiver of the signal knows the code generated, though the spreading effect does reduce the overall power of each channel.

In theory, a pacemaker or ICD should only be accessible by the corresponding manufactures programmer, however, as can be seen in the previous examples of attacks it has been possible to bypass the need for these devices. Fundamentally this is an unavoidable failing with all communication technologies. If you are going to allow wireless connectivity then you must account for unauthorised access attempts, so plan accordingly.

5.2 *Replay Attack*

Home monitoring units send data to and from pacemakers and ICDs when the user is in the vicinity. This data can be captured mid-traffic by utilising the listening functions of a radio antenna, and then it can be replayed back to the device. Since the data or commands it is being sent came from the device originally it may be able to read them, whether the unit accepts this signal is down to the security employed by the receiver.

Since medical implants are commonplace in the UK it is expected that the MICS range could be flooded with signals. These signals clearly do not affect each other however as otherwise they would be subjected to constant replay attacks. Therefore, it can be surmised that some form of unique identifier must be used. If this is the case, then to successfully perform this attack a signal from the same device must be played back to it. If this is not the case then, theoretically any signal from a device of the same type and manufacturer could be used to attack any other.

5.3 Code Injection

Code injection is a generic term that refers to the unauthorised uploading of potentially malicious code [6]. The programming language used can alter however the fundamental techniques remain the same. When malicious code is packaged it is referred to as malware, this is a catch-all term given to computer viruses.

There are various cyber-security platforms and automated software that is specially designed to remove malware, however, if this code is not detected by such tools then it is left to the user to go through the system until it is found. Anti-virus providers and cyber-security agencies typically have in-house experts who specialise in searching for malicious code, once found their clients are notified and a patch to resolve the issue is pushed out. There are many skilled individuals who design malware to perform all sorts of functions such as stealing information, hijacking a device, blackmail purposes or just because they enjoy doing it. Due to the increase in IoT devices and expertise in computer skills, the amount of malware in circulation will exponentially increase.

Pacemakers and ICDs are re-programmable, they have to be to ensure that any issues with the software can be patched. This opens up a possible avenue for attack, if code is accepted from any source then malicious malware could be uploaded to the device instead. Code does not need to be long and complex, if simple commands are accepted then it would be possible to upload a command to download the data, wipe the device entirely or even switch the device off.

5.4 Summary

Radio Frequency has been previously stated as being easily breakable, however, the results from the 2019 work could argue that they are shielded enough to alleviate users concerns. It could be a legal consideration as to why documentation states potential risks of EMI interference, that device manufacturers who implement RF technology must inform the user of potential risk.

The devices used in the tests in 2019 were provided by the NHS, they were standard modern units and as such it is expected that they should have a reasonable defence against hacking. The conclusion of the work was that for the attacks to work, the individual must have expertise and knowledge of both wireless commination (in this instance RF) and the inner workings of the devices being targeted.

6 Forensic Investigation on the Internet of Things and Considerations on 5G Networks

As shown in Sect. 4.1, performing forensic investigation on IoT is complex due to the multiple architectures that investigators have to deal with. In addition, the arrival of 5G makes it more complex because of the massive use of Software Defined Networks (SDNs). Performing forensic investigation on IoT could mean to interact with cloud servers, communication between different VPSs, performing packet analysis between transport networks and SDNs and also analysing infected devices of end users that could violate their rights in terms of data privacy.

The proposed forensic investigation process model is done considering the main components that involve an IoT architecture as mentioned on the previous sections. It has been designed from high to low-level approach allowing forensic investigators to obtain precise information and retrieve a better perception over the detailed components, named processes, stages, sub-stages and principles. It consists of 7 processes, which each one is formed by a different number of stages. This guideline is shown only at its high level, and details of the model can be found at the research from Ibarra et al. [13]. The overview of this model works along with eight concurrent processes regardless of the architecture investigators are interacting with (Fig. 3):

- Preserve Digital and Physical Evidence – Evidence must be retained in its original form and its integrity must be preserved from the opening to the closing stage of the investigation for both physical and digital evidence. It is paramount for investigators to show that evidence has not been altered and if some unavoidable changes were made to report them and justify. IoT networks deal with massive amounts of personal data therefore, the requirement of ensuring privacy during the investigation to assure GDPR compliance. Achieving privacy and integrity of physical and digital evidence ensure a high quality investigation and reliable evidence to present it to a court. The preservation process might involve investigators to prevent people without authorisation to enter or leave the crime scene, system/device/network/VPS isolation to acquire the volatile data and locate suspicious processes running. Preservation also includes the assurance of log files before its removal and a full backup of the imaged system.
- Preserve Chain of Custody – Digital evidence is often prone to be handled by different parties, and in some cases its poor preservation allows courts and defensive members to challenge it with confidence leading to its rejection.

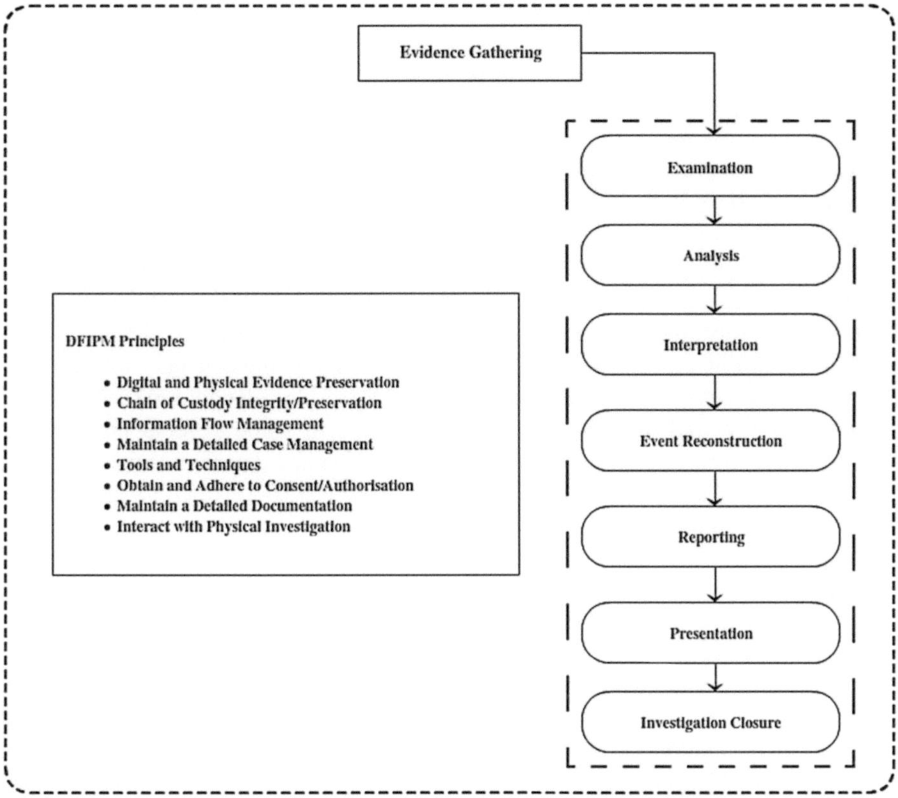

Fig. 3 Guideline for forensic investigation on IoMT

Events are correlated in order to reconstruct the crime scene within a CoC, and this must be considered the main component for any forensic investigation. Potential digital evidence is gathered from threat hunting and incident response (IR) detection stages, processing physical and digital crime scenes, hence the initiation of the CoC and this principle should be observed from the IR detection stage. Proper, accurate and detailed documentation are essential to preserve the CoC as well as supporting evidence such as videos, pictures and drawings. In addition, a reliable CoC demands from investigators to records of the personnel responsible for handling evidence including actions taken with dates and it might require the development of supporting documents that would contribute to the final report prior to its presentation in courts.

- Manage Information Flow – This principle is about the permission for investigators to interact with the variety of laws, languages, etc. appropriately during the entire investigation. One example is the interaction between two investigative entities responsible for the same case, or digital evidence exchange between parties. It can be protected using hashing algorithms such as MD5, SHA-1 or any PKI-based encryption.

- Maintain a Detailed Case Management – It refers to manage wisely the investigation, record and keep track of evidential items, events and crucial forensic findings. Casey [4] points out the importance of this principle as one of the main components of scaffolding to bind all evidence, reports, supporting documentation for the building of a strong case. Likewise, Khatir et al. [19] highlights the effectiveness of an investigation based on strong case management.
- Prepare Tools and Techniques – Forensic investigators must need to use diverse tools and techniques to perform each process during the investigation. This principle is extensively covered by standardised documents such as NIST [17], The International Organisation for Standardisation (2005), (2013), same as technical reporting like the Information Assurance Advisory Council (IAAC) [32]. The known tools can be used for system imaging and data carving i.e. FTK, EnCase, as well as for packet analysis i.e. Wireshark, Tcpdump, Solarwinds. However for IoT devices it is likely to perform some reverse engineering techniques to assess the behaviour of the firmware and determine any malicious code modified against the original with tools such as IDAPro, GDB for instance and the execution of MITM attacks to extract the current firmware from the device depending on the communication protocols developed by the provider. For IP address tracking there are a variety of open source and online tools i.e. ping, nslookup, dig, traceroute, Whois, WhatIsMyIPAddress [34] or IP Location [14].
- Obtain and Adhere to Consent – Any investigation requires authorisation either internal or external. This principle requires from investigating entities to obtain proper consent from: governments, system administrators, users., when carrying out an investigation. Now that GDPR has been implemented across Europe, it is paramount for investigators to execute processes precisely because personal data must not be compromised during an investigation and the protection of people is crucial. In addition, it is possible that users must not allow the retrieval of potential evidence for security reasons that could likely interfere with the performance of investigators. One option is the proposal of a smart contract [21], based on blockchain technologies that allows to perform a secure and reliable forensic investigation.
- Maintain a Detailed Documentation – Activities and actions performed must be logged and documented in detail using comprehensive vocabulary that would allow legal courts to understand the details of the crime executed in order to make fair decisions when the case is presented on audiences. The documentation includes possible changes across the investigation that should be recorded and mentioned during the presentation to justify the actions that investigations performed.
- Interact with Physical Investigation – Even the crime was performed in the digital world, the main component of technology is people. Investigators must interact with people involved in the scene that might witness some unusual event that could contribute to the development of the investigation. However, details should be recorded and authorised by the witness to be presented due to the GDPR regulation. The more supporting evidence investigators collect to present at courts, the stronger and more reliable the CoC gets.

The adoption of IoT must be heavily considered as an important use case in 5G because of the resource constraints that these devices currently have (e.g. e-home, wearable/implantable devices, industrial IoT). It is paramount to consider that 5G networks offer higher download/upload speed rates, and the current cyber attack trend that is currently affecting 4G. Therefore, 5G will offer more efficient execution of attacks especially affecting the most of software-defined layers.

For instance, as shown in the research by Nomikos et al. [28], the communication in 5G is defined by software as well bringing the challenge of creating a Dynamic Radio Access Control Network (DyRAN). Hence, controlling unusual behaviour in this part is important to avoid resource consumption, and this lack of accountability is of course a clear problem for IoT as shown in Nieto et al. [27], which clearly affects 5G networks as well.

Other important feature of 5G is the Device-to-Device (D2D) communication created to increase the coverage of the network e.g. network relays [28]. This could facilitate the set of vulnerabilities and attacks propagated hop by hop leading to possible access to critical parts of the infrastructure i.e. software controllers. As shown in the ENISA 5G security report [3], SDN controllers are prone to attacks to the communication APIs between controllers and between controllers and the SDN elements close to the end user.

One of the most important topics to discuss in 5G is the Mobile Edge Computing (MEC), bringing improvements in terms of data, storage and performance exploiting the latests changes in this new architecture. Therefore, the requirement of working with massive data traffic amounts. Finally, a critical feature is the ability to virtualise network functions and thus, using Network Function Virtualisation (NFV) allows to replace software with more ease compared with hardware based networks. This can allow to isolate attacks immediately by just stopping the service and containing the infected VPS, but on the other side the use of software leads the system to vulnerabilities related to coding errors and the requirement of constant patching.

References

1. Beavers J (2019) Hacking pacemakers: a feasibility study. In: IEEE 12th international conference on global security, safety and sustainability (ICGS3)
2. Beavers J, Pournouri S (2019) Blockchain and clinical trial. Springer. Chapter 11: recent cyber attacks and vulnerabilities in medical devices and healthcare institutions
3. Belmonte Martin A, Marinos L, Rekleitis E, Spanoudakis G, Petroulakis N (2015) Threat landscape and good practice guide for software defined networks/5G. European Union Agency for Network and Information Security (ENISA), Heraklion
4. Casey E (2011) Digital evidence and computer crime: forensic science, computers and the internet, 3rd edn. Elsevier Academic Press, New York
5. Cimpanu C (2018) Hacker might have stolen the healthcare data for half of Norway's Available at: https://www.bleepingcomputer.com/news/security/hacker-might-have-stolen-the-healthcare-data-for-half-of-norways-population/. Accessed 25 Dec 2019
6. Code Injection (2013). Retrieved from https://www.owasp.org/index.php/Code_Injection

7. Denial of Service (2015). Retrieved from https://www.owasp.org/index.php/Denial_of_Service
8. DiGiacomo J (2018) Data beach statistics for 2018 plus totals from 2017 | Revision Legal %. [online] Revision Legal. Available at: https://revisionlegal.com/data-breach/2018statistics/. Accessed 10 Feb 2019
9. Fatal flaws in ten pacemakers make for Denial of Life attacks (2016) Retrieved from https://www.theregister.co.uk/2016/12/01/denial_of_life_attacks_on_pacemakers/
10. Finkle J (2016) J&J warns diabetic patients: insulin pump vulnerable to hacking. Reuters. Retrieved from https://www.reuters.com/article/us-johnson-johnson-cyber-insulin-pumps-e/jj-warns-diabetic-patients-insulin-pump-vulnerable-to-hacking-idUSKCN12411L
11. Focus on: Pacemakers (n.d.). Retrieved from https://www.bhf.org.uk/heart-matters-magazine/medical/pacemakers
12. Halperin D, Heydt-Benjamin TS, Ransford B, Clark SS, Defend B, Morgan W, Fu K, Kohno T, Maisel WH (2008) Pacemakers and implantable cardiac defibrillators: software radio attacks and zero-power defenses. IEEE symposium on security and privacy
13. Ibarra J, Jahankhani H, Kendzierskyj S (2019) Cyber-physical attacks and the value of healthcare data: facing an era of cyber extortion and organised crime. In: Blockchain and clinical trial. Springer, Cham, pp 115–137
14. IP Location (2016) Where is geolocation of an IP address? Available at: https://www.iplocation.net/. Accessed 30 Aug 2019
15. Jack B (2017). Retrieved from https://en.wikipedia.org/wiki/Barnaby_Jack
16. Jamming & Radio Interference: Understanding the impact (n.d.) The institute of engineering and technology. https://doi.org/10.1049/etr.2012.9002
17. Kent K, Chevalier S, Grance T, Dang H (2006) Guide to integrating forensic techniques into incident response. NIST Spec Publ 10(14):800–886
18. Khan S (2017) The role of forensics in the internet of things: motivations and requirements. IEEE Internet Initiative eNewsletter
19. Khatir M, Hejazi M, Sneiders E (2008) Two-dimensional evidence reliability amplification process model for digital forensics. In: Third international annual workshop on digital forensics and incident analysis, pp 21–29
20. Lam B (2017) NHS cyber attack: views from the front line. Pharm J. Retrieved from https://www.pharmcaceutical-journal.com/opinion/qa/nhs-cyber-attack-views-from-the-front-line/20202794.article
21. Lone AH, Mir RN (2018) Forensic-chain: ethereum blockchain based digital forensics chain of custody. SPCSJ 1(2):21–27; Scientific Cyber Security Association (SCSA), 2017 ISSN: 2587–4667
22. Montasari R (2016) The comprehensive digital forensic investigation process model (CDFIPM) for digital forensic practice. PhD thesis, University of Derby
23. Montasari R (2017a) A standardised data acquisition process model for digital forensic. Int J Inform Comput Secur 9(3):229–249
24. Montasari R (2017b) Digital evidence: disclosure and admissibility in the United Kingdom Jurisdiction. In: International conference on global security, safety, and sustainability. Springer, Cham, pp 42–52
25. Morgan L (2018) List of data breaches and cyber attacks in March 2018. [online] IT Governance Blog. Available at: https://www.itgovernance.co.uk/blog/list-of-data-breachesand-cyber-attacks-inmarch-2018/. Accessed 26 Apr 2018
26. New York Post (2016) Yes, pacemakers can get hacked. Retrieved from http://nypost.com/2016/12/29/yes-pacemakers-can-get-hacked
27. Nieto A, Roman R, Lopez J (2016) Digital witness: safeguarding digital evidence by using secure architectures in personal devices. IEEE Netw 30(6):34–41
28. Nomikos N, Nieto A, Makris P, Skoutas DN, Vouyioukas D, Rizomiliotis P, Lopez J, Skianis C (2015) Relay selection for secure 5G green communications. Telecommun Syst 59(1):169–187
29. O'Connor Y, Rowan W, Lynch L, Heavin C (2017) Privacy by design: informed consent and internet of things for smart health. Proc Comput Sci 113:653–658

30. Pacemakers (n.d.). Retrieved from https://www.bhf.org.uk/heart-health/treatments/pacemakers
31. Seals T (2018) Abbott addresses life-threatening flaw in a half-million pacemakers. Retrieved May 19, 2018, from https://threatpost.com/abbott-addresses-life-threatening-flaw-in-a-half-million-pacemakers/131709/
32. Sommer P (2008) Directors' and corporate advisors' guide to digital investigations and evidence. U.K. Information Assurance Advisory Council. Available at: https://www.ucisa.ac.uk/~/media/Files/members/activities/ist/DigitalInvestigationsGuide.ashx. Accessed 30 Aug 2019
33. Terry N (2017) Existential challenges for healthcare data protection in the United States. Ethics Med Pub Health 3(1):19–27
34. WhatIsMyIPAddress (2016) How you connect to the world. Available at: http://whatismyipaddress.com/. Accessed 30 Aug 2019
35. Yuce MR, Islam MN (2016) Review of medical implant communication system (MICS) band and network. ICT Express 2(4):188–194. https://doi.org/10.1016/j.icte.2016.08.010
36. Zetter K (2015) Medical devices that are vulnerable to life-threatening hacks. Retrieved from https://www.wired.com/2015/11/medical-devices-that-are-vulnerable-to-life-threatening-hacks/

Digital Twins for Precision Healthcare

Gabriela Ahmadi-Assalemi, Haider Al-Khateeb, Carsten Maple, Gregory Epiphaniou, Zhraa A. Alhaboby, Sultan Alkaabi, and Doaa Alhaboby

Abstract Precision healthcare is an emerging concept that will see technology-driven digital transformation of the health service. It enables customised patient outcomes via the development of novel, targeted medical approaches with a focus on intelligent, data-centric smart healthcare models. Currently, precision healthcare is seen as a challenging model to apply due to the complexity of the healthcare ecosystem, which is a multi-level and multifaceted environment with high real-time interactions among disciplines, practitioners, patients and discrete computer systems. Digital Twins (DT) pairs individual physical artefacts with digital models reflecting their status in real-time. Creating a live-model for healthcare services introduces new opportunities for patient care including better risk assessment and evaluation without disturbing daily activities. In this article, to address design and management in this complexity, we examine recent work in Digital Twins (DT) to investigate the goals of precision healthcare at a patient and healthcare system levels. We further discuss the role of DT to achieve precision healthcare, proposed frameworks, the value of active participation and continuous monitoring, and the cyber-security challenges and ethical implications for this emerging paradigm.

Keywords Digital healthcare · Smart healthcare · Real-time model · Cyber-physical systems · Ethics

---

G. Ahmadi-Assalemi · H. Al-Khateeb (✉) · G. Epiphaniou · Z. A. Alhaboby
Wolverhampton Cyber Research Institute (WCRI), University of Wolverhampton, Wolverhampton, UK
e-mail: H.Al-Khateeb@wlv.ac.uk

C. Maple
University of Warwick, WMG Group, Coventry, West Midlands, UK

S. Alkaabi
Institute of Informatics and Computing in Energy, Universiti Tenaga Nasional, Kajang, Malaysia

D. Alhaboby
Faculty of Medicine, University of Duisburg-Essen, Duisburg/Essen, Germany

H. Jahankhani et al. (eds.), *Cyber Defence in the Age of AI, Smart Societies and Augmented Humanity*, Advanced Sciences and Technologies for Security Applications, https://doi.org/10.1007/978-3-030-35746-7_8

1 Introduction

Traditionally, the healthcare system was reactive by design. It supported patients after they became symptomatic with the disease rather than providing preventative care. This has changed over time, with the ambitious definition of health as the state of complete wellbeing including the physical, mental, social aspects on top of the biomedical one [1]. The evolving concept of healthcare deviated from focusing solely on the illness towards primary healthcare and health promotion [2]. Healthcare became a multi-level multifaceted complex.

In addition to the change in defining what is 'health', the complexity in healthcare resulted from the interaction of different factors such as the specialities in medicine, communication channels, the health system, and the context in which all of these operate. This complexity has undermined the adoption of digitally-enabled innovations in healthcare and often led to resistance in adaptation or failure in application [3].

The emergence of evidence-based medicine helped to 'regulate' the clinical decision process by healthcare professionals to achieve optimal treatment. In principle, this requires regular development and updates to clinical guidelines based on research findings. However, the guidelines could not replace the human factor in terms of the physicians' experience, and the patients' input [4]. Additionally, patients' needs are often complex and require a personalised management plan, think about chronic conditions and co-morbidities.

Recent years have witnessed a surge in digital health technologies but their adoption into clinical practice is comparatively slow. While there is a profound shift in the way individuals participate in health matters, the transformative benefits of the technological innovations remain to be realised. Therefore, better understanding is needed on how new healthcare technologies meet the needs between patients and healthcare practitioners and how this leverages the quality of care [5, 6].

The concept of Digital Twins (DT) has emerged to enable modelling and the fusion of individual physical artefacts with digital models reflecting their status in real-time. Healthcare, one of the fastest-growing sectors, due to its system complexity has a need to model its services and resources to improve the quality of care, services and patients' outcomes [7].

In this article, the challenges and the role of digital technologies in healthcare are discussed alongside the concept of DT technology in precision healthcare. We discuss the key transformational technologies and examine recent work in proposed DT frameworks. We further discuss the role of DT to improve the delivery of precision health. We cover the cyber-security challenges afterwards. The last part of the article deals with the ethical implications for this emerging paradigm.

2 Defining Precision Healthcare and Digital Twins

2.1 The Cost of Healthcare and Its Challenges

The healthcare lifecycle, a continuum originating at birth and ending at death, is a highly complex ecosystem converged of multiple disciplines which makes it incredibly difficult to view healthcare as a single domain. Globally, there are several approaches to healthcare, some are driven by the private sector (e.g. Switzerland, USA), others include the UK's National Health Service, the social insurance-based system of France, Netherlands and Germany and there is also the Canadian provincial government health insurance.

In the USA, it is widely acknowledged that healthcare costs, at almost 18% GDP in 2011 and forecasted to rise to 20% by 2020, in their current form are unsustainable. Challenging areas include preventative care, increased dependency for the chronically ill for whom coordination is deemed essential for health and function, and excessive use of medication [8]. Medical errors are the third leading cause of death [9], therefore, strategies are required to design safer systems to mitigate the frequency, visibility and consequence of these errors. Canada's healthcare expenditure represents 11.3% of its GDP. The data reported by the Canadian Institute for Health Information shows that in 2016, 16.5% of the population was in the age group of 65 and older with the highest spending of 44.8% on the health expenditure [10]. Within the EU member states, Germany has the highest healthcare expenditure equivalent to 11.3% GDP, but this varies across member states based on a number of factors ranging from disease burden, system priorities and costs. A significant portion of the health expenditure is spent on curative and rehabilitative care while other major categories are health-related long-term care followed by prevention and public health [11]. In the UK, the healthcare expenditure is at 9.6% GDP according to the latest available information from the Office for National Statistics (ONS) with 96% of the government spending related to curative or rehabilitative, health-related long-term and preventative care [12].

Latest data published by the United Nations Department of Economic and Social Affairs show that the global population is ageing with the number of older people set to double to reach 2.1 billion by 2050, overtaking the adolescents and youths. In addition, of the 67 countries surveyed, data indicates that more older people live independently compared to 1990 [13]. Specifically, in the UK by 2040, it is estimated that one in seven people will be aged over 75 [14]. With greater longevity comes increased demand on the healthcare system and increase in complexity of care due to long-term and chronic disease. It is estimated that nearly 50% of all medical resources globally will be used by the elderly [13] with the health threats of those 75 years of age and over attributed to chronic illnesses, respiratory disease, Alzheimer's disease including other forms of dementia, diabetes and heart disease [11]. In 2015 across the European Union 1.2 million people across EU died prematurely that could have been avoided through more effective healthcare [11], with 86% of all deaths in Europe attributed to chronic diseases, and 80%

of those affecting elderly over the age of 65 [14]. Duration, treatability, added complications, prevalence and weakened immune system are some of the facets of chronic diseases, which if combined with independent living within the community create complex medical needs across multiple healthcare disciplines. The World Health Organisation reported that chronic disease threat is increasing, it needs to be better understood and acted upon. Comprehensive and integrated government-led action incorporating existing scientific knowledge is required to overcome this threat [15]. In the UK, the government recognises the importance of technology and innovation in healthcare to transform patients' care with an ambitious vision of Healthtech [16].

2.2 The Role of Digital Technologies in Healthcare

The relentless proliferation of innovations in disruptive digital technologies such as 5G, Edge Computing, Human Augmentation, Artificial Intelligence, Digital Twin combined with big data and substantial computational power will have a significant impact on society over the next decade and offer opportunities to create new digital ecosystems [17, 18]. The arrival of IPv6 and 5G networks could mean that over 50 billion Internet of Things (IoT) devices will be connected by 2020 [19, 20] many of which are medical devices and on-body sensors. IoT enabled Medical Cyber-physical Systems (mCPS), pave the way for the next generation of digital transformation in healthcare. Instead of relying on infrequent visits to hospitals, patients' health monitoring could be used in real-time to empower individuals, facilitate early detection, or manage plans for chronic conditions. Wearable health sensors anticipate a \$650 million global market share by 2020 which should save \$200 billion in healthcare cost over the next 25 years [21]. Furthermore, assistive technologies, life-critical networked medical devices [22] and an increase in real-time data collection create a unique opportunity to enable healthcare professionals to deliver more convenient and accurate healthcare service including remote operations. Smart IoT services will continue to revolutionise how healthcare is delivered and how we manage our health [5, 7, 14, 23, 24]. Technology will help managing large datasets, a key driver for research to improve the outcomes for disease prevention and early detection. The widespread and consistent adoption of disruptive technologies into healthcare fused with other smart sectors such as smart-homes and smart-mobility creates unprecedented opportunities to lower national healthcare spending and improve citizens wellbeing.

2.3 Towards Precision Healthcare

The term "digital health" includes a number of medical technologies, disruptive innovations and communication networks converged inseparably in providing healthcare services. Digital health has not been clearly defined in academic literature

[25] despite several attempts to provide a definition for it [26–28]. The World Health Organisation (WHO) defines digital health as "the use of digital, mobile and wireless technologies to support the achievement of health objectives", the term is used interchangeably in literature with mHealth, eHealth [29], virtual care, telehealth or telemedicine [5]. Despite advances in medical research and improved treatments, the increasing healthcare costs, rising life expectancy and shortage of health workers refocuses the efforts towards disease prevention. The estimated global shortage of health workers will be 12.9 million by 2035 [30]. Preventative medicine is an established field [31], aspects of the vision originate from the Human Genome Project which has enabled deeper understanding of medicine, the underlying disease mechanisms, environment-biology interactions and exploration of complex diseases including diabetes, heart disease, cancer, rare diseases, neurological or developmental disorders leading to personalised diagnosis and treatment [21, 32, 33]. Precision medicine is referred to by other terms including system medicine, P4 or computational system biomedicine. These terms all describe the idea of delivering targeted and the right treatment to the patient when it is required [6, 32–34] (Fig. 1).

2.4 *A Digital Twin for Precision Healthcare*

The concept of precision healthcare draws upon the experiences of other smart sectors. For example in engineering, aircraft engine's health is monitored in real-time by a quantum of sensors, actuators and controllers to prevent failures and

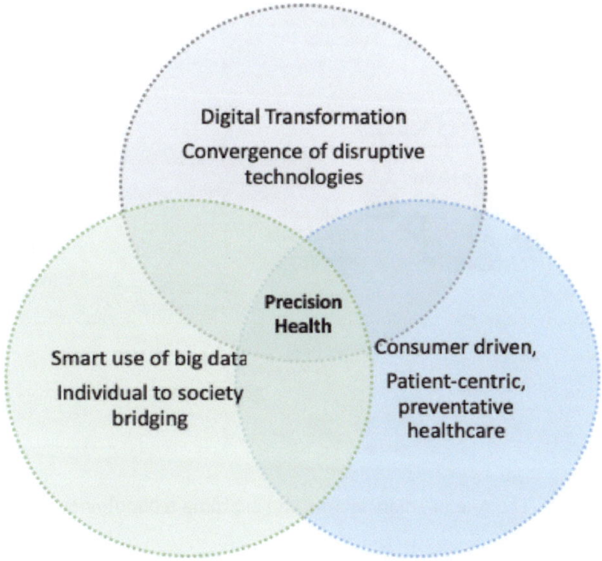

Fig. 1 Physical-Data-Cyber Converged domains driving precision health

repairs are forecasted using "Digital Twins" (DT) [35]. In the manufacturing sector, Industry 4.0 converges the physical and cyber domains through interconnectivity of Cyber-Physical Systems (CPS) to provide a virtual representation of the manufacturing lifecycle [36]. Additionally, gathering information to pinpoint anomaly or deviation from the norm using Artificial Intelligence is a mature concept used in the field of cybersecurity and implemented within anomaly-based intrusion detection systems. Digital Twining, a converged paradigm of cyber-physical-data domains, reflects on the physical artefacts within a virtual, computer-based representation of itself with data passed to it in real-time. The National Aeronautics and Space Administration (NASA) suggests "a Digital Twin is an integrated multi-physics, multiscale, probabilistic simulation of an as-built vehicle or system that uses the best available physical models, sensor updates, fleet history, etc., to mirror the life of its corresponding flying twin" [37]. In engineering, such dynamic computer models are instrumental in predictive analysis, like when to carry out maintenance or when modelling real-world engineering artefacts. In healthcare, the concept of the DT, a "virtual patient", is the same as in engineering. Adopting and adapting this novel engineering practice will elevate healthcare to a different level in disease prevention, early detection of disease, enhancement of patient care, wellbeing and lower the cost of national healthcare. In principle these concepts can be analysed and parallels found in analogy with practice applied in engineering and cyber security of "normal behaviour", "anomalous behaviour", "predictive maintenance", "automation" and "optimisation" [21, 38] (Fig. 2).

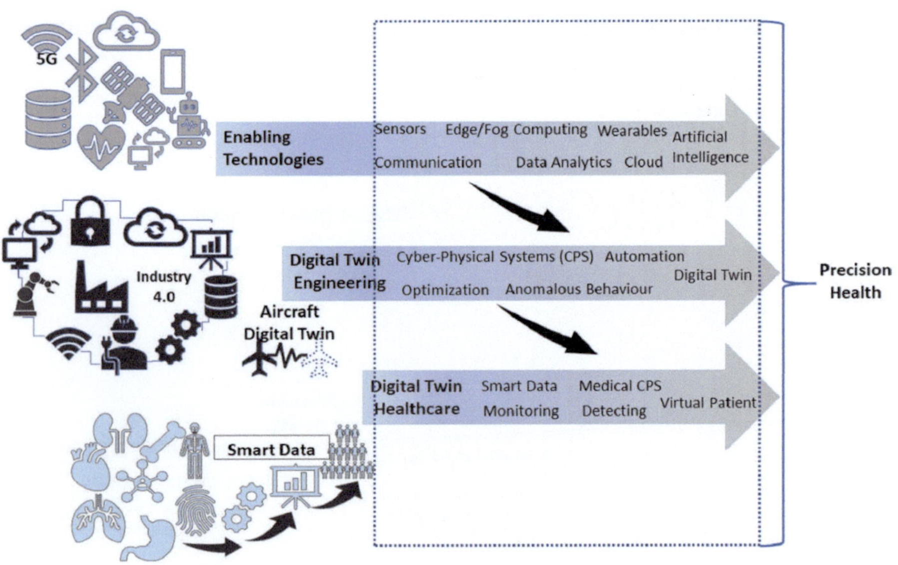

Fig. 2 The role of smart sectors and enabling technologies in delivering precision health

3 Key Enabling Technologies

3.1 *Gartner's Hype Cycle*

Five key trends and a number of dominating technologies emerged in the 2018 Gartner's Hype Cycle all of which are expected to be significant market disruptors. Although some of these are on-the-rise such as 5G, others including AI reached a peak with wider adoption expected within 2–5 years. The period of maturity has been estimated as 5–10 years for Smart Robots, Smart Workspaces, Edge Computing, Digital Twins, Biochips and IoT Platforms [18, 39]. However, frameworks will be required to create the foundation to apply these within the healthcare sector.

3.2 *Transformational Technologies and Associated Applications*

The healthcare industry is undergoing a rapid digital transformation but one of the key challenges of solving the highly complex healthcare ecosystem remains the real-time interaction and convergence of the cyber, physical and natural domains. An effective way to model this challenge is through the concept of DT. Within the current trend of digitalisation, automation, application of AI and big data, Industry 4.0 has paved the way for a systematic integration of IoT and CPS converging physical objects and digital technology to develop and maintain products' lifecycle from design to implementations using the concept of DT [36, 40, 41]. The concept of product lifecycle from marketing models' perspective, dates back to 1967, to the development of a product lifecycle for ethical drugs, which was greatly facilitated by the availability of the relevant data records, whilst the stages of the process were inspired by the biological lifecycle of introduction, growth, maturity and decline [42]. At present, despite utilising modelling, the different data streams still remain isolated and fragmented, therefore meaningless to the smart manufacturing industry. The aim of the current research is to shift the paradigm from focusing on the physical products to their virtual models to solve the problem of fusing the various big datasets across the different stages of the product lifecycle [41]. The exploitation of recent and emerging technologies integrates the physical objects, the virtual models and the data in real-time to create a twin representation of the physical product in a digital space. For example, DT has been applied in Industry 4.0 as virtual factories for as part of the asset management lifecycle [43], production system or IoT lifecycle [44–47]. Likewise, in the aviation industry to build high-fidelity flight models [35], to simulate aerothermal model predictions [48], to detect and monitor aircraft structure damage [49], predictive maintenance and failures.

Findings from the emerging literature, based on their description of the digital health ecosystem, agree on a common ground that the digital healthcare ecosystem is a subset of larger digital ecosystems [25, 50, 51] and a core component of smart cities [52, 53] forming a complex growing network of fragile [54] Cyber-Physical-Natural (CPN) systems centred around people. Digital ecosystems create dynamic digital communities of shared infrastructures, resources and knowledge that with associated applications can define and deliver a set of healthcare services and interactions. Despite the volumes of data that healthcare applications produce and depend on, the datasets remain isolated, significant data artefacts duplicated, with number of providers maintaining multiple Patient Health Records with inefficient sharing practices, lack of motivation to collaborate hindering early detection of diseases or management of chronic illnesses with continued drain on the healthcare resources, thus the quantum of healthcare data is meaningless and does not realise the aims and benefits of precision healthcare [50]. Data collected from IoT connected CPS creates an unprecedented wealth of information but to achieve the true potential of the data, research was that cross-smart sector transferable solutions should emerge from individual smart sectors [55]. For example, combining data for beds planning, with staff rotas and patient flow aided by AI could help manage the inconsistency between demand and capacity in hospitals [7] whilst extending the model to add data from GPs to provide early detection of an epidemic or integrating major incidents data from smart transport could help manage emergency care more proactively and efficiently [56]. Additionally, combining data from Medical Cyber-Physical Systems (mCPS), wearable technology, medical records and external behaviour captured using tracking across smart spaces [14, 24] could revolutionise care of the elderly in the community or those with chronic illnesses.

However, we argue that it is the twin representation of the physical object in digital space aspect of Industry 4.0 and the aviation smart sectors that are of value to precision healthcare. Although the concept of DT is not new and emerged from other smart sectors, its application in healthcare is new and extremely ambitious. Healthcare by its very nature is patient-centric but relies on technology to deliver its services. DT creates a model that converges the patient's physical world with the various datasets in cyberspace in real-time in order to combine the patient-centric nature of healthcare and draw on the benefits of technology to better understand the patient risk and define interventions with applying precision approaches to improve health in the population context [32]. The key components of the DT are common to the smart sectors across multitude of environments:

IoT Platform the driving mechanism of the digitalised ecosystem, which connects the quantum of sensors, actuators and controller from CPS and facilitates network communication between the physical objects. 5G's ubiquitous infrastructure will be a significant driving force for its commercial rollout, which is expected to have a massive impact on society and business bringing about societal and economic opportunities for everyday connected objects and innovative applications across number of smart sectors including smart homes, smart transport, smart grids, smart workplace, smart health and others [57–59].

Edge Computing combining 5G with mobile edge technology, also in its infancy in terms of wide commercial rollout, could provide real-time collaborations, monitoring of patients or even remote surgery [59]. With the explosion of data, devices and interactions, cloud architecture on its own cannot handle the influx of information and processing of data far-removed from its source creating latency and performance issues. With the advancement of IoT driven CPS huge amounts of data will be generated continuously that must be stored, processed and responded securely and cognitively [19]. For example, medical devices are increasingly IoT enabled like Computed Tomography (CT) or medical therapeutic activities in the community and are capable of processing the data at the edge [60, 61].

Cloud That said, cloud technology is a fundamental building block for ubiquitous CPS and services in precision healthcare due to its fundamental characteristics of a distributed, on-demand, scalable and virtualised service. However, there are a few cloud storage platforms specific to healthcare that enable real-time monitoring of patients [24]. The concept of large-scale cloud storage integrated with edge computing and 5G network capabilities can be converged with large numbers of IoT enabled devices amongst themselves, across different smart sectors and also with human users, enabling large scale IoT design and deployment at different abstraction layers [62, 63].

Artificial Intelligence (AI) drives a paradigm shift in healthcare through a widely applied combination of a highly complex algorithm that aims to mimic human cognitive functions in a range of applications and smart sectors. AI, including Deep Learning and Machine Learning, can be applied to a wider range of healthcare data to process complex data structures and enable computers to collect knowledge, thus human intervention in building that knowledge is not required.

CPS the gradual integration of CPS technologies by Industry 4.0 has led the way to enable real-time monitoring of physical activities in virtual space through the networked connectivity of CPS [36] into other smart sectors including urban space in smart cities [64], smart grids [65], smart homes [66], smart workplaces [67], smart transport [68] and healthcare [22] forming safety-critical, intelligent networked systems. **mCPS** are critical connected medical devices that are used increasingly more in hospitals for the provision of quality patient care [22].

The emerging digitalised ecosystems in healthcare will require visionaries, new business strategies and support of innovative technical foundation that enable the physical-cyber-data domains converged and bridged with humans to shift the paradigm from conventional to precision healthcare.

4 Proposed DT Frameworks

Precision healthcare is seen as a significant challenge. To address the design and management of this complexity, we investigate recent works in DT to realize the goals of precision healthcare.

4.1 DT for Better Community Healthcare

The problems that continue to persist in healthcare are the absence of real-time interactions, lack of convergence between medical physical and information systems, absence of active participation and continuous interactive monitoring throughout the elderly person's lifecycle [24]. The panacea to these problems is presented in the concept of Digital Twin for Healthcare (DTH), a novel cloud-based framework for the care of elderly in the community. The study adopts NASA's definition of DT [37] and highlights the complexity of interactions between people, medical devices and the variety of institutions involved in providing healthcare services to elderly. It is argued that due to the current complexity of the healthcare system, introducing DT can achieve greater medical flexibility, reduced medical risk and cost through modelling and simulation with real scenarios, thus gaining better quality and efficiency in disease diagnosis, treatment and prediction. The key phase behind the proposed CloudDTH medical simulation approach is the physical to digital representation using advanced 3D modelling techniques, followed by inclusion of real-time data from external factors like weather, elderly patients' physiological data from wearable monitoring devices, patients' healthcare records and virtual data from digital models. The fused data is stored in the healthcare cloud service. Utilising modelling methods, a virtual model is constructed for fast simulations with use of machine learning algorithm for accuracy of crisis prediction. The study discusses number of DTH applications: an early crisis warning, real-time supervision and scheduling. As the CloudDTH receives real-time data it is put through the virtual model and optimised, the model produces warnings and the scheduling system is based on predictions of the combined data. The viability of CloudDT was tested with 2 volunteers simulating a normal and abnormal heart rate. The experimental process successfully distinguished between the volunteers' needs based on the DTH model and demonstrated the feasibility of individualised medicine using DTH. Furthermore, the crisis warning was simulated using virtual modelling and combined with the hospital scheduling showed promising results. The authors conclude that although most of the research and commercial efforts concentrate on platforms, business models, standard and Health IoT interoperability, DTH is an effective way to solve the physical and cyber convergence and interactions. Whilst [14] are looking at ways to augment traditional clinical health services using IoT devices to help detect early changes in the lives of the elderly in the community, [24] demonstrates that DTH with the knowledge extracted from the real-time data enables the delivery of precision healthcare.

4.2 DT as Part of Intelligent Control and Emergency Planning in Hospitals

A crisis warning system was simulated with promising results by researchers proposing a concept of DTH [24]. Diversifying and evolving the application of DT in healthcare, an emergency unit performance evaluation of current state and major incident intelligent control is researched [56] using Discrete Event Simulation (DES) in the context of a major incident like substantial patients arrival related to an epidemic, as a result of natural disasters like tsunami, earthquake or due to terror attacks. Discrete Event Simulation is a widely researched field for healthcare modelling and although some studies present highly complex models [69] others are more generic and transferrable [70] for wider applicability.

The study highlights the innovative use of DES for modelling and decision-making functions of the framework for use by health-care professionals to enable scenario modelling based on real-time data to create a more efficient patient flow through the emergency unit, reduce patient stay, the demand on resources and increase the number of patients treated. The proposed DT-based modular framework consists of a modular model which is connected to a process analyser tool fed with data from the hospital information system and the patient arrivals forecast including data from the GP network alerts, crisis alerts, patient transfers information and utilisation of other hospital services.

The model was represented using the MedPRO UML-based modelling framework and implemented using the ROCKWELL Arena 14.5. The variables were extracted from the hospital information system, interviews with staff members and observations. The main focus of the study was the process view: patients' care in the emergency and the resources view the healthcare professionals' activities. The viability of the framework was demonstrated on a diverse set of scenarios with predefined key performance indicators. In their concluding remarks, the authors point to the innovative use of DT-based monitoring and control of the hospital emergency unit without disruption to services demonstrating aspects of precision healthcare delivery system through physical and cyber convergence.

4.3 DT as Part of Strategic Planning of Hospital Services

Whilst most DES applications relate to the discrete aspects of healthcare modelling such as clinics or emergency units [71], an innovative DES-base DT concept is proposed to assess and optimise the efficiency of the healthcare delivery systems and evaluate changes thus aid decision making related to staff scheduling, waiting time or appointment problems without disrupting the daily hospital activities [7]. The requirement to remain consistently efficient in the changing healthcare landscape and deal with the inconsistency between demand and capacity create significant challenges [7]. A key challenge highlighted in this study is the increasing demand for health services, therefore increased costs, and it is argued that these changes are due to the growing ageing population and increase in chronic illnesses. The study provided a general modular framework extendible to other healthcare services and simulated four key services for the proof of concepts. The model used the FlexSim HealthCare 3D simulation and modelling tool. To enable accurate and real-time simulation, the input of data was proposed from across hospital information systems, DES and IoT connected ubiquitous computing devices which could be used to track patient flow. The DT model is based on a hospital patient flow which enabled a variety of scenarios to be simulated including patient tracking from admission to discharge in real-time which enables the patient to receive the necessary medication, equipment or operating room at the right time. The framework's methodology feasibility was tested on number of different scenarios. The proof-of-concept shows that improvement of resource usage can be achieved through the concept of DT. Authors in their concluding remarks outlined further development of the proposed approach to handle more complex scenarios.

5 How Can Digital Twin Technology Improve the Delivery of Precision Healthcare

One of the global challenges in public health is the burden of chronic conditions in both developed and developing countries [72]. In medicine, these are a heterogeneous group of diseases characterised by their long duration, frequent recurrence and slow progression. As discussed earlier, the cost of managing chronic conditions, their co-morbidities and complications comprise a burden on health systems worldwide. Further, most of these conditions are subjects for extensive research in which risk factors and treatment options are explored, and this made them a promising aspect in medicine to invest new technologies and promote wellbeing. However, one of the major factors that undermined the use of technology in healthcare is the complexity of patient's needs [3]. This section will focus on managing chronic conditions at different levels and how DT could be hypothetically applied.

The management of chronic conditions is a continuum, it starts at the prevention stage when a risk factor is identified, the pre-diagnosis, the diagnosis and the management stage [73]. The self-management of chronic conditions is an evidence-based approach in healthcare. It implies the involvement of patients in their own care and relies on developing skills, utilising psychological resources, in addition to pharmacological treatment and regular follow up with healthcare practitioners [74]. The management of chronic conditions is complex, at an individual level, healthcare professional and health system levels.

At an individual level, the genetic component is one of the biological determinants of health, which is where precision healthcare mostly relies upon. However, managing chronic conditions is far more complex and it also includes the psycho-social aspects which can vary hugely between individuals [75]. Hence, despite having the same condition, different people have different self-management plans based on the bio-psycho-social aspects of their lives. Thus, the ideal situation in employing DT to improve precision healthcare is to consider these variables, with the acknowledgement that psychological and social aspects, such as lifestyle choices are not easy to measure or monitor.

At a healthcare professional level, the development of evidence-based medicine [4], the adoption of clinical guidelines and the international classification of diseases [76] had brought a relatively common ground of communication among healthcare professionals. However, to provide a personalised care plan the involvement of both patients and healthcare professionals are crucial to feeding into the digitalised form. This could be achieved by maintaining a 'doctor-patient' relationship that is built on trust and informed decisions [77]. This is to encourage patients to share feedback on their own health experiences and their preferences in managing their conditions. For example, if patient x and patient y both have diabetes, and both attend the same clinic, if patient x do not trust the healthcare professional and do not give feedback on the management plan compared to patient y, then the digital twin for patient y will be more personalised and responds to this patient's need.

At a system level, taking into consideration the promises given in precision healthcare and using DT to customise patient care by technology-enabled approaches [38], the journey of a patient with a chronic condition and/or comorbidities through the healthcare system could be transformed. DT could showcase the specific journey for the patient within the healthcare services. This could be achieved through simulating the ideal system navigation, referrals and resource management. It is worth thinking of the implications of precision health and DT application to two established approaches in healthcare, these are 'chronic diseases management' and 'acute case management' programs.

Chronic disease management programs target chronic conditions that are prevalent, with a high cost to manage, and have evidence-based guidelines to follow. National programs for chronic disease management existed in the United States, Austria, Denmark, England, Finland, France, Germany, Italy, Netherlands, and Poland, while regional or private initiatives were also adopted in England, France, Italy, Spain, and Sweden [78]. Hence, these programs are widely implemented, and they rely on modifying the health outcomes by coaching and managing the patients' lifestyle, for example with patients having diabetes, asthma or chronic heart diseases [79, 80]. Disease management programs require a closed-loop of communications among the patients, the treating physician and the disease management nurse. They also require a flow of clinical, behavioural and self-monitoring data within the loop. In such case, precision health and DT could be promising to build a healthcare model where the evidence-based guidelines are embedded in a system that is also encoding the patients' self-monitoring data, behavioural data in addition to the factors discussed under the individual level above.

Accordingly, the best theoretical scenario to apply DT in chronic disease management programs would be a patient with a chronic condition navigating within this healthcare model where the treating physician can have a holistic view that includes clinical and evidence-based information that could be further customised based on each patient's journey. The disease management nurse could be automated and customised to send tailored reminders, educational tips and responses not only to the self-monitoring data but also to the captured behavioural data. Such a model would potentially improve the clinical management outcomes and quality of life of patients with chronic conditions, and potentially avoid complications and their cost upon the system.

Some potential challenges to this application in disease management programs could be related to the variables in real-time that would deviate from the digitalised process [38]. A deviation from a physician side could emerge when the physicians' experience takes over the automated guidelines. From the patient's side, this could be influenced by the patient's complex needs [3] which might influence real-time behaviours, feelings, adherence to the set targets and the utilization of healthcare services.

Another approach is 'acute case management', it deals with acute and expensive cases such as oncology and severe accidents [81]. Applying precision health and DT to acute case management is similar in principle to the application in disease management, however, case management requires many urgent diagnostics and procedures, in addition to costly treatments. To decide the diagnostics to perform, and which procedures and treatment to follow is not simply based on guidelines. It is determined by a complex matrix of guidelines alongside the individual's specific data, such as susceptibility, genetics, or tolerance. Precision health and digital twinning could have a positive impact on this complicated healthcare model, to create a DT for the patient and simulate the best possible care plan which is being promised mainly in oncological research [82]. This would also have an impact on the system navigation and efficient resource management, and most importantly all save time and of unnecessary efforts and trials for such acute or severe cases. Potential challenges to this application could be related to which degree the digital twin could be precise when applying the best care plan on the ground.

A final factor to consider in employing DT in healthcare is the context in which the healthcare professionals and the health system operate, for example, the consideration of minorities, marginalised groups and stereotyping. Examples include the misdiagnosis of mental health illness and misunderstanding of people from ethnic minorities [83]. Accordingly, the design of a DT should consider all of the factors above and avoid exacerbating existing stereotyping and creating a systematically discriminatory process.

6 Why Is It Important to Have an Early Threat Model for DT

6.1 To Facilitate Security-by-Design (SbD) for DT Frameworks

The myriad of connected devices and sensors that exist and are used in solutions across the smart healthcare ecosystem and in other smart sectors, as presented in Fig. 3, could increase the data collection capabilities and the level of automation helped by Artificial Intelligence for precision healthcare [59]. DT represent one of the many concepts of technology-driven digital transformations that are gaining momentum. However, injecting intelligence into old technology, retrofitting machines with sensors to collect data and badge them as smart or collecting and processing large datasets using existing software under the umbrella of cognitive solutions are poor designs that are set to fail. Such practice introduces serious security gaps and a poor approach to cybersecurity. Evidence indicates that the natural desire for cutting-edge technology solutions, cost control, new attractive features are prioritised and security is an after-thought rather than integral to

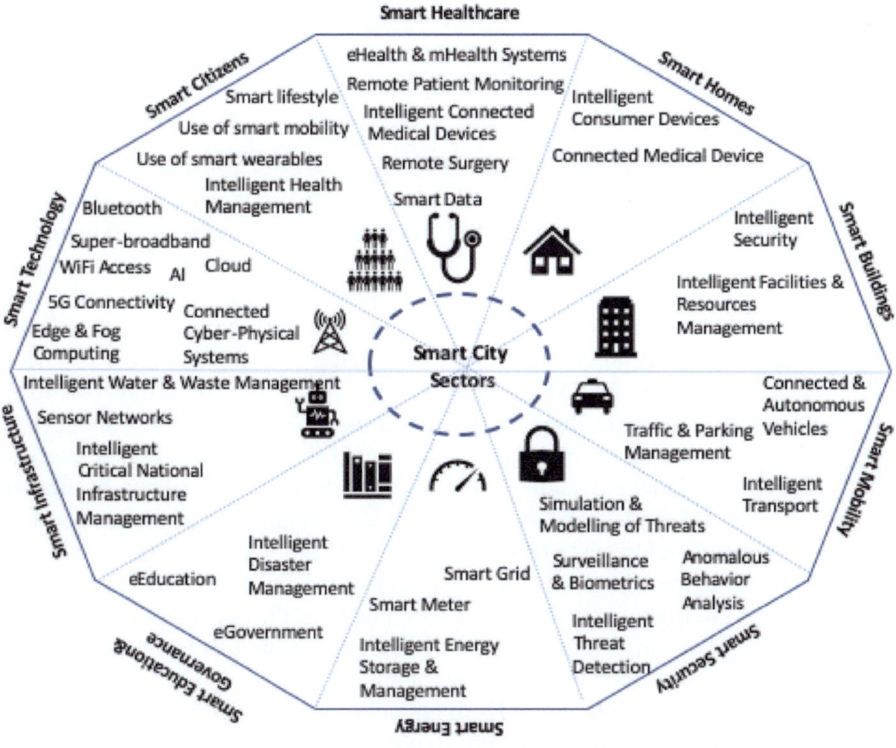

Fig. 3 Smart City sectors impacted by cybersecurity challenges

the framework design [7]. While this paradigm is not specific to the smart healthcare sector [84] there is little evidence of cross-organisational information security sharing, coordination and cybersecurity collaboration [85]. Although DT-based practices in civil engineering provide a good conceptual framework [36], the ubiquitous proliferation of DT across smart sectors from manufacturing to precision healthcare has many complexities introduced by IoT enabled CPS. Cyber-Physical Systems (CPS) are key components of embedded systems and play a critical role in DT, for example, IoT enabled mCPS are transformational to precision healthcare and the data generated from mCPS sensors can be used to learn about patients. But associated risks and vulnerabilities are not well understood, therefore more work is needed to drive SbD which can only be achieved if all aspects, including security, are considered from the outset [86] across all framework tiers. Furthermore, the increase in the interconnectivity and heterogeneity of medical devices and an extensive adoption of disruptive technologies across different sectors in smart cities have inherent cybersecurity threats with a large threat surface and a potential serious impact in the event of a security breach. Therefore, a citizen-centric approach in precision healthcare with a layered security mechanism is needed as part of SbD.

6.2 To Secure Against Inherent and Emerging Threats Through Defence-in-Depth Mechanisms Across All DT Domains

One of the great concerns with IT infrastructure, smart devices and IoT as fundamental supporting elements for DT in precision healthcare is cybersecurity. While collecting, moving and managing colossal amounts of data is underpinned by robust infrastructures, the proliferation of smart technologies and their implementation across smart city sectors are moving too fast for development of standards [87]. If the proposed DT and relevant frameworks for delivery of precision healthcare lack suitable security techniques and methods applied consistently across physical, cyber and data domains the DT concept remains vulnerable by design. A robust DT implementation requires threat modelling of the inherent security challenges of components and enabling technologies that makeup DT, as shown in Fig. 4, many of which are reported in recent studies [88, 89]. Most devices that makeup healthcare applications and DT in delivery of precision healthcare are wireless by nature and involve humans, therefore data privacy and security are major concerns. Whether the application of the sensor devices is wearable or implantable in humans [24] or it is part of intelligent control, emergency planning [7, 56] or used for remote surgery [23], the security threats in smart devices are significant. These threats can be summarised as Boot Process Vulnerabilities, Hardware Exploitation, Chip-Level Exploitation, Encryption and Hash Function Implementations, Backdoors in Remote Access Channels, and Software Exploitation [88]. For example, CPS, key components of DT, attract compromised-key attacks due to many sensors using

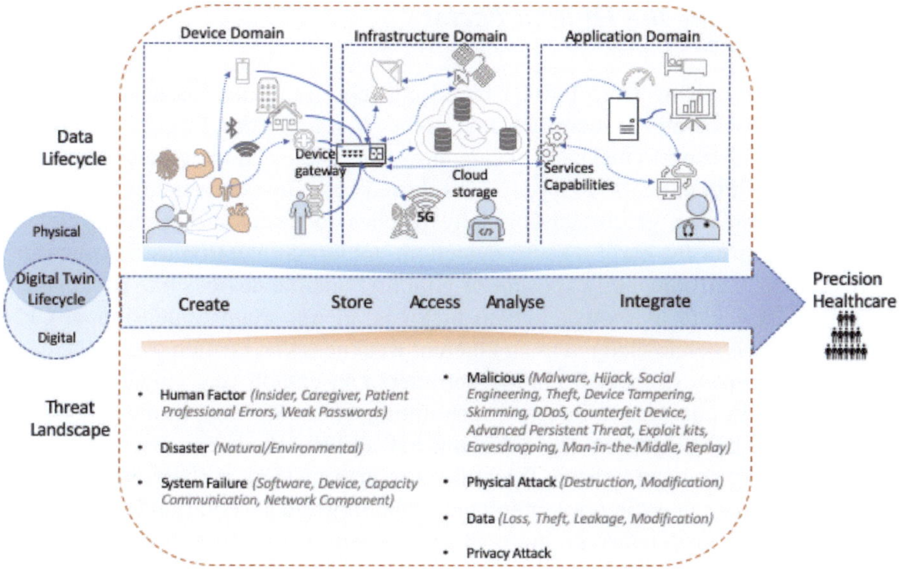

Fig. 4 Inherent and emerging threat landscape across digital twins concept in precision healthcare

cryptographic keys for the handshake protocols which include authenticity checks [90]. But despite the overall system hardening, the supporting infrastructure can be tampered with to embed malicious software or monitoring feature thus creating a backdoor [23, 91, 92] for easier access and possible data or Intellectual Property (IP) theft at a later time. Therefore, security measures to counter the threats must be considered and resolved during the design phases and innovative countermeasures are required to support modern defence-in-depth approach. Sensor designs have few external security features and therefore are prone to physical tampering. In addition to secure storage of data, existing protection methods can be used in securing DT hardware, for example through use of the Trusted Platform Module (TPM) chip, binding the software to the DT hardware. Physical access to the facilities should be restricted and supported by strong multi-factor or biometric authentication. A taxonomy of aspects that should be included in countering security threats include systems, administrative, physical, technical, information and finally healthcare data [24, 93]. Developing a concept of DT for remote surgery using mobile networks includes many different engineering fields including robotics, computer sciences and IT infrastructure development. Other applications of DT in the delivery of precision healthcare interact with different smart sectors like smart building, smart homes or smart transport. Smart cities include automated systems that can introduce and support the intelligent management of precision health components including data, medical equipment and management and supervision of public health [52]. Therefore, delivery of precision healthcare is truly cross-disciplinary and requires significant research with security embedded in its design from the outset.

6.3 To Mitigate the Human Factor

Security issues arising from the DT-based concept are numerous and due to the direct human involvement security is one of the most important aspects to consider. Precision Health aims to make healthcare more accessible and integrates monitoring and diagnosis into patients' everyday lives through use of smartwatches to measure pulse and heart-rate, electrocardiography patches to detect arrhythmias, clothes integrated sensors for early disease detection, epidermal electronics that measure vital signs to environmental exposure or vital implantable devices like pacemakers to name a few [21]. Although human factor is an important dimension in CPN ecosystems, it is acknowledged as an inherent weakness that is often overlooked and significantly underestimated [55, 94]. Intruders can exploit this vulnerability and phishing attacks continue to increase constituting a serious threat in the cyberspace with evidence of seeking out the emerging IoT and smart devices as a target [95]. Furthermore, there is an existent threat from human insiders in the workplace [67, 96] including human error, non-compliance, unauthorised access, fraud or industrial espionage. As a proprietary digital content, DT requires protection in terms of IP [23], therefore security mechanism is required to distinguish normal and malicious behaviours [97].

6.4 To Comply with and Influence the Development the Regulatory Landscape

The regulatory landscape is very diverse and lacks standards and criteria. Local, national and international laws and legal frameworks are a key element of effective public health policy and practice. For instance, precision health will produce more data, but the information gained could identify the patient's disease risk with far-reaching consequences. In the US the Genetic Information Non-discrimination Act 2014 (GINA) prohibits genetic discrimination by employers under the federal law since 2008 but this does not extend to other sectors such as insurers. With the increased value of genomic information in precision healthcare more comprehensive legislation is needed. The Health Insurance Portability and Accountability Act (HIPPA), an organisation-centric regulation, protect health information within the healthcare providers remit but does not extend to cover medical device companies who are not obliged to implement secure communication channels [21]. In the European Union, the General Data Protection Regulation (GDPR), a consumer-centric regulation, enforces data protection by design. Although there are attempts to publish baseline recommendations, due to the complexity and diversity of the IoT-based applications defining baseline security is a major challenge and there is no common approach to IoT security [98]. Beyond IoT security, regulatory concerns in other smart sectors that directly impact the delivery of precision healthcare are yet to be addressed. In the US, the National Institute of Standards and Technology

(NIST) developed guidelines for the Network of 'Things' [99], the Department for Homeland Security published strategic principles for security of IoT [100], the U.S. Department of Health and Human Services Food and Drug Administration (FDA) Center for Devices and Radiological Health covers recommendations on managing post-market cybersecurity vulnerabilities of marketed and distribute medical devices [101] but they are non-binding. The "Internet of Things Cybersecurity Improvement Act of 2017" was introduced to set the minimum set of requirements for IoT implementations [102]. Additionally, the IT security catalogue from ISO contains multi-part standards focusing on Internet of Things (IoT) Reference Architecture namely the ISO/IEC 30141:2018, Security Techniques ISO/IEC 27001:2013, and ISO/IEC 27002:2013, which are concerned with information security management practices. This fragile ecosystem is governed by the fragmentation of standards [98] due to the speed of the technology evolution.

7 Ethical Implications of the Emerging Paradigm

Precision healthcare is an evolving field and many technologies that will support the delivery of its objectives are in early stages of development or are yet to be developed. While technological innovations are potentially transformational, they bring about numerous challenges. The solutions to a novel approach to healthcare are complex, influenced by technological developments, socio-political aspects and different healthcare models. The multifaceted nature of the ethical issues apart from privacy, confidentiality and regulatory aspect should consider social justice, informed consent and marginalised groups. Difficulties in defining a universal meaning for ethics is largely due to the varied interpretations and multi-layered environments. This section will focus on the ethical implications of the emerging paradigm of the novel technological applications of digital twinning in precision healthcare at an individual, healthcare, research and technology levels.

At an individual level, there is an increased enthusiasm to use technologies, wearable sensors or implants to promote and manage health through individual participation. However, this may give rise to issues of equality and social justice, therefore inadvertently counterbalance the desired effect of improving the populations' healthcare [38]. Therefore, a commitment to consistent approach to access precision healthcare is required. Additionally, the use of wearable or implantable health monitoring devices is supported by real-time data collection capabilities and combined with the application of digital twinning technologies produce large datasets about patients including Electronic Health Records, genome sequences and behavioural data. There are significant risks in collecting, transmitting and storing data containing personal information and in practice can be ethically challenging to gather and manage. Lack of transparency and possible malicious use could be detrimental; therefore, it is important to be clear and consistent in the commitment of informed consent and privacy to maintain the patients' trust in the delivery of precision health.

At healthcare level, the data collection process can be very costly and even harmful to the patients, which highlights the question of how to achieve the balance to maximise the benefit but limit the burden. For example, should treatment be repeated, even if it causes risk to patient to gather the data or should it be based on reported patient outcomes? Next, aspects of precision healthcare need to be defined that will be supported using the acquired datasets, for instance, preventative, early diagnosis, therapeutic or chronic conditions. Questions could arise if markets and data monetisation are the ultimate drivers of implementing DT in precision healthcare thus mechanisms would be required for the delivery of precision healthcare for the disadvantaged groups.

At research level, the promise of precision healthcare goes far beyond treating those who are already symptomatic with illness or enabling the individual to take a more active role in management of their health. The capability to proactively prevent disease generates valuable raw data that drives health research and blurs the boundary between care and research. A growing number of organisations including health research, technology, life sciences work collaboratively to develop a common harmonised framework of approaches to enable secure and responsible sharing of genomics and clinical data to enable scientific progress and advancement in medicine in a highly ethical, secure and responsible manner [33]. Researchers who seek to leverage digital technologies research programs in the delivery of precision health should consistently maintain informed consent, autonomy of choice and should handle the data with integrity and transparency.

Finally, at the technology level, digitisation is borderless, and the data flows are global. Whilst the large-scale collection of data brings societal and individual benefits, the large-scale collection, transformation, convergence and aggregation is a substantial regulatory challenge [103]. The ability to collect large amounts of information, have consequences for the patient's safety, freedom and privacy. Therefore, strong regulatory and governance structure is required to ensure that appropriate security-by-design, regulation and audit frameworks are adopted, and transparency of the DT and its storage structure are maintained to safeguard the patient's rights to safety, informed consent and privacy.

8 Conclusion

The aim of this proactive patient care is to pre-empt the disease through preventative medicine and early detection, which could change the societal culture by empowering the individual to prevent their own disease. Traditionally, individual's age, family tree and more recently genetic screening were some of the key aspects of establishing a person's disease risk [21]. However, other factors affecting the person's wellbeing and disease including environmental, demographic, socio-economic or biological in a constantly changing landscape are not detected during routine health screening. Precision healthcare develops the concept of the precision approach and converges the individual to the population capitalising on smart use of big data by understanding and linking the individual-level data with the wider societal context [32]. Thus, the emerging concept of precision healthcare encourages preventative medicine, early detection and monitoring based on the patient's individual risk.

The advent of precision healthcare will see technology-driven digital transformation of the health service that will enable customised patient outcomes through novel and targeted medical approaches. To improve patients' health and wellbeing, the demand for intelligent, data-centric smart healthcare models using technological innovation and artificial intelligence (AI) is increasing. Few studies explored the potential of DT for precision healthcare and proposed frameworks usually linking the physical, cyber and data domains to the current and future needs of healthcare. DT pairs individual physical artefacts with digital models reflecting their status in real-time. Creating a live model for healthcare services introduces better risk assessment and evaluation without disturbing daily activities. The frameworks ranged from modelling and simulation of emergency units' performance in hospitals in the event of major incidents to the optimisation of healthcare delivery systems for the elderly or those with chronic diseases.

References

1. Constitution of the world health organization: Principles., 2005
2. W. H. Organization (1986) The Ottawa charter for health promotion
3. Greenhalgh T, Wherton J, Papoutsi C, Lynch J, Hughes G, Hinder S, Procter R, Shaw S (2018) Analysing the role of complexity in explaining the fortunes of technology programmes: empirical application of the NASSS framework. BMC Med 16(1):66. https://doi.org/10.1186/s12916-018-1050-6
4. Greenhalgh T, Howick J, Maskrey N (2014) Evidence based medicine: a movement in crisis? BMJ 348:g3725. https://doi.org/10.1136/bmj.g3725
5. Bhavnani SP, Sitapati AM (2019) Virtual care 2.0—a vision for the future of data-driven technology-enabled healthcare. Curr Treat Options Cardiovasc Med 21(5):21. https://doi.org/10.1007/s11936-019-0727-2
6. Pritchard DE, Moeckel F, Villa MS, Housman LT, McCarty CA, McLeod HL (2017) Strategies for integrating personalized medicine into healthcare practice. Pers Med 14(2):141–152. https://doi.org/10.2217/pme-2016-0064
7. Karakra A, Fontanili F, Lamine E, Lamothe J, Taweel A (2018) Pervasive computing integrated discrete event simulation for a hospital digital twin, pp 1–6. https://doi.org/10.1109/AICCSA.2018.8612796
8. Berwick DM, Hackbarth AD (2012) Eliminating waste in US health care. JAMA 307(14):1513–1516. https://doi.org/10.1001/jama.2012.362
9. Makary MA, Daniel M (2016) Medical error—the third leading cause of death in the US. BMJ 353:i2139. https://doi.org/10.1136/bmj.i2139
10. C. I. f. H. Information (2018) National health expenditure trends, 1975 to 2018. https://www.cihi.ca/en/health-spending/2018/national-health-expenditure-trends
11. O. E. Union (2018) Health at a glance Europe 2018: state of health in the EU Cycle, Paris. https://doi.org/10.1787/23056088
12. O. f. N. Statistics (2019) Healthcare expenditure, UK Health Accounts: 2017. Office for National Statistics, 25/04/2019, p 27

13. P. D United Nations Department of Economic and Social Affairs (2017) World Population Ageing 2017. https://www.un.org/en/development/desa/population/theme/ageing/WPA2017.asp
14. Bryant N, Spencer N, King A, Crooks P, Deakin J, Young S (2017) IoT and smart city services to support independence and wellbeing of older people, pp 1–6. https://doi.org/10.23919/SOFTCOM.2017.8115553
15. W. H. Organisation (2008) Preventing chronic diseases: a vital Investment.https://apps.who.int/iris/bitstream/handle/10665/43314/9241563001_eng.pdf;jsessionid=F24F4FB022DCC5C9DCF70BAE5BA95C8D?sequence=1
16. U. D. o. H. a. S. Care (2018) Policy Paper: the future of healthcare: our vision for digital, data and technology in health and care. D. o. H. a. S. Care (ed) UK Government
17. Sehrawat D, Gill NS (2018) Emerging trends and future computing technologies: a vision for smart environment. Int J Adv Res Comput Sci 9(2):839. https://doi.org/10.26483/ijarcs.v9i2.5838
18. Gartner (2018) 5 Trends emerge in the gartner hype cycle for emerging technologies, 2018. ID G00363408, Gartner. https://www.gartner.com/document/3886564?ref=TypeAheadSearch&qid=808dc69ca889b4bd3fa85b2e3
19. Rahman MA, Rashid MM, Hossain MS, Hassanain E, Alhamid MF, Guizani M (2019) Blockchain and IoT-based cognitive edge framework for sharing economy services in a smart city. IEEE Access:1–1. https://doi.org/10.1109/ACCESS.2019.2896065
20. Pacheco J, Zhu X, Badr Y, Hariri S (2017) Enabling risk management for smart infrastructures with an anomaly behavior analysis intrusion detection system, pp 324–328. https://doi.org/10.1109/FAS-W.2017.167
21. Gambhir SS, Ge TJ, Vermesh O, Spitler R (2018) Toward achieving precision health. Sci Transl Med 10(430):eaao3612. https://doi.org/10.1126/scitranslmed.aao3612
22. Lee I, Sokolsky O, Chen S, Hatcliff J, Jee E, Kim B, King A, Mullen-Fortino M, Park S, Roederer A, Venkatasubramanian KK (2012) Challenges and research directions in medical cyber–physical systems. Proc IEEE 100(1):75–90. https://doi.org/10.1109/JPROC.2011.2165270
23. Laaki H, Miche Y, Tammi K (2019) Prototyping a digital twin for real time remote control over Mobile networks: application of remote surgery. IEEE Access 7:20325–20336. https://doi.org/10.1109/ACCESS.2019.2897018
24. Liu Y, Zhang L, Yang Y, Zhou L, Ren L, Wang F, Liu R, Pang Z, Deen MJ (2019) A novel cloud-based framework for the elderly healthcare services using digital twin. IEEE Access 7:49088–49101. https://doi.org/10.1109/ACCESS.2019.2909828
25. Iyawa GE, Herselman M, Botha A (2016) Digital health innovation ecosystems: from systematic literature review to conceptual framework. Proced Comput Sci 100:244–252. https://doi.org/10.1016/j.procs.2016.09.149
26. Robinson L, Griffiths M, Wray J, Ure C, Stein-Hodgins JR, Shires G (2015) The use of digital health technology and social media to support breast screening. In: Digital mammography. Springer, pp 105–111. https://doi.org/10.1007/978-3-319-04831-4_13
27. Mellodge P, Vendetti C (2011) Remotely monitoring a patient's mobility: a digital health application. IEEE Potentials 30(2):33–38. https://doi.org/10.1109/MPOT.2010.939453
28. Kostkova P (2015) Grand challenges in digital health. Front Public Health 3:134. https://doi.org/10.3389/fpubh.2015.00134
29. W. T. Organisation (2016) Monitoring and evaluating digital health interventions: a practical guide to conducting research and assessment. ISBN 978–92–4-151176-6, Geneva. https://apps.who.int/iris/bitstream/handle/10665/252183/9789241511766-eng.pdf
30. W. T. Organisation (2014) A universal truth: no health without a workforce.pdf. Geneva. https://www.who.int/workforcealliance/knowledge/resources/GHWA-a_universal_truth_report.pdf?ua=1
31. Terris M (1975) Evolution of public health and preventive medicine in the United States. Am J Public Health 65(2):161–169. https://doi.org/10.2105/AJPH.65.2.161
32. Colijn C, Jones N, Johnston IG, Yaliraki S, Barahona M (2017) Toward precision healthcare: context and mathematical challenges. Front Physiol 8:136. https://doi.org/10.3389/fphys.2017.00136

33. Carrasco-Ramiro F, Peiró-Pastor R, Aguado B (2017) Human genomics projects and precision medicine Gene Ther 24:551, 08/14/online. https://doi.org/10.1038/gt.2017.77
34. Flores M, Glusman G, Brogaard K, Price ND, Hood L (2013) P4 medicine: how systems medicine will transform the healthcare sector and society. Pers Med 10(6):565–576. https://doi.org/10.2217/pme.13.57
35. Tuegel EJ, Ingraffea AR, Eason TG, Spottswood SM (2011) Reengineering aircraft structural life prediction using a digital twin. Int J Aerosp Eng 2011:1–14. https://doi.org/10.1155/2011/154798
36. Negri E, Fumagalli L, Macchi M (2017) A review of the roles of digital twin in cps-based production systems. Proced Manuf 11:939–948. https://doi.org/10.1016/j.promfg.2017.07.198
37. Glaessgen E, Stargel D (2012) The digital twin paradigm for future NASA and US Air Force vehicles, p 1818. https://doi.org/10.2514/6.2012-1818
38. Bruynseels K, Santoni de Sio F, van den Hoven J (2018) Digital twins in health care: ethical implications of an emerging engineering paradigm. Front Genet 9(31). https://doi.org/10.3389/fgene.2018.00031
39. Gartner (2018) Hype cycle for emerging technologies, 2018. ID G00340159. https://www.gartner.com/document/3885468?qid=eaeac87a4acbfd43931fc95&ref=solrAll&refval=224658212&toggle=1
40. Lee J, Bagheri B, Kao H-A (2015) A cyber-physical systems architecture for industry 4.0-based manufacturing systems. Manuf Lett 3:18–23. https://doi.org/10.1016/j.mfglet.2014.12.001
41. Tao F, Cheng J, Qi Q, Zhang M, Zhang H, Sui F (2018) Digital twin-driven product design, manufacturing and service with big data. Int J Adv Manuf Technol 94(9–12):3563–3576. https://doi.org/10.1007/s00170-017-0233-1
42. Cox WE (1967) Product life cycles as marketing models. J Bus 40(4):375–384
43. Sacco M, Pedrazzoli P, Terkaj W (2010) VFF: virtual factory framework, pp 1–8. https://doi.org/10.1109/ICE.2010.7477041
44. Rosen R, Von Wichert G, Lo G, Bettenhausen KD (2015) About the importance of autonomy and digital twins for the future of manufacturing. IFAC-PapersOnLine 48(3):567–572. https://doi.org/10.1016/j.ifacol.2015.06.141
45. Schluse M, Rossmann J (2016) From simulation to experimentable digital twins: Simulation-based development and operation of complex technical systems, pp 1–6. https://doi.org/10.1109/SysEng.2016.7753162
46. Canedo A (2016) Industrial IoT lifecycle via digital twins, pp 1–1
47. Schroeder GN, Steinmetz C, Pereira CE, Espindola DB (2016) Digital twin data modeling with automationML and a communication methodology for data exchange. IFAC-PapersOnLine 49(30):12–17. https://doi.org/10.1016/j.ifacol.2016.11.115
48. Smarslok B, Culler A, Mahadevan S (2012) Error quantification and confidence assessment of aerothermal model predictions for hypersonic aircraft, p 1817. https://doi.org/10.2514/6.2012-1817
49. Bielefeldt B, Hochhalter J, Hartl D (2015) Computationally efficient analysis of SMA sensory particles embedded in complex aerostructures using a substructure approach, pp V001T02A007–V001T02A007. https://doi.org/10.1115/SMASIS2015-8975
50. Qureshi B (2014) Towards a digital ecosystem for predictive healthcare analytics. In: Proceedings of the 6th international conference on Management of Emergent Digital EcoSystems, Buraidah, Al Qassim, Saudi Arabia, pp 34–41. https://doi.org/10.1145/2668260.2668286
51. León MC, Nieto-Hipólito JI, Garibaldi-Beltrán J, Amaya-Parra G, Luque-Morales P, Magaña-Espinoza P, Aguilar-Velazco J (April 27, 2016) Designing a model of a digital ecosystem for healthcare and wellness using the business model canvas. J Med Syst 40(6):144. https://doi.org/10.1007/s10916-016-0488-3
52. Pramanik MI, Lau RYK, Demirkan H, Azad MAK (2017) Smart health: Big data enabled health paradigm within smart cities. Expert Sys Appl 87:370–383. https://doi.org/10.1016/j.eswa.2017.06.027

53. Huang G, Fang Y, Wang X, Pei Y, Horn B (2018) A survey on the status of smart healthcare from the universal village perspective, pp 1–6. https://doi.org/10.1109/UV.2018.8642125
54. Haughey H, Epiphaniou G, al-Khateeb HM (2016) Anonymity networks and the fragile cyber ecosystem. Netw Secur 2016(3):10–18. https://doi.org/10.1016/S1353-4858(16)30028-9
55. Boyes HA, Isbell R, Norris P, Watson T (2014) Enabling intelligent cities through cyber security of building information and building systems, pp 1–6. https://doi.org/10.1049/ic.2014.0046
56. Augusto V, Murgier M, Viallon A (2018) A modelling and simulation framework for intelligent control of emergency units in the case of major crisis, pp 2495–2506. https://doi.org/10.1109/WSC.2018.8632438
57. Chen M, Yang J, Zhou J, Hao Y, Zhang J, Youn C (2018) 5G-smart diabetes: toward personalized diabetes diagnosis with healthcare big data clouds. IEEE Commun Mag 56(4):16–23. https://doi.org/10.1109/MCOM.2018.1700788
58. Oleshchuk V, Fensli R (2011) Remote patient monitoring within a future 5G infrastructure. Wirel Pers Commun 57(3):431–439. https://doi.org/10.1007/s11277-010-0078-5
59. Mattos WD d, Gondim PRL (2016) M-Health solutions using 5G networks and M2M communications. IT Professional 18(3):24–29. https://doi.org/10.1109/MITP.2016.52
60. Rahman MA, Hossain MS, Hassanain E, Muhammad G (2018) Semantic multimedia fog computing and IoT environment: sustainability perspective. IEEE Commun Mag 56(5):80–87. https://doi.org/10.1109/MCOM.2018.1700907
61. Rahman MA, Hossain MS (2017) M-therapy: a multisensor framework for in-home therapy management: a social therapy of things perspective. IEEE Internet Things J 5(4):2548–2556. https://doi.org/10.1109/JIOT.2017.2776150
62. Fortino G, Guerrieri A, Russo W, Savaglio C (2014) Integration of agent-based and cloud computing for the smart objects-oriented IoT, pp 493–498. https://doi.org/10.1109/CSCWD.2014.6846894
63. Nastic S, Sehic S, Le D-H, Truong H-L, Dustdar S (2014) Provisioning software-defined IoT cloud systems, pp 288–295. https://doi.org/10.1109/FiCloud.2014.52
64. Ma Y, Li G, Xie H, Zhang H (2018) City profile: using SMART data to create digital URBAN spaces. ISPRS Ann Photogramm Remote Sens Spati Infor Sci 4:75–82. https://doi.org/10.5194/isprs-annals-IV-4-W7-75-2018
65. Orozco A, Pacheco J, Hariri S (2017) Anomaly behavior analysis for smart grid automation system, pp 1–7. https://doi.org/10.1109/ROPEC.2017.8261614
66. Do Q, Martini B, Choo K-KR (2018) Cyber-physical systems information gathering: a smart home case study. Comput Netw 138:1–12. https://doi.org/10.1016/j.comnet.2018.03.024
67. Ahmadi-Assalemi G, al-Khateeb H, Epiphaniou G, Cosson J, Jahankhani H, Pillai P (2019) Federated blockchain-based tracking and liability attribution framework for employees and cyber-physical objects in a smart workplace. https://doi.org/10.1109/ICGS3.2019.8688297
68. al-Khateeb H, Epiphaniou G, Reviczky A, Karadimas P, Heidari H (2018) Proactive threat detection for connected cars using recursive Bayesian estimation. IEEE Sensors J 18(12):4822–4831. https://doi.org/10.1109/JSEN.2017.2782751
69. Glaa B, Hammadi S, Tahon C (2006) Modeling the emergency path handling and emergency department simulation, pp 4585–4590. https://doi.org/10.1109/ICSMC.2006.384869
70. Sinreich D, Marmor YN (2004) A simple and intuitive simulation tool for analyzing emergency department operations, pp 1994–2002. https://doi.org/10.1109/WSC.2004.1371561
71. Kammoun A, Loukil T, Hachicha W (2014) The use of discrete event simulation in hospital supply chain management, pp 143–148. https://doi.org/10.1109/ICAdLT.2014.6864108
72. Alwan A (2011) Global status report on noncommunicable diseases 2010. World Health Organization, Geneva. http://apps.who.int/iris/bitstream/handle/10665/44579/9789240686458_eng.pdf;jsessionid=1D70E16CE9E288647B273D604E1D8991?sequence=1
73. C. f. D. C. a. P (2019) Chronic diseases: the leading causes of death and disability in the United States. 01/08/2019. https://www.cdc.gov/chronicdisease/resources/infographic/chronic-diseases.htm

74. Richard AA, Shea K (2011) Delineation of self-care and associated concepts. J Nurs Scholarsh 43(3):255–264. https://doi.org/10.1111/j.1547-5069.2011.01404.x
75. Barnes C, Mercer G (2010) Exploring disability, 2nd edn, Cambridge, 2nd ed. Cambridge
76. W. H. Organization (1992) The ICD-10 classification of mental and behavioural disorders: clinical descriptions and diagnostic guidelines
77. Cohn S (2015) 'Trust my doctor, trust my pancreas': trust as an emergent quality of social practice. Philos Ethics Humanit Med 10(1):9. https://doi.org/10.1186/s13010-015-0029-6
78. Gemmill M (2008) Research note: chronic disease management in Europe "employment, social affairs and equal opportunities" unit E1-social and demographic analysis. European Commission Directorate-General
79. Fuchs S, Henschke C, Blümel M, Busse R (2014) Disease management programs for type 2 diabetes in Germany: a systematic literature review evaluating effectiveness. Dtsch Arztebl Int 111(26):453. https://doi.org/10.3238/arztebl.2014.0453
80. Gapp O, Schweikert B, Meisinger C, Holle R (2008) Disease management programmes for patients with coronary heart disease—an empirical study of German programmes. Health Policy 88(2–3):176–185. https://doi.org/10.1016/j.healthpol.2008.03.009
81. Norris SL, Nichols PJ, Caspersen CJ, Glasgow RE, Engelgau MM, Jack L Jr, Isham G, Snyder SR, Carande-Kulis VG, Garfield S (2002) The effectiveness of disease and case management for people with diabetes: a systematic review. Am J Prev Med 22(4):15–38. https://doi.org/10.1016/S0749-3797(02)00423-3
82. Kummar S, Williams PM, Lih C-J, Polley EC, Chen AP, Rubinstein LV, Zhao Y, Simon RM, Conley BA, Doroshow JH (2015) Application of molecular profiling in clinical trials for advanced metastatic cancers. JNCI: J Natl Cancer Inst 107(4). https://doi.org/10.1093/jnci/djv003
83. Suite DH, La Bril R, Primm A, Harrison-Ross P (2007) Beyond misdiagnosis, misunderstanding and mistrust: relevance of the historical perspective in the medical and mental health treatment of people of color. J Natl Med Assoc 99(8):879–885
84. Bajramovic E, Waedt K, Ciriello A, Gupta D (2016) Forensic readiness of smart buildings: preconditions for subsequent cybersecurity tests, pp 1–6. https://doi.org/10.1109/ISC2.2016.7580754
85. Skopik F, Settanni G, Fiedler R (2016) A problem shared is a problem halved: A survey on the dimensions of collective cyber defense through security information sharing. Comput Secur 60:154–176. https://doi.org/10.1016/j.cose.2016.04.003
86. He H, Maple C, Watson T, Tiwari A, Mehnen J, Jin Y, Gabrys B (2016) The security challenges in the IoT enabled cyber-physical systems and opportunities for evolutionary computing & other computational intelligence, pp 1015–1021. https://doi.org/10.1109/CEC.2016.7743900
87. Coppinger R (2016) Design through the looking glass [digital twins of real products]. Eng Technol 11(11):58–60. https://doi.org/10.1049/et.2016.1106
88. Wurm J, Jin Y, Liu Y, Hu S, Heffner K, Rahman F, Tehranipoor M (2017) Introduction to cyber-physical system Security: a cross-layer perspective. IEEE Trans Multi-Scale Comput Syst 3(3):215–227. https://doi.org/10.1109/TMSCS.2016.2569446
89. Wu W, Kang R, Li Z (2015) Risk assessment method for cyber security of cyber physical systems, pp 1–5. https://doi.org/10.1109/ICRSE.2015.7366430
90. Shafi Q (2012) Cyber Physical Systems Security: A Brief Survey, pp 146–150. https://doi.org/10.1109/ICCSA.2012.36
91. Gallagher S (2014) Photos of an NSA "upgrade" factory show Cisco router getting implant
92. Pagliery S (2015) Lenovo slipped 'Superfish' malware into laptops. CNN
93. Al Ameen M, Liu J, Kwak K (2012) Security and privacy issues in wireless sensor networks for healthcare applications. J Med Syst 36(1):93–101. https://doi.org/10.1007/s10916-010-9449-4
94. Krombholz K, Hobel H, Huber M, Weippl E (2015) Advanced social engineering attacks. J Inform Secur Appl 22:113–122. https://doi.org/10.1016/j.jisa.2014.09.005

95. Gupta BB, Tewari A, Jain AK, Agrawal DP (2017) Fighting against phishing attacks: state of the art and future challenges. Neural Comput Appl 28(12):3629–3654. https://doi.org/10.1007/s00521-016-2275-y
96. Kammüller F, Nurse JRC, Probst CW (2016) Attack tree analysis for insider threats on the IoT using Isabelle, pp 234–246. https://doi.org/10.1007/978-3-319-39381-0_21
97. Cheh C, Keefe K, Feddersen B, Chen B, Temple WG, Sanders WH (2017) Developing models for physical attacks in cyber-physical systems. In: Proceedings of the 2017 workshop on cyber-physical systems security and privaCy, Dallas, pp 49–55. https://doi.org/10.1145/3140241.3140249
98. European Union Agency For Network And Information Security (ENISA) (2017) Baseline security recommendations for IoT in the context of critical information infrastructures. https://doi.org/10.2824/03228
99. N. I. o. S. a. T. (NIST) (2016) NIST Special Publocation 800-183 Nentworks of 'Things'. Department of Commerce, USA
100. U. D. o. H. Security (2016) Strategic principles for security the Internet of Things (IoT). US Homeland Security
101. U. S. D. o. H. a. H. S. F. a. D. A. F. C. f. D. a. a. R. Health (2018) Postmarket management of cybersecurity in medical devices
102. Internet of Things (IoT) Cybersecurity Improvement Act of 2017 Standard S. 1691, 2017–2018
103. E. g. o. E. i. S. a. N. T. t. t. E. Commission (2015) The ethical implications of new health technologies and citizen participation. Brussels. https://doi.org/10.2872/633988

5G Cybersecurity Vulnerabilities with IoT and Smart Societies

Yelda Shah, Nishan Chelvachandran, Stefan Kendzierskyj, Hamid Jahankhani, and Radovan Janoso

Abstract 5G, the fifth generation of wireless connectivity, is designed to allow long-distance coverage and stable connections as well as rapid data download and upload. As a result of 5G's the wireless-based technology, the data migration enables a speed of 20 Gbps (Gigabyte per second) through wireless mobile data connections, which simplifies the management of excessive data transmission via 5G. The protocols capability for high quantity data transfer speeds with low latency, compared with the previous generations mobile data telephony makes the protocol ideal for both current IoT and automated systems, as well as enabling the development and further proliferation of more. Data transfer speeds and latency rates have been a bottleneck in the roll out of smart technologies. Despite the relatively high data speeds of 4G connectivity, the availability and development of infrastructure, together with the explosion in the ownership and use of devices utilising the technology, has been a limiting factor in the roll out and use of AI and automated technologies such as driverless vehicles and smart city implementations. Whilst 5G looks to solve these limitations brought by previous generations, there are also drawbacks with 5G. The frequency and narrow wavelength, known as millimeter wave, whilst enabling such high data transfer speeds and reduced latency, also has a very limited distance of effectivity. There is only a very short distance before the signal starts to deteriorate, after which, the deterioration is exponential. 5G signals also cannot penetrate or reflect off of buildings and other obstacles very easily. This means that for a 5G networks implementation to be maximised,

Y. Shah · H. Jahankhani (✉)
Northumbria University, London, UK
e-mail: yelda.shah@northumbria.ac.uk; Hamid.jahankhani@northumbria.ac.uk

N. Chelvachandran · R. Janoso
Open Innovation House, Saidot OY, ESPOO, Finland
e-mail: nishan@saidot.ai; Janoso@saidot.ai

S. Kendzierskyj
Cyfortis, Worcester Park, Surrey, UK
e-mail: Stefan@cyfortis.co.uk

H. Jahankhani et al. (eds.), *Cyber Defence in the Age of AI, Smart Societies and Augmented Humanity*, Advanced Sciences and Technologies for Security Applications, https://doi.org/10.1007/978-3-030-35746-7_9

direct line of sight between the connected device and the relays or radioheads must be maintained, or at least, with as minimal obstruction as possible. A work around to this limitation is through densification and utilising large numbers of small cell radio heads throughout a coverage area. This will require that there is far greater investment and redevelopment in the mobile telephony infrastructure for this strategy to be implemented.

Keywords 5G · IoT · AI · Smart cities · Privacy · Cybersecurity · Millimetre

1 Introduction

The proliferation of technology is now exponential. Developments in technology, the increase in computer power and the reduction of cost, has allowed for greater accessibility, use and implementation of this technology in all sectors and industries. The evolution of smart and autonomous technologies, such as artificial intelligence and machine learning, has enabled traditionally labour intensive data analytics tasks to be conducted, quickly and efficiently. Multiple datasets and data lakes that have been traditionally siloed, are now being utilised and interconnected. This methodology is the foundation and backbone for the concept of smart cities. Not only can the data that is being collected be analysed, but the utilisation of wireless sensors distributed through different types of infrastructure to collect data is also at the forefront of technological development. Both consumer level devices and those used by enterprise and public sector implementations are being used to collect data to allow for "smarter" decisions to be made. For example, a nation's power grid. Various IoT devices at different levels and stages of the infrastructure are being used to help balance the load of the grid and minimise waste. The ability to accurately predict peak times of demand together with times of surplus, can help provide efficiency and manage power production. Alternative storage sources can also be utilised to store power created from renewables or during times of surplus production, so that then can then be used at times of greater demand. Another example is with the deployment of public resources and services, in a smart city or smart neighbourhood construct. Pedestrian numbers and movements through a neighbour can be monitored, the level of waste in litter bins monitored. If there is a spike in the number of people in an area, it could be an indication to an incident or event, similarly if the traffic patterns of pedestrians or vehicles change in an area, this could also indicate an abnormal incident. Real time demand for public transport streamline the use of automated fleets, so that transportation is available when there is a demand, rather than operate a regular timetable when there is little or no demand. However, in all of these instances, a reliable, fast data network to enable the backhaul of large quantities of realtime processing data to the cloud is needed to facilitate the effective use of these mechanisms. It is with 5G, that the inhibitors of smart capabilities will be removed.

It is therefore imperative that with a foreseen utilisation and implementation of smart and automated systems within infrastructure and services, that consideration is taken to ensure the privacy and technology of these mechanisms and systems.

Smart technologies, automated systems, and IoT are all dependant on data utilisation and analyse as the backbone of their functionality, with 5G labelled as being the carrier of the data. Securing both the communications mechanisms and the technologies themselves are key to their safe and secure implementation. It is also critical that such methodology, privacy and security frameworks are utilised to enable and instil trust in the use of these technologies, which whilst will be critical in smart infrastructure, will also be pervasive in both nature and scope of use.

2 The Road to 5G

In 1983 almost all communications possible were wireless based and voice-centric, using analogue systems [22]. From 1983 until 2013 many generation communications were introduced. The first generation (1G) mobile system was an integration of FM radios in analogue systems, since manufacturing digital radio systems were expensive. The former European GSM was later changed to the second generation (2G). Transitioning from voice-centric wireless communication and the change into the digital systems, such as EDGE, GPRS and GSM, where the code division multiple access (CDMA) system, was predominantly used in the USA with a bandwidth of 1.25 MHz. From the end of 1990's, the revolutionary third generation (3G) was introduced into the market by connecting data and voice together. From 2013, the migration from 3G to 4G was a representative transgression from the internet at a lower data rate to the high-speed internet used for mobile videos and higher end multimedia. Both, LTE and WiMAX are part of fourth generation (4G) systems with a bandwidth of 20 MHz [22]. The future of mobile communication includes aspects of efficiency and rapidness. According to Rodriguez [22], the system of the fifth generation (5G) of mobile communication will be marked as the reduction in operational expenses in providers as well as an energy-efficient system, which can simultaneously offer very high-speed and ultra-high capacity data. 5G's mission is to offer stakeholders a smooth communication platform, where a vast number of wireless networks build an infrastructure whilst communicating with each other.

2.1 Technology Architecture of 5G

Multiple research has concluded that approximately four to six technologies are collectively responsible for the existence and the function of the next generation network (NGN), the 5G cellular network. According to many specialists and researchers in this field, White [29] states that the innovative 5G is distinctive in three major features, helping to shape the technology with positive impacts, such as:

- The ability of multi-device connectivity
- Higher speeds and
- Lower latency.

More importantly the yearly subscription to mobile broadband systems shows a rapid increase in the number of individuals using it. With the high amount of data and the connected devices to the internet, the number will eventually increase. In a 20-year global perspective, the number of devices connected to the internet demonstrate an exponential increase and Statista [26] forecasts a growth of 48 billion connected devices from 2019 until 2025; which means by 2025 there will be around 75 billion internet connected devices worldwide.

In order for the billions of IoT devices to interact faster with each other and also with a base station for the response to signals/requests, it requires a high speed and stable internet connection. This enables higher data rates for the purpose of information transfers. Therefore, 5G provides universal connectivity for machines, devices, and humans at various operating spectrum bands as the goal is to develop a newly-created network with enhanced features, such as incorporating the growth of prospectively increasing devices into the new network [2]. Therefore, 5G serves as a cutting-edge technology. But before the potentials and security aspects of the next-generation gets discussed, it is essential to examine the technological components behind the development of 5G.

2.2 *Deeper Dive into 5G Components*

Exchanging data in terms of transmitting and receiving information via a local network could be feasible but the more complex the amount of information transmitted, the more data rates it requires. Khalaf et al. [16] recognise the issue by stating that for instance video streaming needs a wider range of data to be transmitted at the current speed or if possible, utilise higher data rates that requires more bandwidth, which is available at higher frequencies. As a result of this issue, Khalaf et al. [16] intended to design a front-end circuit receiver for use in a 60 GHz-range by using millimeter waves to meet the demands of high data rates and ultimately enable faster high-quality streaming.

Millimeter Wave this part of communication technology is one of the core areas of 5G networks and is expected to offer wireless data transfer by settling for a higher bandwidth [11]. However, the issue which arises from this technological concept is that the transmission distance of this particular wave is known to be limited into 100 m in the atmosphere with regards to its deterioration while the transmission occurs. Ultimately, millimeter waves show a disadvantage in comparison to other wave types, which results in a fair transmission coverage.

The wavelength is the spatial distance between two identical and repetitive oscillation trains within one periodic wave movement. Therefore, the wavelength of electromagnetic waves is the ratio of the speed of light (***c***) and the number of wave oscillation per second ***frequency***, measured in Hertz (***Hz***).

$$\lambda = \frac{c}{f} \tag{1}$$

As mm-waves are located within the spectrum band and have a frequency between 30 GHz and 300 GHz, as the wavelength becomes shorter, the higher the frequency gets [27]. Wavelengths of millimeter waves are approximately 10^{-2} m to 10^{-4} m in amplitude size. Therefore, millimeter waves have an extreme high frequency (***EHF***).

In fact, the selection of frequency is essential in the sense that previous mobile technologies mainly used the lower frequency band. Therefore, 5G is expected to use higher frequencies within the frequency bands. However, higher frequency decays faster than lower frequency and is comparatively more sensitive to signal losses [21]. If both, a lower frequency antenna and a higher frequency antenna were to transmit data at the same power/speed/data rate, the HF- antenna would have a low area coverage, whereas lower frequency has not. As a result, users get higher data rates if the cell size is small. One possible solution for expanding the coverage size for users of high frequency, is the use and implementation of massive numbers of antennas possible due to HF's poor penetration capability, which decays faster. In addition to that, it has a high possibility to reflect from walls. However, lower frequency gets reflected from walls and objects but still has a proper penetration capability [10]. Because the capacity for spectrum in the lower frequency bands for 5G is nearly full occupied, various discussions are present as to what kind of frequency should be used for 5G. It is essential to have very large bandwidth to be able to deliver a massive amount of data rates and for some of these innovative services for which 5G is being advertised, such as IoT and AI. To achieve the utilization of higher bandwidth, mobile telecommunication providers have to set for higher frequencies, which is about 30 GHz and is known as the millimeter waves. As a result, supporting and extensive spectrum bands represent a significant necessity for 5G to handle high data rate demands as well as immense capacities [2].

The 5G network is designed to serve service throughout multiple layers for multiple needs:

- Coverage Layer: < 1 Gigahertz
- Capacity Layer: 1–6 Gigahertz
- High Throughput Layer: 6–100 Gigahertz (generally referring to them as millimeter wave but mm-waves means frequencies above 30 GHz).

Multiple Input Multiple Output (MIMO) is a transmission system for the use of multiple transmitting and receiving aerials to enable wireless communication.

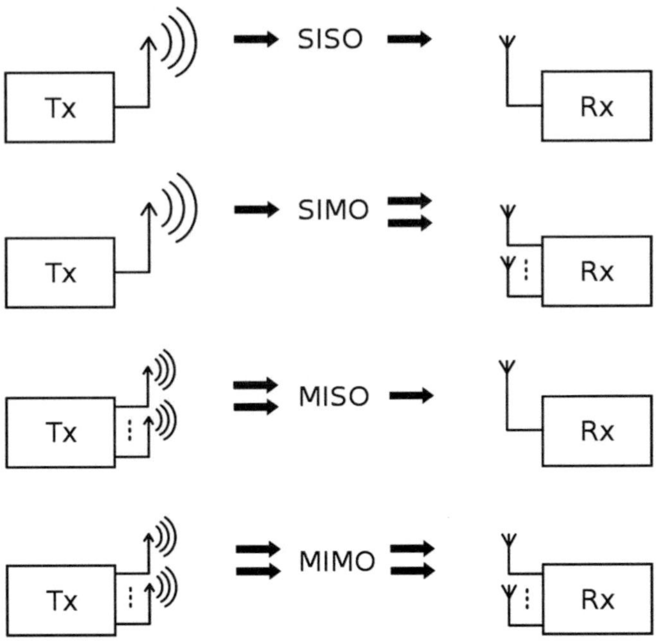

Fig. 1 Various types of transmission channels (Source: Baumgärtner [4])

Figure 1 shows four types of various inputs and outputs:

- ***"SISO"*** is a single input and single output process, which has one antenna each on both, the transmitting and receiving end.
- ***"SIMO"*** is a transmission system, which differs from the previous type with regards to the receiving aerial by having multiple output antennas than just one.
- However, as SIMO ***"MISO"*** is also a mixed system, where a transmitter has multiple antennas, but the receiver has only one antenna.
- A multivariable system, such as ***"MIMO"*** has equal numbers of antennas on both ends (transmitter as well as receiver).

The significant part of 5G's architectural design is the implementation of high numbers of antennas to enhance the reception power. Choosing a high frequency domain for next generation's network shows a positive impact on the latency of the new system as well as the high data rates. Therefore, high frequencies allow for massive arrays of antennas and beamforming supports for higher reception strength and reliability of the network power for multiple devices.

Small Cells One essential part of 5G's architecture are small cells. Small cells are defined as "low-power wireless access points that operate in licensed spectrum" ([22], p. 64). In order to serve high-density urban locations with characteristic properties, such as number of users demanding high data rate capacities, small cells represent an alternative solution resulting in complementing the existing mobile network and densifying the network in crowded areas, such as hotspots. Also, Edfors et al. [8] support the general idea of deploying small cells to promote network densification, by overlooking numerous isolated base stations (BS) and achieve a non-homogeneous network architecture. As a result, small cells are considered to satisfy the architectural requirements for the 5G cellular network. Ge et al. [13] state that in order for the 5G mobile network to be significantly reliant, the number of 5G base stations (BS) need to increase between 40 to 50 base stations per km^2, that is when Ge et al. [13] call 5G an "ultra-dense cellular network. Rodriguez [22] concluded that small cells offer an improvement in many applicative fields, such as in urban and rural areas and in applications for companies and homes, as well as an enrichment of provision in cellular capacity and coverage.

Beamforming In contrast to the current network carrier (4G/LTE), the next generation network (NGN) represent an unique signal coverage.

According to Infineon [15], current network carriers (4G/LTE) transmit antenna signals in an evenly distributed signal around the antennas of the base station (BS) towards the signal-receiving User Equipment (UE). However, these types of antennas loose signal power quickly; the further devices are from the signal-transmitting network cell, which might cause interference and result in weaker communication speed as well as higher latency due to connection to neighbouring cell. However, technology impacts these issues by deploying massive MIMO's with beamforming antennas in the new 5G network.

While the traffic demand is equally high in both networks, the approach taken to rectify the issue is different in the 5G network. The reason for that lies in the beamforming technology. Beamforming allows for the BS's massive MIMO (mMIMO) to send out concentrated signals directed to the user's specific need of data rate. Therefore, with low-powered antennas around a mMIMO-BS, beamforming can tap full capacity, which results also in an increasing number of capacities being focused and delivered directly to each user within the cell without losing signal strength when the number of users is high. Therefore, each user will get its own beam and can get high data rates with very reliable coverage.

Non-Orthogonal Multiple Access (NOMA) demonstrates advantages to the network technology because the segregation of signals provides entry to the base station (BS) [2]. Furthermore, Al-Dulaimi et al. [2] points out the efficient benefit of NOMA over OMA by implying code and power-domains characterising the concept of NOMA, which result in 5G's effectiveness through extensive connectivity and agility through the minimising of latency to a lower threshold.

Multi-Access Edge Computing (MEC) According to 3GPP's (2019) Release 15, section 5.5.2.2.1 mentions that all 5G-enabled services, running on the network, will be supported through a "Service Hosting Environment", which will be an integrated feature in an operator's network. Furthermore, mitigation of issue load on the network will be assured according to Release 15 (Rel-15). Section 5.5.2.2.1 especially points out that the bandwidth and the latency will experience a pressure release to a manageable minimum due to the support of Hosted Services (3GPP, 2019, Release 15). Enabling this concept, leads to compliance with the technical requirements for better user experience.

3 Smart Cities and 5G

One of the strategic purposes of the 5G mobile network is its implementation within the public and service sector, as well as in multimedia. Autonomous driving is one of the core goals that needs to operate on the 5G network. As a result, smart cities and autonomous vehicles are connected to (massive) IoT devices, which ultimately creates the Vehicle-to-Everything (V2X) communication connection. Therefore, intelligent connectivity within cities could have a massive impact on communication overall [25].

City planners, public sector bodies and private entities are striving to utilise smart and automated technologies and IoT in a way to not only streamline services and maximise of efficiency, but to improve the level of service to the citizen and the consumer, to bring an additional convenience and ease in the delivery of services. To this end, AI is being used to bring together and analysis captured data, that traditionally has been kept in silo, dependant on the agency or reason for the collection. The compute capabilities of artificial intelligence and machine learning has enabled large amounts of collected data not only to be unified and analysed, but for patterns and behaviours to be predicted. This means that smarter, more accurate strategic and operational decisions can be made. These capabilities also mean that data can be collected and analysed in real time, having utilised captured historic data to train the algorithmic systems. In the construct of a smart city, an example of this would be to utilise traffic sensor information from traffic lights at junctions and intersections, to monitor the flow of traffic. Changes in the vehicle rate and flow could indicator an accident or incident or show that traditional "rush hour" times of congestion are different in the particular area. The patterns learned in this instance can also govern and advice on future infrastructure improvements to the road network, or when maintenance and construction is required. Coupling this data with environmental data, pedestrian information, timetables for public transportation systems provide masses of valuable data in which resources and services can be effectively and appropriately managed. The realtime data analysis of traffic flow in a city could also help emergency services navigate through the less congested streets to minimise journey times.

The categorisation of IoT can be defined by their uses and implementations as follows:

- **Connected products** — From connected consumer-level coffeemakers to connected industrial pumps, this category enables end-to-end visibility into product-centric operations. It also promises improvements or even transformation around issues like regulatory compliance and product serviceability.

- **Connected assets** — In contrast with connected products, this category involves high-value, long-lived equipment such as aircraft and industrial machinery. Connected assets link production systems with manufacturing and maintenance processes to increase asset uptime and reduce operational and repair costs.
- **Connected fleets** — This category is all about tracking, monitoring, analysing, and maintaining any assets that move — from trucks to ships to construction equipment — wherever they appear in the network. Extracting data from mobile equipment has been difficult and expensive, so the promise here is immense.
- **Connected infrastructures** — From software networks to power grids to buildings, the majority of IoT sensors are likely to end up in connected infrastructures. This category will deliver new forms of digital operational intelligence to transformation physical systems. The goals will be to drive economic growth, improve service, and allow for more effective and efficient operations and risk mitigation.
- **Connected markets** — Markets apply to any activity that involves physical space, from retail centres to farms to cities. IoT can help cities, rural areas, and other markets to optimize use of assets and natural resources; reduce energy usage, emissions, and congestion; and improve efficiency and quality of life.
- **Connected people** — This category focuses on improving work, life, and health by linking people and communities, enabling organizations to evolve into new business models, and delivering better lifestyle experiences.

This of course, requires the proliferation and unification not only of the stored captured data, but also the data that is collected in realtime by IoT and smart devices. It is therefore imperative that the vulnerabilities and potential threat and attack vectors of these devices is considered before their implementation into a high impact system.

Possible attacks on an IoT infrastructure could include:

- Affecting target system behaviour by directly influencing deployed sensors to provide incorrect/faulty readings
- Create sensor impostor – Obtain IoT network access credentials and create (D)DoS attack on existing sensor to inject impostor(s)
- (D)DoS attack on sensor network to disable data collection
- Intelligence – Information collection and related analysis to observe typical patterns
- Disruptions on infrastructure – make grid elements to malfunction to cause either partial of full grid failure
- Modify water processing/ventilation to go outside of safety limits
- Get access to more secure networks/cloud through IoT infrastructure
- Modification of wearable/implanted health devices to cause bodily harm

Considerations also need to be taken into the exploitation of the IoT infrastructure itself, whereby unsecured devices could be infected to form a BotNet, used to attack other systems remote, and to great effect, given the number of potential susceptible hosts on an IoT network.

4 Massive Machine Type Communication (mMTC) and the "Internet-of-Things" (IoT) in 5G

Internet-of-Things (IoT) by definition, is the connectivity between two domains, the virtual and physical space of internet and things as well as the interaction between hardware components and software [17]. In addition to the many definition of the Internet-of-Things this phenomenon, also defined as an international infrastructure, has a massive effect on the Information Society by enhancing interconnection of already existing networks and further developing information and communication technologies (ICT). Another definition the Internet of Things (IoT) is known by is its established collection of objects in the form of a sensor software and electronic installations as well as the connected network to provide those objects data exchange with various addressees, such as other connected smart devices, network operators/producers and service providers [18]. Furthermore, researchers agree that, the internet of things intent is to simplify data exchange amongst various objects and allocate an IT infrastructure [28]. Therefore, 5G demonstrates specifically the enablement of machine-type communication (MTC), such as "Autonomous Driving" and the feasibility of other devices, for instance, "Augmented Reality" and "Virtual Reality" with latency delays of almost 1 ms by a bandwidth throughput ranging between 100 Mbps and 1Gbps, whereas autonomous driving shows a bandwidth throughput of 10 Mbps. Therefore, communication between larger machine-type devices is seen as rising in respect to the potential services enabled by the 5G network.

The Internet of Things could also be defined as objects and/or devices, which could be computing devices that show varying connections in communication protocols [18]. Because the implementation of 5G is currently in the making, those protocols usually use lower power protocols until 5G is implemented, which ultimately is designed to offer long power and steady communication between devices. The enablement of services and technologies on both the existing standard networks as well as the future 5G network, the machine-to-machine (M2M) connection, is a point interest. M2M connectivity can be achieved by the existing networks, typically 3G and LTE/4G. However, with the connective link of larger appliances, such as massive machines to the virtual space, it could be categorised into the area of Industrial IoT. Moore [19] acknowledges similarities between Smart Factories and the Industrial Internet-of-Things because while a smart factory includes cloud computing, complex systems and machine learning, the Industry 4.0 consists of barely less human involvement in automation and manufacturing industry as well as data exchange technologies. Although, IoT connects devices, systems, and other objects with the internet to simplify various types of environments, such as industries, economies, urban infrastructures and households, the Internet of Things transmits a massive amount of data. Atlam and Wills [3] strongly refer to fundamental handling of privacy, safety and security issues that arise from the use and the excessive data flow. The safety of IoT includes the consideration of ethical utilization of IoT and social behaviour contributing to the safeguarding of

IoT technologies, especially a stakeholder's trust in the secure feasibility as well as prohibiting unsuitable risk or physical damage provoked by individual IoT parts and eventually by the whole system [3].

4.1 Network Slicing

As an enabler for applications and services, such as the Internet of things, or massive machine type communications (mMTC), 5G is envisioned to serve multiple areas in technology while enlarging the network infrastructure. Therefore, network slicing allows for various types of devices to get utilized by segmenting the virtual network layers of the base network with its physical characteristics [20].

Furthermore, dataflow among User-End devices and application servers could be carried out through network slicing within flexible network topology by dividing control plane from user plane which results in the separation of data flow from the transmission of user data [23]. In order for the network to be open between slices, El Hattachi and Erfanian [9] suggest defining the requirement for openness so that persistent user experience is given as well as open interfaces among the two main network planes (C- and U-plane) when defining the performance of a 5G network slice and considering the allocation of resources by distributing the scheduler of a Random Access Technology (RAT) throughout the number of network slices. For enabling smartphone use, the first network slice needs to set "fully-fledged" functions across the network. The second layer demonstrated, is designed for supporting device-to-device communication, especially automotive devices. According to El Hattachi and Erfanian [9], automotive use cases can be realized by carrying out functions implemented at the cloud node as well as vertical application due to latency constraints, which needs proper definition of open interfaces to allow such vertical application on a virtual node (cloud edge node). However, El Hattachi and Erfanian's [9] findings point out insufficiency in security, latency as well as in the reliability of a 5G slice countenancing larger devices, such as vehicular use cases, whereas these factors show compliance with the 5G slice for smartphone use cases. Therefore, it is said that 5G network requirements are essential, especially ensuring functionality and control, and enabling secure End-to-End (E2E) operation at any given time [9].

4.2 Cloud Computing and Its Architecture

New network architectures and other use cases establish fundamental concerns for 5G's security, which needs to reconsider a solution to the improvement of security requirements. So called "new cloud virtualization technologies such as software-defined networking (SDN) and network functions virtualization (NFV)" [...] ([24], p.1) are said to create loopholes for vulnerabilities, which undermine the overall security of the 5G network although these network architectures excel flexibility, programmability and openness. SDxCentral [24] goes further by demonstrating system downfalls due to the misuse of management interfaces of an SDN partition to attack either the overall management system or the SDN controller, which ultimately results in a security breach.

In contrary, SDN networks mainly focus on the separation of control plane from data plane by centralising control instead of standardizing network protocols, whereas NFV networks focus on the replacement of certain network functions with software by using cloud computing services [31], which show a significant potential to mitigate CAPEX and OPEX, known as Capital and Operational Expenditures [1]. Lowering these expenditures show a positive benefit in the heterogeneity of 5G services, such as its functionalities and architecture because flexibility of the 5G network is, amongst other things, a key component of the divergent requirements of 5G driven applications [7]. Furthermore, the deployment of cloud services is purely based on network preferences [14]. Efficiency is an advantage feature of cloud computing because it does not own physical infrastructure for the maintenance of services, data and application ran by operators [1]. There are a number of expectations 5G is surrounded of, one of them being the multi-tenancy model.

5 5G and Issues in Privacy and Security

One of the significant features of 5G to consider, is data handling and storing solutions. Huawei [14] points out that "security" as such, remains an indispensable factor for business continuity. Furthermore, Huawei [14] suggests the consideration of applying privacy and security properties from former generations of mobile network to the upcoming mobile network (5G) so that business continuity can be provided. By enormously mitigating the impact of security breaches and understanding the influence that risk factors have, business continuity can be subject to audit through consistent safeguarding [5]. Therefore, maintaining privacy and user rights as well as taking actions tackling cyber issues that are arising from the interconnections of devices through the highly reliable and capable 5G mobile telecommunication service needs more evaluation. As a result, it is now to allocate and examine the impact of key aspects of security issues on this network but also the security gaps affecting other services relying on 5G. Before evaluating various approaches of securing the 5G network and to extract efficient solutions based on the specific risk factors to the network, it is firstly to showcase the essential parts of the 5G network that could lead to a higher probability of network vulnerability. Even current networks (4G/LTE) and also 5G, consist of different properties catering to different services. The Internet of Things shows exposure to numerous vulnerabilities because the technological structure exhibits potential weak spots, although it was developed based on core objectives, such as reliable network connection.

Miller [18] categorically classifies "Theft", "Privacy, "Safety" as well as "Productivity" as the most significant attack types and ultimate risk factors for IoT landscapes (system, network, infrastructure). With the 5G network adding function and enhancement to the reliability and availability of faster wireless service to applications, appliances and other 5G driven technologies, the security issue gains importance and further highlights to 5G. 3GPP's (2019) newest Release 15, introduces the development for additional space for massive connections between devices but also to deliver faster services with reduced latency. Under section 7.3 of 3GPP's (2019) technical specification in Release 15, it is stated that this newest release "builds on the LTE features for Machine-Type Communications (MTC) introduced in Rel-13 and Rel-14 (e.g., low-complexity UE categories M1 and M2, and Coverage Enhancement Modes A and B) by adding support for new use cases and general improvements with respect to latency, power consumption, spectral efficiency, and access control. Although, 5G will be capable to cover high numbers of devices, machines and other appliances, the amount of data retrieved and processed will increase enormously. That is when the confidentiality of vulnerable information may get violated. The risk for users may be immense. As Miller [18] mentioned it, the risk of being affected of theft is especially high with the use of autonomous vehicles because hackers can get access to the vehicle's remote keyless entry system but the possibility of unauthorized access to homes are almost as high.

5.1 Trust in 5G

Elevating the research question to the aspect of "trust", it is to discuss whether 5G can be an entrusted instrument the government and its citizens interact with. It may be a widely discussed aspect, which needs the in-depth outlook it deserves. Therefore, to discuss technological security the importance of "trust" must be evaluated and implemented within this particular technology. Huawei [14] explained that the 4G network provides an insufficient trust model because it already covers an established and bidirectional trust-relationship between "Users" and a "Network", but it does not exhibit a link between "Users" and the specific "Service" technologies (in this case the 4G mobile network) must provide.

With the introduction of the 5G network technology into the mobile communication market there is a mutual but distinct expectation of trust on both public and private side. Fogg and Tseng [12] state that the usability of technology is a crucial factor of trust by which a user's degree of trust is measured by. Moreover, trust as an accumulation of key elements of trust, which comprise factors, such as availability, reliability and privacy, into the definition of trust with regards to the field of technology.

5.2 5G Risks and Vulnerabilities with IoT

When using technologies of different kinds, for instance, mobile phones, laptops, or IP-based/public networks, there is always a danger of personal data being unprotected due to a lack of proper network security. Especially, that many IoT devices, AI and other 5G enabled services are slowly getting implemented into the markets. Amongst other things, 5G's technological specification includes the coverage of 3G and 4G/LTE. Therefore, a vast number of risk components mark critical security challenges for the 5G network.

Power supply depicts a crucial point when assessing risks, the 5G network has on users and the security structure of a nation. Ahmad et al. [1] mentioned the tremendous criticality a collapse of wired power supply systems might have on affecting systems within the network chain, such as data handling and electrical systems, which are integrated into society and were occurred by a security breach.

With consideration of existing mobile communication networks and their specific technical protocols, for instance, HSPDA/HSPA+, GSM and LTE, individuals were gradually introduced to the power and the ability of today's technology. Telecommunication providers are eager to provide profitable services designed around maintaining customers privacy by also fulfilling information security requirements when offering Voice-IP (VoIP), national and international services, such as PBAX, call and messaging services as well as roaming [30]. Therefore, the Internet of Things is exposed to a number of security threats and vulnerabilities. Ahmad et al. [1] point out several major security issues:

1. "Flash network traffic: High number of end-user devices and new things (IoT).
2. Security of radio interfaces: Radio interface encryption keys sent over insecure channels.
3. User plane integrity: No cryptographic integrity protection for the user data plane.
4. Mandated security in the network: Service-driven constraints on the security architecture leading to the optional use of security measures.
5. Roaming security: User-security parameters are not updated with roaming from one operator network to another, leading to security compromises with roaming.
6. Denial of Service (DoS) attacks on the infrastructure: Visible nature of network control elements, and unencrypted control channels.
7. Signalling storms: Distributed control systems requiring coordination, e.g. Non-Access Stratum (NAS) layer of Third Generation Partnership Project (3GPP) protocols.
8. DoS attacks on end-user devices: No security measures for operating systems, applications, and configuration data on user devices".

5.3 Practical Applications of Vulnerabilities in 5G/IoT (Table 1)

Table 1 Security issues in next generation network (5G/IoT)

Security threat	Target on network component	Effect SDN	ED TE NFV	Chnology Cloud	Links	Privacy
DoS Attack	Central control component	X	X	X		
Hijacking Attacks	SDN controller, hypervisor	x	x			
Signalling Storms	5G core network components			X	X	
Resource (slice) Theft	Hypervisor, shared cloud resources		X	X		
Configuration Attacks	SDN virtual switches, routers	X	x			
Saturation Attacks	SDN controller and switches	X				
Penetration Attacks	Virtual resources, clouds	x		X		
User Identity Theft	User information data bases			x		X
TCP Level Attacks	SDN controller-switch communications	X			X	
Man-in-the-middle Attacks	SDN controller-communications	x			X	X
Reset and IP Spoofing	Control channels				X	
Scanning Attacks	Open air interfaces				X	x
Secure Key Exposure	Encrypted channels				X	
Semantic Information Attacks	Subscriber Location				x	X
Timing Attacks	Subscriber Location			x		X
Boundary Attacks	Subscriber Location					X
IMSI Catching Attacks	Subscriber identity				x	X

6 Conclusions

Summarizing all mentioned aspects, demonstrates a vertical shift within the Information Technology industry. Major digital key drivers are the enablement of IoT, AI, cloud, real-time big data and the ultra-high-connectivity. Therefore, the implementation of 5G might bear privacy and security issues. Research has studied possible solutions or implications on compliance and regulative actions. Chandran and Labo [6] introduced a compliance-based approach on effective promotion of ethics within organization. Especially in times where globalization shapes a fundamental foundation for the growth of markets and sustainability of business relations, the role of ethics becomes more important, which results in a development and increased use of code of ethics [32]. Nevertheless, 5G demonstrated to function as a solid and sophisticated network. Its implementation date in 2020 caused massive research and case studies to be undertaken and were made to counteract major risk factors; so that a reliable network can grow and lead the future of technology into a different direction. Furthermore, future works are needed especially for assessing risk, threat and security as well as privacy matters according to diverse use cases. This can be conducted after 5G, with all the services running on the network, is fully initiated into the market and understood by the society. Subsequently, this will require further risk, threat or economic analysis throughout a longer period of time. Therefore, a possible recommendation could be undertaking questionnaires before the implementation of 5G. In this way network providers, but in particular the government, can get an insight of citizens awareness towards technical change and their knowledge in all aspects of AI and the Internet of Things. Because smart cities will gain more relevancy with the implementation of 5G, artificial intelligence processing data to create a pattern for understanding, will be subject of further research. As a result, future work possibly consists of conducting surveys within the automotive industry as well as the health sector, where data processing as well as data privacy act as a major concern. Further observation would be advisable via a thorough bidirectional analysis on the risk factors and cyber security issues impacting the 5G networks.

Ultimately, thorough analysis from different sectors of society can help the government to standardise a security, ethical and data framework of the network and of all services enabled by it. Future questions regarding IoT, could be comparing IoT as it is now, and what may be the issues in 2020, when larger bandwidths are available, and everything is 5G compatible and beyond 5G in terms of 6G and its capabilities. Standardisation is a key question and observation of all risk factors before full rollout of 5G/6G, so that there is less chance of risk to privacy exposure, security breaches and leakage of personable identifiable information.

References

1. Ahmad I, Gurtov A, Kumar T, Liyanage M, Okwuibe J, Ylianttila M (2017) [online] Available at http://jultika.oulu.fi/files/nbnfi-fe201902124647.pdf. Accessed 12 Sept 2019
2. Al-Dulaimi A, Chih-Lin I, Wang X (2018) 5G networks: fundamental requirements, enabling technologies, and operations management, 1st edn. Wiley, New Jersey
3. Atlam HF, Wills GB (2019) IoT security, privacy, safety and ethics. In: Farsi M et al (eds) Digital twin technologies and smart cities, internet of things. Springer Nature Switzerland AG 2020
4. Baumgärtner B (2018) Die Bezeichnungen "SISO", "SIMO", "MISO" und "MIMO" beziehen sich auf den Übertragungskanal. Dessen "Eingang" sind die sendenden Geräte. Entsprechend werden die Empfänger als "Output" des Kanals bezeichnet. [online] Available at https://de.wikipedia.org/wiki/MIMO_(Nachrichtentechnik)#/media/Datei:MIMO_SIMO_MISO_SISO_explanation_without_confusion.svg. Accessed 08 June 2019
5. Calder A, Watkins S (2015) IT governance: an international guide to data security and ISO27001/ISO27002, 6th edn. Kogan Page, London
6. Chandran S, Lobo A (2016) Ethics and compliance in corporations: value based approach. 2016 IEEE International Symposium on Ethics in Engineering, Science and Technology (ETHICS) 1(1):1–4
7. Condoluci M, Mahmoodi T (2018) Softwarization and virtualization in 5G mobile networks: benefits, trends and challenges. Comput Netw 146(1):65–84
8. Edfors O, Larsson EG, Marzetta TL, Tufvesson F (2014) Massive MIMO for next generation wireless systems. IEEE Communications Magazine. pp 186–195
9. El Hattachi R, Erfanian J (2015) 5G white paper: NGMN 5G initiative. NGMN Alliance. [online] Available at https://www.ngmn.org/fileadmin/ngmn/content/images/news/ngmn_news/NGMN_5G_White_Paper_V1_0.pdf. Accessed 25 July
10. Ellingsen SE, Løvholt F, Madshus C, Norèn-Cosgriff K (2017) Simulating low frequency sound transmission through walls and windows by a two-way coupled fluid structure interaction model. J Sound Vib 396(1):203–216
11. Everythingrf (2018) What are millimeter waves?. [jpeg]. [online] Available at https://www.everythingrf.com/community/what-are-millimeter-waves. Accessed 07 July 2019
12. Fogg BJ, Tseng S (1999) Credibility and computing technology. Commun ACM 42(5):39–44
13. Ge X, Mao G, Han T, Tu S, Wang CX (2016) 5G ultra-dense cellular networks. IEEE Wirel Commun 23(1):72–79
14. Huawei (2018) 5G Security: forward thinking Huawei white paper. [online] Available at https://www.huawei.com/minisite/5g/img/5G_Security_Whitepaper_en.pdf. Accessed 13 July 2019
15. Infineon (2019) 5G – The high-speed mobile network of the future. [jpeg]. [online]. Available at https://www.infineon.com/cms/en/discoveries/mobile-communication-5g/. Accessed 28 July 2019
16. Khalaf K, Long J-R, Vidojkovic V, Wambacq P (2015) Data transmission at millimeter waves: exploiting the 60 GHz on silicon, 1st edn. Springer, Berlin/Heidelberg
17. Luoma E, Mazhelis O, Warma H (2012) Defining an Internet-of-Things Eco-System. In: Andreev S, Balandin S, Koucheryavy Y (eds) Internet of things, smart spaces, and next generation networking: 12th international conference, NEW2AN 2012 and 5th conference, ruSmart 2012, St. Petersburg, Russia, August 2012, Proceedings, 1st edn. Springer, Berlin Heidelberg
18. Miller L (2016) IoT security for dummies, INSIDE secure edition. 1st. Chichester/West Sussex: Wiley
19. Moore M (2018) What is industry 4.0? Everything you need to know. The latest news, views and developments in the world of Industry 4.0. [online] Available at https://www.techradar.com/news/what-is-industry-40-everything-you-need-to-know. Accessed 23 May 2019

20. Van der Nagel I (2016) The challenge of a secured 5G network. [online]. Available at https://itnext.io/the-challenges-of-a-secured-5g-network-2d0b283c9619. Accessed 27 July 2019
21. Nesterova M, Nesterova Y, Nicol S (2018) Evaluating power density for 5G applications. 2018 IEEE 5G World Forum (5GWF), pp 347–350
22. Rodriguez J (2015) Fundamentals of 5G mobile networks, 1st edn. Wiley, Chichester/West Sussex
23. Sawall A (2018) 3GPP: Standardisierung von 5G ist fertig. [online] Available at https://www.golem.de/news/3gpp-standardisierung-von-5g-ist-fertiggestellt-1806-134992.html. Accessed 10 Aug 2019
24. SDxCentral (2019) What are the top 5G security. Challenges. [online] Available at https://www.sdxcentral.com/5g/definitions/top-5g-security-challenges/. Accessed 17 Aug 2019
25. Seeburn K (2019) 5G and AI: a potentially potent combination. [online] Available at http://www.isaca.org/Knowledge-Center/Blog/Lists/Posts/Post.aspx?ID=1146#Comments. Accessed 19 Aug 2019
26. Statista (2016) Internet of things (IoT) connected devices installed base worldwide from 2015 to 2025 (in billions). [jpeg]. [online] Available at https://www.statista.com/statistics/471264/iot-number-of-connected-devices-worldwide/. Accessed 05 July 2019
27. Techplayon (2017) Wavelength, frequency, amplitude and phase – defining waves!. [jpeg]. [online] Available at http://www.techplayon.com/wavelength-frequency-amplitude-phase-defining-waves/. Accessed 07 July 2019
28. Weber RH (2010) Internet of things – new security and privacy challenges. Comput Law Secur Rev 26(1):23–30
29. White G (2018) IPoC: a new core networking protocol for 5G networks. [online] Available at: https://www.cablelabs.com/ipoc-a-new-core-networking-protocol-for-5g-networks. Accessed 05 July 2019
30. Yesuf AS (2017) A review of risk identification approaches in the telecommunication domain. [online] Available at https://www.researchgate.net/publication/314392917_A_Review_of_Risk_Identification_Approaches_in_the_Telecommunication_Domain [PDF] Conference paper. Conference: the 3rd international conference on information systems security and privacy – ICISSP. Accessed 15 August 2019
31. Zhang Y (2018) Network function virtualization concepts and applicability in 5G networks, 1st edn. Wiley, New Jersey
32. Zinner S (2014) Codes of ethics move into the "third generations". 2014 IEEE International Symposium on Ethics in Science, Technology and Engineering 1(1):1–3

Part III
Technology of Cyber Attacks

Blockchain, TTP Attacks and Harmonious Relationship with AI

Stefan Kendzierskyj and Hamid Jahankhani

Abstract Blockchain and decentralised distributed ledger technologies are being viewed as a mechanism to provide further protection and enhance the security of data by using its properties of immutability, auditability and encryption whilst providing transparency amongst parties who may not know each other; so, operating in a trustless environment. It's true that blockchain has its roots in cryptocurrency applications and is still evolving for that purpose in the financial sector, but many other organisations across different industries are beginning to see the non-crypto use cases where this mechanism to record data that cannot be changed or reversed or apply as smart contracts (as a way to time-stamp transactions between parties) is becoming extremely relevant and purposeful. A variety of industry sectors, besides Finance, has undertaken the use of these distributed technologies and beneficial attributes of blockchain from the healthcare and pharmaceutical, real estate, retail and supply chain, legal and publishing. Organisations have flexible options to run blockchain as permissionless (anyone can join), permissioned (where those need to be invited) or hybrid (a consortium type) and whether data should be held on-chain or off-chain. With industry entering its fourth industrial revolution (Industry 4.0) the addition of blockchain as a complimentary technology has its place and there are some industries very suited to the significant impact it may bring. Also, the advances of Internet of Things, Machine Learning and Artificial Intelligence has meant more pressures on potential impacts to data and the ripple effects that cyber-attacks may cause. This has also become complicated, as cyber-attacks have become much more sophisticated over recent years with the different configuration types and various industry sectors have suffered from a range of these different attack vectors, resulting in some devastating outcomes. These have manifested in

S. Kendzierskyj (✉)
Cyfortis, Worcester Park, Surrey, UK
e-mail: Stefan@cyfortis.co.uk

H. Jahankhani
Northumbria University London, London, UK
e-mail: Hamid.jahankhani@northumbria.ac.uk

H. Jahankhani et al. (eds.), *Cyber Defence in the Age of AI, Smart Societies and Augmented Humanity*, Advanced Sciences and Technologies for Security Applications, https://doi.org/10.1007/978-3-030-35746-7_10

the shape of ransomware, malware, manipulation methods, phishing and spear-phishing. Whilst data breaches are a serious incident, in most organisations, there is a growing concern regarding attacks that are designed to have a more destructive effect such as the Ukraine cyber-attack in 2015 that resulted in a shutdown of the power grid or the WannaCry ransomware attack in 2017 that caused widespread chaos with healthcare institutions unable to carry out any tasks since access to data/systems was unavailable.

Keywords Blockchain · TTP · Cyberattacks · Decentralised · DLT · AI · IoT

1 Introduction

Cyberattacks and its frequency regarding data loss within organisations is becoming too familiar and continually evolving in many types of attacks and motivations behind them. Whilst companies undertake activities to mitigate attacks and lessen its success ratios, it is considered that a centralised network or trusted third party (TTP) presents an obvious security risk. This is due to all trust placed with an entity that is trusted by all other entities in that community or ecosystem to perform and be the authority to set functions. The TTP may provide a service or number of services and is trusted to store data which may be personal/sensitive or have a mission critical purpose (for example banking details/data). Once a TTP has been compromised then the risk of data loss, identity theft or other issues such as destructive malware, ransomware, etc., undertakes a path where there are many impacts, not just to the TTP entity in loss of services, revenue, credibility but also to individuals (such as identity theft or destructive malware, etc).

Whilst this chapter discusses alternatives of TTPs, such as blockchain, it does not suggest to invalidate TTP use cases since that is not in scope and actually many organisations are looking at ways to ensure that threat landscape of TTPs is further diminished by the right technology infrastructure, training of employees (phishing still remains the most used method to enter an organisation), etc. However, with the right business use cases, blockchain can be very applicable with its attributes to be utilised to further secure data privacy but offer transparency. Blockchain may also work in harmony with TTPs so that a hybrid model can be deployed where perhaps data is more sensitive or mission critical.

2 Blockchain and Centralised Systems

2.1 Decentralised or Centralised Systems

It is beneficial to understand what is meant by centralised and decentralised systems. To put it simply in a centralised system the authority is given to one entity which could be an enterprise or individual that oversees all the transactions. So, this centralised entity would require all data, communications, information, etc., to enter and leave through a central hub or number of private servers (an example such as Google or Facebook). Blockchain is a decentralised distributed ledger with no single central authority and transactions are executed amongst multiple parties where peer-to-peer interaction drives the network, hence no single central entity that data would pass through (BitTorrent is an example of a peer-to-peer network). This means that decentralised platforms can allow for more privacy since information does not go through one point or entity source and will pass through a number of points. In a centralised network all actions and information are continually monitored and there is a potential for privacy to be compromised. In certain organisations such as Facebook, information is consented in the terms and conditions and allows a degree of being able to package up the individual's information and then a loss of control over a user's data is happening. Using a decentralised system therefore allows a stronger method to protect your identity/personal information.

2.2 *What Is Blockchain*

Blockchain is said to be one of the most disruptive computing paradigms after the Internet; Swan [17]. The technology has advanced more than just a means of financial and economic trading and looks to provide a consensus of trust which can be where transactions require storing by multiple parties who may be unknown and untrusted. Over recent years different arms of industry are starting to harness the technology to take advantage of its benefits, attributes and methods of application. The concept of blockchain was first circulated as a white paper, Satoshi [14], and created using a pseudonym under Satoshin Nakamoto. Originally, developed for intention as a cryptocurrency (Bitcoin) and is often referred to as Blockchain 1.0 for classification in the use of cryptocurrencies, Swan ([17], p. ix). Blockchain 2.0 refers to smart contracts which go beyond just cash transactions and blockchain 3.0 goes further in the way of alternative use in healthcare, government, intellectual property, etc.

Blockchain technology has been quoted as allowing records to be 'shared by all network nodes, updated by miners, monitored by everyone, and owned and controlled by no one', Swan ([17], p.1). Blockchain is based on a decentralised system; a distributed ledger database where sequential inventory of transactions has identical copies shared and maintained by numerous individuals over network nodes. The multiple parties hold the consensus over the data and its validity rather than one individual. Under certain blockchain setup it also offers a protection against attack since more than 51% of the blockchain network would need to be compromised. Engelhardt [3] summarizes this immutability (a key component of blockchain) as each record in the chain contains exact information on creation, the cryptographic signature in the preceding record in the chain and any arbitrary data. The signature (a hash which is a unique record identifier) has the cryptographic sequence of a particular length, as alphanumeric, which uniquely determines the digital entity. This is how, if compromised by changing a previous record, the break in the chain would be identified. Records cannot be removed, only added by the approval of the consensus (this is one of the consideration points to match GDPR as the 'right to be forgotten'). With adding encryption to each block, only keepers of the private keys have access (the pubic key is an openly visible key, but the private key only unlocks the data permissible on the blockchain). Auditability is another important aspect as transparency allows all stakeholders to see the data; and something that would help solve improving the transparency of data. When new transactions take place between parties it is broadcast to the network for all to see and the network miners will verify it. It's this collective verification that will only allow the block to be added and how the 'trustless' method works and not relying on a single entity to be the only authority to verify (as is the centralised models). There can be some limitations such as rate of transactions and block size but is not a compelling issue since for industry records type blockchain may just be used for authenticating rather than holding data on the chain itself [10].

2.3 Types of Blockchain

There are different types of blockchain, and stakeholders need evaluate not just technology requirements but also the business model requirements which can have a host of questions that needs understanding to decide who can enter the blockchain community setup for a particular scenario. Understanding these types will help organisations decide what type blockchain technology should be deployed and how should they allow permissions/authentication as depending on what type of data it is. These generally can be classified into three main types of Public, Private and Consortium (or hybrid) blockchains but can have options of many consensus mechanisms; these are the algorithms that set in place how the network will operate (consensus examples are discussed in Table 1).

- **Permissionless Blockchains:** Also referred to as public blockchains and allow anyone to participate with no restrictions on reading/submitting transactions. All network nodes are unknown but take part in the consensus process. Examples of these are Bitcoin and Ethereum. However, public permissioned blockchains are restricted to those allowed to enter but anyone can read/submit transactions.
- **Permissioned Blockchains:** These private blockchains will restrict access and who may enter the network of nodes and transactions are only validated by those recognised as authenticated on the ledger; essentially the network belongs to an entity or organisation. Private permissioned are usually totally restricted such as Bluemix by IBM or Rubix by Deloitte as example.
- **Consortium Blockchains:** Approved entities validate requirements. An ecosystem would find this type blockchain more suited since all parties will have a common aim in deciding what process of data workflow should be included, etc., and so only where a particular group participates in the consensus process.

Table 1 Blockchain consensus mechanisms

Blockchain consensus mechanisms/algorithms	Characteristic description of mechanisms
Proof of Work (PoW)	PoW was the first consensus algorithm that was created and used for cryptocurrency purposes such as Bitcoin. This works on the principle of heavy computational calculations, commonly referred to as mining, and therefore assumes nodes performing these calculations are not malicious to attack the network. These calculations are complicated mathematical puzzles and the answer to the mathematical puzzle is a hash. Once one node inside the blockchain network solves or identifies this value the block is then broadcast to the rest of nodes to verify and mutually agree that the value is correct. Presuming this is a correct value then all nodes must update their blockchains. The immutable factor is that all newly created blocks are linked to the previous blocks and provides the cryptographic resistance that gives the security of blockchain. It is also a provision against tampering since more than 51% of the network (or hashing poser of all nodes) would have to be under the malicious control to affect a tampering outcome. The benefit of defending a Denial-of-Service attack is there due to time and cost to undertake a DoS attack that renders it not cost effective to undertake (a 51% attack requires enormous computational power). However, there is an overhead to consider for miners since the more computational power to solve puzzles means more success (but this equates to costly equipment to run complex algorithms. **Pros:** Its existence has been stable since inception in 2009 **Cons:** Uses up a lot of energy and can be slow, vulnerable to 51% attack **Examples:** Bitcoin, Litecoin, Ethereum
Proof of Stake (PoS)	The main advantages of POS are efficiency (low electricity consumption and reduced hardware costs) and security (increased computational complexity and skill required by adversaries). Also, there is an element of randomisation that makes the network more decentralised and in PoS there are no miners and so means no need to release new coins for a reward to miners thus keeping the coin more stabilized. PoS has the same objectives as PoW in terms of reaching consensus, but how that is achieved as a method is different and PoS usually will use transaction stakes as a reward (rather than cryptocurrency as rewards for miners). The PoS algorithm essentially makes use of a pseudorandom method to select a node to be the validator of the next block. Participating users are required to lock-in a certain amount of coin stake and the size of the stake will determine the ratio of that node being selected as the next 'validator' to forge the next block as well as the age of the stake and wealth of the node. However, it is equally important that favouritism is not encouraged merely based on node wealth and so more techniques are utilised such as having coin age selection (calculated as number of days the coins are held as stake by the volume of coins staked) and randomised block selection (these are nodes with a combination of low hash value but with highest stake). Once a block is created the coin age is reset to zero and the node will require waiting for 30 days before possibility to sign another block. **Pros:** Expensive to mount an attack and more energy efficient than PoW **Cons:** Nothing at Stake theory and only richest stakeholders have the opportunity to have consensus control **Examples:** Peercoin, Nxt

(continued)

Table 1 (continued)

Blockchain consensus mechanisms/ algorithms	Characteristic description of mechanisms
Delegated proof of stake (DPoS)	DPoS stakeholders delegate the hashing capability to a group of nodes, referred to as witnesses, so that they will be responsible for achieving consensus regarding generation and validation of new blocks and are rewarded. This voting system has proven to be efficient, fast and helps achieve autonomous cooperation. The voting power is proportional to the number of coins participants hold. Any inappropriate behaviour or inefficiencies can mean expulsion. DPoS mechanisms are said to be highly scalable with its ability to take on more processing of transactions when comparing to PoW and PoS. **Pros:** Energy efficient and fast **Cons:** Validator concerns as those with high stakes can vote themselves in as validators **Examples:** EOS, BitShares
Proof of Authority (PoA)	PoA mechanisms require block validators to stake their reputation instead of coins, so leveraging the value of identities. The PoA consensus mechanism is a highly scalable system and has been linked to use cases in supply chain of custody scenarios or where there is a large logistical requirement. Usually also better suited to private blockchains due to performance benefits. Some conditions are required to be present: Validators are required to confirm their real identities. Computationally expensive to become a validator; investment of money and stake of reputation ensures less chance of dishonest behaviours. A uniform method to select validators. PoA has been depicted as a mechanism that sacrifices decentralisation and is a good method to provide efficiency in centralised methods. Other limitations could be that as the identity of validators has to be revealed there is a sacrifice of privacy and third-party manipulation; in such cases a malicious person outside the network can know who to approach and potentially corrupt to disrupt the PoA system. **Pros:** Fast and efficient **Cons:** Deemed to be weighted towards centralisation **Examples:** POA.Network, VeChain
Proof of Elapsed Time (PoET)	Validators in PoET mechanisms will wait a random amount of time for every new block the validator will create. The leader of the new block is the first person to finish waiting. It relies on a particular CPU instruction from Intel called Software Guard Extensions (SGX) that allows to run the trusted execution of programs in a protected environment. This ensures it satisfies PoET requirements. This consensus algorithm is less resource intensive that say PoW as the miner's processor can remain more dormant and allocate to other tasks during the specified time so increasing efficiency capabilities. Since it does not require a mathematical puzzle to be solved is what gives rise to the efficiency since it uses the randomized timer to select the block leader. However, some weaknesses in SGX technology are said to be identified since it relies on specialised hardware and in some ways goes against the principle of decentralised models since there is total reliance on a third party as the consensus model is built on Intel equipment. **Pros:** Participation cost is low and transparent on how leader is legitimately selected **Cons:** Specialized hardware is required and a reliance on the provider, Intel (almost what blockchain wants to move away from regarding third party trust) **Examples:** HyperLedger Sawtooth

(continued)

Table 1 (continued)

Blockchain consensus mechanisms/ algorithms	Characteristic description of mechanisms
Proof of Importance (PoI)	PoI has some similarities to PoS as nodes need put forward an amount of currency to create blocks and creation of a block is in proportion to some score. The difference is in PoI there are more variables in the score such as vested amount of currency to create blocks; net transfers or total spent in last 30 days, number/size of transactions, node importance score. So, productivity is a key aspect not just coins accumulated. **Pros:** stake evaluation and importance assumed in network **Cons:** minimum stake of 10,000 XEM is required for vesting **Examples:** NEM
Practical Byzantine Fault Tolerance (PBFT)	PBFT establishes a practical Byzantine state machine replication and looks to operate even if there are malicious nodes in the system. Nodes are sequentially ordered with a leader node allocated as the primary node whist providing backup nodes (or secondary nodes). These secondary nodes are able to transition to a primary node, which is useful as a backup in case of a failure. PBFT has a function rule that no more than one third of the nodes can be malicious or adversaries. Consensus is broken down into four phases where a request is sent to the leader node and in turn the leader node broadcasts to the backup nodes (this assumes that all backup node is treated equally, Byzantine General's problem). Primary and secondary nodes perform the task and a reply is sent to the client. The client checks all the multiple identical replies and final identical reply is usually checked by honest nodes and approved to either accept or reject. PBFT is more efficient if the number of nodes is smaller as there is an overhead caused by high communications that increases as more nodes join to the network. It is also vulnerable to a Sybil attack where one entity controls many identities. However, as it does not require the PoW type transactions can make also a reduction is energy usage. **Pros:** High transaction throughput **Cons:** Exponential message count as nodes are added **Examples:** Zilliqa
Raft	Raft is a consensus that was developed out of Stanford University and handles, even in the situation of failures, the issue of having multiple servers agree on a shared state. Raft can also manage replicated logs and elects a leader in the cluster who in turn is responsible to accept client requests and manage log replication to other servers. Every server exists in three states, and can act as leader, follower or candidate. Under normal operational circumstances there is only one leader and the rest of the server's act as followers (in this case they are passive) and usually sends a heartbeat message to inform followers of its existence. Followers have a timeout and expect to receive a heartbeat so timers are reset otherwise status will change to candidate to proceed to leader election. **Pros:** Simple model that provides development implementation in multiple languages **Cons:** Mostly for private and permissioned networks **Examples:** Quorum

In terms of differences it will mostly lie or have impact in the aspect of decentralisation and how the technology handles the data considerations. In the case of public blockchains these offer complete decentralisation but whereas hybrid or consortium blockchains will be partly decentralised. For private blockchains, as they are mostly controlled by one entity or organisation that set the 'rules', can be a similar concept to centralisation as the group of users are a closed group.

For transaction processing purposes, organisations need to evaluate on blockchain's technical and confidential considerations. They need look at how best to execute access to data and can be dependent on what type of data it is, if it is sensitive, etc. The data storage methods can be as follows:

1. **On-chain**: data is stored on the blockchain structure.
2. **Off-chain**: access links are saved on blockchain and act as authenticated indicators to data stored in other centralised networks/databases.
3. **Hybrid**: having a mixture of the above with some standard data sets stored directly on the blockchain (beneficial for immediate permissioned use) and other access to off-chain data links.
4. **InterPlanetary File System (IPFS):** A protocol/network that allows peer-to-peer hypermedia storing and sharing of content, held across a distributed file system (such as BitTorrent). Network nodes store content it is interested in but with indexing information so it can be intuitive as to where content is stored. When requests are made to look up/search content, the network will request the nodes storing the content behind a unique hash to provide it. Mentioned by many to be the replacement to HTTP and the web of tomorrow.

2.4 Consensus Structures

Consensus is characterised as a general agreement of the state blockchain is in. This agreement is fundamental if a transaction recorded on blockchain ensures that it upholds non-tampering, is viewed as correct and no malicious activity has taken place. The consensus protocols are the primary rules of a blockchain and the algorithm acts as the mechanism that the rules can be adhered to. So essentially the algorithm is there to advise the steps to take, so compliance is achieved with the end result expected.

There are differing consensus protocols and algorithms used that are dependent what may be suited to deploy but fall under Byzantine Fault Tolerance and Leader-based types. Table 1 depicts some of the more popular mechanisms (there are many more possible consensus models) and it is clear that some types favour particular scenarios such as PoA favouring to achieve more scalability and throughput by losing some of its decentralisation capability or how cryptocurrency networks require decentralisation as the priority.

Whilst Table 1 indicates some popular consensus mechanisms there are more that can be available such as:

Proof of Reputation, Proof of Capacity, Proof of History, Proof of Stake Velocity, Proof of Burn, Proof of Identity, Proof of Activity, Proof of Time, Proof of Existence, Proof or Retrievability, Stellar Consensus, Proof of Believability, Directed Acyclic Graphs, Mokka.

2.5 Benefits of Blockchain

Blockchain is set to disrupt industry and help transform the traditional methods and business models. The benefits of blockchain are more clearly understood now by industry and besides just technical understanding it is helpful if approached from understanding current issues faced by their specific industry and reviewing how the beneficial attributes of blockchain can be applied. Many organisations from healthcare, government, manufacturing, pharmaceutical, financial, media/publishing, etc., have already undertaken successful pilots and achieved higher returns than anticipated. Surveys undertaken in industry indicate heavy support for blockchain and this will be due to severe problems associated around interoperability, security, data integrity, privacy issues and other challenges in centralised networks.

A white paper which assisted Government with market research, [5] advises that 'blockchain promises to put privacy and control of data back in the hands of citizens'. However, the cost for example in the healthcare industry, has risen dramatically over the years and for some countries it is becoming a crisis that is growing in terms of expense. Over the last 10 years there has been an increase on health expenditure by 60%, The World Bank [18]. This is compounded with the issue in developed countries with the population becoming more aged and therefore likely to worsen. In the healthcare sector, blockchain is looked upon as the technology to provide a solution in revolutionising the re-use of data to control huge masses of healthcare data that is anonymised and give way to new research and innovations. So, it not only protects the privacy of patients, but removes the expensive middle layers and escrow services that mediate; so, connecting all stakeholders without this expense, Engelhardt [3].

Another interesting aspect of blockchain benefits is, due to its nature of decentralisation it may offer better protection against types of cyber-attacks, such as the WannaCry ransomware attacks in 2017, since blockchain would need to be simultaneously attacked by numerous sites; Mattei [8]. There needs to be a more secure layer given to protect users as the rise in malicious outsider attacks for identity theft is becoming one of the most common types and causes of cyberattacks.

The following section gives similarity use cases across many industry sectors and references healthcare in places as a specific industry example due to the potential of many use cases that could be applied.

Interoperability, Transparency and Immutability Traditionally a lot of data is still held in silos and managed under a centralised or trusted third party (TTP) and the current challenges are the accountability in those silos. In healthcare, medical records are a good example of silo data and how these content details that make up the identity of an individual consisting of all data points spread through one's journey in life, are stored. Clearly interoperability is an issue that does not just make it inefficient but also potential to tampering, loss of data (accidental loss by those without privileges to access). There is also limited control of data ownership whilst the patient whom is the centre of all has the least ownership or control. The unseen challenge is also the serious levels of data breaches and damage of data loss to the users through identity theft. Blockchain can eliminate these data silos and provide a more coherent and seamless integrated data model that can control access through cryptographic methods, but where authenticated, make medical encounters between disparate parties more accessible. Transparency is also a key point as to help auditability become more visible to all. As blockchain is a distributed ledger where all participants share the same documentation (rather than individual copies) and has to be applied through a consensus mechanism (as per Table 1). If tampering was to occur then the majority of the network would need to collude in this and alteration of all subsequent records. It's something that does not take place and although privacy and security are also important aspects, so too is transparency and is given as equal emphasis.

Privacy and Security the confidentiality, integrity, availability and audit (CIAA) is subject to a lot of pressurised issues both internally through non-malicious behaviours as, for example accidental loss of data, and to outside vectors such as targeted malicious behaviours for the purpose of identity and data theft. Industry is trying to tackle this with the day to day traditional structures (the typical network security, compliance, Intrusion Prevention and Detection Systems, training, etc.,) that help mitigate risk and have a continual cycle of lessons learned. But through an additional layer of blockchain, it can offer enhanced security with encryption and would increase the integrity with the use of a decentralised and distributed ledger system. With data encrypted, more complex permission settings and the most suited consensus mechanism, will offer good controls for a secure authenticated data interchange.

Time Stamping and Chronological Ordering If content and data are held in silos there is more chance of corruption, fraud, traceability, audit trail and so on. One of the benefits blockchain can offer is its undeniable immutability. If even blockchain is used only for its authentication part to allow permission to data stored in a 'data lake' (a centralised data storage held in cloud environments) then the fingerprint of all activities is held in chronological order (smart contracts) and cannot be tampered. Depending on the types of blockchain, then parties can make use of its smart contracts to help provide greater efficiency and integration, all time-stamped and ordered to give the immutability factor.

Procurement and Contract Process Using smart contracts properties of blockchain will help relieve a lot of the complex processes, negotiations and supply chain issues by streamlining to give efficiency and reduce cost. This can provide automated supplier contracts and analytics to maximise productivity and control.

Traceability For supply chain this is a cost saving, efficient and key aspect to provide transparency over the chain of custody and in some cases offset counterfeit processes, products or records/transactions. With the immutability, time-stamping, and proof of records end-to-end gives industry that confidence that the business model is untampered, single version of the truth and audit trail at any given point.

3 Trusted Third Parties (TTPS)

3.1 The Risk of Trusted Third Parties

Over time the world has become accustomed by freely accepting an entity to manage and preside over sometimes mission critical data such as identity, bank/finance details, medical records and so on. It had even arrived at the point where users of main centralised platforms such as Facebook, Instagram, Google, did not fully understand the impact of the personal data and what ownership rights they have and fragile position they may face if the data is used in a certain way (without their knowledge or presumed consent) or when successful attacks cause data breaches and data theft. In fact, it was not until over recent years and the exponential growth in cyberattacks has the individual become so concerned. With so much data theft the impact of this to the individual and what is sold in the dark web is not fully known. Entities and organisations who act as the custodian of data, guarantee the safety and privacy of data, in good faith, but even with best intentions, security protocols/compliance, training and technical defence systems still cannot protect data theft.

3.2 Threat Landscape and Cyberattack Types of TTPs

The variation of cyberattacks and sophistication has dramatically increased over recent years and given TTPs more landscape to cover and threat model; since this is now not just limited to traditional networks, but also exposure across the Internet of Things (IoT), cloud and edge computing. It is interesting to see the evolution/development of new attack methods, but entry success relies on behavioural patterns of victims to activate old and reliable techniques such as phishing or spear-phishing attacks and allow that initial entry by an attacker, where further and more destructive attacks take place. For example, using advance persistent threats (APTs), ransomware or ability to create a back door to conduct many attacks or reconnaissance.

The popular types of cyberattacks can be reviewed in Table 2 below as to give indication to some challenges many organisations and individuals are facing.

Table 2 Cyberattack landscape

Types of cyberattack	Main common characteristics
Phishing	A general social engineering email type attack that may not be specifically targeted for an individual in mind but more of a blanket type of attack in the opportunity to attain a response and obtain a foothold into an organisation. Usually the base level entry for attacks that then go onto to produce other type attacks such as Advance Persistent Threats. These typically could be randomly generated to produce mass volume of emails.
Spear phishing	A phishing attack as an email that is specifically aimed at a particular individual and are executed by individual attackers and not randomly generated. The emails appear genuine as to appear from the recipients own organisation or one that is known to be trusted either as organisation or personally.
Whale phishing	Another type phishing attack that is organised in such a way as to appear as a high profile executive such as CEO or CFO and designed to acquire important information due to the privileges access one would expect these position to hold. Usually a monetary request is behind the attacker's motive.
Ransomware	Ransomware is a malware attack that can cause organisations wide and disruptive damage (such as WannaCry, 2017) or individual victim's data being blocked. The malware blocks access to the data and threatens to either publish sensitive information, delete it or not provide a decryption key to unlock it unless a ransom is paid. Sometimes even when a ransom is paid there was no intention to release with a decrypt key and was intended as a destructive malware.
DDoS	A Denial-of-Service (DDoS) attack has the objective to cause organisations mass disruption by and causing inaccessible services by overwhelming the intended target with traffic. This could cause anything from system slowdown to crashing if also the intention was to flood the network with so much traffic. The design of the attack causes business, employees and users of the service mass inconvenience and damage to reputation of the organisation. Sometimes the primary objective of a DDoS attack appears just this mass disruption but may be used to mass other cyberattack motives. As there are many DoS & DDoS types the popular ones as follows TCP SYN flood attack, Teardrop attack, Smurf attack, Ping of Death attack and Botnets.
Botnets	A Botnet is used to spam and launch DDoS through a network of hijacked computers and devices which is infected with the botnet malware and remotely controlled by the cyber attacker. Commands can be transferred not just as traditional command and control but through peer-to-peer networks.
Trojans	A Trojan (much the same as the concept of a Trojan Horse in Greek mythology) hides and masks itself in what appears as a useful program to the user unaware of the malware that sits in the program. Sometimes the purposes of a Trojan are to open a back door to the attacker to perform a host of other cyberattacks or perform other activities such as listening to sensitive information or sit in the network undetected with necessarily performing any attack and when his information task is achieved will leave.
SQL injection	Structured Query Language Injection (SQLI) attack although an older type of attack is still used and a dangerous attack to expose access data. The attack uses malicious code to manipulate backend databases or admin control and gain access to data that will be sensitive company data and personal data. SQLI is viewed as a serious threat to an organisation with loss of credibility to its customer base and worse if records are stolen, deleted or altered in any way.

(continued)

Table 2 (continued)

Types of cyberattack	Main common characteristics
Cross-Site Scripting (XSS)	Similar to SQLI attacks and where attackers run scripts in the target's browser using third-party web resources. Malicious JavaScript into the website's database. The attacker's malicious payload, which is part of the HTML that the victim is accessing on the website, is transmitted to the victim's browser and victim's cookie is sent to the attackers' server. Session hijacking is the most likely outcome, but more complex consequences can occur where attackers can log keystrokes, control the victim's terminal or reconnaissance.
Man-in-the-middle	A dangerous cybersecurity breach attack that the attacker can perform a number of options from simply eavesdropping on communications, to intercepting and amending data as by changing the requested key with his own and so assuming a 'trusted' identity. In many occasions it might just simply be eavesdropping and gathering intelligence and when the objectives are complete to leave the network undetected. The type of MiTM attacks can be classified as session hijacking, IP Spoofing and Replay attack.
Drive-by attack	A malware distribution attack usually found on insecure websites where a cyber attacker can plant malicious script into PHP and HTTP pages. When a visitor comes to the site malware is installed through a script or can redirect the victim to a website that the attacker is in control of. Given the name of drive-by attack, as the victim merely has to visit a site to become affected by the malware script. The danger usually is unawareness by the victim as attacks are completed silently.
Advance Persistent Threats (APTs)	Complex and usually a more sophisticated cyberattack as it involves a number of attacks but initially may start as a phishing attack to gain entry and move through the network to gain credentials etc., where reconnaissance is built to determine how the attacker requires the attack to play out: It could be destructive malware, cause massive damage to critical national infrastructure and so on.
Wiper attacks	Wiper attacks have been used in conjunction with APT attacks and ransomware and have a particularly destructive nature for data and disk wiping
Malvertising	A malware attack that uses advertising to hide malicious code where victims assume adverts to be safe. A number of consequences can unfold when the victim succumbs to a successful malvertising attack which can be anything from system damage, remote control or data access which may be sensitive data. The malware can dictate payloads to be set off at given times (these would be preinstalled programs).
Password attack	Attackers use a suite of tools such as sniffers, dictionary attacks, and cracking software programs to assist to decrypt and obtain passwords without authorisation. Still an attack that can work due to the user deploying weak or easy passwords (organisations now, as part of cyber defence strategy, usually insist on strong passwords).
Zero day exploit	The term zero day comes about due to a software/hardware/firmware vulnerability being discovered by attackers before manufacturers or other know about it or security professionals can fix it.
Insider attack	As the term indicates that attacks or malicious behaviour are performed inside the network by authorized individuals that will have access to the systems/network. Since most of the focus will be on external attacks this may be harder to detect although more is being looked at regarding behaviour and artificial intelligence software to help detect.

Phishing and Ransomware

Phishing has been around for some time and until users stop clicking on unknown senders of emails or enabling programs in attachments, the attackers will still create successes and is one of the more popular ways for attackers to enter a network. Steer [16] notes a more dangerous threat where the email is more targeted for specific individual employees and known as spear-phishing which will have objective aims and why it can have more serious consequences since the attackers have built a plan on social engineering and what the desired end result should be. Attackers now search social media to make an email look convincingly genuine to mask the tracks of the ransomware. This 'familiarity' of appearance to a user makes this method a very dangerous entry point. Even newer dangers of malvertising no longer require users to 'click' links to activate the malwares.

In 2016, phishing attacks rose by 65% with an increase over 2015 mentioned by Sharma [15] and set to increase further again with organisations more in the sights of the attackers than individuals. But as social media brings many benefits, so this also offers attackers a wealth of additional information that gives rise to a more sophisticated method and likely chance of a victim opening an infected email/program.

Regarding ransomware, there has been significant increase in the volume and diversity of ransomware attacks over the last few years (see Fig. 1 for growth details). Attack formations have become more complicated, progressing from script

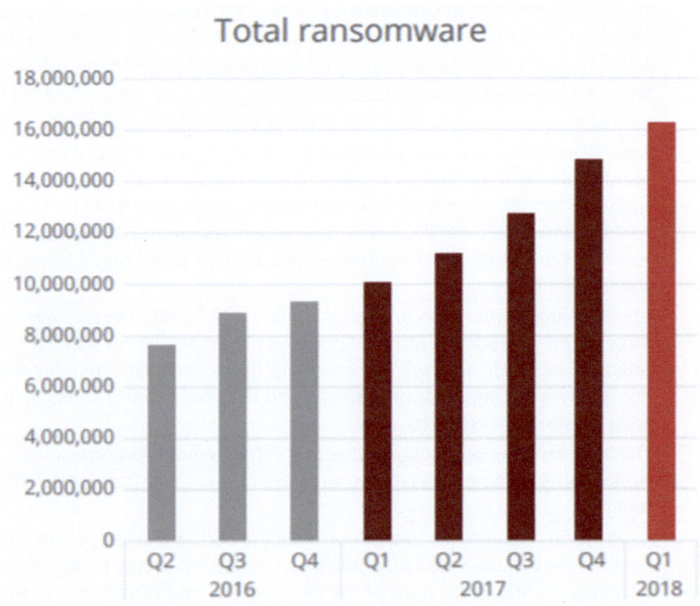

Fig. 1 Ransomware increases (McAfee Labs 2018)

kiddies to crime-for-hire (Ransomware-as-a-Service) and sophisticated hackers, as Parent and Cushack [12] mentions 'a new layer cake in cybercrime' has arisen. However, attack success relies on the victim acting as the catalyst to initiate the chain of events. Ransomware is sophisticated and malicious software that blocks the victim's access to their files or can threaten to publish or make available the victim's data (based from cryptovirology). What usually happens next is the request for a ransom to be paid to allow access to the victim's files again. Over the last 5 years the extent and complexity of ransomware has increased, and some organisations have taken to paying the ransoms through digital currencies such as Bitcoin. The victim's dilemma is there is no sure way to know if after paying a ransom whether the attacker will release the decryption key. There are some victims that have paid the ransom and were successful in accessing the data once again. Others have paid but the cyberattackers did not release the decryption keys and others where they received the decryption keys but were found to be not working. There is always the possibility as well that after paying the initial ransom the cyber criminals may continue to ask for additional ransom. Without doubt, attacks are on the increase and even 3 years ago was noted: [4] wrote that GCHQ and the Ministry of Defence warned the government of a serious cyber-attack and that the National Cyber Security Centre (NCSC) was a step in the right direction.

Attackers are using more manipulative means by deploying sub-domains, hidden URLs (link hidden under plain text) misspelt URLs (bought domains that look similar to the real site e.g. google.com) and IDN homograph attacks. McAfee [9] labs report indicates a rise in malware scripting techniques, using JavaScript, VBScript, PHP, Powershell, etc., to distribute including in use of ransomware (e.g. Nemucod ransomware using PHP and JavaScript). The main use case for attackers taking this route is evasion and reason why this recent technique has seen a sharp rise over last 2 years, McAfee [9].

Yaqoob et al. [19] discuss how sensors and IoT devices are expected to reach 26 billion by 2020 deployed across from wearable devices, connected smart homes, cars, health, utility, etc. On the one hand it is great technological advancement but as Yaqoob et al. [19] mention that over 70% of IoT devices have vulnerability to attacks and with the growth of ransomware it is inevitable that IoT offers a vast surface attack area. In tandem to technological advancement the preservation of confidentiality, integrity and availability (CIA) are important and safer ways to protect IoT from malware which should take priority.

Ransomware can also be penetrated using botnets inside IoT networks, activated through phishing and compromise the whole IoT network and attackers can inject ransomware-as-a-service into the cloud and the ransomware may then appear as a 'legitimate service' to the user. More research shows the extent of IoT potential issues with thermostat hacking (turning the temperature up), Flocker ransomware (locking the smart TV as malware is embedded in fake movie), Android Simplocker affecting Android wearable devices and some tests on practicality of ransomware through smart bulbs.

There has also been a sharp increase in using malvertising to propagate malware, specifically ransomware [13]. This new type of technique threatens the digital marketing many organisations has legitimately enjoyed but now find users may block Ads for fear of malwares that can potentially be deployed. Users need not even click on the malicious Ads and can be infected just by visiting the website. This is in itself a new technique that provides infection without user interaction (e.g. Astrum exploit using Flash vulnerability). Palmer [11] reported a few UK Universities that were hit with this ransomware but very hard to consolidate extent of attacks as many education and government type organisations resist to openly discuss in fear of negative publicity.

The future surface attack area for TTPs has taken a new threat and dimension. Mansfield-Devine [6] view the outlook as heading towards ransomware distributed through mobile and the cloud. This can be a big concern to contain since a user can upload an infected file to the cloud and unknowingly provide a wide dispersal ratio to the rest of the community (users confusingly assume it's a legitimate 'safe' service to download). Also, more and more enterprises are seeking cloud and SaaS based models in turn believing this to be safer than on premise models and why more analysis of cloud security needs to happen. Whilst it will be an ongoing education process to help enforce the human not to click an unauthorised link or carefully observe before clicking it still does not solve the problem of how to control the data access, storage and in the event of cyber breach give the malicious attackers more hurdles with added layers such as blockchain and encryption. Also, with blockchain there is further research on using artificial intelligence (AI) to detect malicious outsider attacks and its patterns which gives further protection levels to help blockchain stay more protected.

3.3 Blockchain and TTPs

The idea of there being no intermediaries such as centralised entities or TTPs is not what is being advocated here. Clearly, there is still a requirement for TTPs and also blockchain itself has some negative connotations. Under a decentralised system and for example, regarding a cryptocurrency wallet, losing your password is pretty much a point of no return since there is no way to retrieve it if there was no backup of writing it down, etc. But what could be a possibility that may work, and benefit is a blend of both operating in synergy. Certainly, this is already being applied in many industry use cases.

Take for example the interoperability in Cyber-Physical systems (CPS). These are a range of software networks, communications, sensors that interconnect into the real world, between physical and virtual states. In many industries these often involve mission critical processes so, for example, in manufacturing they would process real-time information to industrial machines, supply chain or perhaps in the energy sector provide smart grids which would consist many controls, sensors and equipment working in conformity. Smart grids work more intelligently to add resilience to the power network by detecting where the outage may be and contain it; so, avoiding the blackouts or power surges that could cause major disruption if power is not available and a domino effect to other such as banking/finance, healthcare, etc. It is also driven by the fact that more physical systems controlled by embedded software are connecting and interacting with other systems via the internet. Regarding blockchain, Dong et al. [2] explore blockchain as the trusted environment to support interactions that CPS function on in an energy grid. It looks at three parts; blockchain distributed data storage, IoT sensing technology and cloud-based delivery systems and how they can interact together and support the physical infrastructure (the generation, transmission and delivery). Blockchain supports the CPS with its decentralized data storage to give traceability and non-tamperable qualities and it is not relying on itself to be a single point of failure as is the case with TTPs). Also, the smart contracts allow programmable execution of work and authorization for auditability purposes. The interactions between IoT and blockchain make it flexible as to how the data can be stored (cloud architecture sits on top of IoT and then blockchain) so dependent on data if on-chain or off-chain. Dong et al. [2] make a good point of the SCADA (systems used to monitor plant/manufacturing equipment) vulnerabilities and where data is centrally stored which is highly vulnerable to cyber attackers and propose operating a blockchain based grid data protection mechanism. CPS architecture must consider not only the individual subsystems but a system of systems approach to cyber security and is developing aggressively in IoT for industrial and IoMT in healthcare and new products launched are already possessing that data sharing and networking capability.

Real time computing (RTC) is already being used in many applications, especially mission critical systems such as autopilot systems in aircrafts and anti-lock brakes in vehicles. In such mission critical systems, due to latency and time sensitivity, systems running real-time operating systems (RTOS) process data right at the edge of the system and almost instantaneously decide on the next course of action. As a next elevation of security blockchain could be deployed with RTC in mind. There can be many scenarios where this is of good use. For example, in supply chains where achieving high performance and efficiency could be used in harmony with blockchain to provide transparency, immutability and security it offers.

For IoT, it is heavily reliant on cloud computing due to the amount of data collected, however does all this data need to be collected and sent out to the cloud? Processing data at the edge can prove to be as effective as processing in the cloud as demonstrated in systems with real-time processing capabilities. Cloud computing provides a more resilient computing platform due to its distributed physical nature however there are still vulnerabilities. This is why research and pilots are being assessed, as Dong et al. [2] mentions on the benefits of integrating blockchain with IoT and cloud computing in the energy sector of critical national infrastructure (CNI) as various decisions can be taken as to how to store the data and where permissions should be allowed.

Shifting mind-sets from a cloud based perspective to performing real-time processing at the edge of the network will not only reduce threats caused by infrastructure system vulnerabilities but will also address time sensitive systems and potentially could be used to address the issue of privacy which is a much talked about threat to end users. There is no suggestion that real-time processing at the edge should replace cloud computing but rather the two technologies should complement each other or even integrate with the third technology of blockchain. Managing data efficiently at the edge using real-time processing should allow tighter and more granular controls to be put in place to manage the type of data that should be pushed out into the cloud or stored elsewhere if on blockchain. Big data algorithms and machine learning can then be employed to further manage data effectively in the cloud or across blockchain, improving edge endpoint reactions and access and with blockchain certainly take the risk away from cloud computing, fog and RTC .

4 Harmonious Relationship of Blockchain and Artificial Intelligence

The two technologies of blockchain and artificial intelligence (AI) appear to have set on a path of convergence, [1]. This has ranged from technological benefits of AI deploying more efficiency over mining capabilities with optimisation of energy consumption and methods of federated learning to provide that leaner processing in the form of data sharding (simply meaning data portioning and separating larger databases into smaller components).

Blockchain, with its record keeping attributes, can provide more coherent understanding of decisions made and impacts of machine learning. Analysis is made easier if these records can be traced. So, a machine learning algorithm can be subject to continual changes as its learning process dictates that efficiency in pathway changes as it learns in real-time. However, as these algorithms get more intelligent through machine learning it also becomes more difficult to look at how conclusions were arrived at. There is the trust factor in these conclusions; meaning potential algorithm bias or ways to ensure the algorithm had not deviated so far as to be corrupted. Therefore, it will become more and more important to ensure an audit trail is there, can be tracked and alert or flag up inconsistencies. It's clear blockchain offers the immutability, time-stamping and can be used for the audit trail purpose and therefore can complement the attributes that AI offers. Data is secured by blockchain but also enhances this relationship in its audit capacity, so the full journey of machine learning and algorithm changes are understood and provide the transparency, which is becoming a hot topic of discussion.

Data science and its benefits offered in many industries is also helping to accelerate the relationship of blockchain and AI. For example, take the field of genome research and the significant factor that AI plays here in terms of machine learning and not just applied to human health but also to agriculture and animal husbandry, Marr [7]. AI of course brings that element to research to make analysis, faster, accurate and with many options to deviate outcomes. This of course makes gene technology truly exciting for developing precision and personalised medicine and AI can help overcome many previous barriers by offering modelling, predictions based on specific individual's genes. It's now easy to see how then blockchain can be complimentary to AI in this area, since there needs to be a strict access protocol, audit trail, traceability and all data analysis secured, so there is no selective reporting or possibility for bias views to the data. There would be very sensitive individual personal data and the security and privacy that blockchain offers ensures there is that high level of security to deter any cyberattacks that can cause data leaks or accidental exposure of data. Transparency is also important to be able to verify the basis of operation within the business and to the individuals that data is created from.

There would be many more examples in all sectors of industry where the complementary positioning of AI ands blockchain are clear to all.

5 Conclusions

This chapter explains the current necessity of trusted third parties (TTPs) and the role cloud based organisations operate under but also demonstrates the risk of exposure to data leaks by cyberattacks and impacts caused by the many different variations when TTPs are breached. Blockchain attributes of immutability, traceability, auditability, securing privacy and offering transparency, offers possibilities of adding another layer of protection to sensitive data. This technology can give the single version of the truth and full audit control on a chain of custody. Also, the interesting convergence of blockchain and AI is positioning many opportunities to further develop faster data analysis and provide a methodology to ensure authentication, protocols and so on are adhered to.

References

1. Corea F (2017) The convergence of AI and Blockchain: what's the deal? Medium. Available at: https://medium.com/@Francesco_AI/the-convergence-of-ai-and-blockchain-whats-the-deal-60c618e3accc. Accessed 10 Aug 2019
2. Dong Z, Lou F, Liang G (2018) Blockchain: a secure, decentralized, trusted cyber infrastructure solution for future energy systems. Mod Power Syst Clean Energy 6(5):958–967. Available at: Doi: https://doi.org/10.1007/s40565-018-0418-0. Accessed 10 Aug 2019
3. Engelhardt M (2017) Hitching healthcare to the chain: an introduction to Blockchain technology in the healthcare sector. Technol Innov Manag Rev 7(10):22–34. Available at: https://doi.org/10.22215/timreview/1111. Accessed 10 Aug 2019
4. Huchins M (2017) Britain knew the risks, says ex-military chief. The Times, 16 May, pp 14–15. Available at: https://www.thetimes.co.uk/article/britain-knew-the-risks-says-ex-military-chief-hz2lpfvfl. Accessed 14 Aug 2019
5. IBM Global Business Services (2016) Blockchain: the chain of trust and its potential to transform healthcare. [Online]. Available at: https://www.healthit.gov/sites/default/files/8-31-blockchain-ibm_ideation-challenge_aug8.pdf. Accessed 10 Aug 2019

6. Mansfield-Devine S (2016) Ransomware: taking businesses hostage. Netw Secur 2016(10):8–17. [Online]. Available at: https://doi.org/10.1016/S1353-4858(16)30096-4. Accessed 1 Aug 2019
7. Marr B (2018) The wonderful ways artificial intelligence is transforming genomics and gene editing. Forbes. [Online]. Available at: https://www.forbes.com/sites/bernardmarr/2018/11/16/the-amazing-ways-artificial-intelligence-is-transforming-genomics-and-gene-editing/#70e819a642c1. Accessed 10 Aug 2019
8. Mattei TA (2017) Privacy, confidentiality, and security of health care information: lessons from the recent WannaCry cyberattack. World Neurosurg 104:972–974. Available at: https://doi.org/10.1016/j.wneu.2017.06.104. Accessed 10 Aug 2019
9. McAfee (2017) McAfee labs threat report. [Online]. Available at: https://www.mcafee.com/uk/resources/reports/rp-quarterly-threats-sept-2017.pdf. Accessed 1 Aug 2019
10. Mettler M (2016) Blockchain technology in healthcare: the revolution starts here 2016 IEEE 18th International Conference on e-Health Networking, Applications and Services (Healthcom). Available at: DOI: https://doi.org/10.1109/HealthCom.2016.7749510. Accessed 18 Aug 2019
11. Palmer D (2017) ZDnet. Available at: http://www.zdnet.com/article/this-malvertising-campaign-infected-pcs-with-ransomware-without-users-even-clicking-a-link/. Accessed 6 Aug 2019
12. Parent M, Cusack B (2016) Cybersecurity in 2016: people, technology, and processes. Bus Horiz 59(6):567–569. [Online]. Available at: https://doi.org/10.1016/j.bushor.2016.08.005. (http://www.sciencedirect.com/science/article/pii/S0007681316300829). Accessed 6 Aug 2019
13. RiskIQ Research (2017) Business Wire. [Online]. Available at: http://www.businesswire.com/news/home/20170131005420/en/Malvertising-Rises-132-2016-2015-RiskIQ-Research. Accessed 6 Aug 2019
14. Satoshi N (2008) Bitcoin: a peer-to-peer electronic cash system 2008 [Online]. Available at: https://bitcoin.org/bitcoin.pdf. Accessed 18 Aug 2019
15. Sharma U (2017) Blockchain in healthcare: patient benefits and more [Online]. Available at: https://www.ibm.com/blogs/blockchain/2017/10/blockchain-in-healthcare-patient-benefits-and-more/. Accessed 18 Aug 2019
16. Steer, S. Defending against spear-phishing, Comput Fraud Secur, 2017, 8, 2017, 18–20. [Online]. Available at: https://doi.org/10.1016/S1361-3723(17)30074-X. (http://www.sciencedirect.com/science/article/pii/S136137231730074X). Accessed 18 Aug 2019
17. Swan M (2015) Blockchain: blueprint for a new economy. O'Reily Media Inc, Sebastopol
18. World Bank (2015) World development indicators: Health expenditure per capita (current US$). The World Bank [Online]. Available at: https://data.worldbank.org/indicator/SH.XPD.PCAP. Accessed 18 Aug 2019
19. Yaqoob I et al (2017) The rise of ransomware and emerging security challenges in the Internet of Things. Comput Netw 129(2):444–458. Available at: https://doi.org/10.1016/j.comnet.2017.09.003. Accessed 6 Aug 2019

Protecting Privacy and Security Using Tor and Blockchain and De-anonymization Risks

Stilyan Petrov, Stefan Kendzierskyj, and Hamid Jahankhani

Abstract The huge increase in data usage and the rapid development of new technologies such as cloud, IoT, and has also led to the exponential increase in cyber threats online. Anonymity and privacy services have equally seen an exceptional growth rate since the introduction of Blockchain and Tor network, as more individuals demand anonymous services away from the traditional centralised offerings, but also seek more security and privacy. This chapter will review quantitative analysis undertaken to critically evaluate Tor and Blockchain as emerging technologies, by an in-depth comparison of their security and privacy properties. Further analysis is undertaken by utilising network and data points that highlight the necessity of urgent deployment of innovative methods to protect users' anonymity utilising Blockchain application over the Tor network. By undertaking experimental analysis, it is possible to determine Tor packets from common packets and raises the question on possibilities of cyberattacks leading to loss of personable identifiable information (PII) and de-anonymization.

Keywords Tor · Blockchain · Cybersecurity · Anonymity · Privacy · Cyberattack · Onion services · PII · Encryption · Security

1 Introduction

With everyday life having seamless interaction with activities online we are living in an era where most of the criminal activities are also being performed online. This has meant organisations and individuals are facing an urgent necessity to

---

S. Petrov · H. Jahankhani (✉)
Northumbria University London, London, UK
e-mail: Stilyan.petrov@northumbria.ac.uk; Hamid.jahankhani@northumbria.ac.uk

S. Kendzierskyj
Cyfortis, Worcester Park, Surrey, UK
e-mail: Stefan@cyfortis.co.uk

H. Jahankhani et al. (eds.), *Cyber Defence in the Age of AI, Smart Societies and Augmented Humanity*, Advanced Sciences and Technologies for Security Applications, https://doi.org/10.1007/978-3-030-35746-7_11

acquire new security solutions and services. The changing landscape of cyberattacks that compromise personal identifiable information (PII) particularly in critical national infrastructure, such as in sectors of Healthcare, Finance, Defence, Civil Nuclear and so on, required the development of more secure technologies that also respect privacy. This has led to some organisations seeking further emerging technologies such as blockchain and also individuals using the Tor network as way of ensuring anonymity. As an innovative and highly sophisticated technology, Tor anonymity network attracts developers, researchers and criminals to utilise it in various "good and bad" [19] ways, but also due to core reasons – to remain private and secure anonymity. Thus, Tor delivers strong protection techniques to secure the confidentiality, integrity and availability (CIA) of the data, and the real identity of the individual, however, it represents vulnerabilities leading to leakage of PII [30]. Therefore, there are opportunities to use other emerging solutions such as blockchain that can protect data, enhance security and provide transparency.

According to Al Jawaheri et al. [1], in the contemporary world, there is an increasing need to improve security and privacy over the Internet due to the swift rising of threats. To protect the online transactions, a secure and decentralized peer-to-peer network such as blockchain can be a desirable technology to benefit both society, individuals and organisations. The demand for blockchain from government, organisations, industry, and supply chain is growing due to blockchain attributes such as immutability, privacy, traceability, audit, etc., that provide a trustful distributed network without meeting the weaknesses that centralised authorities hold (or trusted third parties). Nevertheless, the growing number of malicious attempts and widening attack surfaces against blockchain are extremely concerning especially with its increasing adoption and popularity.

Scientists and developers are working together for a superior and secure future with attention driven towards Smart Cities, IoT, Big Data, Quantum Computing, Biotelemetry and so on. So, it is highly desired for the adoption of Tor and blockchain as emerging technologies to deliver a more trustworthy mechanism, that offers security, privacy, and anonymity and which safeguards users' data/assets, and can prevent any de-anonymization possibilities.

2 The Motivation Towards Anonymity and Tor Network

Nurmi [42] stated that anonymity and pseudo-anonymity are nothing more than privacy of identity. Under anonymity, Nurmi refers that the user manages who can view their identity, so placing control over the one he or she wishes to know the identity. One of the earliest examples for online anonymity was in 1993 when Johan Helsingius ran an anonymous email service called anon.penet.fi. The function of these services is to provide anonymous accounts that act as a proxy or corresponded between real email addresses and pseudonymous addresses. However, the weak anonymity is exposed by stripping identifying headers from outbound re-

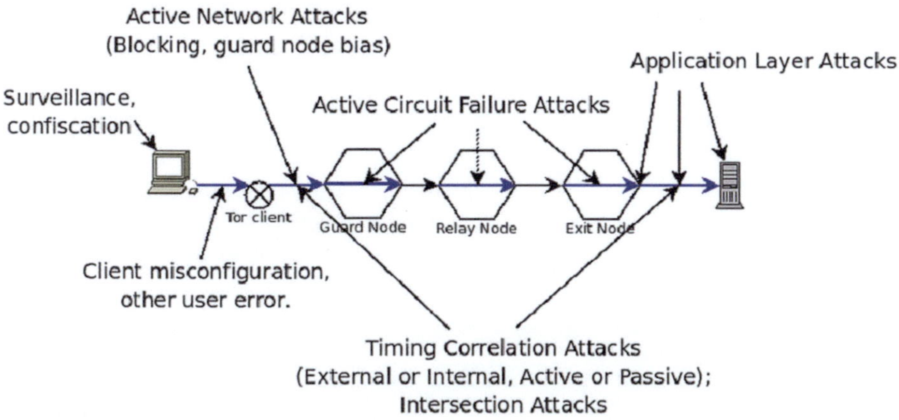

Fig. 1 Tor de-anonymization attacks [42]

mailed messages and the limitations of the services in terms of security, closed the Helsingius remailer service; Nurmi [42].

Further, summarized by Nurmi [42] revealed four categories under which a user can lose its anonymity in the Tor network (see Fig. 1):

- Comprehensive operation security can lead to failure
- Attacks intended towards the onion server software's that are not intended to be adopted on Tor.
- End-user vulnerabilities such as weak passwords or unpatched systems/applica tions.
- If entry and exit relay are condemned it may incur traffic and timing correlation attacks.

However, the Nurmi [42] research was not focused on the end-user vulnerabilities in configuration and settings. Moreover, after many experiments and analyses on the Tor network and the onion services, Juha Nurmi claimed in the conference that Tor network possesses a unique feature – as in the more users there are, then the harder the de-anonymization attack would succeed.

In 2012, the "highly-ranked" National Security Agency (NSA) employee, Edward Snowden, revealed to the world, highly confidential secrets about the USA surveillance and admitted that the NSA has been spying on citizens all over the world [25]. In 2012, the NSA published documentation called "Tor Stinks", declared that de-anonymization of Tor users was not just possible to be accomplished but also a workable conception (see Fig. 2). Moreover, the NSA stated that they will never be able to de-anonymize all Tor users, however, they could be able to reveal an identity only on a small fraction of them.

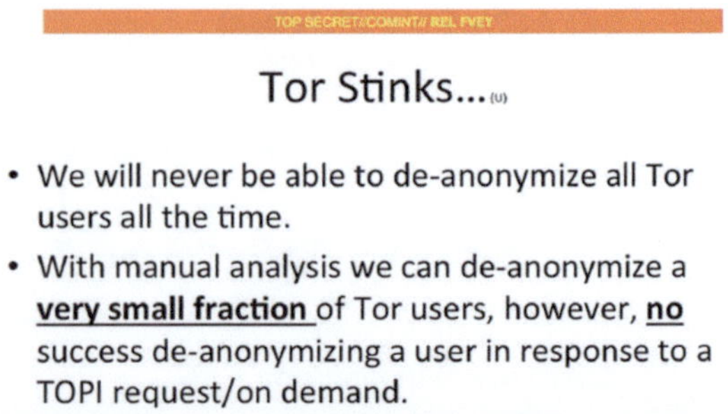

Fig. 2 NSA document 'Tor Stinks'

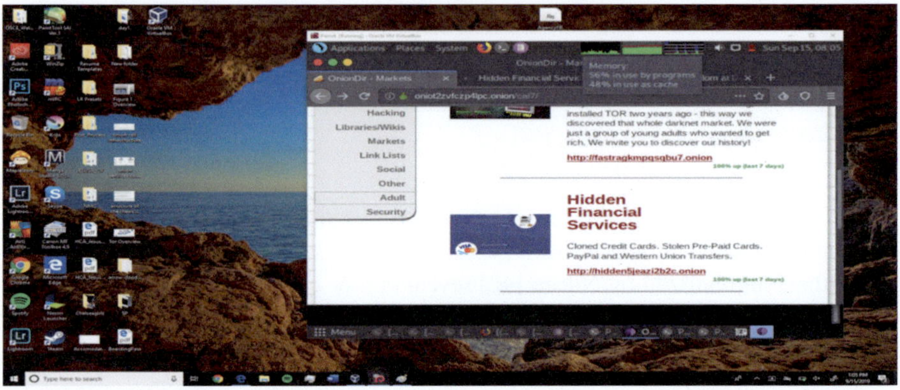

Fig. 3 Tor services rendered by criminals

2.1 *Tor Origins and Overview*

A recent technical report from Al Jawaheri et al. [1], mentions that Tor is utilised mostly from those that want to hide their anonymity such as journalists, whistleblowers, censorship fighting organizations, sensitive topics researchers and of course, criminals. Nevertheless, as a freely usable portal for illegal activities such as buying drugs, weapons, downloading child-abuse videos and even a place where you can hire an assassin, the majority of Tor is legal, stated by Moore and Rid [38]. Figures 3 and 4 however display more of the 'darker' side of Tor.

2.1.1 What Is Tor?

Tor is the most broadly used anonymous network, daily accessed by more than four million people and designed by volunteer-run Onion Routers (ORs), hereinafter called relays/nodes and provides peer-to-peer traffic encryption by using the 128 bit AES in counter mode (still not compromised) and checksums for integrity checking. Moreover, the three relays, linking to each other with the path algorithm selection, form a Tor circuit. According to the Döpmann et al. [24], the circuits represent a

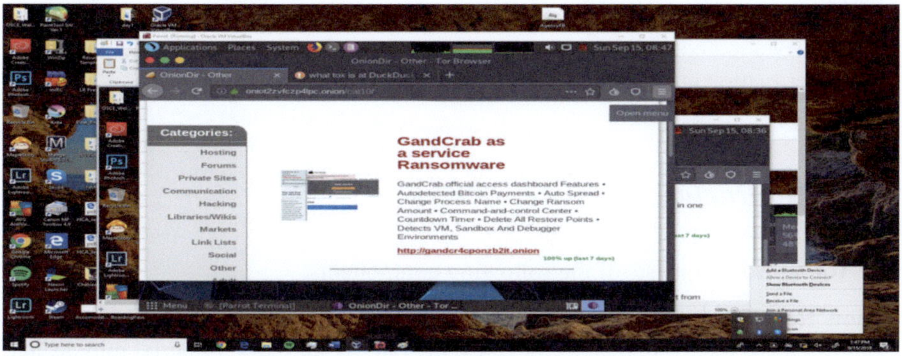

Fig. 4 Tor ransomware and malware as a service offering

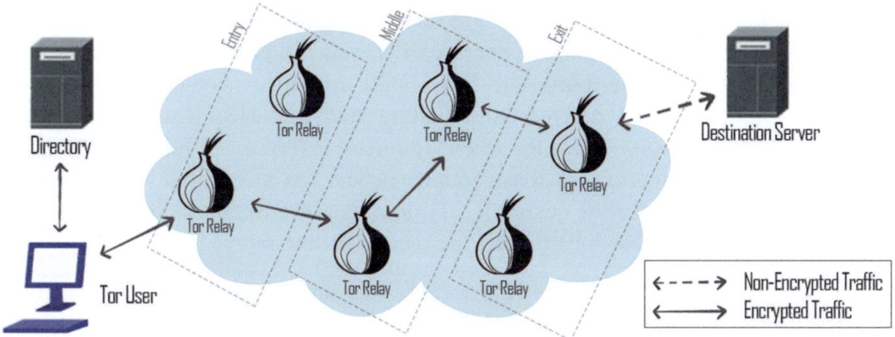

Fig. 5 Tor network infrastructure

Country	Autonomous System	Consensus Weight	Advertised Bandwidth	Guard Probability	Middle Probability	Exit Probability	Relays	Guard	Exit
(86 distinct)	(1106 distinct)	99.9929%	48459.64 MiB/s	99.9508%	100.0521%	99.9667%	6524	3003	992
Total		**99.99%**	**48459.64 MiB/s**	**99.95%**	**100.05%**	**99.97%**	**6524**	**3003**	**992**

Fig. 6 The number of currently running relays. (Derived from TorProject [48–51])

cryptographically secured tunnel where adds or strips one layer of encryption on the Relay Cell (Fig. 5).

An analysis in 2019 of the Tor network shows at that time it consisted of 6524 relays (routers) from which 3003 of them are guards and 992 are exit relays (Fig. 6).

A volunteer means a person allows their PC to receive and pass traffic over the Tor. This is a significant design strategy, developed with the origin of the network, which means increasing the number of the volunteer-run relays means the more secure, anonymous and sophisticated against an attack on Tor network. Additionally, Tor is a developing technology and over the years acquired new features in terms of anonymity, privacy and reliability, and looks to prevent de-anonymization attack vectors. In a research paper, Biryukov and Pustogarov [8, 9] justified that due to the live-development and low-latency properties of the Tor network, the traffic and timing correlation attacks cannot be entirely eradicated.

2.2 Tor Structure

The Tor network consists of the following components:

Onion Routers Referred also to ORs; they form the spine of the Tor network. Controlled by the volunteer users' they operate as an entry (guard), middle and exit nodes. These create a commonly three-hop Tor circuit where the traffic is enveloped in n layers of encryption with n unique session key and sent to all relays in it. Each node designs a descriptor (explained in the section How Onion Services Works?) which contains its own bandwidth, IP, public key, circuit negotiation, exit policies, etc. about the hidden services [2], and sends it to the Tor Directory Authorities (DAs). Furthermore, the DAs build consensus documents based on consensus algorithm and delivers them along with the HSs descriptors to the Directory Servers (DSs). When the client (later on referred as Onion Proxy (OP) wants to obtain information for these routers he/she has to connect to the DS and extract up-to-date data for the currently usable nodes. On the period of writing this chapter, there are nine DSs which keep the records of the relays [44].

Onion Proxy (Client) An end-user, accepted as an Onion Proxy (OP), who is utilising the Tor browser as an entry point to the network. The OP regularly contacts to the DSs so they do get a list of operated ORs and their descriptors. Also, to defend clients against route detection and reconnaissance attacks, both the records and the consensus algorithms are constantly updated on each hour to deliver a coherent and reliable image of the network to all clients; so as to provide a current and consistent picture of the network to all clients. Consensus documents are published precisely once an hour and descriptors are real-time updated as their contents change.

Directories The Directory Authorities (DAs) send the consensus document and the OPs descriptors to Directory Servers (DSs), thus they deliver information of the current state of the circuit. Moreover, the DSs store the descriptor and the OPs public key in the so-called Distributed Hash Table (DHT).

Relay Cell Onion relay sends, in anonymous reason, 512 bytes or recently 514 bytes cells size traffic on the Tor network which its payload (actual data) is encrypted with the Elliptic Curve Cryptography. This public key cryptography is presently resilient under attacks. Döpmann et al. [24] stated that if we look in-depth in the transport layer in security terms of Tor keeping in mind that the cells are exchanged between the relays with the TLS/SSL TCP connections. Figure 7 below, reviews the relay cell and the cell architecture.

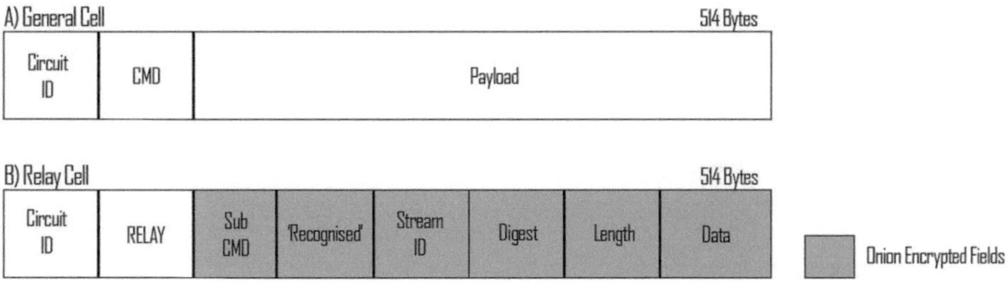

Fig. 7 Tor relay cell architecture

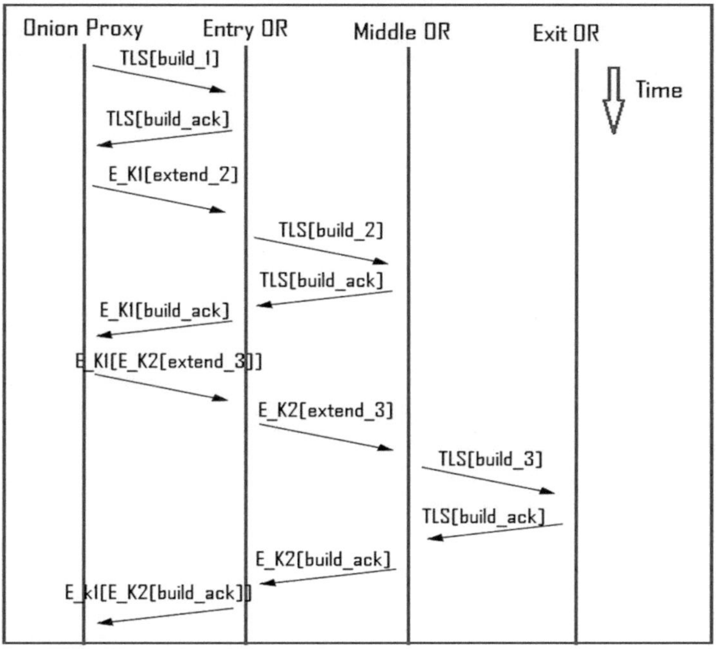

Fig. 8 Tor circuit creation, Salo [45]

When a user launches a Tor browser and attempts to access the onion service, the Create Cell is going through the circuit of relays that includes end-to-end encryption from 128-bit AES cipher in counter mode, checksums for integrity checking and asymmetric session keys (X25519); Saleh et al. [44]. So, to establish a new circuit (see Fig. 8), a Create Cell is sent to the first node on the path, which usually consist of three relays (see Fig. 5 on Tor infrastructure).

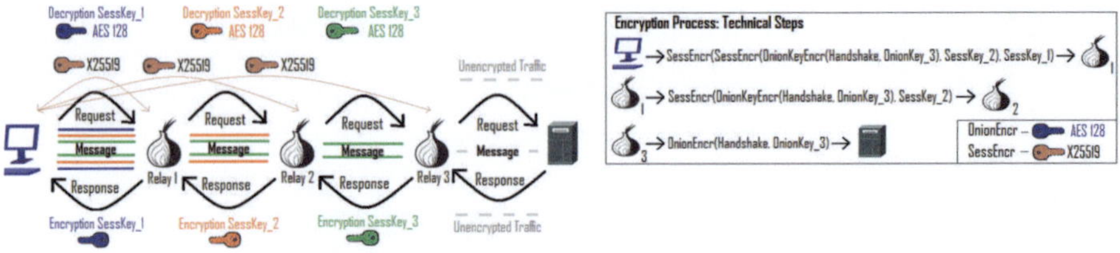

Fig. 9 Tor traffic encryption process

The Create Cell contains a circuit ID in its header and the onion key from senders' side in the payload section with the SessKey of each of the relays. This is called the Diffie-Hellman key exchange which is using TLS connection. Once a node receives the cell, it sends back a key and a signature confirmation that it received the cell through the DH process. Finally, a session key is acquired and the circuit between the first node and the client is established, [24].

2.3 How Does Tor Work?

The Tor user (OP) connects to the DSs to derive a current status of the Tor network and its relays information (public addresses, exit policies, bandwidth and so on). For the purposes of this chapter and explanation, it can be summarized in several technical steps how a three-hop onion circuit, consists of well-suited exit policies, is built and how it operates (Fig. 9).

Currently, v3 onion protocol is using 128-bit AES symmetric key with ECDHE asymmetric key exchange algorithm.

The data encryption process and key exchange algorithm performed by the Tor protocol over on a relay cell are represented in steps below.

Request

1. The OP sends through the DH Key-Exchange the so-called temporarily Session Key to the first (entry) node.
2. Relay 1 decrypts the *Create Cell* using *SessKey1* and sends it to the Relay 2.
3. Relay 2 destructs the *Create Cell* and constructs a *Relay Cell*. Moreover, it generates *SessKey2* and sends it to the Relay 3.
4. Relay 3 decrypts the *Create Cell* using *SessKey3* and sends it to the intended destination address known from the origin OP request.
5. The last (exit) relay decrypts the message with the onion (handshake) key and redirects it to the server.

Response

1. Upon receiving *response* from the server, the Relay 3 encrypts the *Create Cell* with *SessKey3* and sends it to the Relay 2.
2. Then, Relay 2 encrypts it with *SessKey2* and sends it to the Relay 1.

3. Relay 1 encrypts the response with *SessKey1* and sends it back to the OP. Thus, it strips the layers encryption with its private key and *PublicKey3* to see in plain text the response.

Generally, the requests are one layer of public (onion) key encryption and several layers (depends on the built circuit) of session (symmetric) key encryption. On the other hand, the response is using only symmetric key encryption without a public key.

Overview of Onion (Hidden) Services Nurmi [42] stated that several Tor protocols between the client and a server can make a location obscured. Thus, onion services are TCP-based network connections which are accessed through the Tor browser and use not human-meaningful () *.onion* top-level domain (TLD) name. The major purposes behind onion services are robustness and anonymously of servers, access-control protection and hiding the real individualities of onion services administrators [15]. According to the statistics derived from https://metrics.torproject.org/, there are more 80,000 onion services (see Fig. 10) on the Tor network, nevertheless, only a minor portion of these are web services [42].

Undoubtedly, huge amounts of the onion websites publish illegal activities such as child abuse pictures and video sharing or perform drug market services [41]. Therefore, for these reasons the Tor network and its anonymity are regularly criticized. According to the TorProject [48–51], another fundamental property, that makes the services more secure and obscure, is that accessing the onion services the Tor traffic remains in Tor network. There is no exit node in the circuit which prevents many de-anonymization attacks such as monitoring the unencrypted traffic, running

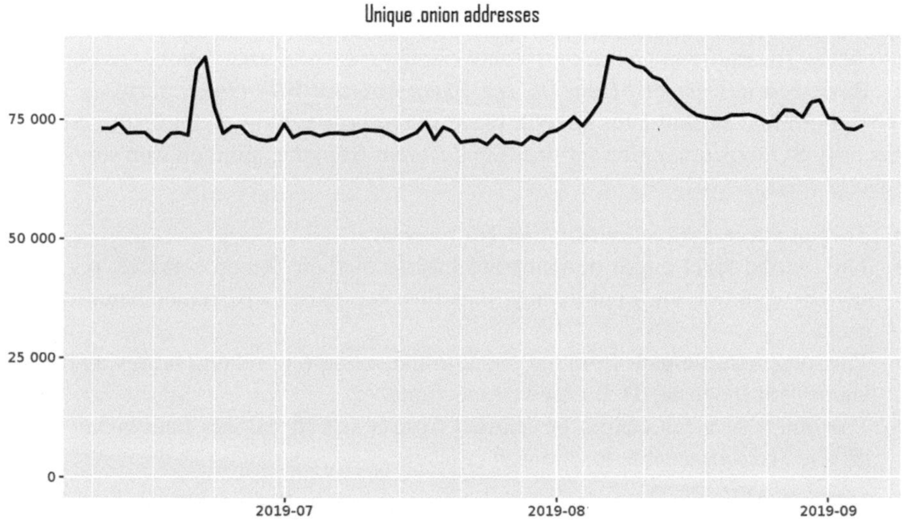

Fig. 10 Statistics of the active onion services: TorProject [48–51]

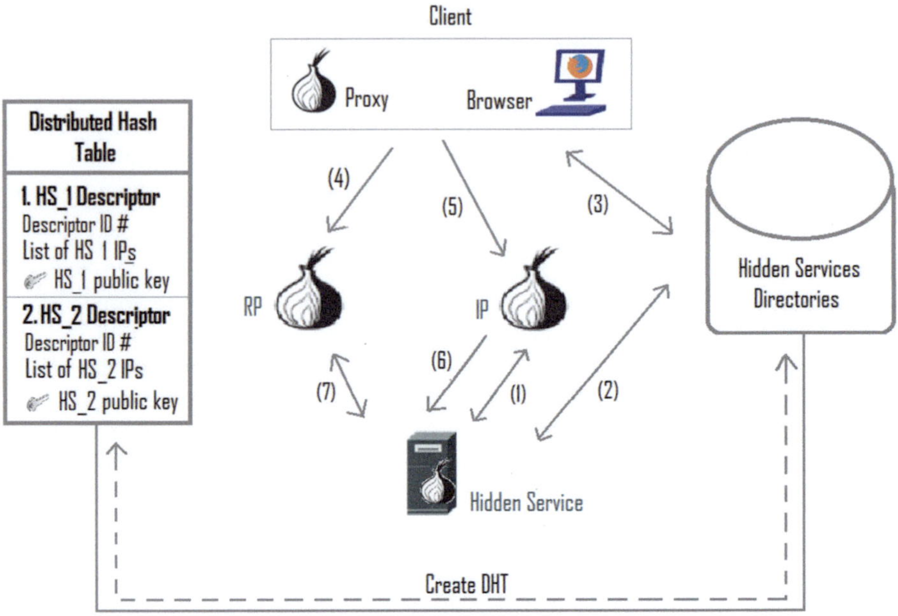

Fig. 11 Establishment of connection between Tor client and onion service

malicious relays and so on. Moreover, the onion domain address represents a hash function of public key which delivers a security and anonymity in client-server TCP communication Derdge [25]. A unique protection measurement developed by TorProject [48–51] which provides pure end-to-end security is the additional IPs and RP relays that extends the circuit with three more hops, and moreover, does not allow the Tor traffic to go out of the network (peer-to-peer encryption, see Fig. 11).

Reviewed in a paper *"How Do Tor Users Interact With Onion Services"* from Winter et al. [56], that later on took place in the proceedings of the 27th USENIX Security Symposium, onion services are different from the common web services in several ways:

- They are accessible only from the Tor browser.
- The second-level onion domains are hashes of their (onion services, e.g. web server, webmail, etc.) public key, thus they are hard to be remembered by the users.
- The longer traffic path from the client to the server (six relays) brings up higher latency for the usual TCP-based connections.
- The onion services cannot be learned from a search engine, thus makes them private and untraceable for the web crawlers.

A comprehensive evaluation, formed of steps, details the typical communication establishment process between the client and an onion service:

1. Bob produces a public key pair *"to identify himself as a service"* [22]. Thus, he randomly selects three IPs (with addresses of base64 string) as contact points and builds circuits to each of them.
2. Bob advertises the IPs and information about his onion service on the HSDirs, which on the other hand creates a *"key-value lookup system with authenticated updates"* [23] called Distributed Hash Table (DHT).
3. Alice heard about Bob's onion address, either from Bob or from a friend (since the onion services are not published in the search engines) and thus, she extracts more details about the Bob service from the DHT table [11].
4. Alice chooses a random OR as a Rendezvous Point (RP) which serves as a meeting point for the connection between her and Bob. Thus, she builds a circuit to the RP and establishes a "one-time secret" cookie so do recognize when Bob connects [35].
5. Alice builds a circuit to one of the Bob's IPs in order to announce herself, her RP, the rendezvous cookie and the beginning of the DH handshake. The IP sends the message to Bob [37].
6. If Bob wants to talk with Alice, he connects to her RP by building a circuit to it, and sends a message contains the rendezvous point (as verification about himself), the other half of the DH handshake, and the hash of a session key which they now share [2].
7. The RP connects Alice's and Bob's circuits, but none of them has information about each other. Then, Alice sends a begin cell to Bob's OP which connects to his onion service.

According to the AlSabah and Goldberg [2], the descriptors are "assembled-based self-reported information by the relays" that consist of different metadata such as public cryptographic keys for relays authentication and "contact details" of the assigned IPs; and thus support sub-protocol versions for each relay, and Marquez [34] stated that new descriptors are generated every 24 h. The computation of descriptors was announced in 2013 by Biryukov et al., as below:

descriptor-id = H(public-key-id || secret-id-part)
secret-id-part = H(descriptor-cookie || time-period ||replica-index).

2.4 Security and Privacy in Tor

According to TorProject [48–51], the combination of encryption parameters such as stream cipher, public key cipher, Diffie-Hellman protocol, and a hash function assure users anonymity and privacy in accessing the onion services. Furthermore, there are other fundamental security methods as follows:

Forward Secrecy Referred as Perfect Forward Secrecy (PFS), it is a fundamental security technique which delivers guarantee that the session keys wouldn't be compromised even if the server's private key is stolen. Introduced in the Second-Generation Onion Router research report by Dingledine et al. [22], the main idea behind this method is to safeguard past sessions against upcoming compromises of secret keys.

Router Selection Algorithm Initially, it's chosen randomly for the path of routers that the traffic will go through but after research and further developments it implemented four key properties:

1. one router cannot exist more than once in a circuit, and no two routers belong to the same class B subnet
2. DAs appoint the so-called "flags" on the ORs where each flag is based on the routers "performance, stability and roles in the network" [2]
3. due to the de-anonymization attacks that happened between the client and the first node, the TorProject developed a selection between three "entry guard nodes" that would be assigned for period of 30–60 days, and moreover, utilised for all circuits within that time
4. the consequent selection of an OR depends proportionally on its offered bandwidth. [23].

In 2015, Alsabah and Goldberg described this process as highly comprehensive. Tor utilises a cascade buffer architecture to govern cells traveling through the circuit. It works when an OR receives a cell (either from other OR or OP, or from the server) it then directs the TCP connection from its output buffer to a subsequent OR input buffer. Then, the cell is encrypted or decrypted (based on the direction of the traffic), and placed on the First-In, First-Out queuing process. Furthermore, a scheduler is utilised to extract cells from the queuing mode to the 32 KB output buffer. Lastly, the cell traffic is sent to the kernel TCP send buffer which proceeds it to the next OR or OP.

Blocking Resistance Due to censorship, there are some countries such as China and Iran where the government extract the OR information from the DAs and then block these routers to prevent access to Tor. Therefore, TorProject proposed a distinctive type of router called a "bridge" which can be utilised by the OP to act as first (entry) node in the circuit [2]. As claimed in Wang et al. [54] the main difference between them and entry guards are that the "bridges" are not listed in the DAs to prevent from de-anonymization and enumeration attacks. Except IP addresses blocking, the governments are attempting to block the traffic flows too. So do prevent this, Tor bridges utilise so-called "pluggable-transport" extensions called *obsf4* that hide their traffic and camouflage it to appear as another protocols or applications (such as WhatsApp, SMTP, HTTP).

Traffic Throttling It is a method that limits the bandwidth on the Tor network. According to Alsabah and Goldberg [2], it delivers to the ORs the ability to regulate the congestion and traffic overload. Furthermore, they stated that there are some "levels of throttling" to threshold the data accessed on the network and as follows:

- Rate-Limiting: ORs adopt rate-limiting algorithm which allows them to control the amount of the bandwidth that one OR can spend on Tor.
- Circuit Window: Circuits in the Tor network initially adopt "*end-to-end window-based flow control algorithm*" [2] which constrains the amount of the data at all stages the cells going through. Thus, the algorithm establishes a circuit window of 1000 cells between OP and exit relay. Moreover, it works as each time the data is sent, the window size is lessening by 1, until it reaches 0. Furthermore, when the OP or exit relay receives 100 cells, it sends an acknowledgement cell to any end of the circuit. This cell is called *circuit_sendme*, and when is sent to all ends, the OR increases the window size by 100 cells.
- Stream Window: TCP stream flow algorithm which acts similar to a Circuit Window. This time the size of the window is 500 cells and any time 50 cells are received by the OR cause stream window size increments by another 50 cells until it reaches 500 [2].

2.5 Advanced Security and Privacy in Tor

Descriptors Protection The onion services descriptors are secured by two layers of encryption:

1. delivers confidentiality against the adversaries who do not know the public key address of the certain onion service;
2. provides client authentication and safeguards against the users that do not have valid credentials.

Public Key Cipher Tor is utilising RSA with 1024-bit keys for authentication and key exchange during SSL/TLS circuit establishment. According to the live-term period they have, there are three types of RSA keys:

- A long-term "Identity key" utilised to digitally sign documents, and to determine the node entity.
- A medium-term "Onion key" which decrypts encryption layers when circuit extension is requested. This key is rotated each time the path is extended.
- A short-term "Connecting key" utilised to establish a TLS connection. This key is required to be rotated/regenerated at least once a day.

Stream Cipher Tor Project is utilising AES-128 in counter mode, with IV 0 bytes, or AES-256.

Digital Signature Curve25519 group and the Ed25519 signature formats protect the integrity and authenticity of the message by adding tamper-resistance [48–51].

- Curve25519 is a medium-term *ntor* "Onion key" utilised to control RSA onion key handshakes when circuit expansion is requested. According to TorProject [48–51], the combination of both keys successfully protects from stolen or tampered key pairs during exchange.
- Ed25519 handles three different sub-keys: (1) A long-term "master identity key" which is utilised to sign the (2) key. Lovecruft et al. [33] published a v3 onion specification protocol document, accessed on https://gitweb.torproject.org/torspec.git/tree/guard-spec.txt, that this key must never be changed and for security measurements "*it must be kept offline*"; (2) A medium-key "signing key" that signs practically everything else in the Tor network. It is signed by the "master identity key", kept online and a new one should be produced regularly; (3) A short-term "link authentication key" utilised to verify the link handshake. These keys are signed by the medium-keys and are generated regularly as well.

DH Key Exchange It is utilises generator (g) of 2 and for the modulus (p) it is utilises 1024 safe prime.

Hash Function TorProject [48–51] declared that they are using SHA256 and SHA3–256 in someplace.

2.6 Advantages and Disadvantages of Tor

As was mentioned earlier, that Tor is the most widely utilised anonymous tool, and extensive research papers are there on the advantages and disadvantages of using it. A Tor survey paper from Saleh et al. [44] very effectively summarizes the pros and cons of the Tor network and gives a side by side comparison on blockchain (see Table 1).

Although Tor is quite a new technology, it has undergone tremendous transformation over recent years. Together researchers and developers are using its open-source code to work towards creating more protection methods that prevent Tor users from de-anonymization. However, the development of the technologies and often utilising newly made protocols and applications, hide huge security challenges caused by broadening the attack surfaces [16]. Table 2 summarizes current attack vectors and how they affect the different class of Tor structure.

Moreover, the growing usage of the onion services makes it more an attractive 'bite' for the hackers. According to the Biryukov and Pustogarov [8], Bitcoin transaction through a Tor browser is not safe and secure method. Since the Tor network is utilised daily from more than 3–4 million users and there are more than 100,000 Bitcoin transactions, it is visible the huge scope of scalability and use cases of both technologies. Therefore, it is obvious that even a small design weakness in one of them can lead to tremendous issues and losses of identities, money and maybe more serious outcomes such as loss of life [1].

Table 1 Advantages and disadvantages in Tor and Blockchain

Technology	Pros	Cons
Blockchain	Decentralization	Scalability & storage
	Distribution	51 % majority attack
	Security	Prone to be compromised
	Privacy	
	Integrity	
	Reliability	
	Availability	
	Transparency	
	Cost-efficiency	
	Fault-tolerance	
	Audit & accountability	
	Stability	
Tor	Anonymity	Prone to zero–day attacks
	Security	Low-latency
	Privacy	Monitored and track by law authorities
	Open-source code	Deployment of incompatible with anonymity applications
	Prevent censorship	
	Free software	
	Support *.onion* domains	
	Compatible with Blockchain & VPN	

The botnet attack weaknesses are evaluated in the research from Nicholas in 2014. According to the research, there is still a huge challenge which Tor developers urgently need to address or solution.

3 Blockchain

Within recent years, blockchain has been tremendously developed and become a broadly utilised technology. According to the blockchain survey from Deloitte [21], more than 53% of the responders claimed that in 2019, blockchain became an urgent priority for their organisation. However, these are appetite statistics for the hackers who broaden and strengthen their attack vectors. The number of stolen wallets and compromised online digital systems keep growing simultaneously with growing the importance and the usage of Blockchain. So, the facts are represented in statistics of the BCSEC on the blockchain attack events in 2018, according to which, more than 2 billion dollars of economic losses caused by blockchain security weaknesses (Fig. 12).

Blockchain technology is still in the early stages of fast improvement, and therefore its security is behind the speed and level of the expansion of the new

Table 2 Displays the attack vectors and how they are associated to the Tor systems

	Threat	Client	Onion service	Tor	Application	Relays
Tor structure	Denial of service		✓	✓	✓	
	Low-resources routing	✓		✓		✓
	Botnet			✓		
Peer-to-Peer system	Cell flooding DoS attack			✓		✓
	Tor cells manipulation attack	✓				✓
	Traffic and timing correlation attack	✓			✓	✓
	P2P information leakage	✓	✓			
	Off-path MitM	✓	✓			✓
	Sniper attack	✓				
	Congestion attack		✓	✓		
	Exploiting routing algorithm	✓	✓	✓		✓
	Tor guard selection attack	✓				✓
	Bridge discovery	✓	✓	✓		
Application	Torben	✓	✓		✓	
	Replay attack	✓			✓	
	Shaping			✓	✓	

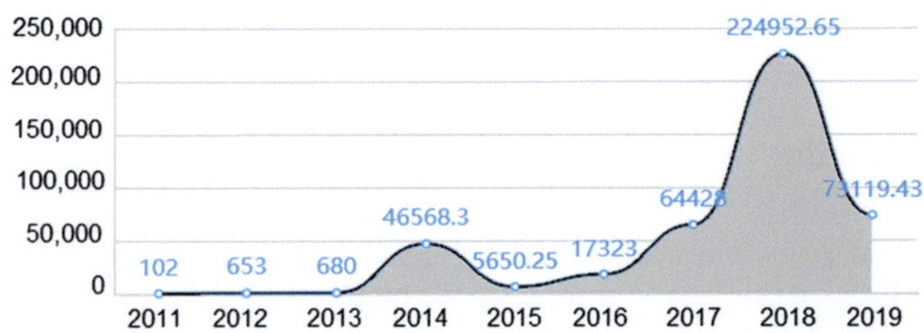

Fig. 12 Economic losses caused by blockchain security weaknesses (ten thousand dollars) [55]

century innovation and techniques. The risks could occur either from external or internal applicants. The increase in the popularity of blockchain obliges new requirements in terms of security and privacy protection of the transactions. Moreover, since the attack surface against it became more complex, it sets new challenges to the current security solutions utilised by blockchain. Compromising of transactions in transit, breaking authentication mechanisms, user account theft, and so on are the reasons of insistent establishment and development of new security and privacy solutions. The hash function SHA-256 and the encryption algorithm Elliptic Curve Cryptography utilised in blockchain are still safe until quantum computing development [55]. Moreover, NIST [40] announced what could be the impact of this innovative technology on the most utilised common cryptographic algorithms.

3.1 Blockchain Development

Blockchain is a current solution of secure computing in a live-network system without a Central Authority (CA). Not just organizing the transaction records into a hierarchical method but also secure storing them by using hash and cryptographic algorithms. The main features behind the design of the blockchain architecture are *"decentralization, tamper-resistance, safety and reliability"* [54].

Although the concept of cryptography-chained blocks was introduced first in 1992 and the usage of Merkle trees as an effective enhancement of the hash chain evaluated by Bayer, Haber and Stornetta, the Bitcoin was announced in 2009 as a peer-to-peer, distributed digital system that adopted blockchain as a secure ledger for its online transactions. For more than 10 years Blockchain developed from digital currency (Blockchain 1.0), then moved to smart contracts (Blockchain 2.0), and currently Blockchain 3.0 as a high security-level innovative technology.

According to Zhang et al. [57], the annual profits of blockchain-based applications world-wide barely get to the $2,5 billion in 2016 while it is expecting to reach 20 billion by 2025, with an annual growth rate of 26%. Furthermore, in their research it is stated that governments have published white papers and technical reports of blockchain to contribute positively with the distribution and enhancement of the new century technologies. A report issued in 2016 from the UK chief scientific adviser Sir Mark Walport [53], focuses on analysis and planning the better use cases for the future of distributed ledger technologies. Moreover, the European Central bank published papers on the advantages of using distributed ledger for secure transactions. Financial institutions, consulting companies and IT vendors such as Citibank, HSBC, Microsoft, IBM, Cisco, etc., claimed that blockchain is working as a secure and distributed ledger that stores all online transactions "effectively, persistently, and in a verifiable manner". Therefore, there is an increase in investments and researches of the three main next-generation technologies such as blockchain, artificial intelligence (AI) and big data [20].

3.1.1 What Is Blockchain?

Moubarak et al. [39] stated that a blockchain is a very fascinating technology which is rapidly being adopted and still under development from industries such as the Internet of Things, healthcare, smart energy, retail, etc. Blockchain is a "secure, distributed, peer-to-peer environment" [39] operating on the concept of storing and sharing transactions which can be reviewed and read/write from almost

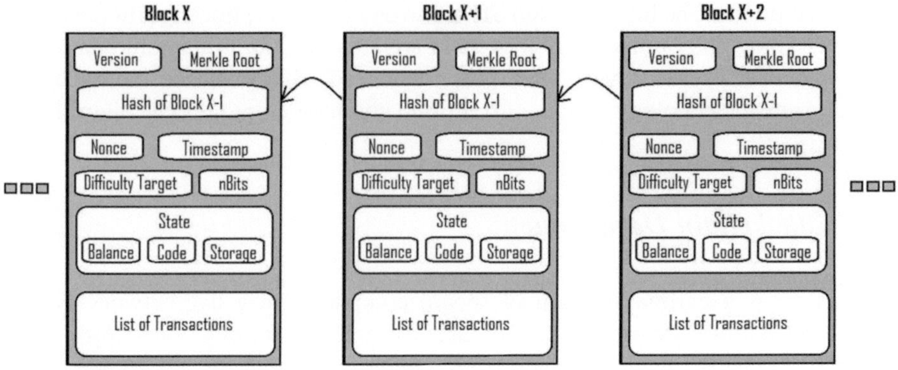

Fig. 13 Blockchain structure

anyone who participates in this network chain. Moreover, the features such as usage of cryptographic schemes and consensus algorithms cause fault tolerance and reliable platform for decentralized and trustful sharing sensitive information, such as transactions, over the Internet. Figure 13 gives an overview of the structure and components of Blockchain.

Precisely, one block in the chain consists not only of transaction records, but also keeps the hash value of its block + the hash value of the previous block that serves as a cryptographic bond between both the blocks. Wang et al. [55] stated that the integrity is guaranteed so when a block is added to a chain it cannot be altered or compromised due to its connections to the all blocks in the network.

Blockchain components can further be described as.

- ***Version*** – Identifies the suite of block validation procedures to follow [54].
- ***Merkle Root –*** It is a hash function computed from the sum of all transactions within the certain block. "*Its main purpose is to calculate the hash of a block from a hash of his sons*" [46].
- ***Nonce –*** It is a 4-byte field that begins with 0 and grows each time the hash has been calculated. Moreover, it is a variable incremented by the consensus algorithm. Utilised to prevent double-spending attacks [31].
- ***Timestamp –*** A record of the current time in seconds since January 1, 1970 [52].
- ***Difficulty Target*** – Represents how difficult the current target makes it in claiming the validity of a certain block. Moreover, the hash is sized in bits, where the lower the target in bits is, the more difficult it is to compute a corresponding hash.
- ***nBits*** – Denotes the goal maximum of a valid block hash [46].

3.1.2 Limitations of Blockchain

Blockchain is very secure and consistent; however, it has not been immune. Since it is regulated by computers, and also nowadays potentially misunderstood or perhaps utilised unnecessarily, blockchain hides some limitations and misconceptions [40]. This section highlights some of them below:

- Since the technology is distributed, it means that the transactions and accounts of everyone on the blockchain network are visible.
- The fact that there is no central authority means that there is no one you can call in case of something wrong happen.
- The security and consistency of the system hang on the blockchain code and its math functions.
- In terms of ownership, they are not completely decentralized, announced by NIST in their internal report *"Blockchain Technology Overview"* in 2018. According to them, the permissionless (public) networks are controlled mainly by blockchain users, publishing nodes and software developers. Meanwhile, the permissioned networks (private/consortium) are usually configured and run by an owner.
- The use of blockchain doesn't address the habitual cyber-security risks that necessitate precise and practical risk management process. Several of these "habitual" threats implicate human interaction. Consequently, a vigorous cyber-security program is crucial for protecting the participant's sensitive information, especially since the hackers could be the application developers or common users with privileged access.

3.2 Blockchain Security and Privacy Challenges

Baker and Steiner [5] published a paper that critically deliberates the challenges of blockchain in regard to its security and privacy. Nevertheless, they focused mostly on the attacks towards the application level of particular consensus algorithms.

According to the Saad et al. [43], blockchain is a modern live-development technology where everyday developers and analysts are working together to minimize the challenges and prevent the attack surfaces on blockchain distributed ledger system.

Since the requirements for mining (adding) a new block in this consensus algorithm depends on the stake (deposit) propose. Firstly, announced by Eyal and Sirer [27], and then cited by Saleh et al. [44], the mining process could lead to highly sophisticated attack surfaces against blockchain. Table 3, summarizes current attack vectors and how they affect the different class of systems.

Furthermore, numerous studies have been undertaken in order to classify and expose the attack surfaces, the limitations and weaknesses of blockchain networks; however, with the emergence of new technologies and with the development of

Table 3 Attack vectors associated to blockchain systems (adversaries attack methods and their corresponded target systems)

	Attacks	Blockchain	Miners	Mining pools	Application	Users
Blockchain structure	Orphaned blocks	✓	✓	✓		
	Forks	✓				
Peer-to-Peer system	BGP hijacks		✓	✓		✓
	DNS hijacks		✓	✓		✓
	Majority attack	✓	✓		✓	
	Selfish mining	✓	✓	✓		
	DDoS attack	✓	✓	✓	✓	
	Eclipse attack		✓			✓
	Timejacking attack		✓	✓	✓	
	Consensus delay		✓	✓		✓
	Finney attack		✓	✓		✓
Application	Wallet theft				✓	✓
	Blockchain ingestion	✓				
	Double-spending	✓				✓
	Cryptojacking				✓	✓
	Smart contract DoS	✓			✓	✓
	Reentracy attack				✓	✓
	Replay attack	✓		✓	✓	✓
	Overflow attack				✓	✓
	Balance attack				✓	✓

blockchain into version 3, the attack landscape highly increases which creates new security and privacy challenges [44].

Additional challenges for blockchain developers appeared when GDPR has been taken place in 2018. Moreover, there has been concerns and questions of how the data is controlled, stored and utilised, and what are the roles and responsibilities of blockchain users and third parties.

3.3 Blockchain Advanced Security and Privacy

According to Zhang et al. [57], the combination security practices such as Hash chain, Merkle tree, digital signature with consensus mechanisms, helps blockchain minimize the attack surface of Bitcoin. Hash Chained Storage Is a blockchain method that contains of two blocks – Hash Pointer and Merkle Tree.

Hash Pointer It is a cryptographic hash result indicating the address of the data stored in the blockchain. The most utilised cryptographic hash function in blockchain nowadays is SHA-256 due to its pre-image resistance (one-way function) and collision resistance security properties [40]. Moreover, they stated that the hash function is 32 bytes represented as a 64-hexadecimal string. So, blockchain ID addresses represent 34 characters hash of the public key. Zhang et al. [57] summarized two core purposes of utilising the hash pointers:

1. assess if the data has tampered;
2. the link between the blocks.

Each block only knows the address of its predecessor block. The hash of the stored block is freely verified by blockchain users to prove whether or not the block is tampered. This main feature prevents the adversary to write on a block because they have to change the hash pointers of the all previous blocks in the curtain chain. According to Anwar [3, 4], this feature delivers "an extra layer of protection and prevent any type of violations". Thus, an anniversary wouldn't be able to forge the data in the beginning of the chain where the genesis (origin, main) block operates.

Merkle Tree Expressed as a binary search tree, it is utilising hash pointers in order to link "tree" nodes together [57]. It operates on the concept of "parent–children" nodes where the parent nodes are on the top level of pair of children nodes that are on the lower-level. The Merkle Tree algorithm creates a new data node for each two lower level nodes which consist of the hashes of both nodes. This process is iterating until it reaches the genesis block. The main advantage of this blockchain technique is preventing information from tampering by following down utilising the hash pointers. That's true due to the reason that the adversary has to change all blocks to the bottom of the tree which is easy to be recognized and determinate by the Blockchain security teams. Moreover, it verifies rapidly and efficiently the membership of genuine nodes by representing them in a "root" tree.

Digital Signature Blockchain relies on research results of cryptography, which the heart of the security and privacy methods which delivers the CIA of the data blocks. Saleh et al. [44] announced three main components that forms a digital signature scheme:

1. key generation algorithm which creates two keys – first one private, utilised to sign a message and kept privately, the second one is public, utilised to verify if the message has signature signed with its corresponding key.
2. The next component is the signing algorithm. It generates a digital signature on the input of the message by utilising private key.
3. The last component is the verification algorithm. It combines the digital signature, a message and its public key, and verifies the signature utilising the public key. As an output, this algorithm returns a Boolean value. According to the Saleh et al. [44], a secure and precise digital signature is the one which is verifiable and effectively unforgeable.

Elliptic Curve Digital Signature Algorithm (ECDSA) Hanke et al. [29] stated that until the time of quantum computing is not broadly used, the ECDSA is still safe and will remain as a secure encryption algorithm for a few years. It is a successor of elliptic curve *"secp256k1"*, delivering 128 bits of encryption and it's been proven as robust against forgery attacks [14].

Public Keys The most relevant responsibility of the public keys is to maintain the legitimacy of the message by using the (Public Key Infrastructure) PKI. It operates as the one who writes the message signs it with his/her private key and then sends it to the receiver who uses the senders' public key to verify the message. Thus, the public keys are linked to the identities of entities and stored in the CA.

Consensus Due to its decentralization to add a new block to the chain, each participant has the choice whether or not to add to their own copy of the blockchain ledger. The main purpose of it is to search an agreement upon a single state from at least 51% of the network and to prevent dishonest nodes and malicious behavior [57].

There are various security and privacy techniques utilised by blockchain to keep the anonymity of the transactions and to protect the users' identity.

1. **Anonymous Signatures**

The most typical types of these signatures are Group and Ring signatures:

Group Signature Cryptographic scheme established in 1991 by Chaum and Heist. Within a group, any of the participants can sign a message for the group with his/her private key and then any member of this group can check and verify with the groups' public key whether or not the message is signed by a particular group member. The group signatures require a group manager to act as an authority that setup a group by adding and removing participants, handling events and so on. Therefore, these signatures apply in the consortium blockchain networks. Recently, the massive Chinese data exchange provider called Juzix [32] employed group signatures on their platform to enhance the current security measurements.

Ring Signature Zhang et al. [57] announced two main differences between the ring and group signatures:

1. There is no central authority which means even on a case of debate the real identity of the member cannot be extracted;
2. any member can create a "ring" by him/herself without additional requirements and resources.

As the number of the participants using ring signatures in one group increase, it becomes harder for a hacker to reveal the real identity of the participant. Since there is no signature manager, it applies in the public Blockchain networks.

2. **Homomorphic Encryption (HE)**

It is a strong cryptographic method which executes certain types of computation directly on the ciphertext, instead first on the plaintext mode. This technique addresses the limitation of the privacy protection on the public Blockchain and "delivers ready access to the encrypted data for auditing and other purposes" [57].

3. **Mixing**

As mentioned earlier, Blockchain is not anonymous, therefore, to prevent the leakage of information due to the association of pseudo-address and user's identity, a so-called "mixing" method system is developed. It functions as a random exchange of user's coins with another user's so doing obfuscation of the ownership of the account. Moreover, Zhang et al. [57] described two types of mixing services and evaluated their properties.

Mixcoin Established by Bonneau et al. in [13], it guarantees anonymous payment in Bitcoin. It works as mixing all users coins simultaneously and moreover, utilises an accountability mechanism so do detect stealing from the wallets.

Joint Payment CoinJoin. MaxWell [36] proposed this method as a concept of "joint" payment. The last is done by joining together transaction payments from different users in order to reduce the possibility of linking between the transactions and users. This happens as users negotiate with which transactions they want to combine with the central trusted servers. However, a single compromise over this central authority can lead to the disclosure of all logs and transactions that users joint with.

4. **Non-Interactive Zero-Knowledge Proof (NIZK) Proof**

NIZK is an enhanced privacy-protecting technique announced in the 1985 by Goldwasser, Micali and Rackoff. Its main idea is an interaction between the certifier and verifier in the Blockchain with zero-knowledge about each other. Additionally, it means that a user privately generates input and broadcasts it on the network for an output proof, without revealing any disclosure information about themselves. Thus, the other users interact and trust the output they perceive without any knowledge or information about the owner.

Advanced and highly effective application of zero-knowledge is presented in 2013 by Bitansky et al., called zero-knowledge Succinct Non-interactive Argument of Knowledge (*zk-SNARK*) proof which nowadays serves as core security and privacy technique adopted from *Zcash* protocol introduced in 2013 by Ben-Sasson et al.

Ethereum adopted *zk-SNARK* proof verification and called it *"baby" ZoE* from Zerocash over Ethereum [57]. Thus, this contract accepts a user to store non-disclosed amount of ETH units without revealing any information, but only by showing "a serial number" as assurance in front of a Merkle tree.

5. **The Trusted Execution Environment (TEE) Based Smart Contracts**

An extremely useful method for enhancing confidentiality and integrity is a segregated execution environment which successfully prevent other software's and OSs to tamper or learn the condition of the application currently running on it. TEE finds application in the Intel Software Guard eXtensions (SGX) based on which Cheng et al. [18] introduced a trustworthy platform for confidentiality-augmented smart contract execution. Moreover, proposed by Zyskind, Nathan and Pentland in 2015 "Enigma ledger" utilises TEE to create a smart contract with the decentralized scoring algorithm ENIGMA [26].

3.4 Advantages and Disadvantages of Blockchain

Built with high complexity, delivers high trust, decentralized security and privacy online environment with minimum processing fees and without any errors. Blockchain technology not only resolves the issues in centralized systems however, it is a "gold" mine invention for the industries and governments around the world. Therefore, in Table 4 are some pros and cons of adoption of the distributed ledger platform, blockchain, where Moubarak et al. [39] presented a general comparison of the main features that both emerging technologies are utilising.

4 Methods to De-anonymise Tor Packets

Looking at the interest area of privacy and security of anonymous networks, more than 50% of Tor research is focused on the de-anonymization of the anonymous

Table 4 Basic comparison of Tor to Bitcoin DLT system. Derived from [39]

	Tor	Bitcoin
Main function	Anonymity system	Trustless digital cash system upon Blockchain
Control of network	Non-profit foundation (TorProject)	Community
Number of nodes	Fixed/limited	Open
Actors	Relays, bridges, peers	Peers
Network type	Distributed	Distributed
Incentivized	No	Yes
Anonymity mechanisms	Yes	No
Routing actors	Relays, bridges	No
Confidentiality	Yes	Yes
Integrity	Yes	Yes
Authenticity	Yes	Yes
Scalability	Limited	Yes
Trust	Yes	Yes
Faulth-tolerance	Yes	Yes
Data storage	No	Yes
Content Addr. networks	DHT	Node hash
Client P2P connections	Yes	No

network. Approximately 25% of the study papers are covering path selection mechanisms. Around 25% studies are related to the performance examination and enhancement mechanisms of Tor network. Moreover, only 9% of research papers have been able to identify the real IP address of the onion services. More than 90% of the de-anonymization Tor studies conducted attacks over its inherent vulnerabilities. However, insufficient research focuses on the exploited autonomous systems, servers, flag cheating and also, on the compromised blockchain applications and mechanisms after being implemented in the Tor network. A mindmap represents the security, privacy and anonymity related studies, shown in Fig. 14.

4.1 *Experiment Example to Determine Tor Packets from Common Packets*

A typical configuration to undertake an experimental environment to test anonymity can be setup as and suggested as following:

1. Kali Linux and Parrot Security virtual disk image (VDI) files were run under VirtualBox software, where the first refers to HOST and second refers to USER;
2. "*Host-only*" and "*Nat*" network adapters were assigned in the both Kali and Parrot OSs. The first network adapter provides internal (virtual) communication between both VMs, while the second network adapter delivers access to the WAN in order to be installed the needed software's and to be accessed the Tor Browser;
3. Under Kali Linux OS, the project created and configured onion web service hosted on a *.onion* domain and accessible only through Tor;

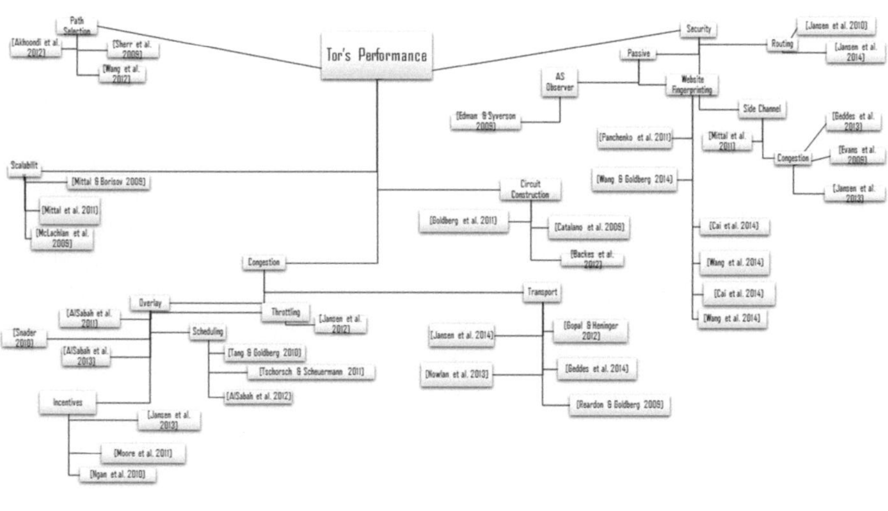

Fig. 14 Mind-map of relative research studies

Table 5 Hardware utilised for the study

Hardware	Virtual OS	RAM	Network
Asus GL553V	Kali Linux 2019.3 (64 bit)	8 GB	Up to 50 Mbps (Limited)
Microsoft Surface Pro 2	Parrot Security 4.6 (64 bit)	4 GB	Up to 50 Mbps (Limited)

Table 6 Software utilised for the study

Software	Version
Oracle VM VirtualBox	5.2 (64 bit)
Tor (Proxy)	8.5.5 (64 bit)
Wireshark (Network Sniffing)	3.1.0 (64 bit)
PcapXray (Packet Analyse and Data Extraction)	3.0 (64 bit)

4. The Tor Browser, Wireshark and PcapXray to be suitably installed and configured on the HOST;
5. The Tor Browser also installed on the USER in order to connect to the hosted *.onion* website.

Typical low-end hardware configurations and examples as per Tables 5 and 6 that are enough to undertake an experiment to determine Tor packet from common packets.

For the approach to test and experiment, then 5 possible approaches can be considered:

- HSDir memory extraction
- Onion address bruteforcing
- Primary data collection
- Network sniffing
- Data analysing and extraction

For the purposes of this chapter the research was based on primary data collection, network sniffing and data extraction. The incentive behind this choice can be seen as follows:

The HSDir memory extraction will be a reliable and successful method if there is sufficient network bandwidth and more resource power in order to obtain "HSDir" flag which allows to extract the memory and conduct additional data analyses.

Bruteforcing could be a successful method to derive an onion service address only performed over v2 protocol (16 characters string). However, it is impracticable to attempt bruteforcing over v3 addresses (56 characters string) due to time-consumption and lack of computation power.

Network sniffing is a method to capture packets through the network. This approach is reliable due to non-additional security settings set in the experiment environment. Moreover, it acts as a standpoint where further analysis could be achieved.

Data analyzing and extraction are valid methods if the previous approaches, as gathering knowledge by collecting primary data and network sniffing, are undertaken. Utilising sophisticated forensic tools, it's possible to extract details such as communication and device information from the packet. Although TorProject developed and implemented numerous amounts of anonymity and obfuscation techniques, this approach is able successfully to distinguish Tor packets from the common traffic.

Network Sniffing The object is to detect Tor network sniffing in order to find out is it possible to distinguish a Tor packet (default port 9051) from the commonly met protocol ports such as http (80), https (443), SMTP (25), etc. Here, the HOST marunningne is ran and Wireshark to capture Tor incoming packets from the USER.

Data Analysis and Extraction After successful capture of the Tor packet, it is then saved with a *".pcap"* extension. Thus, the HOST utilised a network forensic tool named *PcapXray* which details the captured packet by designing a network diagram and extracts data such as device identification, important communication and so forth.

Result Analysis The method clearly can detect and classify whether a packet is a Tor packet or not. These findings deliberate security and privacy concerns that should be further addressed in order to protect personable identifiable information (PII). Further research on how to prevent packet leakage and loss of PII could be looked at to develop a multi-layer solution of Tor and blockchain.

This chapter disclosed that it is possible to distinguish a Tor packet from the common traffic without much effort and utilising low-cost hardware. However, the study can be utilised as a tool for further researches into the Tor users' security. More than 80% of the onion services are still utilising v2 protocol [34], where the encryption methods are quite weak and perhaps could be broken. The v3 protocol is still in development, thus further researches can analyse its cryptographic algorithms and decide whether or not they are well applied. Because of its cryptographic advanced techniques and sub-protocols, it's recommended for urgent adoption of v3 Tor protocol where v2 is still utilised. However, the users are struggling to remember or keep a record of all these 56 characters in the address bars. Thus, this can lead to new security gaps and vulnerabilities such as phishing and reverse engineering attacks. Therefore, it is a necessity to develop a v3 custom names generation tool which creates much safer and easier methods for those who utilise onion services.

The following are typical stages to discover the Tor packet and to help visualize the stages in such an experiment as previously explained (Figs. 15, 16, 17, 18, 19, 20, 21, 22, 23).

Fig. 15 Access the Onion website and view the circuit established during the connection

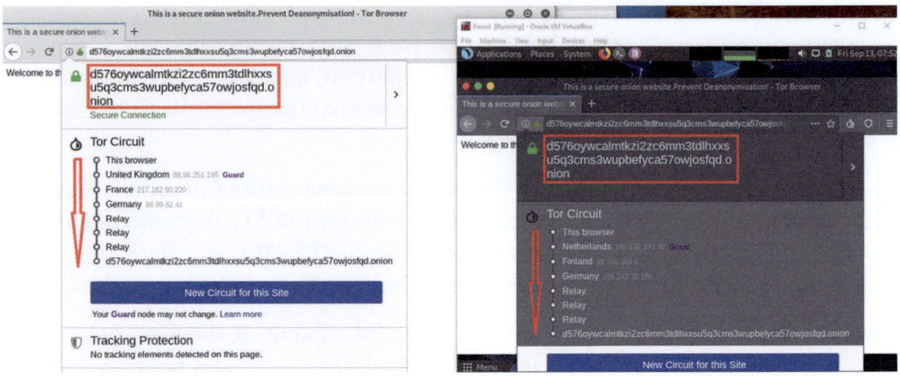

Fig. 16 Access the Onion website from the both VMs

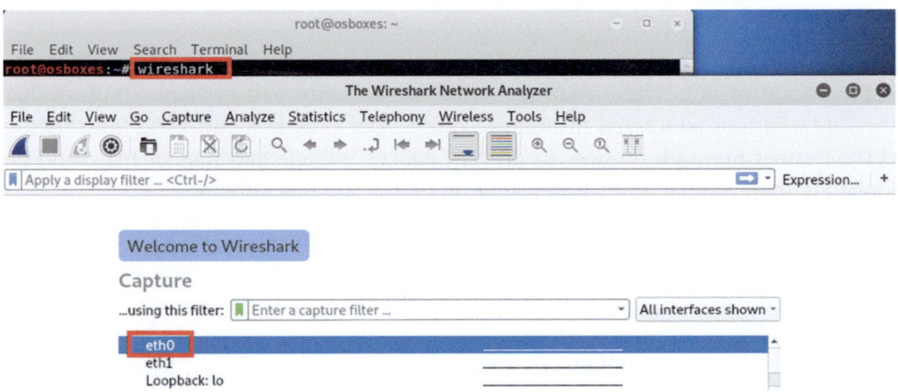

Fig. 17 Run Wireshark network sniffing tool on Eth0 Nat interface to capture the incoming packets

5 Conclusions

This chapter looked at the privacy and security of anonymous technologies such as Tor and blockchain and in particular their privacy and security concerns, gaps or vulnerabilities. The experiment and studies takes forward utilised classification, quantification and comparative evaluation of multiple study papers covering Tor and Blockchain. It appears, in the best of knowledge, that there was not any other research papers that perform a deep and comprehensive analysis in the state of

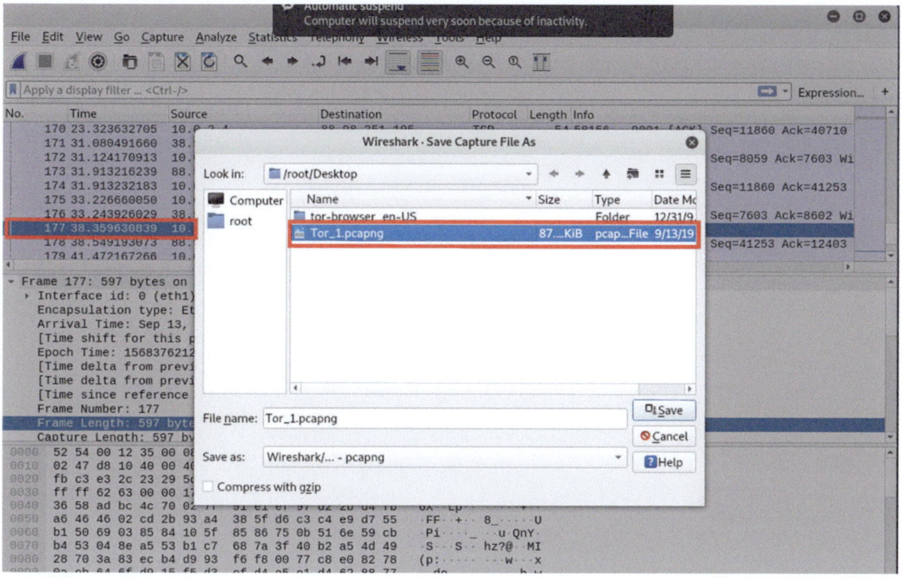

Fig. 18 Capture data packet and save it as .pcapng file

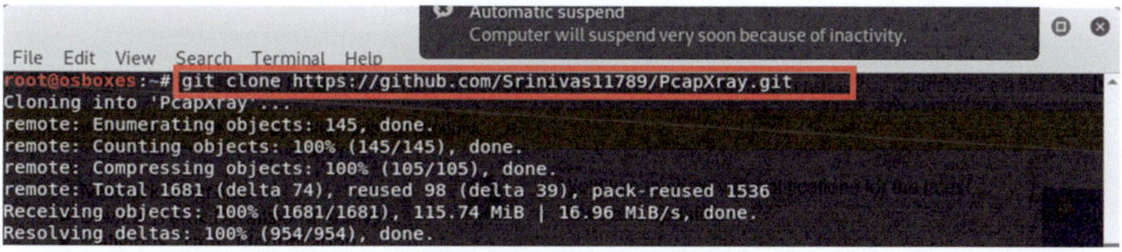

Fig. 19 Clone PcapXray forensic tool repository from Github.com

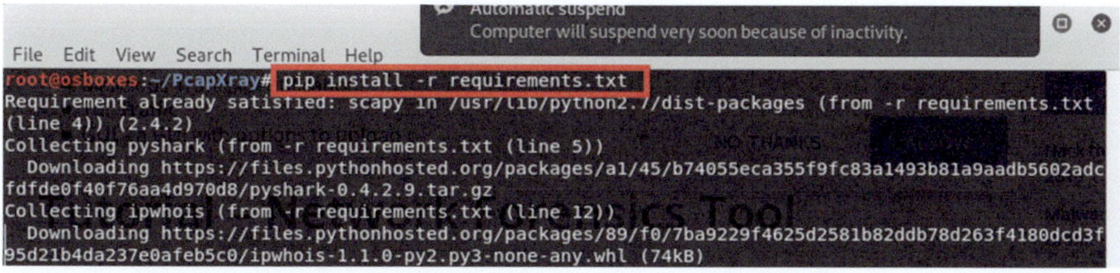

Fig. 20 Install the PcapXray Requirements.txt file by utilising python package installation utility – Pip

security in both the emerging technologies of Tor and blockchain. Although they assure security, privacy, and particularly Tor – anonymity, the attack surfaces that lead to user's de-anonymization and compromise of CIA assets are increasing at an alarming rate. However, it's expected that various lessons gained from the

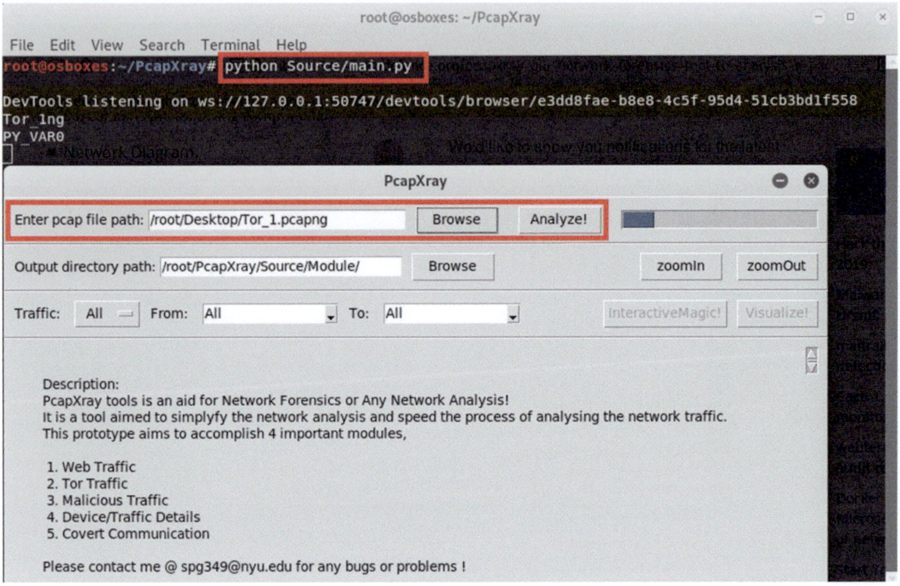

Fig. 21 Run the PcapXray tool with the python command and browse the path location to the Saved .pcapng file

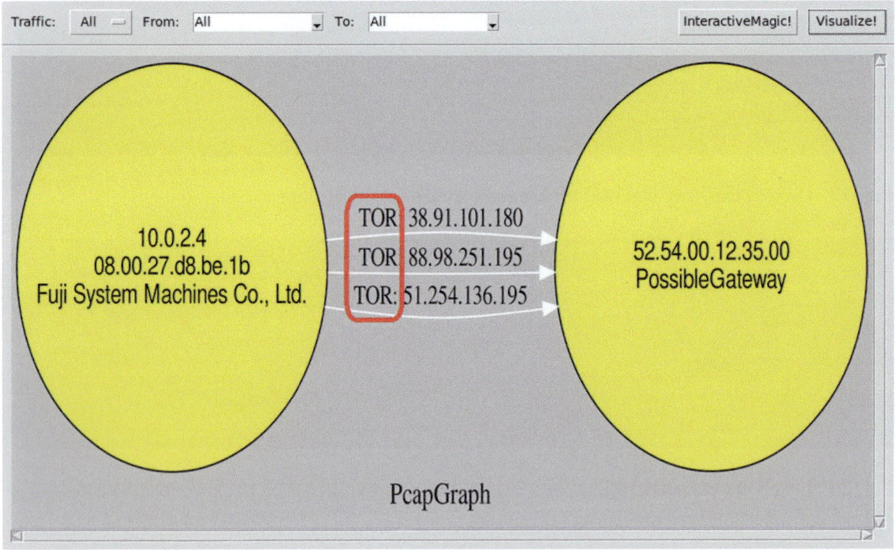

Fig. 22 PcapXray visualization of the Scanned .pcapng file due to extract the data from it

experience and usability design of the public Internet could be applied to the onion services. A smart home novel approache can be considered where the combination

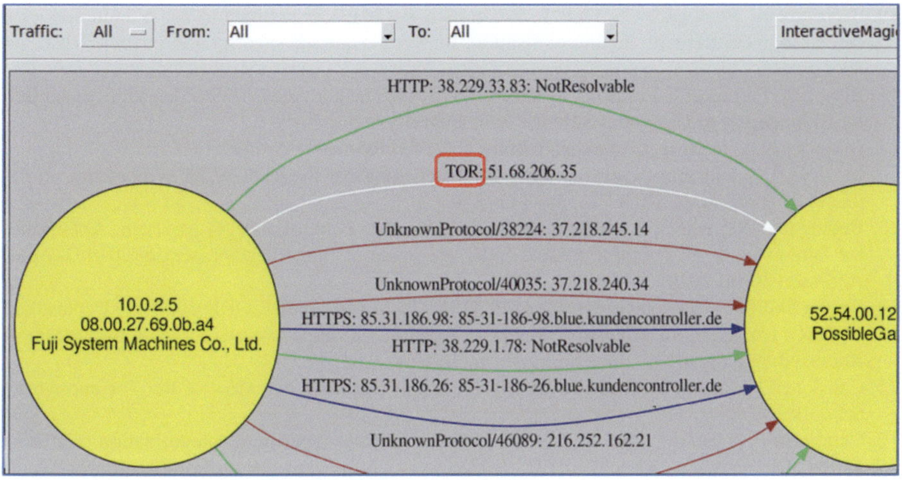

Fig. 23 PcapXray visualization after the anonymity technique "obsf4" was switched on

of IoT, Blockchain and Tor will work together for a better and secure future. Thus, a platform which reliably and successfully implements the advanced security properties and techniques utilised in Tor and Blockchain could be the new "panacea" against cyberattacks and potential leakage of PII.

References

1. Al Jawaheri H, Al Sabah M, Boshmaf Y, Erbad A (2019) Deanonymizing Tor hidden service users through Bitcoin transactions analysis. [online] Available at: https://arxiv.org/pdf/1801.07501.pdf. Accessed 2 Sept 2019
2. AlSabah M, Goldberg I (2015) Performance and security improvements for Tor: a survey. [ebook] Available at: https://eprint.iacr.org/2015/235.pdf. Accessed 25 July 2019
3. Anwar U (2017) Blockchain: anonymisation techniques within distributed ledgers
4. Anwar H (2018) Consensus algorithms: the root of the Blockchain Technology. [online] 101 Blockchains. Available at: https://101blockchains.com/consensus-algorithms-blockchain/#2. Accessed 18 Aug 2019
5. Baker J, Steiner J (2015) Blockchain: the solution for transparency in product. [online] Provenance. Available at: https://www.provenance.org/whitepaper. Accessed 22 Aug 2019
6. Bayer D, Haber S, Stornetta W (1992) Improving the efficiency and reliability of digital time-stamping. [ebook] Available at: http://citeseerx.ist.psu.edu/viewdoc/download?doi=10.1.1.71.4891&rep=rep1&type=pdf. Accessed 1 Aug 2019
7. Ben-Sasson E, Chiesa A, Genkin D, Tromer E, Virza M (2013) SNARKs for C: verifying program executions succinctly and in zero knowledge. [ebook] Available at: https://eprint.iacr.org/2013/507.pdf. Accessed 28 July 2019
8. Biryukov A, Pustogarov I (2015) Bitcoin over Tor isn't a good idea. IEEE Symp Secur Priv 2015:122–134
9. Biryukov A, Pustogarov I (2015) Proof-of-work as anonymous micropayment: rewarding a Tor relay? [ebook] Available at: https://eprint.iacr.org/2014/1011.pdf. Accessed 15 Aug 2019

10. Biryukov A, Pustogarov I, Thill F, Weinmann R (2013) Trawling for Tor hidden services: detection, measurement, de-anonymisation. In: Symposium on security and privacy. IEEE, pp 80–94
11. Bissoli A, Farinacci F, Prosseda A, Veterini S (2017) Deanonymize Tor hidden services master. In: Engineering of Computer Science. Web Security and Privacy
12. Bitansky N, Canetti R, Chiesa A, Tromer E (2013) Recursive composition and bootstrapping for SNARKs and proof-carrying data. In: Proceedings of the 45th ACM symposium on the theory of computing, STOC'13, pp. 111–120
13. Bonneau J, Narayanan A, Miller A, Clark J, Kroll J, Felten E (2014) Mixcoin. Anonymity for Bitcoin with accountable mixes. [pdf] Available at: https://eprint.iacr.org/2014/077.pdf. Accessed 20 Jul 2019
14. Brown D (2000) The exact security of ECDSA, Technical report CORR 2000-54. Department of C&O, University of Waterloo, Waterloo. Available at: https://www.cacr.math.uwaterloo.ca. Accessed 19 Aug 2019
15. Çalışkan E, Minárik T, Osula A (2015) Technical and legal overview of the Tor anonymity network
16. Cambiaso E, Vaccari I, Patti L, Aiello M (2017) Darknet security: a categorization of attacks to the Tor network. [ebook] IEEE. Available at: http://ceur-ws.org/Vol-2315/paper10.pdf. Accessed 2 Sept 2019
17. Chaum D, Heist E (1991) Group signatures. In: Davies D (ed) Advances in cryptology – EUROCRYPT'91. Springer, Berlin/Heidelberg, pp 257–265
18. Cheng R, Zhang F, Kos J, He W, Hynes N, Johnson N, Juels A, Miller A, Song D (2018) Ekiden: a platform for confidentiality-preserving, trustworthy, and performant smart contract execution
19. Choudhary D (2018) The onion routing-the good and the bad. [pdf] Available at: https://www.researchgate.net/publication/327867486_The_Onion_Routing-The_Good_and_The_Bad/citation/download. Accessed 12 Aug 2019
20. Cuzzocrea A (2017) Multidimensional mining of big social data for supporting advanced big data analytics. In: 2017 40th international convention on information and communication technology, electronics and microelectronics (MIPRO). Springer, Opatija, pp 1337–1342
21. Deloitte (2019) Deloitte's 2019 global Blockchain survey. [online] Available at: https://www2.deloitte.com/content/dam/Deloitte/se/Documents/risk/DI_2019-global-blockchain-survey.pdf. Accessed 25 Aug 2019
22. Dingledine R, Mathewson N, Syverson P (2004) Tor: the second-generation onion router. Naval Research Lab, Washington, DC
23. Dingledine R, Mathewson N, Syverson P (2005) Challenges in deploying low-latency anonymity (DRAFT). [ebook] Available at: https://pdfs.semanticscholar.org/29d7/36eed9e71d4b1b0781ff30c7cecb1d6b7fa8.pdf?_ga=2.152482289.999236869.1568448972-1206205724.1551784981. Accessed 28 Aug 2019
24. Döpmann C, Rust S, Tschorsch F (2018) Exploring deployment strategies for the Tor network
25. Dredge S (2013) What is Tor? A beginner's guide to the privacy tool. [online] the Guardian. Available at: https://www.theguardian.com/technology/2013/nov/05/tor-beginners-guide-nsa-browser. Accessed 14 Jul 2019
26. ENIGMA (2019) Enigma – securing the decentralized web. [online] Enigma. Available at: https://enigma.co. Accessed 11 Sept 2019
27. Eyal I, Sirer EG (2013) Majority is not enough: Bitcoin mining is vulnerable. [online] Available at: https://arxiv.org/pdf/1311.0243.pdf. Accessed 5 Aug 2019
28. Goldwasser S, Micali S, Rackoff C (1985) STOC'85 Proceedings of the seventeenth annual ACM symposium on theory of computing. In: STOC '85 proceedings of the seventeenth annual ACM symposium on theory of computing. [online] New York, pp 291–304. Available at: https://doi.org/10.1137/0218012. Accessed 1 Sept 2019
29. Hanke T, Movahedi M, Williams D (2018) DFINITY technology overview series consensus system. In: 2018 proceedings of technology overview series. DFINITY, Stiftung

30. Ibarra J, Jahankhani H (2018) Cyber-physical attacks and the value of healthcare data: facing an era of cyber extortion and organised crime
31. Jesus E, Chicarino V, Albuquerque C, Rocha A (2018) A survey of how to use blockchain to secure internet of things and the stalker attack. In: Security and communication networks
32. Juzix (n.d.) [online] Available at: http://www.juzix.io/index_en.html. Accessed 12 Aug 2019
33. Lovecruft I, Kadianakis G, Bini O, Mathewson N (n.d.) guard-spec.txt – torspec – Tor's protocol specifications. [online] Gitweb.torproject.org. Available at: https://gitweb.torproject.org/torspec.git/tree/guard-spec.txt. Accessed 5 Aug 2019
34. Marquez J (2018) Tor: hidden service intelligence extraction. [pdf] Available at: https://pdfs.semanticscholar.org/76cd/e9c9fc3bb18e0c2b4fdbe023df07db1de9a2.pdf. Accessed 31 Aug 2019
35. Mathewson N, Wilson-Brown T, Johnson A (2017) Tor proposal 288: privacy-preserving statistics with Privcount in Tor (Shamir version). [online] gitweb.torproject.org. Available at: https://gitweb.torproject.org/torspec.git/tree/proposals/288-privcount-with-shamir.txt. Accessed 7 Aug 2019
36. Maxwell G (2013) CoinJoin: Bitcoin privacy for the real world. [online] Available at: https://bitcointalk.org/index.php?topic=279249. Accessed 29 Aug 2019
37. Monk B, Mitchell J, Frank R, Davies G (2018) Uncovering tor: an examination of the network structure. In: Security and communication networks.
38. Moore D, Rid T (2016) Cryptopolitik and the Darknet. Survival 58(1):7–38
39. Moubarak J, Filiol E, Chamoun M (2017) Comparative analysis of blockchain technologies and TOR network: two faces of the same reality? In: 2017 1st cyber security in networking conference (CSNet). [online] Rio de Janeiro, pp 1–9. Available at: https://ieeexplore.ieee.org/document/8242004. Accessed 7 Aug 2019
40. NIST (2018) Blockchain technology overview. Internal Report 8202. [online] NIST. Available at: https://doi.org/10.6028/NIST.IR.8202. Accessed 5 Sept 2019
41. NSA (2012) Tor Stinks https://edwardsnowden.com/docs/doc/tor-stinks-presentation.pdf. Accessed 5 Sept 2019
42. Nurmi J (2019) Understanding the usage of anonymous onion services. [pdf] Available at: https://tutcris.tut.fi/portal/files/18769092/TUNI_nurmi.pdf. Accessed 2 Sep 2019
43. Saad M, Spaulding J, Njilla L, Kamhoua C, Shetty S, Nyang D, Mohaisen A (2019) Exploring the attack surface of Blockchain: a systematic overview. [ebook] Available at: https://arxiv.org/pdf/1904.03487.pdf. Accessed 23 Jul 2019
44. Saleh S, Qadir J, Ilyas M (2018) Shedding light on the dark corners of the internet: a survey of Tor research. J Netw Comput Appl 114:1–28
45. Salo J (2012) Recent attacks on Tor. [ebook] Available at: http://www.cse.hut.fi/en/publications/B/11/papers/salo.pdf. Accessed 29 Aug 2019
46. Sayadi S, Rejeb S, Choukair Z (2018) Blockchain challenges and security schemes: a survey
47. TorProject (2015) The Tor browser. [online] Cdn.ttgtmedia.com. Available at: https://cdn.ttgtmedia.com/rms/pdf/Hiding%20Behind%20the%20Keyboard_Ch%202.pdf. Accessed 7 Sept 2019
48. TorProject (n.d.) The Tor project | privacy & freedom online. [online] Available at: https://www.torproject.org/. Accessed 22 June 2019
49. TorProject (n.d.) Torproject's git repository browser. [online] Gitweb.torproject.org. Available at: https://gitweb.torproject.org/. Accessed 18 July 2019
50. TorProject (n.d.). Tor's protocol specifications. [online] Available at: https://gitweb.torproject.org/torspec.git/tree/tor-spec.txt. Accessed 15 July 2019
51. TorProject (n.d.) Welcome to Tor metrics. [online] Metrics.torproject.org. Available at: https://metrics.torproject.org. Accessed 15 July 2019
52. Vidrih M (2018) What Is a Block in the Blockchain? [online] Medium. Available at: https://medium.com/datadriveninvestor/what-is-a-block-in-the-blockchain-c7a420270373. Accessed 20 Aug 2019

53. Walport M (2016) Government Office for science annual report 2016–2017. [online] Available at: https://assets.publishing.service.gov.uk/government/uploads/system/uploads/attachment_data/file/642307/Government_Office_for_Science_Annual_Report_final_v2_16-17.pdf. Accessed 3 Sep 2019
54. Wang H, Zheng Z, Xie S, Dai H, Chen X (2018) Blockchain challenges and opportunities: a survey. Int J Web Grid Serv 14(4):352
55. Wang H, Wang Y, Cao Z, Li Z, Xiong G (2019) An overview of Blockchain security analysis. In: Yun X et al (eds) Cyber security. CNCERT 2018. Communications in computer and information science, vol 970. Springer, Singapore
56. Winter P, Edmundson A, Roberts L (2018) How do tor users interact with onion services? [ebook] Available at: https://arxiv.org/pdf/1806.11278.pdf. Accessed 8 July 2019
57. Zhang R, Xue R, Liu L (2019) Security and privacy on Blockchain. [ebook] Available at: https://arxiv.org/pdf/1903.07602.pdf. Accessed 30 Aug 2019
58. Zyskind G, Nathan O, Pentland A (2015) Enigma: decentralized computation platform with guaranteed

Experimental Analyses in Search of Effective Mitigation for Login Cross-Site Request Forgery

Y. Shibuya, K. Mwitondi, and S. Zargari

Abstract Advancements in web applications and on-line services continue to stimulate business growth and other applications across the globe. Alongside these developments are the increasing cyber security risks and vulnerabilities, inevitably entailing mitigations. Web application vulnerabilities are security holes, which attackers may attempt to exploit, hence potentially causing serious damage to business, such as stealing sensitive data and compromising business resources. Since web applications are now widely used, critical business environments such as internet banking, communication of sensitive data and online shopping, require robust protective measures against a wide range of vulnerabilities. This work explores remediation methods – HTTP header verification, tokenisation and challenge-response authentication of vulnerabilities against login CSRF attacks. Experiments comprising of nine test cases with the three mitigation methods and three vulnerabilities are conducted to identify whether exploitation of vulnerabilities was able to bypass a mitigation method and how the mitigation behaved in web applications of virtual environments. Using techniques and specific scripts of simulated web applications, three mitigation methods are mapped to the exploitation of the three vulnerabilities in different settings in search of an optimal solution. Results indicate that the HTTP header verification was not successful in protecting users from clickjacking exploitation, while it was successful in protecting against XSS and CSRF attacks. Further, exploitation of the three vulnerabilities bypassed the tokenisation mitigation and XSS attacks were prevented by challenge-response authentication, although exploitation of clickjacking and CSRF defeated the mitigation. The significance of these results lies in the fact that different methods are effective or ineffective in different conditions and therefore no single solution can be considered

---

Y. Shibuya (✉)
Cybereason, Tokyo, Japan
e-mail: yuki.shibuya@cybereason.com

K. Mwitondi · S. Zargari
Faculty of Science, Technology and Arts, Sheffield Hallam University, Sheffield, UK
e-mail: k.mwitondi@shu.ac.uk; s.zargari@shu.ac.uk

H. Jahankhani et al. (eds.), *Cyber Defence in the Age of AI, Smart Societies and Augmented Humanity*, Advanced Sciences and Technologies for Security Applications, https://doi.org/10.1007/978-3-030-35746-7_12

as most appropriate for web applications. The study concludes that best practices can be sought through empirical and experimental studies, via which observation and analysis of behaviours of different solutions under different scenarios of attacks are conducted. Such experiments, designed to bypass mitigations, provide insights into robust and appropriate implementation approaches and, in the era of Artificial Intelligence and Big Data, they should be routinely and automatically conducted.

Keywords Challenge-response authentication · Clickjacking · Cross-Site Request Forgery (CSRF) · Cross-Site Scripting (XSS) · HTTP header verification · Login CSRF · Tokenisation

1 Introduction

Login Cross-Site Request Forgery (CSRF) implementation methods constitute an interesting focal point in the cyber security research community and several mitigation methods have been discussed highlighting a range of benefits and drawbacks [38]. Evidence in the literature shows that Hyper Text Transfer Protocol (HTTP) header verification, tokenisation and challenge-response authentication are effective solutions for CSRF, whilst exploitation of clickjacking, cross-site scripting (XSS) and CSRF in a non-login page can hedge the effectiveness of the mitigation under certain conditions [21]. However, the effectiveness and limitations of these solutions have not sufficiently been researched. This topic is critical, especially for web developers who need to ensure that an implemented control is as secure as possible. This study focuses on methods of implementing HTTP header verification, tokenisation and challenge-response authentication, as effective protection means against login CSRF attacks. Experiments were conducted to identify whether exploitation of vulnerabilities was able to bypass a mitigation method and how the mitigation behaved in web applications of virtual environments. The experiments comprised nine test cases with the three mitigation methods and the three vulnerabilities. The results indicate that the HTTP header verification was not successful in protecting users from clickjacking exploitation, while it was successful in protecting against XSS and CSRF attacks.

Further, exploitation of the three vulnerabilities bypassed the tokenisation mitigation and XSS attacks were prevented by challenge-response authentication, although exploitation of clickjacking and CSRF defeated the mitigation. The chapter is organised sections as follows. Section 1 focuses on the background and motivation of the study, research problem and objectives. Literature review is covered in Sect. 2 followed by the study methodology in Sect. 3, covering project experiment designs including specific scripts and mitigation validation techniques. Results and Analysis are presented in Sect. 4 followed by concluding remarks in Sect. 5, which highlights future research paths.

1.1 Motivation

Advancements in web applications and on-line services continue to stimulate business growth and other applications across the globe. Alongside these developments are the increasing cyber security risks and vulnerabilities, inevitably entailing mitigations. Web application vulnerabilities are security holes, which attackers may attempt to exploit, hence potentially causing serious damage to business, such as stealing sensitive data and compromising business resources. Since web applications are now widely used, critical business environments such as internet banking, communication of sensitive data and online shopping, require robust protective measures against a wide range of vulnerabilities. This work was motivated by the need to contribute to such protection.

1.2 Research Problem and Objectives

The main problem of this work is to uncover the effectiveness of implementing HTTP header verification, tokenisation and challenge-response authentication as mitigation methods for Login CSRF. It seeks to reveal the effectiveness by identifying their limitations by challenging the mitigation in a non-login experimental page using attacks such as clickjacking, XSS and CSRF. We set the following study objectives.

1. To analyse the current study about Login CSRF and its mitigation methods to understand advantages and drawbacks of each solution.
2. To assess the three solutions effectiveness for Login CSRF based on simulated web applications.
3. To identify the factors that may degenerate effectiveness of the mitigation methods.
4. To make recommendations of efficient implementation approaches of the three mitigations with consideration of uncovered restrictions.

2 Literature Review

There has been significant progress in addressing flaws in web applications such as CSRF as reported by Farah et al. [14]. CSRF can cause a user on a legitimate website to unintentionally perform undesirable actions, such as password change, money transfer and online shopping with the user's privilege [30]. CSRF can be exploited via several forms. For instance, attackers may exploit CSRF vulnerabilities on the login page to force legitimate users to log in as attackers when target users are deceived to send valid HTTP requests which attackers craft and include the attacker's credentials [33]. Without knowing an attacker's username or password, a deceived user will submit a valid login form via a crafted HTTP request intended to exploit login CSRF vulnerability. When a login CSRF attack is successful, sensitive information such as credit card data might be extracted if tricked users leave the data on web pages. Several solutions have been suggested to protect against Login CSRF attacks, reporting advantages and disadvantages of different solutions over others. Let us start by looking at the login CSRF.

2.1 CSRF and Login CSRF

CSRF vulnerabilities have been detected on famous websites including Skype, Netflix, Google and Ali Express [32]. By exploiting this flaw, an attacker forces a victim to perform unwanted actions with the target's privilege through a forged request containing parameters necessary for a specific action to accomplish. For example, a target user tricked with a non-technical method, such as phishing email attacks and social engineering, consciously or unconsciously executes attacker's malicious Hypertext Mark-Up Language (HTML) or Javascript codes included on a web page or an email attachment [30]. Typical CSRF attacks can be performed through an embedded image tag or link in an email or web page to automatically send a forged request [27]. Due to these characteristics, web applications may allow CSRF attacks to be successful if a website does not appropriately verify that a request is sent with user's consent. Login CSRF, a variation of CSRF [33], exploits CSRF vulnerability on a login page. With this vulnerability, a target is forced to log into a web application with an attacker's account, being deceived by a forged request to follow a malicious link containing the request with hidden attacker's credentials to automatically and unintentionally respond. If a target is unaware that he or she logs in as an attacker, the target is likely to store personal data or keep traces on the attacker's account. After the target user explores web pages using attacker's credentials and leaves personal information on the attacker's account, the attacker will be able to collect sensitive information, such as and credit card information. Sensitive bank account information of a target on an e-government website was able to be available for an attacker by exploiting Login CSRF vulnerabilities [36]. Contrary to CSRF, with which an attacker exploits a user's privilege, Login CSRF does not require the victim to have an active session. Because attacker's credentials are utilised in this attack, different methods should be necessary for mitigation of Login CSRF. Research on Login CSRF has provided three major solutions for vulnerability- HTTP header verification, Tokenisation and Challenge-response authentication.

2.2 *Mitigation for Login CSRF*

The Login CSRF vulnerability can be remediated by mitigation methods including HTTP header verification, Tokenisation and Challenge-response authentication according to the research. Different solutions have different techniques with advantages and limitations as mitigation as outlined below.

2.2.1 HTTP Header Verification Protection with Its Benefits and Restrictions

HTTP header verification can mitigate Login CSRF with HTTP header fields including header fields "Referer" and "Origin", which can verify a source of a request [36]. Because a "Referer" header value contains the previous page of a request, in order to identify a request source, a server can distinguish a legitimate login request from an attacker's forged login request. Sudhodanan et al. [36] show that since the "Referer" header identifies the source of a request, it is possible to reject the request unless it is originated from a trusted Uniform Resource Locator (URL). Because a proper request should have the "Referer" value of a particular URL, a forged request to exploit Login CSRF can be blocked with verification of the "Referer" header in Login CSRF attack scenarios. In fact, use of the "Referer" header is, in general, widely accepted on the client side because the field is denied by only small percentage around between 0.05% and 0.22% of HTTPS (HTTP Secure) applications, over which login requests ought to be communicated [23].

Since it can be assumed that Login CSRF exists on the login page, in which the request is encrypted with HTTPS, this result implies that the "Referer" field is relatively a reliable indicator of Login CSRF attacks because the header is unlikely to be removed. Despite the effectiveness of the protection, the "Referer" has disadvantages, mainly relating to privacy issues, as the field includes users' previous requests [12]. Thus, while the solution may be effective, the issue of user privacy must be carefully addressed. To address this issue, the original header is regarded as another effective field for the protection, dictating a source domain with which an HTTP request initiates without containing the entire path [40]. Because the "Referer" header can be improved by the "Origin" header, as the "Origin" sends a server only a protocol, a domain and a port over "POST" requests, the "Origin" header mitigates the drawback of privacy [6]. That is, by identifying a source server of a request whilst concealing request details, the "Origin" header overcomes the disadvantage of the "Referer" header.

However, implementation of the "Origin" header presents limitations to block Login CSRF attacks in many situations. For instance, a web application with the web domain of "bank.com" cannot block CSRF attacks from "broker.com/forum/" if "bank.com" accepts requests from the login page "broker.com/accounts/" because the "Origin" header excludes a path after the web domain [9]. In this scenario, a Login CSRF attack may be feasible if a malicious script is embedded in the forum page because the web application "bank.com" permits any accesses from "broker.com". A web application verifying the Origin header to prevent Login CSRF may have to create a separate subdomain for the login page and non-login pages, such as a forum page. Thus, it is also necessary to mitigate CSRF vulnerabilities in the non-login pages to implement the Origin header as protection. In summary, although the Referer header and the Origin header have been expected to allow only a legitimate request to prevent Login CSRF, the headers have issues including privacy and may be vulnerable to attacks in which CSRF vulnerabilities within a non-login page are exploited to bypass verification.

2.2.2 Tokenisation with Its Benefits and Restrictions

Tokenisation is another mitigation method for preventing Login CSRF. In this case, a web server generates an unpredictable token parameter as a secret random token and stores the value within a login request with which legitimate users can be identified [9]. Because of this characteristic, tokenisation can be effective for Login CSRF attacks. Login CSRF can be protected by tokenisation because a server can refuse illegitimate requests from attackers who attempt to exploit the vulnerability by evaluating a hidden validation token which is unassociated with a session and generated with a random value [39]. Several technologies, such as double submit cookies, can realise the implementation of the random token to block against Login CSRF. Double submit cookies, for example, can mitigate Login CSRF because a random value is generated as both a cookie and a hidden parameter to compare the two values when a login form is submitted [11]. Double submit cookies are effective because the two values must be identical. Unless the value of the cookie is identified, attackers virtually cannot predict a valid token to include within a forged request because the cookie value cannot generally be read or changed by an attacker from a different domain [30, 31]. Despite its effectiveness as a mitigation method, tokenisation has its limitations. For instance, a token can be stolen when an attacker persuades users to send a request with a token to attacker's web frame [9]. This exploitation may be feasible when the attacker's web content is injected into a target website to extract a valid token. An example of this exploitation is given by OWASP [30], that the token system can be defeated by XSS vulnerabilities because XSS attacks with XML HTTP "Request" can be exploited to sniff a token from a response to insert a malicious request. Tokenisation can be a secure protection for Login CSRF provided relevant vulnerabilities do not exist on a target web application.

2.2.3 Challenge-Response Authentication with Its Benefits and Restrictions

Challenge-response authentication is another mitigation method for Login CSRF because the request value generally cannot be guessed by attackers. The method requires users to send another secret value in addition to a username and a password.

The difference between tokenisation and challenge-response authentication is that a user needs to identify and send a value in challenge-response authentication while a token is automatically communicated in tokenisation [30]. CAPTCHA authentication, a method of implementing challenge-response authentication, requires a user to send an unpredictable token text or audio message displayed on the screen in addition to user's credentials [20]. Because this token is unshared with the other users and valid one time only, it is virtually impossible for attackers to guess the token. A study by Moradi and Moghaddam [25] regarding CAPTCHA explained the validity of CAPTCHA as mitigation of Login CSRF because attackers are unable to specify and send the correct value since the CAPTCHA image can be viewed by only legitimate user. Similar to CAPTCHA, OWASP [30] assures the one-time token is also an effective strategy for the defence. The one-time token can be realised by asking users to send a token which is valid only one time and communicated to users via email or a mobile application. Both these methods are expected to block Login CSRF attacks because the token is mostly unpredictable. However, some researchers have claimed that the authentication is inappropriate because attackers can bypass the mitigation in many situations. For instance, Moradi and Moghaddam [25] proved that a clickjacking attack, via which users are tricked to click an invisible link or button to send information to attackers, can be a decisive factor for bypassing Login CSRF. Exploitation of clickjacking may persuade a target user to solve a challenge of the authentication and send attacker's credentials [25]. Login CSRF attacks can be successful even if challenge-response authentication is implemented to mitigate the vulnerability when an attacker attempts to bypass the mitigation exploiting a clickjacking vulnerability.

2.3 Vulnerability Attacks Which Can Bypass the Mitigation for Login CSRF

As evidenced by research, the mitigation methods for Login CSRF including HTTP header verification, tokenisation and challenge-response authentication may be defeated if attackers exploit vulnerabilities of clickjacking, XSS or CSRF in a non-login page. These three vulnerabilities are discussed in detail with respect to attack characteristics and to identify how they can be exploited for the Login CSRF attack.

2.3.1 Clickjacking and Exploitation of the Vulnerability

The purpose of clickjacking exploitation is to craft appearance of a web page and force users to execute an activity which the attacker intends. Attackers exploit clickjacking vulnerability on a web application to persuade users to execute an invisible button, form or link on a page whose appearance is crafted to deceive the victims [37]. An attacker prepares a web page which loads a visible button and a transparent button covered by the visible button to persuade a target user to click the invisible button although the target user seemingly clicks the visible button [37]. Based on this theory, a malicious web page which is apparently harmless includes an invisible target web page which is vulnerable to clickjacking, using an "iframe" HTML tag [29]. Clicking the visible button in this page can cause a malicious request which the target user does not intend to execute. A practical example attack is to force a user to send a hidden "Like" with an "iframe" tag to a post on a social networking system Facebook post although a victim believes to click an attractive message "CLICK HERE TO WIN AN IPAD" [34]. A tricked user would click the button of the message although the target user, in fact, sends "Like" of the hidden request. Malicious hidden requests change their forms, depending on functionalities on target web applications.

In order to exploit clickjacking to bypass Login CSRF mitigation by forcing a target user to log in as the attacker, an attacker may prepare a fake page containing a target login page with the "iframe" tag. The login page of the target web application, framed in the crafted page, is invisible to a visiting user whilst the attacker's crafted page appears on a client web browser. A typical attack exploiting Login CSRF and clickjacking on a target web application will induce the user to click a visible button or link on the crafted page. Clicking the button causes a victim's web browser to send a request with the attacker's credentials to the target login page, forcing the user to log in as the attacker.

2.3.2 XSS and Exploitation of the Vulnerability

Cross-Site Scripting or XSS is a popular web application vulnerability. Exploiting an XSS flaw of insufficient validation and encoding of user inputs and outputs, malicious codes injected by attackers execute on a target web page [2]. XSS may enable an attacker to attain sensitive data through malicious codes, such as Javascript, HTML, Flash, VBScript or ActiveX [10]. Attacker's arbitrarily crafted codes can execute when XSS is exploited. This happens when special characters for script languages, in addition to general letters, can be included in inputs and outputs with inappropriate validation. With this exploitation, a victim's web browser is forced to execute attacker's crafted scripts for purposes of session hijacking, defacing websites, leading users to malicious pages, and impersonating users [41]. For XSS to be exploited, the attacker's script is sent to an XSS-vulnerable page in the target web application to force a victim's web browser to execute the malicious code [31]. Although exploitation effects may depend on functionalities which are vulnerable to XSS in target web applications, exploiting XSS can generally result in serious damage on target web applications and possibly users, because attackers can perform malicious codes including data theft or spoofing. In an example attack exploiting an XSS vulnerability of a form in a blog page, a malicious Javascript code

<scipt> window.location="http://evil.com/?cookie=" + document.cookie </script>

can be injected by the attacker into the blog page [2]. When a victim user visits the blog page, the injected script will execute in the victim's browser to send the victim's cookie value controlled by "document.cookie" to the attacker's website "http://evil.com/" [2]. Because of insufficient filtering of an attacker's input which includes a "*<script >*" code, the victim web browser executes the malicious code as a legitimate embedded script controlled by the web application. Similar to this example attack, attackers will exploit Login CSRF and XSS by embedding a Javascript code to an XSS-vulnerable functionality to force a target user to send attacker's credentials and log in as attackers. The malicious script should contain a request with an attacker's credentials to a login page.

2.3.3 CSRF in a Non-login Page and Exploitation of the Vulnerability

Exploitation of CSRF can, in general, result in target user's unintentional execution of functionality on a web application due to insufficient authenticity verification of the action on the server-side. The victim may be seriously damaged if the action involves sensitive data such as password change or on-line financial transactions. In an example scenario of bank transfer, if a web application accepts 100 dollars transfer with a GET request of, say

http://netbank.com/transfer.do?acct=PersonB&amount=$100,

confirming a money receiver with a parameter "acct" and the amount of transferring money with a parameter "amount", an attacker would modify a request to

http://netbank.com/transfer.do?acct=AttackerA&amount=$100

and send a target user an email with the link of

<a href="http://netbank.com/transfer.do?acct=AttackerA&amount=$100">Read more!</a>

by convincing a target logged-in user to click the link in order to send "100 dollars" to the malicious user "Attacker" [17]. Instructed by the attacker, innocent users would click the link after they log in into the web application. The money transfer to the attacker will complete if the server fails to sufficiently verify that the request is intentionally performed by the users. An attacker may exploit a CSRF vulnerability on a non-login page to accomplish Login CSRF by providing a malicious script in a vulnerable form. An attacker may prepare a malicious code which contains attacker's credentials in a CSRF-vulnerable page to force a target user to execute the malicious request. This attack will be feasible in a web application which authenticates users with only a username and a password and does not verify user's intention of the action including confirmation that login requests are sent from the legitimate login page.

Although CSRF is apparently similar to clickjacking and XSS attacks, it differs in terms of attack scenarios and necessity of user's authentication. Firstly, CSRF requires different user interaction from clickjacking for the exploitation as it forces a victim browser to execute an action without user's direct interaction, whilst the clickjacking vulnerability hijacks a user action through a forged request between a web browser and a legitimate web page by inducing a target user to perform the action [24]. An attacker simply injects a forged request containing a legitimate request to exploit CSRF although target users are persuaded to perform certain actions by clickjacking exploitation through forged appearances containing fake requests. The difference between CSRF and XSS is the necessity of user authentication for successful exploits. Authenticated users which the server trusts are generally targeted in CSRF attacks to execute an attacker's crafted script requested to a vulnerable web application [4]. Targeted actions in CSRF generally require privileges which authenticated users can provide. In contrast, XSS is caused by insufficient validation of user inputs whose responses are just returned to clients, often resulting in the execution of malicious scripts in other users [5]. In other words, XSS does not need target users to be authenticated as long as the vulnerability exists in user inputs with inappropriate filtering. Thus, exploitation of CSRF may consider user's authentication including active sessions, whilst target users do not have to be authenticated in order that XSS can be exploited.

2.4 Summary and Limitation

We have seen in this section that different strategies for Login CSRF mitigation have their own advantages and disadvantages. HTTP header values of "Referer" and "Origin" can recognise legitimate requests by identifying where a login request is originated, although the Referer field can partially reveal users' privacy, and the Origin header may not prevent a malicious CSRF attack originated from a page within the same web domain. Tokenisation, with which tokens are used to distinguish legitimate access from malicious requests, may be effective in protecting against Login CSRF. Challenge-response authentication provides a user-interactive token for authentication to protect against Login CSRF, but it cannot protect against Login CSRF attacks if attackers exploit the clickjacking vulnerability. Limitations of these mitigation methods are identifiable when the vulnerabilities of clickjacking, XSS and CSRF on a non-login page are exploited. Clickjacking can be exploited to deface an appearance of a webpage for the Login CSRF attack. Malicious codes with attacker's credentials are sent by exploitation of XSS and Login CSRF. Often, target users unintentionally execute malicious scripts caused by CSRF in a non-login page so that attackers exploit Login CSRF.

Despite the discussion of benefits and drawbacks of the three mitigation methods, few researchers have described how each of the mitigation methods should be applied to protect against Login CSRF. In other words, although the Login CSRF vulnerability needs to be remediated due to its significance by the discussed strategies, it seems that the current research has not sufficiently discussed effective implementation approaches of the mitigation methods for Login CSRF while preventing the three attacks from being exploited to bypass the mitigation. This topic is significant mainly because different methods are effective or ineffective in different conditions and therefore no single solution can be considered as most appropriate for web applications. Consequently, best practice can be sought through empirical and experimental studies, via which observation and analysis of behaviours of different solutions under different scenarios of attacks are conducted. Results from such experiments, designed to bypass mitigations, will typically result provide insights into robust and appropriate implementation approaches. In the next exposition, we describe an enhanced implementation methodology, based on experiments to observe the effectiveness of the solutions affected by exploitation of the vulnerabilities.

3 Methodology

This section describes the adopted study methodology including data collection, implementation techniques as well as specific scripts of simulated web applications. The implementation strategy is to map the three mitigation methods and the exploitation of the three vulnerabilities in different settings in search of an optimal solution.

3.1 Data and Experiment Scenarios

The experiments were undertaken through the following procedures. Firstly, a basic simulated web application was established and the Login CSRF vulnerability on this basic web application was remediated with the three mitigation solutions – that is, HTTP header verification, tokenisation and challenge-response authentication. In order to observe the effectiveness of each of the mitigation methods without influences of the other two solutions, each of the mitigation methods is implemented on a separate web application containing identical structure and web pages to the basic application. Focus then turned to implementation techniques of the three mitigation methods on three separate web applications. The three attack factors of Clickjacking, XSS and CSRF in the blog page were then exploited to bypass each of the mitigation and breach Login CSRF.

Nine scenarios with all combinations of the three solutions and the three attacks were set in the experiment. Attack scenarios and malicious scripts in the test cases were customised with consideration of characteristics of a mitigation method and a performed attack. Finally, the experiments observed the effectiveness of the mitigation method in each scenario including results of whether each of the attacks could bypass each of the solutions. The basic exploitation scripts of these three attacks on experiment web applications and bypassing attempts were collected as useful data for further analysis.

3.2 *Implementation Strategy*

This section describes details of implementation techniques and the scripting of a simulated web application, the three mitigation methods for Login CSRF and the three attacks bypassing the mitigation to fully provide experiment methodology.

Although non-technical attack procedures including effective email contents to deceive target users are significant for the attacks in the experiments to be successful, the experiments in this work focused on the technical feasibility of attacks and mitigation in the implementation of the experiment test cases and the results analysis.

3.2.1 Web Application in the Project Experiment

An in-house web application is established and controlled to observe the effectiveness of the three mitigation strategies for Login CSRF including HTTP header verification, tokenisation and challenge-response authentication in the presence of the three involving factors clickjacking, XSS and CSRF on a non-login page. Figure 1 is a screen flow diagram of the established web application, containing four pages; Login (index.php), Menu (menu.php), Blog (blog.php) and Search (search.php).

Pages of the web application are coded in HTML, Cascading Style Sheets (CSS) for the style of the document and programming languages of Javascript and PHP (Hypertext Processor) version 5.4.16. Users must enter a username and a password in the login page (index.php) to be authenticated to access authenticated pages including the menu page and the blog page while authenticity is not required for the access to the search page. The Login CSRF vulnerability resides on the login page of this basic web application because the server-side script of the login page of the web application verifies the user's authenticity with solely a combination of an entered username and password in the database. Authenticated users are redirected to the menu page (menu.php) and from there they can access, post and leave a comment on the blog page (blog.php). Posted comments are displayed on the same page, but the page includes a CSRF vulnerability, which allows an attacker to leave a malicious script in the forum, forcing a target user to execute a malicious code with a request of the attacker's credentials toward the login page. The search page (search.php) is accessible from the login page without authentication and user inputs are displayed through a search box in the search page. XSS is exploitable in this page to a victim's web browser to execute a malicious code because filtering of user

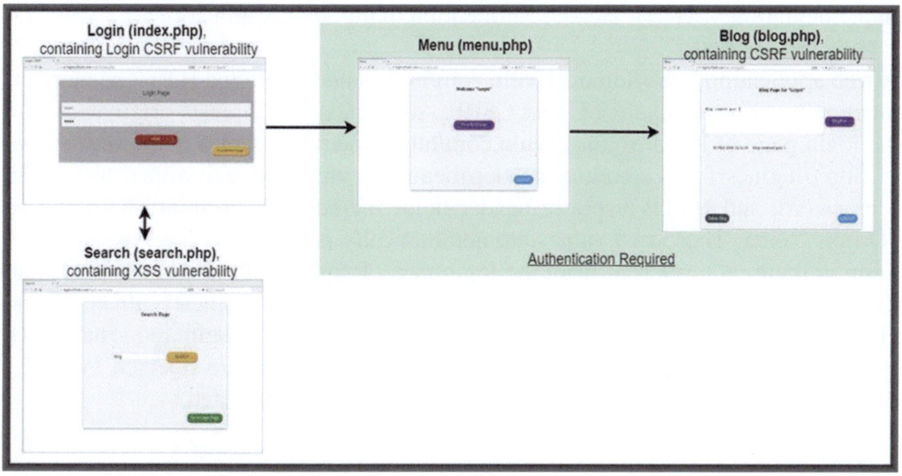

Fig. 1 Screen flow diagram of the basic web application

Table 1 Experimental web applications, remediation methods and URLs

	Remediation method of login CSRF	URL
Web application 1	HTTP header verification	http://logincsrfvuln.com/http-header/
Web application 2	Tokenisation	http://logincsrfvuln.com/token/
Web application 3	Challenge-response authentication	http://logincsrfvuln.com/ch-res/

inputs is insufficient. Account names used in the experiments are “target” as the victim and “maluser” as the attacker. The user “target” is forced to log in to the web application as the user “maluser” when Login CSRF exploitation is successful.

3.2.2 Implementation of the Three Mitigation Methods

In this section, technical implementation of the three mitigation methods HTTP header verification, tokenisation with double submit cookie and challenge-response authentication with Google’s reCAPTCHA are described. Firstly, the Login CSRF vulnerability in the experiments web application is remediated with three different solutions. The three mitigation methods are applied on three separate web applications http://logincsrfvuln.com/http-header/, http://logincsrfvuln.com/token/ & http://logincsrfvuln.com/ch-res/ respectively. Table 1 describes the mitigation methods and the URL of the experiment web applications. Implementing one solution for Login CSRF in a separate application will ensure that no other two mitigation techniques affect the implemented solution to observe the effectiveness when the vulnerabilities are exploited. The three applications contain identical structure excluding the applied mitigation for Login CSRF.

Implementation of HTTP Header Verification in the Web Applications

A web application http://logincsrfvuln.com/http-header/ with the HTTP header verification is protected against Login CSRF. Authentication requests originated from the login page of http://logincsrfvuln.com/http-header/index.php to the menu page of http://logincsrfvuln.com/http-header/menu.php are validated with a username, a password, and HTTP response header fields including a "Referer" field and an "Origin" filled. These two values are automatically provided by a server-side program in login requests sent from the login page. Required HTTP header fields and their values for legitimate login requests are "Origin: http://logincsrfvuln.com" and "Referer: http://logincsrfvuln.com/http-header/index.php". Login requests which contain incorrect values of either or both of the two fields are rejected even if a correct combination of a user account and a password is requested.

Implementation of Tokenisation in the Web Applications

The Login CSRF vulnerability on a web application http://logincsrfvuln.com/token/ is remediated with tokenisation, which validates a one-time token value included in a login request. The tokenisation implemented on this web application is "double submit cookie" because of its high security-that is, a cookie and a parameter of an HTTP request send a pseudorandom token which is validated on the server to verify whether the two values are identical [30]. The tokenisation technology is believed to be secure because the valid token cannot be obtained by an attacker on a different web domain resource [13], our experiments follows this principle.

Users accessing the login page on http://logincsrfvuln.com/token/index.php on which tokenisation is implemented, are provided with a cookie named "CSRFtoken" which is randomly generated with 32 character-lengths on the login page. This cookie is difficult to guess due to its randomness and length. Additionally, a hidden request parameter named "CSRFtoken" is provided to users on the login page. This hidden parameter is generated with the same value of the cookie on the server-side. Users need to send the two identical values of a valid "CSRFtoken" cookie and a valid "CSRFtoken" parameter because the server validates the values in addition to a username and a password. Users fail to be authenticated if the two values are different or invalid. For successfully authenticated users, a cookie "PHPSESSID" for session management is separately issued on the menu page http://logincsrfvuln.com/token/menu.php in addition to the "CSRFtoken" cookie. The cookie and the hidden value are newly generated or reissued in every access to the login page.

Challenge-Response Authentication in the Web Applications

A web application whose path is http://logincsrfvuln.com/ch-res/ is protected against Login CSRF with "Google's reCAPTCHA". The authentication asks the user to solve a challenge to separate requests from humans and automated programs

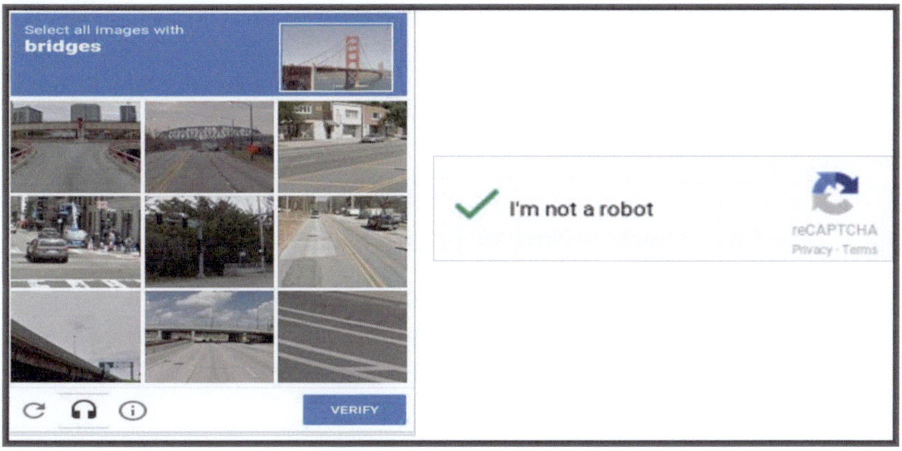

Fig. 2 An example of reCAPTCHA visual challenge

and identify malicious bot requests [1]. Because automated programs can hardly solve challenge questions as answers change in every request, it is believed that only legitimate users who can identify questions will send correct answers. The Login CSRF vulnerability on this web application with reCAPTCHA is difficult to exploit unless attackers can manipulate login requests to force target users to send correct answers. The experiment web application is equipped with reCAPTCHA version 2, which is the most stable version as of June 2018. Google's reCAPTCHA in the experiment web applications requires users to click the checkbox (right hand side panel in Fig. 2) which is shown on the screen and situationally answers a visual (or alternatively an audio) question (left hand side panel), the answer to which changes with every access. Users can send a valid value of the reCAPTCHA value when a checkbox becomes checked after answering a question. In other words, a valid value of the reCAPTCHA is not generated until challenge questions are fully answered. When a challenge question is correctly solved, the parameter named "g-recaptcha-response" is sent to the web application. Thus, the web application receives the three parameters "username", "password" and "g-recaptcha-response" via a login request form on http://logincsrfvuln.com/ch-res/index.php and allows users with the correct values to access authenticated pages.

In order to make the most of reCAPTCHA and follow a suggested implementation method by the vendor, the experiment web applications connect to an external Google's resource. In the backend of reCAPTCHA authentication, the experiment web applications send two parameters "secret" and "response" to a Google API (Application Programming Interface) page https://www.google.com/recaptcha/api/siteverify. A web application with reCAPTCHA needs to send a parameter named "secret" with the secret key and a parameter named "response", along with the user's reCAPTCHA response token to the Google page, which returns a response of a JSON object to validate a user's reCAPTCHA token. The "secret" value in the experiments is stored in a location which is inaccessible to remote users. The "response" parameter in the experiments is a value of "g-recaptcha-response" which a client submits to the web applications, generated when a user successfully solves a challenge. If a response from the Google page is "true", the Challenge-response authentication is successfully validated.

3.2.3 The Three Attacks to Mitigate Login CSRF Protection

The capacity of the three attacks – clickjacking, XSS and CSRF in a non-login page – to defeat the three mitigation methods for Login CSRF are described in this section. In particular, details of the attack scripts and exploitation flow are described.

Exploitation Technique of Clickjacking in the Experiments

In the experiments, the attacker frames the login page of the target web application within an attacker's crafted page http://malsite.com/clickjacking.php, exploiting the clickjacking vulnerability. Deceived by an attacker's instruction with non-technical methods, typically via email, the target user in the experiments is supposed to enter fake codes which are actually the attacker's credentials on an attacker's malicious web page http://malsite.com/clickjacking.php, resulting in a success of the Login CSRF attack. In the first step of the attack exploiting clickjacking, the target user firstly logs into the web application through the legitimate login page with the user's own credentials as "target", led by an attacker's fake instruction. The target then opens a web page http://malsite.com/clickjacking.php crafted by the attacker and frames the login page of the web application. This malicious page is authorised by the attacker with a different domain. In order to exploit clickjacking, the target login page needs to be framed into an invisible iframe with the "opacity" property [18].

Although the malicious page is apparently indifferent to authentication, the user can enter a username and a password which can be sent to the login page on this crafted page because the login page is invisibly framed with the "opacity" property of "0", controlled by a CSS script. The value of the "opacity" property is intentionally set to "0.3", which is more than "0", for the visual explanation in this project. Additionally, the "z-index" property, which is also controlled by a CSS script, of the framed login page is set to "2" so that client's manipulation of the frame login page is prioritised to the control on the attacker's fake form page. On the malicious page which appears to be legitimate, the tricked user types the attacker's credentials on the two fields disguised as legitimate inputs which are instructed by the attacker. When the user clicks the fake button "Send the codes", a login request with the attacker's credentials is transferred to the legitimate login page, logging the user to the target web application as the attacker "maluser". Finally, the user is instructed to go back to the web application without being aware that the user would explore the web application as the attacker "maluser".

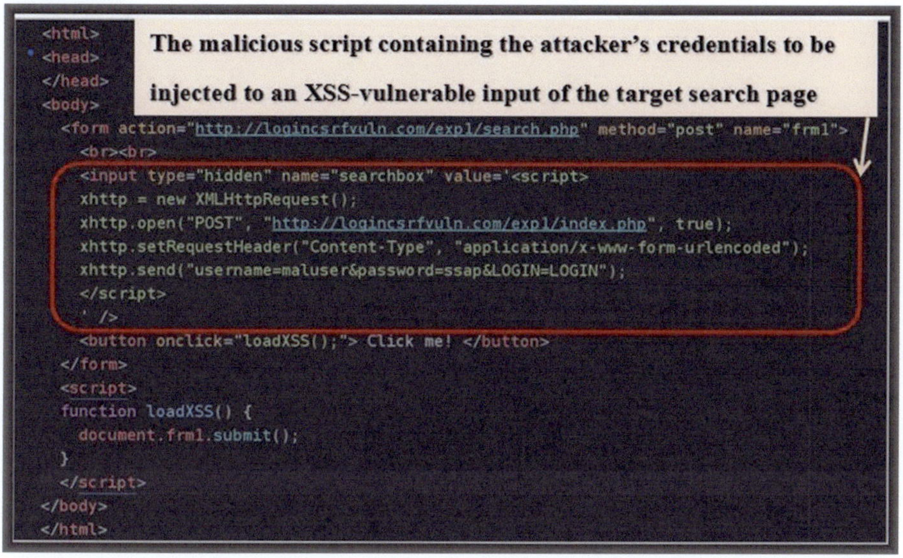

Fig. 3 The source code of the malicious page to exploit XSS in the search page

Exploitation Technique of XSS in the Experiments

Exploiting an XSS vulnerability in the search page of the target applications, the attacker can force the target user "target" to send an arbitrary request with the attacker's credentials and to log into the target web application as the attacker "maluser". When the reflected XSS, a variation of XSS, is exploited, a user is tricked by means of non-technical methods, such as an attacker's crafted email, to click a malicious link, send a malformed form or access an attacker's web page [31]. A reflected XSS attack exploiting Login CSRF in the experiment web applications follows a procedure in which the target user is persuaded to execute a malicious script included an attacker's page http://malsite.com/XSS.php. This malicious web page is supposed to be on the attacker's server. The malicious script to exploit the XSS vulnerability in the search page has the source code of Fig. 3, including an "XMLHttpRequest" object.

The "XMLHttpRequest" object is injected to this malicious code to send an arbitrary crafted login request through the XSS vulnerability. Manipulating an XMLHttpRequest object enables an HTTP request to open a page and transfer a request, establishing cross-site requests [26]. The XMLHttpRequest object in the malicious page creates a request with an attacker's username and password to the login page http://logincsrfvuln.com/exp1/index.php. When a target user clicks the button "Click me!" on the malicious page "http://malsite.com/XSS.php", the victim's web browser loads the target search page with the malicious code including the XMLHttpRequest object to send the attacker's credentials to login onto http://

Fig. 4 The injected malicious script as a rogue comment by the attacker

logincsrfvuln.com/exp1/index.php. Consequently, the user logins as the attacker "maluser" without consciously sending a login request because the login page accepts valid credentials of the attacker "maluser".

Exploitation Technique of CSRF in a Non-login Page in the Experiments

The attacker can exploit a CSRF vulnerability in the target blog page in the experiments to perform a Login CSRF attack. Firstly, the attacker creates a malicious script containing the "XMLHttpRequest" object in a blog page comment, such as the code of Fig. 4.

Similar to the exploitation code of the XSS vulnerability in the search page, the injected malicious script automatically sends the attacker's credentials to the login page with the "XMLHttpRequest" object when the target user visits the blog page without the target user's consent. The attack to exploit the CSRF vulnerability in the blog page carries out a stored CSRF attack technique. In the stored CSRF exploitation, a variant of CSRF attacks, an attacker's crafted request is injected in a trusted page on which authenticated users can leave and read comments including a blog page or an online discussion site [28]. Exploitation of stored CSRF vulnerability in the blog page forces the authenticated victim to perform an attacker's intended action to execute the malicious script and login as "maluser". Despite an innocent appearance of the blog page after injection of the malicious code, when the user logs into the web application with the user's own credentials and visits the blog page, the malicious script automatically loads, forcing the victim web browser to send the parameters of the attacker's credentials to the target login page. Thus, the current user account of the target user is converted to the account "maluser" although the user does not know the attacker's username and password.

4 Results, Analyses and Suggestions of Effective Implementation

This section presents results and analyses from nine experiment test cases of the three solution approaches, based on the attack types described above. The analyses are designed to lead to effective implementation approaches for future applications.

4.1 Implementation, Results and Analyses

Table 2 summarises the results of the experiments in the nine cases, showing whether an attack factor was successfully defeated by the mitigation method or not.

The following sub-sections provide detailed explanations of Table 2, covering attack procedures and results in each of the scenarios. The investigation focuses on important parameters to bypass the mitigation methods discussed above.

4.1.1 HTTP Header Verification as the Mitigation and Clickjacking the Attack

Exploitation of the Clickjacking flaw on the login page of a web application http://logincsrfvuln.com/http-header/ on which HTTP header verification is implemented was successfully able to bypass the mitigation for the Login CSRF vulnerability because the "Referer" field value and the "Origin" field value transferred via the framed login page were accepted as valid. A successful attack in this scenario underlies the example attack described above – exploiting clickjacking by framing the target login page http://logincsrfvuln.com/http-header/index.php. Similar to this example, an attacker's page http://malsite.com/http-header/clickjacking.php contained the target login page with an "iframe" tag with the "src" attribute of http://logincsrfvuln.com/http-header/index.php. If a target user "target" was persuaded to type the attacker's "maluser" username and password on the attacker's page without knowing that the login page was framed in the malicious page, the submitted request was accepted on the server, logging the target user into the web application

Table 2 Summary of the experiment results

		Attack		
		Clickjacking	XSS (search page)	CSRF (blog page)
Mitigation	HTTP header verification	Bypassed	Not bypassed	Not bypassed
	Tokenisation	Bypassed	Bypassed	Bypassed
	Challenge-response authentication	Bypassed	Not bypassed	Bypassed

as “maluser”. Analysis of HTTP headers within the successful attack request has identified that a valid “Referer” value of “Referer: http://logincsrfvuln.com/http-header/index.php” and a valid “Origin” value of “Origin: http://logincsrfvuln.com” were transferred to the target web application because the framed login page sent its login submit form to the target web application along with a rogue request of the attacker’s page to the attacker’s server. Thus, clickjacking defeated the mitigation of HTTP header verification because the two legitimate HTTP header field values were generated with the target login page framed within the attacker’s crafted page.

4.1.2 HTTP Header Verification as the Mitigation and XSS as the Attack

The mitigation of HTTP header verification prevented an attack of the XSS vulnerability via the search page http://logincsrfvuln.com/http-header/search.php from defeating the defence because the “Referer” field was unable to be forged to a valid value by the attack. Several attacks in this scenario were attempted to bypass the mitigation by forging the “Referer” field to “Referer: http://logincsrfvuln.com/http-header/index.php” with an attacker’s page http://malsite.com/http-header/XSS.php. The malicious page was basically coded with the script described in section “Exploitation technique of XSS in the experiments” although the target login URL was customised. The modification of the HTTP header was necessary because the server automatically added the “Referer” field value of http://logincsrfvuln.com/http-header/search.php when the malicious login requests were sent from the search page to the login page, whilst the “Origin” field contained a valid value of http://logincsrfvuln.com since the attack request initiated within the same domain. However, none of the attacks, including this attack, in the experiments was able to change to the valid “Referer” value. In an analysis of HTTP headers in an attack login request, the “Referer” field remained the search page path http://logincsrfvuln.com/http-header/search.php even despite an attempt of changing the value with an additional code designed to add an HTTP header field with an arbitrary value, to the exploitation code. Thus, since the field “Referer: http://logincsrfvuln.com/http-header/search.php”, was rejected by the HTTP header verification on the server-side, exploitation of the XSS in the search page failed to defeat the mitigation of HTTP header verification.

4.1.3 HTTP Header Verification as the Mitigation for CSRF the Attack

HTTP header verification as mitigation for Login CSRF protected against the attacks in the experiment, exploiting the CSRF flaw in the blog page http://logincsrfvuln.com/http-header/blog.php due to the failure of transferring a valid “Referer” field value to the target web application. Similar to the test case of XSS exploitation, “Referer” remained the path of the blog page http://logincsrfvuln.com/http-header/blog.php in the attacks although a valid value with “Origin: http://logincsrfvuln.com” was provided for “Origin” through several attempts in this test case. Based on the attack script explained in the section “Exploitation technique of CSRF in a non-login page in the experiments”, this exploitation code was attempted to force the target user “target” to send the attacker’s (“maluser”) credentials when the target user visited the blog page with “Referer” set “Referer: http://logincsrfvuln.com/http-header/index.php”. Investigation of HTTP headers in the attacks revealed that “Referer” was overwritten to “Referer: http://logincsrfvuln.com/http-header/blog.php”, on which the XSS vulnerability was exploited even if an attack code was attempted to retain “Referer: http://logincsrfvuln.com/http-header/index.php” in this field with a code aimed at inserting an arbitrary HTTP header. The “Referer” field was eventually provided by the server-side and modified to the original path from which a request to the login page initiated. Due to the failure of “Referer” modification, exploitation of the CSRF vulnerability in the blog page could not bypass HTTP header verification.

4.1.4 Tokenisation as the Mitigation for Clickjacking Attack

Tokenisation mitigation for Login CSRF was defeated by exploitation of the clickjacking vulnerability on the web application http://logincsrfvuln.com/token because a valid combination of a “CSRFtoken” cookie and a “CSRFtoken” request parameter was sent in maliciously. Framing the login page http://logincsrfvuln.com/token/index.php in an attacker’s page with an attacker’s rogue guide of the target user contributed to the success of this attack scenario. The attacker’s page http://malsite.com/token/clickjacking.php framed the target login page http://logincsrfvuln.com/token/index.php with a transparent appearance by the code of ‘<iframe src=“http://logincsrfvuln.com/token/index.php”>’.

The target user “target”, who was deceived to enter the attacker’s (“maluser”) credentials through the malicious page, successfully forcing them to an authenticated page. Investigation of the successful attack has identified that a “CSRFtoken” cookie and a “CSRFtoken” parameter were generated with the same value of 32 random characters and provided with the user when a request to the target login page was established, following the legitimate procedure on the target web application. Because the two valid values were transferred to the target web application when the forged form submit was sent on the attacker’s page, the malicious login request was accepted on the target server, resulting in the success of exploitation of the clickjacking vulnerability to bypass tokenisation.

4.1.5 Tokenisation as the Mitigation and XSS as the Attack

The Tokenisation mitigation for Login CSRF in the web application http://logincsrfvuln.com/token was defeated by a script which obtained a valid “CSRFtoken” value and sent both the attacker’s (“maluser”) credentials and the token value when XSS in the search page was exploited. Several attacks were attempted in this scenario. From some of the failed attacks, a malicious script

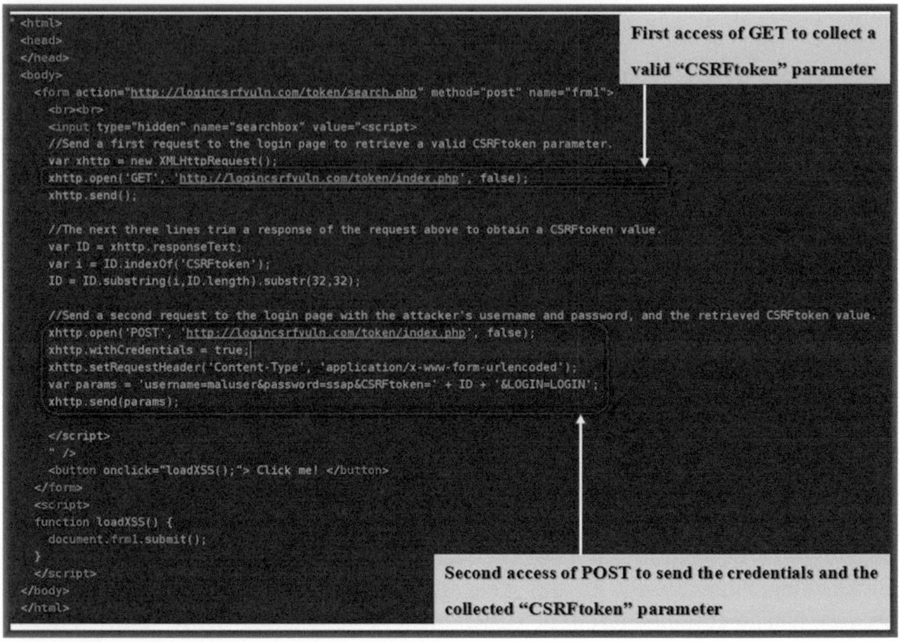

Fig. 5 Malicious script for the successful XSS exploitation

exploiting XSS needed to read a valid “CSRFtoken” parameter value and include it as the “CSRF” parameter with the login request. For a web application on which a cookie and a request parameter share a CSRF token, XSS can be exploited to collect a valid CSRF token parameter through a GET request with the “XMLHttpRequest” script and insert the value into a malicious attack POST request [3]. Based on this theory, the script in Fig. 5 was able to bypass the mitigation by collecting a “CSRFtoken” parameter in the first GET request in the login page and embedding it in the second POST request with the attacker’s (“maluser”) credentials. In this case, the target user visited the attacker’s malicious page and executed the malicious script and navigated to the authenticated pages because both the “CSRFtoken” cookie and a “CSRFtoken” with the same values were included within the malicious login request. It can be inferred that the malicious script defeated the mitigation because the access simulated a login procedure of a legitimate user’s access- the GET request is equivalent to a call of a login page, and user credentials with a retrieved token are sent in the following POST request, which can occur when users click the button to login after they enter usernames and passwords in the login page.

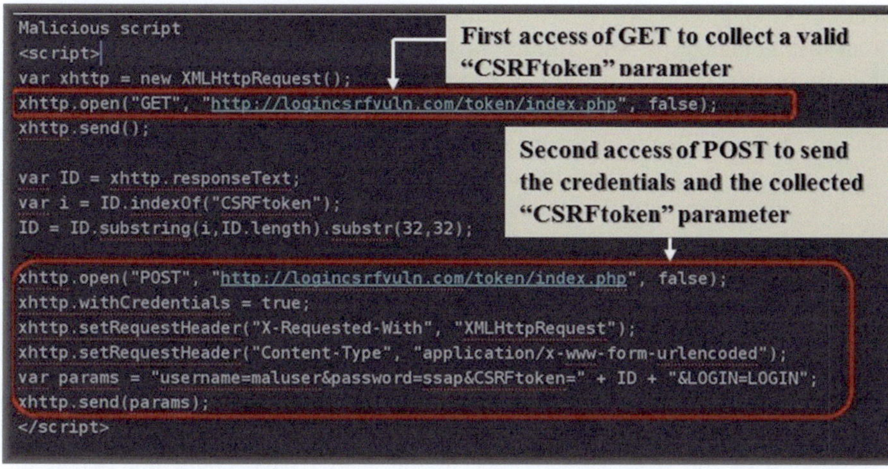

Fig. 6 A malicious script as a blog comment defeating tokenisation

4.1.6 Tokenisation as the Mitigation and CSRF as the Attack

A malicious script injected into a CSRF-vulnerable input in the blog page, which collected a valid "CSRFtoken" parameter and sent the attacker's credentials along with the token value was able to bypass the Tokenisation mitigation on the web application "http://logincsrfvuln.com/token". Similar to the exploitation of the XSS vulnerability in the search page, described in the Sect. 4.1.5, two requests to the login page, described in Fig. 6 were injected to the blog page by the attacker "maluser".

The POST request with the attacker's credentials and a valid "CSRFtoken" parameter value collected in the preceding GET request was sent to the login page upon execution by the user. In a successful attack, the victim user obtained a "CSRFtoken" hidden parameter with a 32 random character value generated by the server along with a "CSRFtoken" cookie with the same value in the first GET request. This value was embedded in the second POST request as well as the credentials of "maluser". Eventually, the Login CSRF attack exploiting CSRF in the blog page was successful because the cookie and the parameter with the same value were valid and identical.

4.1.7 Challenge-Response Authentication for Mitigating Clickjacking Attack

The attacker "maluser" would be able to perform a Login CSRF attack on the web application http://logincsrfvuln.com/ch-res, the challenge-response authentication, if the "target" was deceived to solve a challenge of Google's reCAPTCHA on an attacker's page framing the target login page. In this case, the target login page including Google's reCAPTCHA was framed in an attacker's web page, controlled by an *iframe* tag script. The triumph of this attack can be proved by the fact that the login request contains a valid "g-recaptcha-response" value, which is validated in the challenge-response authentication on the server-side, due to a successful solution of the challenge. Because of a valid parameter, the server accepted as a legitimate login request. Thus, the clickjacking vulnerability may enable the target user to be forced to send the attacker's credentials and a valid challenge-response authentication value to the target login page given that the target user is lured to a malicious page framing the login page.

4.1.8 Challenge-Response Authentication as the Mitigation and XSS Attack

Exploitation of the XSS vulnerability in the search page could not defeat the challenge-response authentication as mitigation for Login CSRF because a crafted request injected into the malicious script was unable to contain a valid "g-recaptcha-response" parameter by solving a challenge. It was necessary for a code exploiting the XSS vulnerability to include a valid value of the "g-recaptcha-response" parameter for the Login CSRF attack to be successful. After several exploitation attacks were performed in the scenario, it was found that malicious scripts exploiting XSS could not automatically solve Google's reCAPTCHA authentication on behalf of the user. Since a valid parameter is available when a user connects to a Google's resource by solving a challenge, any attempts of modifying the "g-recaptcha-response" parameter in the attempted attacks were eventually rejected as an invalid value which was not legitimately collected through the Google's resource. In this experiment, it was concluded that the exploitation was unfeasible because a challenge solution procedure of reCAPTCHA could not be included in the rogue attack script, or a malicious request could not be crafted so that the target user would solve a challenge to send a valid "g-recaptca-response" parameter with the credentials of "maluser".

4.1.9 Challenge-Response Authentication as the Mitigation and CSRF Attack

The "target" was forced to login as the "maluser" by exploitation of CSRF in the blog page of the web application on which Login CSRF was protected by the challenge-response authentication, if the victim was deceived to solve a challenge included in an attacker's injected malicious comment in the blog page. In this experiment, the attacker injected a malicious script as a blog comment in Fig. 7, but in this case, a client could send the attacker's username "maluser" and the password as hidden parameters.

In this example attack, the user "target" was instructed to enter a rogue parameter as the activation code and solve a challenge of reCAPTCHA. The malicious submit

```
Ch-res Malicious script <br><br>
<!-- Google recaptcha code -->
<script src="https://www.google.com/recaptcha/api.js"></script>
<form action="index.php" method="POST">
<input type="hidden" name="username" value="maluser" />
<input type="hidden" name="password" value="ssap" />
<input type="text" name="code" placeholder="Type the code here" style="height:30px;width:300px;"/>
<!-- Google recaptcha code -->
<div class="g-recaptcha" data-sitekey="6LdIjlUUAAAAAJ4wrNAe1wzsHC2JXKhngSEPbg7R"></div>
<input type="submit" value="Click here to activate the code" style="height:30px;width:300px;">
</form>
```

Fig. 7 Malicious script injected into the forum on the blog page

form contained two hidden parameters – "username" and "password" including, fake code and reCAPTCHA. If the target user solved a challenge and submitted the code by clicking the fake button, the forged request was sent to the login page, forcing "target" to login to the web application as "maluser". Consequently, the three required parameters were accepted as valid, enabling exploitation of CSRF in the blog page to successfully defeat the Challenge-response authentication as mitigation for Login CSRF.

4.2 *Effective Implementation*

With comprehensive analysis of the experiment results in the previous sections and investigation of traffic of the attacks in the nine test cases, this section proposes effective implementation approaches of the three mitigation methods.

4.2.1 Efficient Implementation of HTTP Header Verification

In order for the HTTP header verification to efficiently protect against exploitation of the Login CSRF vulnerability, a web application should separate a different web domain for a login page from the other non-login pages. It should remediate clickjacking vulnerabilities by permitting only the web domain of a login page to frame resources, and eradicate XSS and CSRF vulnerabilities on a login page. The HTTP header verification with an "Origin" header can be effectively implemented on a web application with a different web domain for a login page from non-login pages to prevent exploitation of XSS and CSRF on non-login pages. The "Origin" header enables clients to access HTTP resources across different web domains for the purpose of Cross-Origin Resource Sharing (CORS) [16].

The "Origin" header can be an effective protection for Login CSRF if web servers can recognise a legitimate request origin page by the field or a web domain, and reject requests from disallowed web domains so that the "Origin" field is difficult to bypass by the performed attacks with XSS and CSRF. An implementation example is a web application with a web domain "login.logincsrfvuln.com" for the login page and a separate domain "web.logincsrfvuln.com" for the other pages. Users are, in this example, required to be authenticated on a login page of the web domain "login.logincsrfvuln.com". Apart from the login page, the non-login pages including the search page and the blog page have a different domain "web.logincsrfvuln.com". This web application can protect against the Login CSRF attack exploiting XSS in the search page and CSRF in the blog page if the server accepts login requests with an HTTP header "Origin: login.logincsrfvuln.com" to identify that requests are originated from the login page. Because the attacks will contain the "Origin" field of the domain "web.logincsrfvuln.com", exploitation of the vulnerabilities will fail since the web application accepts login requests with an "Origin" field of "login.logincsrfvuln.com".

Another effective implementation approach of HTTP header verification is to set a Content-Security-Policy (CSP) HTTP header to remediate the clickjacking and prevent an attacker's crafted page from framing the login page. CSP restricts resources allowed within web applications to prevent attacker's inline codes and malicious functions such as "eval" from executing and control resources inclusion with white-listing filtering [8]. CSP controls loadable resources of client web browsers by defining URI directives including "frame-ancestors" to clarify a resource in which an authoritative content can be framed with an "*iframe*" tag [35]. Because the CSP header field with the directive "frame-ancestors" should be able to permit frameable resources by whitelisting domains, the clickjacking vulnerability can be resolved, whilst the two different domains for a login page and the non-login pages in the first suggested implementation approach of HTTP header verification are also able to be implemented along with this HTTP header. The setting stops remote attackers from exploiting clickjacking vulnerability, because the login page can be loaded with an "*iframe*" tag only within pages of authorised subdomains "logincsrfvuln.com" including "login.logincsrfvuln.com" and "web.logincsrfvuln.com".

Finally, XSS and CSRF must be resolved in order for the HTTP header verification with the "Referer" field to protect against the Login CSRF attack. This is because exploitation of the two vulnerabilities can bypass the "Referer" verification if they exist in a login page. As noted above, exploitation of XSS on the search page and CSRF and on the blog page could not bypass the HTTP header verification because the "Referer" field remained the request source path despite attempts of modification of the value to the login page path. These outcomes indicate that the two vulnerabilities on a login page will defeat the "Referer" field verification because the attacks are accepted as legitimate since the attack requests are originated from a login page. For example, if a login page has XSS vulnerability within a search functionality same as the search page, the flaw can be exploited for Login CSRF with an attack code in an attacker's resource with the techniques discussed above. A user tricked to execute the code will send a login request with the attacker's credentials and a valid "Referer" field of the login page path, resulting in being navigated to an authorised page. This outcome is due to the fact that clients send login parameters with "Referer" of a request source, which will be the login page if XSS vulnerability exists in the login page, as proved in the Sect. 4.1.2.

4.2.2 Efficient Implementation of Tokenisation

Tokenisation would become an efficient solution for Login CSRF if a web application resolves a clickjacking flaw with a single web domain control provides only legitimate users with valid tokens by creating unpredictable tokens and prevents authenticated users from updating their tokens. Tokenisation mitigation can be effective for web applications which remediate the clickjacking vulnerability to prevent illegitimate framing of a login page while managing a single web domain to minimise the authoritative domain control, from analysis of the result in the Sect. 4.1.4. The clickjacking flaw can be remediated with the "X-Frame-Options", an HTTP header field. An "X-Frame-Options" HTTP header allows only particular resources to be included within a frame [15]. In particular, an "X-frame-options" HTTP header field with the value of "SAMEORIGIN "can protect against the Login CSRF attacks by resolving the clickjacking flaw because the header prevents external resources from framing any pages of the web application. This header can be set to one of the three values; "deny" to prohibit any pages from framing a target resource, "same-origin" to permit pages of the same domain to include the page, or "allow-from" to specify domains whose pages can be rendered in a target page [34].

Web applications with a single domain can protect against clickjacking attacks with "X-Frame-Options" with "deny" or "SAMEORIGIN". Unlike the HTTP header verification, the suggested approaches for tokenisation implementation, does not require web applications to divide a domain into subdomains. Less effort may be required for web applications with one web domain to remediate clickjacking and implement tokenisation than the HTTP header verification because only another additional HTTP header is necessary for the effective solution for this vulnerability. As an implementation example of the header field, remote attackers are unable to frame the login page with the domain "logincsrfvuln.com" if the web application has an HTTP header "X-Frame-Options: SAMEORIGIN", allowing pages with only the same domain to include the login page. Tokenisation can be implemented to protect against Login CSRF for web applications with a single domain, remediating clickjacking with "X-Frame-Options". Another effective implementation method of tokenisation is to prevent illegitimate accesses from obtaining valid tokens to deal with the exploitation of XSS as explained in Sect. 4.1.5. A cause for the successful attack of the XSS exploitation is that the malicious code was able to identify a token name with a valid value from a response of access to the login page. In order that servers can authenticate particular permission for a user or a machine with a token, tokens are mostly issued with random alphanumeric letters so that they are not compromised by brute-force attacks: for instance, some JWT (JSON Web Token) as CSRF tokens are encrypted with cryptography such as HMAC (Hash-based Message Authentication Code) [22]. While this fact explains quality of token values, randomness or encryption of tokens may enhance security of the token name to protect malicious codes from collecting a token via XSS exploitation, from the experiment result.

A method of providing only legitimate accesses with valid tokens is to encrypt the name of an additional hidden parameter so that attackers cannot collect a valid token name with a value to include the parameter within a malicious code exploiting the XSS vulnerability. An example of the implementation is to generate a hidden parameter of an encrypted name by Advanced Encryption Standard (AES) with an arbitrary secret key. Through a symmetric encryption method AES, a string can be encrypted by a 128-bit input text and some transformation rounds with a round key [7]. In this example, login requests include this parameter with the fixed token value. A login request in this example generates five parameters for authentication; "username", "password", "CSRFtoken" for Tokenisation, an encrypted parameter as an additional token parameter and "LOGIN" to submit a login form. The server generates and includes a valid token of an encrypted name along with credentials and "CSRFtoken" for legitimate requests. It is considerably difficult for remote attackers to include a valid token within an XSS-exploitation code because the parameter with an encrypted name is virtually unpredictable by clients unless both the passphrase for the encryption and the encryption method are disclosed, which should be confidential and managed by only the server. In summary, XSS exploitation can be blocked by tokenisation with unpredictable parameter name to prevent illegitimate login accesses.

From analysis in Sect. 4.1.6, tokens should be issued only when non-authenticated users request from a login page as an efficient employment of tokenisation because authenticated users are unlikely to login to web applications without logging out applications or losing valid sessions. The main target group of CSRF attacks are authenticated users because web servers generally trust authenticated users, believing that they send only necessary requests controlled by browsers [4]. On the other hand, non-authenticated clients will send login requests from a login page whilst authenticated users also re-login on a login page after they logout of web applications. Because of these two facts, login request procedure including generation of valid tokens for CSRF protection should be processed for only non-authenticated users without valid session cookies to prevent CSRF in non-login pages from being exploited to bypass the mitigation. An implementation example in the experiment web applications is to validate whether accessing users are authenticated on the login page by identifying the session status with a cookie "PHPSESSID". A "CSRFtoken" cookie and a "CSRFtoken" parameter with the same value should be included in requests from non-authenticated users without session cookies "PHPSESSID". In contrast, a "CSRFtoken" cookie should not be provided to authenticated users who already own session cookies "PHPSESSID" and execute the malicious script in Sect. 4.1.6 by visiting the blog page to which the attack script is injected. This control will prevent deceived users in the blog page from being provided with a valid "CSRFtoken" cookie, leading to failure of the Login CSRF attack exploiting CSRF in the blog page. Therefore, by prohibiting authenticated users from obtaining valid CSRF token cookies, Tokenisation will be efficiently implemented against Login CSRF attacks.

4.2.3 Efficient Implementation of Challenge-Response Authentication

Challenge-response authentication as mitigation for Login CSRF should be implemented by accepting authentication requests from only authorised domains of a login page, remediating clickjacking vulnerabilities to prohibit external resources from framing login pages and producing unpredictable answers and validation tokens of challenges. That is, web applications should permit challenge-response authentication scripts to be included within only authorised login pages by separating domains of login pages from non-login pages to prevent authentication codes from being embedded in non-login pages to exploit CSRF vulnerabilities. Because the successful attack in Sect. 4.1.9 can be attributed to insufficient validation of request origin domains of pages in which users solve challenges, illegitimate login requests due to the exploitation of CSRF in non-login pages will be prevented if challenge-response authentication blocks access from non-login pages. As an implementation example, consider Google's reCAPTCHA, which has a feature of controlling authorised domains. Because the Google's reCAPTCHA API key is connected to users' particular domains and their subdomains, the hostname validation can protect against illegal use of reCAPTCHA by unauthorised resources. Google's reCAPTCHA enables web applications to restrict the use of the code by domain names. In this implementation example, the domain of the login page may be assigned to "login.logincsrfvuln.com" while the domain name for non-login pages is "web.logincsrfvuln.com" to divide domains, similar to the example in the Sect. 4.2.1. Because the login requests can be considered to be originated from the login page with the domain "login.logincsrfvuln.com", only this domain should be permitted by the reCAPTCHA code for this application in the setting of the Google's reCAPTCHA website. Due to this setting, the malicious code to exploit CSRF in the blog page is protected because the validation procedure on the Google's side of request origin domains denies reCAPTCHA authentication accesses from the blog page with the domain "web.logincsrfvuln.com". Validation of domains of challenge-response authentication requests will increase the reliability of the authentication system because the authentication codes will not be sent by exploitation of vulnerabilities including CSRF in non-login pages. This validation of web domains should be followed by remediation of clickjacking. Clickjacking vulnerabilities need to be remediated with the "Content-Security-Policy" to prohibit unauthorised domain resources from framing a login page to prevent exploitation of the clickjacking vulnerability from contributing to the success of Login CSRF attacks. The results in the Sect. 4.1.7 in which framing a login page defeated the solution suggest that web applications adopting challenge-response authentication mitigation, must resolve clickjacking vulnerability. In order to be compatible with the hostname validation with separation of domains for login pages and non-login pages, an effective solution will be to implement the CSP HTTP header with "frame-ancestors" to whitelist a web domain of a login page.

Finally, challenge-response authentication should require a parameter of a challenge solution to be complex enough to be protected against brute-force attacks, as suggested by the results in Sect. 4.1.8. Google's reCAPTCHA protected against the XSS exploitation attack because the challenges are considerably difficult for an automated script to provide correct answers, and the authentication code "g-recaptcha-response" is a lengthy parameter with mixed characters of number, lowercase and uppercase.

Similar to the CSRF token in tokenisation, solutions for challenges should be difficult to guess with brute-force attacks since insufficient complexity of a challenge would enhance the attack successfulness. A vulnerable challenge-response authentication may require users to enter a four-digit one-time code which the server generates. This four-digit number code is also a required parameter value along with credentials. Remote attackers exploiting XSS vulnerability on the search page may be able to successfully force the target user to login to the web application as "maluser" with a malicious script which repeatedly attempts to send a different one-time code with the credentials until the valid token is found. This attack can be complete within practical time if the code can be the number between 1000 and 9999 after the victim user executes the code through the attacker's malicious page. In contrast, the one-time parameter with the large number and various characters will prevent attackers from successfully exploiting the vulnerability because of the considerable amount of estimated time consumption. Thus, a complex token parameter for challenge-response authentication to prevent it from being easily predicted will increase the effectiveness of the mitigation for Login CSRF.

5 Concluding Remarks

The chapter discussed the severity of Login CSRF attacks, via which forces target users to login to a web application as malicious attackers to obtain sensitive information of the targets. The experimental analyses lead to a number of suggestions on attaining efficiency in implementing the three solutions – HTTP header verification, tokenisation and challenge-response authentication, as mitigation for Login CSRF, on the backdrop of the mitigation methods' advantages and limitations. The main question of "effective implementation of the three solutions for Login CSRF" was raised because the mitigation methods for Login CSRF can be defeated by exploitation of vulnerabilities such as clickjacking, XSS and CSRF on non-login pages.

5.1 Meeting Study Objectives

The findings imply that several implementation approaches to mitigation need to be considered to block Login CSRF attacks regardless of the exploitation. The experiments consisted of web applications in which the three mitigation methods were implemented to observe their behaviours when exposed to clickjacking, XSS and CSRF attacks on non-login pages. Three web applications, each of which remediated the Login CSRF vulnerability by HTTP header verification, Tokenisation or Challenge-response authentication, were tested to analyse how exploitation of the three vulnerabilities affected the mitigation including the feasibility of defeating the solutions. The outcomes in the experiments show that clickjacking exploitation defeated HTTP header verification, all of the attacks bypassed tokenisation and challenge-response authentication was defeated by the clickjacking and CSRF attacks on a non-login page.

The study performed experimental analyses in search of effective implementation approaches to Login CSRF mitigation. The four objectives set in Sect. 1.2 were all achieved and, based on the findings, efficient implementation of the three methods of attack were proposed. It was established that mitigation of HTTP header verification can be effectively implemented if clickjacking is remediated with separate web domain controls on a login page and non-login pages, and both XSS and CSRF are removed from a login page. Tokenisation should be able to protect against Login CSRF with a single web domain control with elimination of clickjacking, and tokens are secured with unpredictability by encryption of the token parameter name and provided with only non-authenticated users. Finally, the effectiveness of challenge-response authentication can be enhanced if the challenge codes are valid only within authoritative login pages. It is also shown that clickjacking is resolved with a multiple-domain control and challenge solutions are unpredictable so that illegitimate users cannot obtain correct answers.

5.2 Future Directions

This chapter will not have sufficiently identified and analysed concerns regarding the effective implementation of Login CSRF mitigation, for many reasons, particularly the sampling limitations, i.e., insufficient comprehensiveness of attack exploitation and defence. Further, and related to that factor, cybersecurity is a highly dynamic area of study, new attacks and defence methods come and go. For further study in Login CSRF, it can be suggested that future research may be discussed with different attack factors such as social engineering and brute-force attacks to identify more benefits and drawbacks of the mitigation methods than they are discussed in this paper. Social engineering can be discussed from non-technical attacks to clarify when this attack can be exploited to bypass the mitigation for Login CSRF. Social engineering attacks such as malicious web links and conversation are often involved in exploiting of CSRF vulnerabilities [30]. Although this paper mostly dealt with technical attacks and implementation, the topic can be discussed with consideration of non-technical methods in terms of how the attack can deceive target users for the exploitation. In addition, brute-force attacks may also be considered in future research to especially identify the strength of tokens and challenges. As a straightforward and powerful method of decoding, brute-force attacks are considered to bypass most encryption if time and an attacking machine power are sufficient because the attacks attempt all possibilities to defeat passwords or tokens [19]. The foregoing steps can be adopted from a general Big Data approach, establishing shared data repositories for multiple data attributes on attacks and defences.

References

1. Abubaker H, Salah K, Al-Muhairi H, Bentiba A (2015) Cloud-based Arabic reCAPTCHA service: design and architecture. In: 2015 IEEE/ACS 12th international conference of computer systems and applications (AICCSA), pp 1–6. https://doi.org/10.1109/AICCSA.2015.7507189
2. Acunetix. (2012). Cross-site Scripting (XSS). Acunetix. Retrieved from: https://www.acunetix.com/websitesecurity/cross-site-scripting/
3. Acunetix (2014) CSRF and XSS – brothers in arms. Acunetix. Retrieved from https://www.acunetix.com/blog/articles/csrf-xss-brothers-arms/
4. Alvarez E, Correa B, Arango I (2016) An analysis of XSS, CSRF and SQL injection in colombian software and web site development. In: 2016 8th Euro American conference on telematics and information systems (EATIS), pp 1–5. https://doi.org/10.1109/EATIS.2016.7520140
5. Baojiang C, Baolian L, Tingting H (2014) Reverse analysis method of static XSS defect detection technique based on database query language. In: 2014 ninth international conference on P2P, parallel, grid, cloud and internet computing (3PGCIC), pp 487–491. https://doi.org/10.1109/3PGCIC.2014.99
6. Barth A, Jackson C, Mitchell J (2008) Robust defenses for cross-site request forgery. In: Proceedings of the 15th ACM conference on computer and communications security. https://doi.org/10.1145/1455770.1455782
7. Bin Liu BM, Baas BM (2013) Parallel AES encryption engines for many-core processor arrays. IEEE Trans Comput 62(3):536–547. https://doi.org/10.1109/TC.2011.251
8. Calzavara S, Rabitti A, Bugliesi M (2016) Content security problems?: evaluating the effectiveness of content security policy in the wild. In: Proceedings of the ACM conference on computer and communications security, pp 1365–1375. https://doi.org/10.1145/2976749.2978338
9. Czeskis A, Moshchuk A, Kohno T, Wang HJ (2013) Lightweight server support for browser-based CSRF protection. In: Proceedings of the 22nd international conference on world wide web, pp 273–284. https://doi.org/10.1145/1455770.1455782
10. Dayal AM, Ambedkar N, Raw R (2016) A comprehensive inspection of cross site scripting attack. In: 2016 international conference on computing, communication and automation (ICCCA), pp 497–502. https://doi.org/10.1109/CCAA.2016.7813770
11. Detectify AB (2017) Login CSRF. Detectify AB. Retrieved from https://support.detectify.com/customer/en/portal/articles/1969819-login-csrf
12. Ding C (2010) Login cross-site request forgery defence: technical report. Retrieved from http://students.ecs.soton.ac.uk/cd8e10/paper/INFO6003_Technical_Report_Login_Cross_Site_Request_Forgery_Defence_Chaohai_Ding.pdf
13. Dorneanu V (2016) Some words on CSRF and cookies. Retrieved from http://blog.dornea.nu/2016/01/26/some-words-on-csrf-and-cookies/

14. Farah T, Shojol M, Hassan M, Alam D (2016) Assessment of vulnerabilities of web applications of Bangladesh: a case study of XSS & CSRF. In: 2016 sixth international conference on digital information and communication technology and its applications (DICTAP), pp 74–78. https://doi.org/10.1109/DICTAP.2016.7544004
15. Ferry EO, Raw J, Curran K (2015) Security evaluation of the OAuth 2.0 framework. Inf Comput Secur 23(1):73–101. Retrieved from https://search-proquest-com.lcproxy.shu.ac.uk/docview/1786146054/fulltext/94C0FC45FC024D1APQ/1
16. Hothersall-Thomas C, Maffeis S, Novakovic C (2015) BrowserAudit: automated testing of browser security features. In: Proceedings of the 2015 international symposium on software testing and analysis, pp 37–47. https://doi.org/10.1145/2771783.2771789
17. Imperva (2018) Cross site request forgery (CSRF) attack. Imperva. Retrieved from https://www.incapsula.com/web-application-security/csrf-cross-site-request-forgery.html
18. Jain J (2015) Clickjacking, Cursorjacking & Filejacking. Retrieved from https://resources.infosecinstitute.com/bypassing-same-origin-policy-part-3-clickjacking-cursorjacking-filejacking/
19. Jo H, Yoon J (2015) A new countermeasure against brute-force attacks that use high performance computers for big data analysis. Int J Distrib Sens Netw 11(6):1–7. https://doi.org/10.1155/2015/406915
20. Karthika S, Devaki P (2014) An efficient user authentication using captcha and graphical passwords – a survey. Int J Sci Res (IJSR) 3(11):2319–7064. Retrieved from https://pdfs.semanticscholar.org/da60/282b6be853f01082c23734533e4c96aff5d5.pdf
21. Kavitha D, Chandrasekaran S, Rani S (2016) HDTCV: hybrid detection technique for clickjacking vulnerability. In: Dash S, Bhaskar M, Panigrahi B, Das S (eds) Artificial intelligence and evolutionary computations in engineering systems, Advances in intelligent systems and computing, vol 394. Springer, Cham
22. Krapf L, Knobloch G, Antipa D, Leonardo C, Sanso A (2017) U.S. patent no. 9,774,622. U.S. Patent and Trademark Office, Washington, DC
23. Manaswini N, Sahoo P (2016) CSRF attacks on web applications. Int J Adv Comput Tech Appl (IJACTA) 4(1):194–197. Retrieved from http://www.ijacta.com/index.php/ojs/article/view/51/41
24. Miessler D (2008) The difference between CSRF and clickjacking. Retrieved from https://danielmiessler.com/blog/the-difference-between-csrf-and-clickjacking/
25. Moradi H, Moghaddam H (2015) Strategies and scenarios of CSRF attacks against the CAPTCHA forms. J Adv Comput Sci Technol 4(1):15–22. https://doi.org/10.14419/jacst.v4i1.3935
26. Mozilla (2018) Using XMLHttpRequest – web APIs|MDN. Mozilla. Retrieved from https://developer.mozilla.org/en-US/docs/Web/API/XMLHttpRequest/Using_XMLHttpRequest
27. Nagpal B, Chauhan N, Singh N (2014) Cross-site request forgery: vulnerabilities and defenses. I-Manager's J Inf Technol 3(2):13–21. https://doi.org/10.26634/jit.3.2.2778
28. Nagpal B, Chauhan N, Singh N (2016) Additional authentication technique: an efficient approach to prevent cross-site request forgery attack. I-Manager's J Inf Technol 5(2):14–18
29. OWASP (2016) Testing for clickjacking (OTG-CLIENT-009). OWASP. Retrieved from https://www.owasp.org/index.php/Testing_for_Clickjacking_(OTG-CLIENT-009
30. OWASP (2018a) Cross-site request forgery (CSRF) prevention cheat sheet. OWASP. Retrieved from https://www.owasp.org/index.php/Cross-Site_Request_Forgery_(CSRF)_Prevention_Cheat_Sheet
31. OWASP (2018b) Cross-site scripting (XSS). OWASP. Retrieved from https://www.owasp.org/index.php/Cross-site_Scripting_(XSS)
32. Pellegrino G, Johns M, Koch S, Backes M, Rossow C (2017) Deemon: detecting CSRF with dynamic analysis and property graphs. Retrieved from https://arxiv.org/abs/1708.08786
33. Sentamilselvan K, Lakshmana S, Ramkumar N (2014) Cross site request forgery: preventive measures. Int J Comput Appl 106(11):20–25. Retrieved from: http://citeseerx.ist.psu.edu/viewdoc/download?doi=10.1.1.800.3853&rep=rep1&type=pdf

34. Shahriar H, Haddad H, Devendran V (2015) Request and response analysis framework for mitigating clickjacking attacks. Int J Secur Softw Eng (IJSSE) 6(3):1–25. https://doi.org/10.4018/IJSSE.2015070101
35. Stamm S, Sterne B, Markham G (2010) Reining in the web with content security policy. In: Proceedings of the 19th international conference on world wide web, pp 921–930. https://doi.org/10.1145/1772690.1772784
36. Sudhodanan A, Carbone R, Compagna L, Dolgin N, Armando A, Morelli U (2017) Large-scale analysis & detection of authentication cross-site request forgeries. In: 2017 IEEE European symposium on security and privacy (EuroS&P), pp 350–365. https://doi.org/10.1109/EuroSP.2017.45
37. Takamatsu Y, Kono K (2014) Clickjuggler: checking for incomplete defenses against clickjacking. Privacy. In: 2014 twelfth annual international conference on security and trust (PST), pp 224–231. https://doi.org/10.1109/PST.2014.6890943
38. Vasilomanolakis E, Mühlhäuser M (2019) Detection and mitigation of monitor identification attacks in collaborative intrusion detection systems. Int J Netw Manag 29(2):1099–1190. ISSN: 1055-7148
39. Vrindamol P, Neena VV (2015) Detection and prevention of clickjacking and cross site request forgery. Int J Adv Res Sci Eng 4(Special Issue 01):55–64. Retrieved from https://pdfs.semanticscholar.org/eaab/9e476edeeeea0290875ceeb51c19e9572f6c.pdf
40. Yadav P, Parekh C (2017) A report on CSRF security challenges & prevention techniques. In: 2017 international conference on innovations in information, embedded and communication systems (ICIIECS), pp 1–4. https://doi.org/10.1109/ICIIECS.2017.8275852
41. Yusof I, Pathan A (2014) Preventing persistent Cross-Site Scripting (XSS) attack by applying pattern filtering approach. In: The 5th international conference on Information and Communication Technology for The Muslim World (ICT4M), pp 1–6. https://doi.org/10.1109/ICT4M.2014.7020628

Attack Vectors and Advanced Persistent Threats

Sergio F. de Abreu, Stefan Kendzierskyj, and Hamid Jahankhani

Abstract Advanced Persistent Threats (APTs) are destructive and malicious cyber-attacks aimed at high profile, high value targets with clear objectives in mind with a range of desired outputs. In most cases, these threat groups are state sponsored which makes them extremely well financed, organised and resourced. The attack payloads range from data exfiltration and theft to the undermining of critical national infrastructure. These attacks differ from the typical cyberattacks in several different ways but a key differentiation is their patient "low and slow" approach to prevent detection. This approach, although slow, has been very successful and in many cases, detection is years after initial infection. Many of the attacks detected today, have been over a decade in the making. Most concerning is the fact that traditional defence mechanisms have been unsuccessful at detecting these attacks and so how successful will these methods be against a new generation of attacks? The earliest recording of an APT is probably "the cuckoo's egg". An attack in the 1980s in which a West German hacker infiltrated a series of computers in California and over time stole state secrets relating to the US "Star Wars" program. The hacker then sold the information to the Soviet KGB. Although at this point in time, cyber defence was not a government sponsored military department, it raised awareness of just how powerful this threat could be. Since then, worldwide attacks in the private and public sectors have grown exponentially and today, all governments have cyber warfare units.

Most APT attacks are state sponsored; however, this does not mean that attacks are limited to government entities. Far from it. These attacks affect individuals, companies, corporations and governments globally. Attacks can and do encompass a multitude of sophisticated techniques and affect not only the traditional LAN/WAN

S. F. de Abreu · H. Jahankhani
Northumbria University London, London, UK
e-mail: Hamid.jahankhani@northumbria.ac.uk

S. Kendzierskyj (✉)
Cyfortis, Worcester Park, Surrey, UK
e-mail: stefan@cyfortis.com

H. Jahankhani et al. (eds.), *Cyber Defence in the Age of AI, Smart Societies and Augmented Humanity*, Advanced Sciences and Technologies for Security Applications, https://doi.org/10.1007/978-3-030-35746-7_13

environments but could also contaminate new generation networks such as mobile 5G networks, vehicular ad hoc networks (VANET) and Internet of Things (IoT) to name but a few. Dealing with these attacks is challenging, most attacks take years to be discovered and traditional detection mechanisms have been woefully inadequate. The age of machine learning and artificial intelligence has brought significant improvement to the detection challenges faced. These fields allow us to look for far more than attack signatures and characteristics. They allow us to look for patterns of behaviour through massive data quantities at speeds previously unimaginable.

Keywords Advanced persistent threats · APTs · Malware · Machine learning · Artificial intelligence · Threat actors · Cyberattacks

1 Introduction

In June 2010, a cybersecurity researcher named Sergey Ulasen, discovered a malicious computer worm. This worm, codenamed Stuxnet, is thought to have been in development since at least early 2005 and is still regarded as one of the most sophisticated APTs ever seen. Stuxnet's purpose was to sabotage the Iranian nuclear program and reportedly ruined almost one fifth of Iran's nuclear centrifuges causing enough physical damage to the infrastructure to set the entire program back 4 years [6]. This malicious worm was part of what we now know and call an Advanced Persistent Threat (APT).

An APT can be described as a prolonged persistent cyber-attack in which access to a network is achieved but remains undetected over a long period of time. The attackers go to extraordinary lengths to avoid detection. The threat infiltrates the network of choice using a multitude of different attack vectors and once access is gained, advanced methods are used to avoid detection while increasing their foothold on the overall network. These attacks are then used to exfiltrate data, control systems and in some cases destroy infrastructure.

The complexity and cost of APTs suggests that in the vast majority of cases the attacks are specifically targeted, well-funded, resourced and patient which has led to a general consensus that they are state sponsored. According to a recent review of top threat actor groups and the countries they operate from [20], North Korea, Russia and Iran currently list in the top three.

It is widely accepted that the Stuxnet worm was part of an APT attack engineered by both American and Israeli intelligence, although this was never officially confirmed by either country the fact remains that this attack very successfully and significantly damaged the Iranian Nuclear program without the need for any physical military involvement.

Another APT codenamed Duqu, a derivative of Stuxnet suspected of either being created by the same organisations or at least a group with access to the original source code was discovered in 2011. This APT's payload was not to directly cause any damage but rather to gather information specifically around industrial control systems. One of the vital parts required in a sophisticated attack involving different phases of attack.

Traditional attacks tend to try achieve immediate and fast access to a target. The attack is carried out and once the objective is met, the attacker leaves with no clear plan or intention of returning. While APT's often use many of the same techniques to infiltrate a target network, their primary focus is to avoid all detection systems, gain a foothold and begin to spread across the network to ensure that if a compromised node is detected, they still have access to the network via one of the other infected nodes. This allows them to spread slowly and quietly ensuring that they go undetected while they go about their intended attack. A successful attack will not necessary mean that they will leave, if undetected, they will keep their foothold to either use at another point or even sell off to another adversary.

It is important to remember that the threat of APTs wouldn't be restricted to the traditional LAN/WAN network environment but could also be utilised on any type of network. This would include both Internet of Things (IoT), and Vehicular Ad Hoc networks (VANET) infrastructures posing a serious threat and risk to any network.

2 Advance Persistent Threats (APTs)

2.1 What Is an APT

An APT could be defined as a series of both basic and advanced malicious techniques and methods used in conjunction to build an attack which not only grants an attacker access to a victim network but expands and maintains access over a long term to ensure that as much valuable data and malicious damage can be done with the minimum chance of detection.

The attacks differentiate themselves from traditional threats in that:

- The attackers are highly organised, sophisticated, determined and operated by a well-resourced group.
- The targets are specific.
- The purpose is strategic.
- The approach is one of repeated attempts, stays low and slow, adapts to resist defences and is generally long term.

The National Institute of Standards and Technology (NIST) defines an APT as:

> An adversary that possesses sophisticated levels of expertise and significant resources which allow it to create opportunities to achieve its objectives by using multiple attack vectors (e.g., cyber, physical, and deception). These objectives typically include establishing and extending footholds within the information technology infrastructure of the targeted organizations for purposes of exfiltrating information, undermining or impeding critical aspects of a mission, program, or organization; or positioning itself to carry out these objectives in the future. The advanced persistent threat: (i) pursues its objectives repeatedly over an extended period of time; (ii) adapts to defenders' efforts to resist it; and (iii) is determined to maintain the level of interaction needed to execute its objectives. [17]

2.2 The Actors

The vast majority of APT attacks are state sponsored. Looking at currently identified and tracked APTs, their objectives and the groups known to have orchestrated them, and it quickly builds up a picture of the top 6 countries in which the actors operate from, namely:

- North Korea
- Russia
- Iran
- India
- Russia
- China

In a 2018 report by AlienVault [20], the top ten most reported active threat actor groups and their locations were as follows in Table 1:

The Lazarus group, also known to united states intelligence as "Hidden Cobra" is widely accepted to be sponsored and controlled by the North Korean government. This group's primary focus are attacks within the financial markets. One of their campaigns nicknamed "FASTCash" was responsible for large amounts of theft from ATMs in both Asia and Africa with an attack, which started in 2016, and is still ongoing. In 2018, the US department of homeland security (CISA) issued an alert to this effect. On the 10th of April 2019, CISA released another alert attributed to the Lazarus group [7]. This alert details a piece of malware which has the ability to connect to a command and control server in order to transfer stolen files from an infected network.

The Malware, known as "Hoplight" masks traffic between the victim and the remote server by acting as several proxy applications.

Table 1 10 most reported APTs

Rank	Advanced persistent threat	Location
1	Lazarus Group	North Korea
2	Sofacy	Russia
3	Muddy Water	Iran
4	Oil Rig	Iran
5	Patchwork	India
6	Energetic Bear	Russia
7	Kimsuky	North Korea
8	APT 15	China
9	Stone Panda	China
10	Turia	Russia

According to the alert, “The proxies have the ability to generate fake TLS handshake sessions using valid public SSL certificates, disguising network connections with remote malicious actors.” [7]. North Korea’s backing for the Lazarus group falls outside of the typical state sponsorship for the purpose of espionage and intellectual property theft. The objective of this group is purely financial gain, which when one looks at the severely isolated and cash starved state, it is clear why this group is so critical.

The Sofacy group also known as Fancy Bear is highly suspected of being sponsored by Russian military intelligence. In 2018 an indictment by Robert Mueller, the United States special council looking into Russian Interference in the United States 2016 presidential election, identified the Sofacy group as two GRU (Main Directorate of the General Staff of the Armed Forces of the Russian Federation) units knows as Unit 26165 and Unit 74455.

This group has been operating since around the mid 2000s and specifically targets government, military and security organisations.

One of the groups attributed attacks was an attack on German parliament in 2014. Specifically, the government’s “Informationsverbund Berlin-Bonn” (IVBB) network, which is a separate and private network used by the Chancellery and Federal Ministries. Ironically, this network was setup separately from other public networks to ensure an added layer of security.

The Dutch Government also accused the group of data theft from the Organisation for the Prohibition of Chemical Weapons (OPCW) in The Hague and most recently and famously, this group has been specifically mentioned in ties to the 2016 American election meddling investigation. Their primary target is and has always been NATO member states.

Clear actor identification can be challenging. Various vendors and intelligence agencies often name the threat actors differently which can lead to some confusion within the market. Some naming conventions are designed to create a mythological or figurative emotion, others are just naming tags given for the sole purpose of identification, yet others are just named after specific malware that that was used in an attack. A further key reason for differences is that threat actors could occasionally join and then split up causing further confusion on the actual threat actor responsible.

An example of the varied naming conventions could be the APT group “Comment Crew” [10]. This Chinese group, attributed to the second Bureau of the People’s Liberation Army (PLA) is named “Comment Panda” [5] by reseller Crowdstrike, “PLA Unit 61398” [8] by reseller IRL, “TG-8223” by Dell Secure Works “APT 1” [10] by reseller Mandiant and even “brown fox” by reseller iSight. These differences in naming can be confusing and there are calls for standardisation but it’s just not that simple. There are technical and “people” reasons why certain vendors use certain naming conventions.

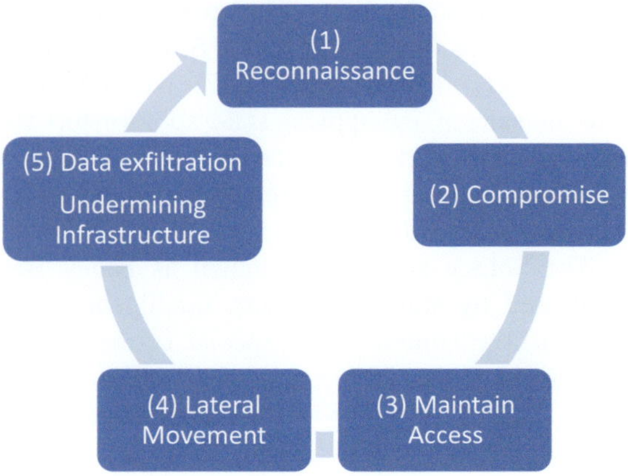

Fig. 1 Typical APT Lifecycle

2.3 APT Lifecycle

The typical APT lifecycle can be split into several different phases (see Fig. 1). Although various researchers break down the steps differently [4] ([22, 29], they all essentially break an attack down into five distinct steps.

Reconnaissance

Once attackers have identified the target and a strategy for attack, they need to research the target so that they are completely familiar with the people, systems and processes that are used. This reconnaissance would typically include both physical and passive cyberattacks in an effort to gather as much information as possible.

The people aspect of the reconnaissance would not necessarily only be staff but could include contractors, vendors and partners. These reconnaissance missions often employ large numbers of researchers and can involve a significant amount of time and cost and are almost always passive to ensure no red flags are raised. If the Stuxnet attack on the Iranian nuclear reactors is reviewed, it can be understood that the attackers had expert knowledge of the internal systems used and critically the Siemens programmable logic controllers (PLCs) used on the centrifuges within the facility. This is no small feat and would have involved significant research and knowledge.

With this knowledge, attackers would then need to identify an initial entry point to the network. This point would not only be the easiest path to entry but also the point where an attack would stand the best chance of going undetected. Wherever possible, multiple points would be targeted to ensure success.

Compromise

In this phase, the attacker crafts an attack with the sole purpose of infecting a victim's machine. This is commonly in the form of a socially engineered attack with spear phishing and watering hole attacks being the preferred route [22], but could really be any available resource to the attacker. The attack could even come indirectly through a third party which is trusted by the victim.

Again, in the case of Stuxnet, it is suspected that an infected removable disk storage unit inadvertently plugged in by a staff member was used to distribute the attack. [6]. Analysis of Stuxnet shows us that four zero-day exploits were built into the malware. This is a massive number in comparison to all other APT attacks. The attacks are well crafted and designed to bypass traditional Intrusion Detection Systems (IDS) while the exploits used are often zero-day attacks that any proactive level of patching would not help to prevent [2].

Internal staff are often regarded as the most cost-effective way to infiltrate the network and this is seen by the amount of attacks targeting end users directly.

Maintaining Access

Maintaining access and lateral movement are really the two phases which set an APT apart from other typical opportunistic type attacks.

Once the attacker has managed to compromise an internal system, in almost all cases its vital that a back door is installed to continually maintain a level of access to the infrastructure. To do this, a Remote Access Trojan (RAT) is installed on the victim machine/s as described by [4, 22].

Once the attacker has created the backdoor to the network, they would then proceed to compromise several other machines thereby ensuring that access can be maintained even if one of the compromised systems are discovered or indeed just taken offline. The RAT will then make a connection to an external Command and Control server (CnC). This CnC server then dictates to the RAT what should be done on the victim machine/s. This would explain how [16, 22] the connection from the RAT to the CnC server will in almost all cases be initiated from the RAT outwards to the CnC. This is done to help hide the traffic and bypass typical security controls, as most networks are configured to be far more lenient on outgoing connections than incoming traffic.

Lateral Movement

APTs operate in a "low and slow" method, gaining access slowly and carefully and spreading their connectivity from within the network.

In this phase, the attacker would be able to perform internal scans to map out traffic routes and other hosts within the network segment. Details of the environment, systems, functions and processes are discovered, both hardware and software vulnerabilities, unprotected network resources and additional access points to the network are mapped. Although internal scans could be detected, the lateral movement is often not, due to the use of compromised valid credentials already obtained as detailed by [22]. Since an APT's main goal is to gain access and remain in the network for an extended amount of time, every method and technique used is built around avoiding detection. One example of the techniques used in an attack is operation Aurora, also known as Hydraq or the Google hack attack. This attack originating in China [9], used an old obfuscation technique called spaghetti code to help keep itself hidden from network protection mechanisms. This was originally recognised as an inefficient and unstructured way of coding which was highly discouraged but was used to great success when the coders were after exactly that effect.

Moving laterally within a network allows the attacker to access and infect further endpoints over time using the elevated privileges gained in earlier steps to access targeted data/systems.

Data Exfiltration
This is the final stage and the objective of the attack. However, this stage does not have to only be about data exfiltration; it could be about undermining critical aspects of the targeted infrastructure as described by [17]. Data exfiltration mentioned by [22] and collaborated by [16] and could be executed in many different ways:

- Encrypted or clear data could be exfiltrated to the CnC server(s). This could be done from one or multiple victims to either one or multiple CnC servers. The advantage of exfiltrating data in an encrypted format would make it even harder for intrusion detection and data loss prevention (DLP) systems to detect the data loss.
- Although data could possibly be exfiltrated all in one go but with the intention of longer-term access to the victim needing to be maintained, very low and slow levels of data leakage would help prevent being detected, successfully exfiltrating data and maintaining access for future use.
- Steganography is a technique that could be used to insert the data into an image which could be displayed as a .jpg file as was the case in the *Duqu* APT [34]. This would appear as normal day to day typical use by a user which would be very difficult to identify as anything malicious [14].
- Physical human intervention could be used to gather the exfiltrated data from a defined location. One way this could be accomplished would be a technique called "dead letter box".

A recent example of successful data exfiltration is represented by the Equifax data leak in 2017 [12, 23] in which 147 million customers sensitive personal information was leaked.

3 Attack Examples

3.1 How Did They Do It?

Looking at two examples, Stuxnet and Lazarus Group, of well-known and successfully implemented APT attacks, we can analyse exactly how these attacks were carried out in each of the five phases to build a complete picture.

3.1.1 Stuxnet

One of the most sophisticated and precise APTs ever detected. This attack was very precisely aimed at Iran's Nuclear plant, Natanz (see Table 2 for attack phases and its descriptions).

Table 2 Stuxnet attack phases and descriptions

Attack phase	Description detail
Reconnaissance	The Stuxnet worm was targeted at very particular and specific Siemens Programmable Logic Controllers (PLCs). The worm was so well written, it required absolutely no intervention from any internal staff to work. A simple plugin to a USB drive was all that was necessary. To achieve this level of functionality the attacker would have to have detailed information of the network, infrastructure and centrifuges.
Compromise	The Natanz plant was air gapped from the internet. It was not possible to attack it directly from the internet however it is widely accepted that the Stuxnet worm was introduced into the plant via a USB key. It is not known whether this was done accidently by staff or deliberately.
Maintain access	Stuxnet was targeted directly at certain logic controllers controlling centrifuges within the plant. It was so specific that while it was programmed to spread from machine to machine, it was coded to search for certain hardware components and if they were not found, no action at all was taken. The worm would lie dormant taking no further action. Additionally, the worm was designed to self-destruct on the 24th of June 2012. In most cases, APT's establish a connection to the outside world by installing a remote access trojan (RAT) on the machine, however in the case of Stuxnet the attackers knew that it would not be possible for a RAT to communicate with the outside world once deployed so the worm had to be completely self-sufficient and run without waiting for any external instructions. An incredibly hard task to accomplish.
Lateral movement	This worm was specifically written to spread at a rapid pace using four in-built zero-day attacks to ensure that it would be able to achieve its target. Although traces of Stuxnet were found on systems all over the world, the biggest concentration of infections were all over Iran. It's important to consider that the worm would take absolutely no action on any machine that didn't have the correct Siemens controller software on it.
Data exfiltration undermining infrastructure	The payload was to destroy centrifuges in the plant. To achieve this, Stuxnet made the centrifuges spin dangerously fast for a short period of time but critically had already infected the monitoring systems within the plant to not detect this change. Although engineers could hear that the centrifuges were spinning dangerously high, the control systems indicated that all was within normal parameters. About a month later Stuxnet then slowed the centrifuges down dramatically for around 50 minutes, again with all control systems showing the plant running within perfectly normal operation parameters. The dangerous repetition of this caused over 1000 centrifuges (around 20%) at the plant to collapse.

3.1.2 Lazarus Group – Financial Threats

Founded in 2009, the Lazarus Group, a very active North Korean sponsored threat group best known for their attacks specifically targeted around financial gain. They attack the world cryptocurrency exchanges, financial institutions and banks Although this is not their only attack profile. Below is a high-level look at one of their most recent attacks on a Chilean organisation called Redbanc (see Table 3 for attack phases and its descriptions).

3.2 Detection Challenges

The sophisticated nature of APT's means there are significant challenges in detecting them. At every stage of their typical lifecycle, everything possible is done to avoid detection.

The reconnaissance is detailed, well-funded and passive to avoid any means of detection while the compromises take any and all approaches necessary from physical infiltration to cyber hacking. In most cases, multiple zero-day attacks are utilized to prevent being detected by traditional intrusion detection systems (IDS) [6], also rendering both system patching and signature based anti-virus and malware detection useless [18]. Messmer [19] and Kruegel [24] argue that even *Sandboxing*, an often used and preferred malware detection method can by bypassed by skilled and well-funded adversaries using methods such as, environment-specific-techniques, human-interaction-techniques, VMware-specific techniques, and configuration-specific-techniques. Using these detection avoidance techniques has led to a 200% rise in malware capable of evading detection [19]. The persistent nature of these attacks means that even in cases where a completely isolated system is enforced, the victim could still be physically compromised by being influenced into plugging a removable media drive into an internal system (USB drop attack) [30].

As previously discussed, maintaining access to the victim is a key aspect of the persistence of an APT. Data exfiltration or undermining the infrastructure can only happen when the correct targets are identified and compromised. This process can take a significant amount of time hence the need for access to be maintained. This is accomplished using external CnC servers which use various techniques to maintain access to the victims while avoiding detection. These methods as described by ([1, 6] include but are not limited to:

- Remote Access tools (RAT) which are often used in day to day business use and make use of a server and client agent.
- Social Networking sites that the victim's machine goes to which could put control information into blog posts and status messages
- TOR Anonymity Networks which by their very nature are designed to hide services and traffic.

Table 3 Lazarus group attack phases and descriptions

Attack phase	Description detail
Reconnaissance	Redbanc is a Chilean company whose business is responsible for all interconnectivity between the ATM infrastructure in the country. To gain access to the network, attackers created a front company and posted a job opening on LinkedIn for a developer position within the company. At that point, they were not sure who would apply for the job. An employee of Redbanc saw the posting and applied for the position.
Compromise	Once the employee had applied for the job, the group arranged a video conference interview over Skype™ and in that interview was asked to download and run a file that would help with the recruitment process seen below in Fig. 2 [11].
	The file appeared to generate a standard job application form, but this file called ApplicationPDF.exe was in fact a Microsoft Visual C#/ Basic .NET (v4.0.30319)-compiled executable file which infected the employee's computer with a piece of malware called PowerRatankba. This malware, allowed the attackers to gain information about:
	The hardware
	Operating system
	Running processes
	RPC and SMB file shares
	Computer name
	User name
	Proxy settings
	Through this compromise, the attackers were able to get further reconnaissance of the target and decide if the other stages of attack would be of value to them. The attackers clearly decided that this was a desirable target.
Maintain access	As well as feeding back information about the target computer, the malware constantly reports on the status of its own remote connection the attacker. The malware gives the attacker the ability to delete the malware from the victim machine, modify and replace ps1 and VBS files, send data to a chosen destination server and download an executable to run via PowerShell. This is archived through its support for several different commands [26].
Lateral movement	The ability to upload further executables from the attacker to the victim gives the attacker many different opportunities to not only maintain access but also spread infection through the network. With the reconnaissance information gained in step one, the attacker knows the machine type, operating system and running processes on a standard staff desktop thereby giving them vital information on the standard company installation profile. Information on running processes is extremely valuable as it allows the attacker to build a profile on any security measures and software running on the machines. This includes specific firewall and anti-virus tools.
	In the case of the attack on Redbanc, infection spread to a significant number of machines.
Data exfiltration	Exact financial losses are not clear as Redbanc has never released any information regarding this however, other attacks by the Lazarus group on ATM infrastructure in Asia and Africa are well documented.

(continued)

Table 3 (continued)

Attack phase	Description detail
Undermining infrastructure	The joint FBI, DHS and Treasury US-Cert technical alert report details the FASTCash scheme used against ATMS. "FASTCash" *schemes remotely compromise payment switch application servers within banks to facilitate fraudulent transactions. The U.S. Government assesses that HIDDEN COBRA actors will continue to use FASTCash tactics to target retail payment systems vulnerable to remote exploitation."*
	"According to a trusted partner's estimation, HIDDEN COBRA actors have stolen tens of millions of dollars. In one incident in 2017, HIDDEN COBRA actors enabled cash to be simultaneously withdrawn from ATMs located in over 30 different countries. In another incident in 2018, HIDDEN COBRA actors enabled cash to be simultaneously withdrawn from ATMs in 23 different countries."
	As previously mentioned, the US government defines the Lazarus group as Hidden Cobra.

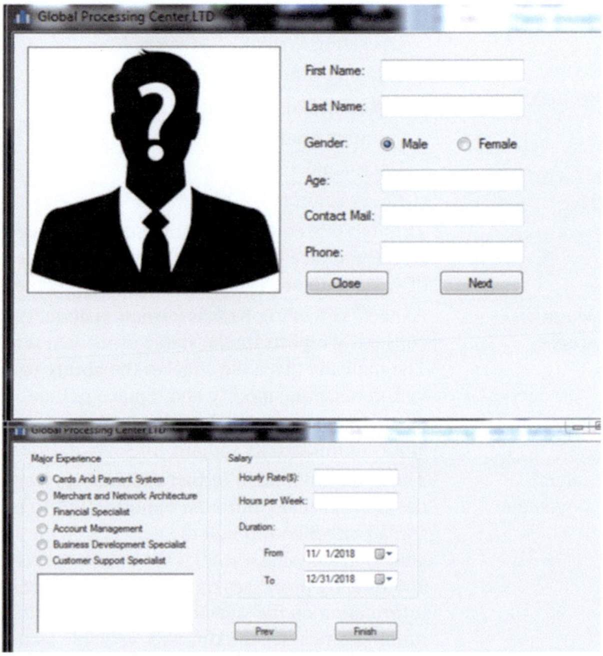

Fig. 2 Redbanc fake job application

The ability to move laterally is arguably the most dangerous phase of the attack and almost certainly the most time consuming. In this phase, the attackers remain undetected by often making use of built in Operating System (OS) features and utilities whose use cases would not look out of the ordinary to any security software. By using these in-built tools, internal reconnaissance would allow the adversary to obtain information about additional systems, network structure, network drives, security software used and network security detection systems. A key part of this phase would be the ability to harvest user credentials, particularly those with elevated access rights. The use of authorised access credentials would generally not flag as suspicious to the internal systems unless accounts were used in multiple locations at the same time. Data exfiltration can be accomplished using low and slow techniques like DNS tunnelling as described by [28]. This technique when done slowly and making use of custom coding is very difficult to detect. Exfiltrated data is compressed to limit the size as much as possible. The data is additionally encrypted using SSL/TLS to restrict the type of scanning that can be performed masking the data and the communication channel. The use of TOR networks is often used to accomplish this.

There are three factors that any successful APT requires:

- The attacker must have the ability to execute their malicious code on a machine(s) within the target environment. This would include individual vehicles in VANET
- The attacker requires the ability to CnC the machine(s) on the target environment and this ability has to be maintained. There must be the ability to get messages in and out of the target network.
- Lateral movement requires that the attacker is visible. If they have valid network credentials, this is hard to detect but they will be visible.

3.3 How Do We Detect APT's Today

As discussed at length already, there are significant challenges with APT detection however significant research on this problem has been done and researches have discussed various different detections methods to deal with this issue.

3.3.1 Network Sensors

Bhatt et al. [3] argue that effective detection of APTs is only possible with network sensors which can detect all attack facets. Further to this [27] finds it is necessary to continuously monitor and analyse features of a TCP/IP connection. These include:

- Number of transferred packets
- Total count of the bytes exchanged
- Duration of TCP/IP connections
- Information on the number of packet flows

Bhatt et al. [3] suggests a method for detection is to install sensors in each layer of the network. All alerts and logs would then be collected and stored. Correlation of data for each layer could then be performed and this would assist in identifying attacks in progress. An issue highlighted with this approach is the sheer number of logs which are typically generated in all the layers of attack. Hale [13] and

MacDonald [21] point out that in a typical network of 100 hosts, one can expect around 100GB of logs and alarms a day. If we consider a typical network with varying node density, mobility and a constant increase in users, analysing this volume with current methods would be extremely challenging. Another proposed technique used to detect attacks is honeypots.

3.3.2 Honeypots

Jasek et al. [15] propose a system of detecting APTs using honeypots, a system or network of systems (honeynet) whose sole purpose it is to attract attackers and then record their activities. The proposal makes use of high and low interaction honeypots as well as separate honeypots on production systems. Jasek et al. [15] argues that traditional honeypots are limited in that they are passive and wait for the attacker. It proposes making use of an agent which is installed and directs the attacker to the honeypots. The engagement is a 5-step process as follows:

1. Connect the system of Honeypots to the production environment using low and high interactive honeypots and activated agents.
2. The attacker compromises a production client and, in their reconnaissance, discovers shared resources on other systems (honeypots)
3. The attacker gains access to the honeypot systems and compromises them.
4. The attacker collects data from the compromised systems and honeynets and sends the information out to the CnC server externally.
5. The administrator detects the compromise from the honeypot systems and the traffic outflow.

With the attacker activity logged and monitored, the administrator(s) is then fed this information. The administrator is then theoretically able to apply rules and procedures to defend against the attack on the production environment (Fig. 3).

While honeypots unquestionably increase our understanding of malicious network activity and provide an interesting option for detection of malicious activity, there are several issues that are raised with the use of honeypots. Questions around the legality and privacy of honeypots exist; collection and monitoring of user information, malicious or not could fall foul of privacy laws. Sokol et al. [32] highlights privacy issues within the European Union (EU) while [25] addresses the

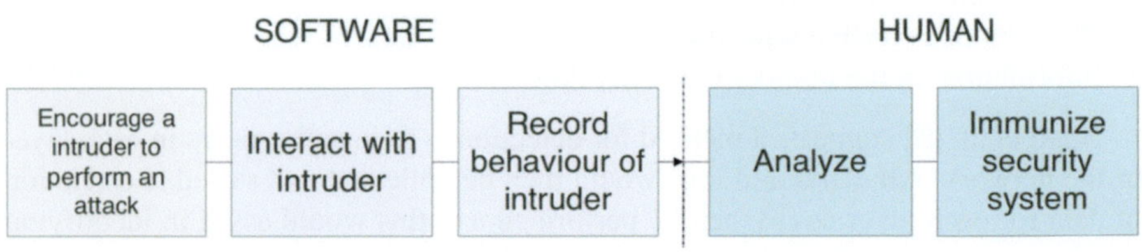

Fig. 3 Honeypot interaction model

same concerns from the legal jurisdiction of the United States of America (USA). There are also concerns around the risk of honeypots and how an attacker realising that a honeypot is being used could then compromise the honeypot in such a way as to attack, infiltrate or harm other systems or organisations [33]. Another prominent proposed detection method is that of machine learning (ML).

4 Machine Learning and Artificial Intelligence

4.1 Current Detection Methodologies

Typical security mechanisms do not adequately address APTs in in this new highly mobile, varied and complex ad-hoc type network world. It is impractical to think that human intervention and detection skills could solve the challenges presented in such a complex and completely ad-hoc network especially when one considers that in certain cases no input or information is available about the attack at all. In such cases unsupervised Machine Learning techniques (ML) are seen as a solution which could deal with this threat. Machine learning techniques can generally be split into two different approaches. Artificial Intelligence (AI) and Computational Intelligence (CI) [35] AI techniques have their roots in traditional methods like statistical modelling while CI techniques are most commonly based on nature-inspired methods that are used to deal with challenges that classic methods are unable to solve. CI methodologies include but are not limited to evolutionary computation (genetic algorithms), fuzzy logic, artificial neural networks (ANN), artificial immune systems (AIS) and swarm intelligence (SI). "*AI handles symbolic knowledge representation, while CI handles numeric representation of information*" [35]. Although it's not always easy to distinguish the boundary between these two broad categories. Hybrid methods are possible and sometimes proposed but generally speaking are used independently of each other.

Fractal dimension-based machine learning is one such possibility proposed by Siddiqui et al. [31]. The authors present a correlation algorithm which makes use of fractal dimensions to detect APT based anomalous traffic patterns with high accuracy and reliability using a feature vector obtained through the processing of TCP/IP session information.

The feature vector selected is based on two metrics:

- Total data packets transferred during a single TCP session
- The duration of a complete TCP session.

The researcher's analysis of TCP data concludes that APT traffic consists of a small count of data packets in a short or long-lived TCP session, whereas normal internet traffic exhibited patterns of a large amount of data packets in a short duration. This is consistent with the APT low and slow exfiltration method already discussed.

The basic requirement of the algorithm is an accurately labelled reference dataset of the features. Each data point is classified as anomalous by comparing the correlation fractal dimensions of the corresponding dataset.

The algorithm first calculates the correlation fractal dimension of the attack and normal reference datasets separately, and then forms a prototypical measure for each class. To classify new input samples, the methodology computes the correlation fractal dimension of the new samples with the reference data set and compares that, to the prototypical measures of the normal and attack data sets. The class for which there is a minimal change in the fractal dimension, indicates that, the point belongs to the particular class. This can also be regarded as finding the similarity index of the new sample and choosing the class to which the input is most similar. This methodology has proven more effective at reducing both false positives and false negatives.

Paredes-Oliva et al. [27] has proposed a novel scheme which also makes use of ML techniques to detect anomalies in traffic patterns. The authors make use of a combination of both frequent item-set mining and decision tree ML techniques to accomplish this and while not directly looking at APTs, such classification would detect anomalies which could then be classified as required. The authors argue that most anomaly detection systems differentiate between normal traffic and anomalies but they do not distinguish different anomaly types which is a key focus of the proposal. The authors first analyse a large set of flows for one or more flow features in common. This is called frequent item-set mining (FIM). An example of this would be a typical network scan; this will produce many separate flows with the same source IP address and destination port. After applying FIM, the result would be one frequent item set with two items: the scanner IP address and the scanned port number. The scheme then builds a decision tree to classify the FIMs as benign or anomalous. Once this process is complete, the anomalies could then be classified by specific type. Figure 4 visually illustrates this process.

Using this methodology, the authors were able to simultaneously monitor two high volume 10Gb/s links and maintain a classification accuracy of 98%.

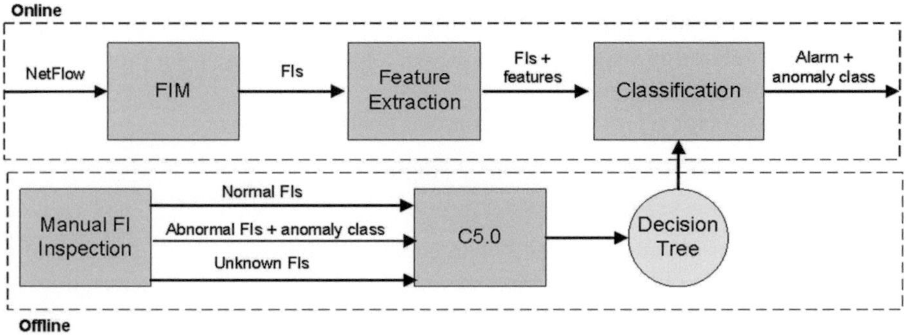

Fig. 4 Anomaly detection system overview [27]

This opens up the question of how does a machine learning classifier begin to identify an attack?

4.2 Attack Visualisation

If we take a standard dataset of benign network traffic and then randomly inject several APT attacks into it, we have the opportunity to analyse these flows and visualise just how the attacks integrate into the traffic.

Taking five separate attacks approximately 5 Mb in total size and injecting this into a 4.4GB standard benign network traffic dataset, we can extract each bidirectional data flow and analyse several attributes of the flows. Breaking these streams down results in 137 APT data streams amongst 7703 benign data streams. A total of 1.78% of the total data.

If we then extract some of the individual attributes of the streams such as:

- Flow duration
- Total forwarded packets (per flow)
- Total backward packets (per flow)
- Maximum forward packet length
- Minimum forward packet length
- Mean forward packet length
- Flow Bytes per second
- Flow packets per second
- Backward packets per second
- Standard packet length
- Down/Up ratio
- Average packet size
- Backward segment size average
- Average forward Bytes/b
- Label (Manually labelled as attack or benign).

It is then possible to view how these attributes are seen by a machine learning classifier. We do this by using WEKA, an application written by the university of Waikato which has built a collection of machine learning algorithms on a single platform to simplify the task of data mining using machine learning classifiers.

Figure 5 is how this data analysis displays in WEKA. The red dots are the benign data streams while the blue dots are the attack data sets. This very clearly highlights the characteristics of the typical low and slow APT data transmission. The duration of flows is much lower over the entire time period under analysis. This, as discussed, is one of the methods used by APTs to avoid detection by traditional intrusion detection systems.

A further illustration of this can be seen in Fig. 6 where average packet sizes are illustrated by grouping them by size over the same duration. A large percentage of the APTs are recorded in the lowest packet data size hidden amongst benign data

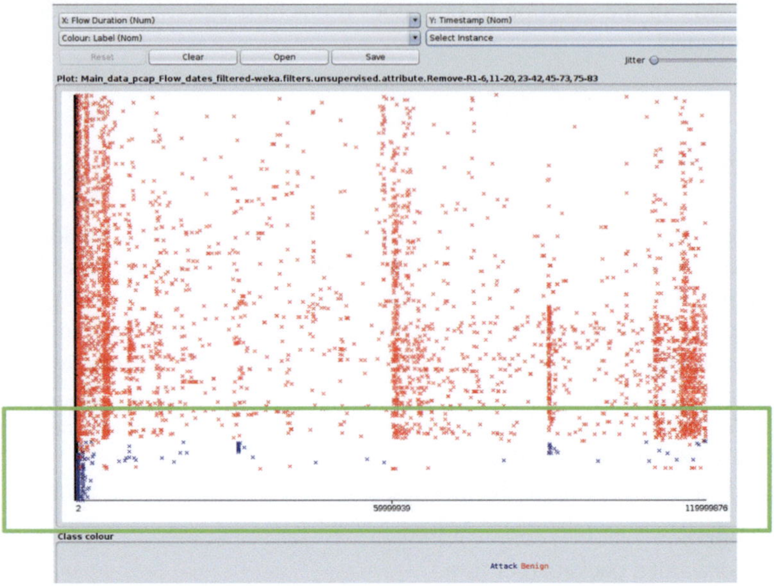

Fig. 5 Visual representation of flow duration typical of APTs

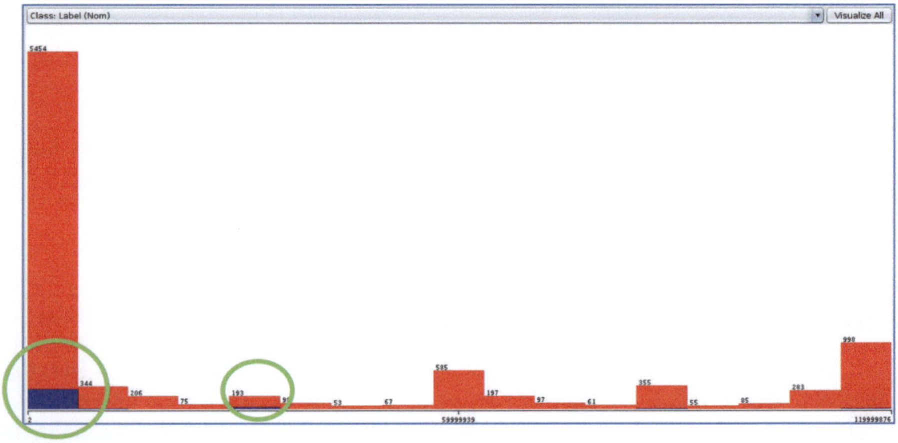

Fig. 6 Visual representation of packet size grouping

flows of the same nature. This grouped with the short flows shows just how data is transmitted, slowly over short periods and small sizes making it very difficult to detect.

4.3 Analysis

Although extremely challenging to detect, there are techniques which can be utilised that give a higher chance of detection. The attacks are sophisticated and well-crafted and often include components that traditional intrusion detection systems (IDS) do not detect.

Too many techniques are passive and look for particular signatures which only work when the attack types have been identified before. To add to this, the volume of data and logs created on a standard corporate LAN/WAN network is staggering. The ever-increasing quantity of data really does make detection a case of finding a needle in a haystack and a fact that attackers rely on.

One successful technique in this detection challenge is searching for suspicious behaviour but the key to this is that it has to be done in the absence of a baseline. One cannot simply analyse a network, assume it's clean and then create a benchmark based on that to analyse future traffic. Fundamentally, it can never be assumed that a network is clean and free from contamination. Applications vary greatly and there is a constant introduction of new and upgraded network components which create an ever-changing network traffic profile.

Honeypots, as mentioned earlier in this chapter, might help to detect an attack but this is a passive approach that doesn't allow for real time analysis and detection and can be extremely difficult to implement in a sophisticated network architecture. They do however help to build an overall knowledge of attacks which in turn helps to identify characteristics that attacks might have in common.

APTs use a combination of techniques and methodology to attack a victim and these will vary depending on who the victim is. Equally, successful defence against this type of adversary will require a combination of differing techniques. A one shoe fits all approach will not work and a consolidated approach will produce better results.

5 Conclusions

Advanced Persistent Threats are an attack type which cannot be underestimated and must be taken seriously. They are hard to detect, prevent, and if infected, to remove. No industry is immune from attack and APT is agnostic to any organisation type.

Reconnaissance of the target is detailed and effective and because most attacks are state sponsored, they are well funded and resourced. The attacks in themselves are specific, with clear objectives in mind.

Attacks are patient and run through several different phases from reconnaissance, compromise, lateral movement and eventually payload delivery. These attacks can take years to deliver their complete payload and all the while, the victim is completely unaware that they are infected. From intellectual property and financial theft to critical infrastructure destruction, the threat is real and applies to all industries and network types and this ‘low and slow’ type attack is what makes this highly dangerous.

When considering the threats, landscape and attack types, attack consequences could be life threatening and devastating. An example of this could be a well-orchestrated attack on an autonomous vehicles VANET where a vehicle is taken over and maliciously used, but there are other attacks on VANET we could consider of a less severe nature where a vehicle could be infiltrated and the cars inbuilt microphone used for handsfree communication compromised, allowing the attacker to listen and record all conversations within the car over an extended period of time. This could be a source of invaluable information to the attacker.

Detection of these attacks using traditional techniques and intrusion detection systems is extremely challenging. A well-crafted attack making use of zero-day exploits used in conjunction with detailed knowledge of the target’s internal systems as in so many recorded cases can infect a network for years.

Real time Identification of suspicious behaviour in large data volumes can successfully be accomplished by systems which implement some form of machine learning classifiers. Human detection alone is impossible. While various detection methodologies have been researched, it is clear that the key lies in the accuracy of the detection and on how refined the classifiers are and how they are adapted to the data type. It is critical to keep false positive results as low as possible to avoid confusion. Artificial Intelligence might allow these classifiers to keep adapting and developing their algorithms as threats advance in this area and continued research in AI and ML may prove to provide beneficial outcomes.

References

1. Adair S, Deibert R, Rohozinski R, Villeneuve N, Walton G (2010) SHADOWS IN THE CLOUD: investigating cyber espionage 2.0|online safety & privacy|computer security. [online] Scribd. Available at https://www.scribd.com/doc/29435784/SHADOWS-IN-THE-CLOUD-Investigating-Cyber-Espionage-2-0#. Accessed 14 June 2018
2. Ben-Asher N, Gonzalez C (2015) Training for the unknown: the role of feedback and similarity in detecting zero-day attacks. Proc Manuf 3:1088–1095
3. Bhatt P, Yano E, Gustavsson P (2014) Towards a framework to detect multi-stage advanced persistent threats attacks. In: 2014 IEEE 8th international symposium on service oriented system engineering
4. Brewer R (2014) Advanced persistent threats: minimising the damage. Netw Secur 2014(4):5–9
5. Cdn0.vox-cdn.com (2014) crowdstrike-intelligence-report-putter-panda.original.pdf. [online]. Available at http://cdn0.vox-cdn.com/assets/4589853/crowdstrike-intelligence-report-putter-panda.original.pdf. Accessed 8 Sept 2019
6. Chen P, Desmet L, Huygens C (2014) A study on advanced persistent threats. In: Communications and multimedia security. Springer, Aveiro, pp 63–72
7. CISA Cyber Infrastructure (2019) MAR-10135536-8 – North Korean Trojan: HOPLIGHT|CISA. [online]. Available at https://www.us-cert.gov/ncas/analysis-reports/AR19-100A. Accessed 23 Sept 2019

8. Council on Foreign Relations (2019) Connect the dots on state-sponsored cyber incidents – PLA unit 61398. [online]. Available at https://www.cfr.org/interactive/cyber-operations/pla-unit-61398. Accessed 15 Sept 2019
9. Ferrer Z, Cebrian Ferrer M (2016) In-depth analysis of Hydraq – in-depth_analysis _of_hydraq_final_231538.pdf. [online] Paper.seebug.org. Available at https://paper.seebug.org/papers/APT/APT_CyberCriminal_Campagin/2010/in-depth_analysis_of_hydraq_final_231538.pdf. Accessed 6 Sept 2019
10. Fireeye Mandiant APT1 Report (2016) APT1: exposing one of China's cyber espionage units – mandiant-apt1-report. [online]. Available at https://www.fireeye.com/content/dam/fireeye-www/services/pdfs/mandiant-apt1-report.pdf. Accessed 18 Sept 2019
11. Flashpoint (2019) Flashpoint – disclosure of Chilean Redbanc intrusion leads to Lazarus Ties. [online]. Available at https://www.flashpoint-intel.com/blog/disclosure-chilean-redbanc-intrusion-lazarus-ties/. Accessed 6 Sept 2019
12. Gressin S (2017) The equifax data breach: what to do. [online] Consumer Information. Available at https://www.consumer.ftc.gov/blog/2017/09/equifax-data-breach-what-do. Accessed 7 Aug 2018
13. Hale B (n.d.) Estimating log generation for security information event and log management. [online] Content.solarwinds.com. Available at http://content.solarwinds.com/creative/pdf/Whitepapers/estimating_log_generation_white_paper.pdf. Accessed 9 June 2018
14. Hussain M, Wahab A, Idris Y, Ho A, Jung K (2018) Image steganography in spatial domain: a survey. Signal Process Image Commun 65:46–66
15. Jasek R, Kolarik M, Vymola T (2013) APT detection system using honeypots. [online] Pdfs.semanticscholar.org. Available at https://pdfs.semanticscholar.org/2f8e/f5890c39579bc9648158b710a1ef2b8366db.pdf. Accessed 12 July 2018
16. Jiang D, Omote K (2015) An approach to detect remote access Trojan in the early stage of communication. In: 2015 IEEE 29th international conference on advanced information networking and applications
17. Joint Task Force Transformation Initiative (2011) Managing information security risk. [online] Nvlpubs.nist.gov. Available at https://nvlpubs.nist.gov/nistpubs/Legacy/SP/nistspecialpublication800-39.pdf. Accessed 4 July 2018
18. Keragala D (2016) Detecting malware and sandbox evasion techniques. [online] Sans.org. Available at https://www.sans.org/reading-room/whitepapers/forensics/detecting-malware-sandbox-evasion-techniques-36667. Accessed 10 June 2018
19. Kruegel C (2015) Evasive malware exposed and deconstructed|USA 2015|RSA conference. [online] Rsaconference.com. Available at https://www.rsaconference.com/events/us15/agenda/sessions/2022/evasive-malware-exposed-and-deconstructed. Accessed 6 June 2018
20. LLC L (2018) Threat actors and exploits top ten lists of 2018|LIFARS, your cyber resiliency partner. [online] LIFARS, your cyber resiliency partner. Available at https://lifars.com/2018/11/threat-actors-exploits-top-ten-2018/. Accessed 19 Sept 2019
21. MacDonald N (2012) Information security is becoming a big data analytics problem. [online] Gartner.com. Available at https://www.gartner.com/id=1960615. Accessed 9 June 2018
22. Marchetti M, Pierazzi F, Colajanni M, Guido A (2016) Analysis of high volumes of network traffic for advanced persistent threat detection. Comput Netw 109:127–141
23. McCandless D (2018) World's biggest data breaches & hacks – information is beautiful. [online] information is beautiful. Available at http://www.informationisbeautiful.net/visualizations/worlds-biggest-data-breaches-hacks/. Accessed 10 Aug 2018
24. Messmer E (2013) Malware-detecting 'sandboxing' technology no silver bullet. [online] network world. Available at https://www.networkworld.com/article/2164758/network-security/malware-detecting%2D%2Dsandboxing%2D%2Dtechnology-no-silver-bullet.html. Accessed 17 June 2018
25. Mokube I, Adams M (2007) Proceedings of the 45th annual southeast regional conference. ACM, New York, pp 321–326

26. Paganini P (2019) Experts link attack on Chilean interbank network Redbanc NK Lazarus APT. [online] Security Affairs. Available at https://securityaffairs.co/wordpress/79929/breaking-news/chilean-research-redbank-lazarus.html. Accessed 6 Sept 2019
27. Paredes-Oliva I, Castell-Uroz I, Barlet-Ros P, Dimitropoulos X, Sole-Pareta J (2012) Practical anomaly detection based on classifying frequent traffic patterns. In: 2012 Proceedings IEEE INFOCOM workshops
28. Raman D, De Sutter B, Coppens B, Volckaert S, De Bosschere K, Danhieux P, Van Buggenhout E (2013) DNS tunneling for network penetration. In: Lecture notes in computer science. Springer, Cham, pp 65–77
29. Rashid P, Ramdhany D, Edwards M, Kibirige S, Babar D, Hutchison P, Chitchyan D (2014) Detecting and preventing data exfiltration. [online] seculanc_data_exfil_report. Available at https://www.lancaster.ac.uk/media/lancaster-university/content-assets/images/security-lancaster/seculanc_data_exfil_report.pdf. Accessed 10 June 2018
30. Scaife N, Carter H, Traynor P, Butler K (2016) CryptoLock (and Drop It): stopping ransomware attacks on user data. In: 2016 IEEE 36th international conference on distributed computing systems (ICDCS)
31. Siddiqui S, Khan M, Ferens K, Kinsner W (2016) Detecting advanced persistent threats using fractal dimension based machine learning classification. In: Proceedings of the 2016 ACM on international workshop on security and privacy analytics – IWSPA'16
32. Sokol P, Míšek J, Husák M (2017) Honeypots and honeynets: issues of privacy. EURASIP J Inf Secur 2017(1):1–9
33. Spitzner L (2002) Honeypots: tracking hackers. Addison-Wesley, Boston
34. Virvilis N, Gritzalis D (2013) The big four – what we did wrong in advanced persistent threat detection? In: 2013 international conference on availability, reliability and security
35. Zamani M, Movahedi M (2015) Machine learning techniques for intrusion detection. [online] Arxiv.org. Available at https://arxiv.org/pdf/1312.2177.pdf. Accessed 21 Dec 2017

Artificial Intelligence in Protecting Smart Building's Cloud Service Infrastructure from Cyberattacks

Petri Vähäkainu, Martti Lehto, Antti Kariluoto, and Anniina Ojalainen

Abstract Gathering and utilizing stored data is gaining popularity and has become a crucial component of smart building infrastructure. The data collected can be stored, for example, into private, public, or hybrid cloud service infrastructure or distributed service by utilizing data platforms. The stored data can be used when implementing services, such as building automation (BAS). Cloud services, IoT sensors, and data platforms can face several kinds of cybersecurity attack vectors such as adversarial, AI-based, DoS/DDoS, insider attacks. If a perpetrator can penetrate the defenses of a data platform, she can cause significant harm to the system. For example, the perpetrator can disrupt a building's automatic heating system or break the heating equipment by using a suitable attack vector for a data platform. This chapter focuses on examining possibilities to protect cloud storage or data platforms from incoming cyberattacks by using, for instance, artificial-intelligence-based tools or trained neural networks that can detect and prevent typical attack vectors.

Keywords Artificial-intelligence-based applications · Artificial intelligence · Cloud service · Data platform · Attack vectors

1 Introduction

Artificial intelligence is a major buzzword nowadays and is considered as the new "oil" of the future with the potential for great societal impact. AI has been under research for many decades, and it was originally presented as a novel way to mimic the cognitive functions of the human brain. AI has the capacity to process vast amounts of data, it has far-reaching applications, and it has been used in armed

P. Vähäkainu (✉) · M. Lehto · A. Kariluoto · A. Ojalainen
Faculty of Information Technology, University of Jyväskylä, Jyväskylä, Finland
e-mail: petri.vahakainu@jyu.fi; martti.lehto@jyu.fi; anjuedka@jyu.fi; anniina.m.t.ojalainen@jyu.fi

H. Jahankhani et al. (eds.), *Cyber Defence in the Age of AI, Smart Societies and Augmented Humanity*, Advanced Sciences and Technologies for Security Applications, https://doi.org/10.1007/978-3-030-35746-7_14

forces, construction, education, healthcare, space exploration and transportation around the world. In the healthcare sector, AI has succeeded in providing accurate diagnoses to prevent skin cancer, treatment recommendations, and provided surgical aid. In the field of smart buildings, AI can assist in finding anomalies and providing future forecasting in order to reduce maintenance costs.

Artificial intelligence can be defined as a system that thinks and acts rationally thinks and acts in such a way as to mimic rational humanlike behavior [24]. AI is a combination of information technology and physiological intelligence, which can be computationally used to reach goals defined. Intelligence is the ability to think through memory formation, pattern recognition, adaptive decision-making and experimental learning. Artificial Intelligence can make machines behave like humans and even surpass them in efficiency [50].

Artificial intelligence can be applied to a range of fields, such as healthcare, predictive building maintenance, military and cybersecurity. Cybersecurity provides the means to access the data and the data stored on them. Effective cybersecurity controls provide a cyberspace infrastructure, which is reliable and resilient. Lacking or absent controls lead to an insecure cyberspace. According to Bayuk et al. [7] cybersecurity applied to prevent, detect and recover from damage to confidentiality, integrity, and availability of information in cyberspace. In order to use all these factors, people, processes and technologies are utilized.

Smart buildings can be seen as a cyber-physical system (CPS) in which smart sensors automatically measure usage, functions, and variables describing the state of a building [70]. Energy, electricity, and water consumption, inside temperature, humidity, and other relevant variables are examined and used to automatically adjust, for instance, the heating system of a smart building. A building can be considered smart, even if only some of these variables are measured.

Cyber-physical systems provide a way to gather relevant data through smart sensors. The data has to be stored privately, securely, and it has to be available. Cloud services enable data replication and strategic storage on multiple servers spread across various geographical locations [72]. Replication improves data availability, reliability, and ensures fault tolerance. Smart services, such as AI-assisted data-intensive automatic heating adjustment system, can be developed using the stored data. In order for such services to work, they need working business models and replication functions to operate globally.

Cybercriminals are constantly looking for new ways to exploit vulnerabilities. Data gathered and stored into cloud storage, or distributed data platforms need to be secured. Cybercriminals today are able to leverage sophisticated attack vectors, including artificial-intelligence-based attacks, in determining exploitable vulnerabilities. These days, cybercriminals can utilize even more sophisticated attack vectors such as artificial intelligence-based attacks in looking for vulnerabilities they can exploit. An insider threat can be a significant threat to the system. An insider threat causes one of the most if even the most significant threat to the system. A malicious insider familiar with an organization's security practices, data, and computer systems can circumvent security controls to gain access to the system and the data. This motivates the need to research novel ways to detect exploitable vulnerabilities and prevent these high-risk cyberattacks.

IoT devices can be used for the collection of data to be stored on cloud services. Artificial intelligence and machine learning can be used in optimizing data usage efficiency. This is why this chapter is organized as follows: Section 2 examines the basics of artificial intelligence and machine learning. Section 3 deals with the basics of cyberspace and cybersecurity. Section 4 introduces smart buildings and services. Section 5 examines cloud services and data platforms. Section 6 presents common cloud vulnerabilities and attack vectors, including DoS and DDoS attacks, IoT based attacks, and insider threats. These attack vectors were selected because they came up repeatedly in scholarly sources. In addition, these vectors are both related to smart homes and cloud services. Section 7 discusses countering cloud cyberattacks and Sect. 8 concludes the chapter.

2 Artificial Intelligence and Machine Learning

Artificial intelligence (AI) and machine learning (ML) are disciplines with algorithms capable of learning representations from data. ML is a subset of AI [57]. ML contains the research areas of deep learning (DL) and deep reinforcement learning (DRL). Relations of these fields can be seen as overlapping circles. AI refers to systems that simulate human behavior. ML refers to systems capable of adapting themselves based on the situation. DL refers to the actual size of structure of the ML model. This applies to DRL as well, but DRL is mainly known for how the ML model learns. Learning is based on an action-feedback loop.

Artificial Intelligence is used in cases which humans consider time-consuming or tedious, or when an AI model can be trained faster than programming an explicit solution. Tasks in which AI and ML algorithms have succeeded particularly well include image recognition, image classification, image generation, and natural language processing. They have also been used for social media monitoring, marketing, predictive health monitoring, robotics, fraud detection [4]. Burnap and Williams [11] used ML for hate speech detection from Twitter and Zhao et al. [83] used artificial neural networks (ANN) to predict building energy usage.

One common feature among these types of models is the need to train the algorithms that need to be trained (such as supervised learning) first before the actual use. Supervised, unsupervised, and DRL methods are used widely. Supervised methods refer to cases where there are pre-labeled data for the training of the algorithm. Unsupervised refers to cases in which the ML algorithm estimates these labels itself. DRL is a special case of training algorithms because it uses feedback in order to learn and that feedback can come from a human expert or from the surrounding system.

In general, algorithmic learning happens based on data inputs and the desired output. Therefore these can be abstracted into a functional representation: f(input) = output. In the case of neural networks (NN), which are extremely popular in ML and AI research, when the algorithm learns, the training changes hidden values based on the results of an activation function for each node of the neural network. Training continues until the model has reached a sufficient level of accuracy. Accuracy is calculated with the minimizing function, which calculates the differences of predictions of the ML algorithm and the given true values.

According to Ghahramani [33], training these algorithms means that they learn models that represent part of the data or the behavior of the data. Another common feature is that AI solutions tend to be data-intensive systems. For example, using the NSL-KDD dataset Potluri and Dietrich [62] trained their DNN model in parallel to accelerate the learning of different attack types. They did not have enough data for all attack types, which resulted in decreased classification performance on those attack types. In order to perform well, DL models tend to need lots of quality data and GPU time. Currently, there is a trend among researchers to find ways to lessen the number of data samples.

3 Cyber Space and Cyber Security

The word Cyber comes from the Greek word κυβερεω (*kybereo*), which means to direct, guide, and control. Cyber refers to the digital world, which includes the surroundings and being present in our daily lives. In the year 1984, William Gibson's Neuromancer novel connected the words" cyber" and" space." Defining cyberspace is still a challenging task. Cyberspace is described in United States Cyberspace Policy Review as an interdependent network of information technology infrastructures and includes the internet, telecommunications network, computer systems, and embedded processors and controllers in critical industries. Typical usage of the term also refers to the virtual environment of information and interactions between people [22]. Based on literacy review cybersecurity can be connected to cyberspace as follows:" cybersecurity is the organization and collection of resources, processes, and structures used to protect cyberspace and cyberspace-enabled systems from occurrences that misalign de jure from de facto property rights" [21].

There is no universally accepted definition of cybersecurity, but the term is broadly used in literacy. Even though there is no universal definition of cybersecurity, a description of the word should bind human and information system component together. Cybersecurity can be defined as a range of actions taken in defense against cyberattacks and their consequences and includes implementing the required countermeasures. Cybersecurity is built on the threat analysis of an organization or institution. The structure and elements of an organization's cybersecurity strategy and its implementation program is based on the estimated threats and risk analyses. In many cases, it becomes necessary to prepare several-targeted cybersecurity strategies and guidelines for an organization. [50, pp. 3–29].

European commission defined cybersecurity in the Cybers Security Strategy of the European Union as the safeguards and actions that can be used to protect the cyber domain both in the civilian and military fields, from those threats that are associated with its interdependent networks and information infrastructure or that may harm them. Cybersecurity strives to preserve the availability and integrity of the networks, infrastructure, and the confidentiality of the information contained therein [30].

Original Martin C. Libicki's model of cyberspace [52] consisted of a three-layer model: semantic, syntactic, and physical. We created and enhanced our unique Libicki's model by adding the cognitive and the service layer into the original Libicki's model for it to better describe the cyber environment concerning a smart building concept discussed in this chapter.

The physical layer is the first layer, which consists of physical components of an information network. The physical layer includes all the equipment necessary to send, receive, store, and interact with and through cyberspace. The hardware and devices concerned are e.g., cables, routers, switches, transmitters, receivers, computers, and hard disks. The layer acts as a bridge between the physical layer and the syntactic layer.

The syntactic layer uses protocols and software to send, receive, store, format, and present gathered data through the physical layer. The syntactic layer divides into sub-layers by using, for example, OSI-model (Open System Interconnection). The syntactic layer is responsible for interaction between the devices connected to the network.

The semantic layer contains all the information and gathered datasets from smart building's IoT sensors and stores them into data storage, such as data warehouses, located on cloud services (data platforms). All the data stored needs to be secure, and current information security goals should be followed. Those information security goals being: confidentiality, integrity, availability of information, authenticity, accountability and non-repudiation, and reliability [9].

The service layer includes digital smart services that implement data gathered from smart building's IoT sensors. Digital smart services can be, for example, smart lock, automatic heating adjusting system, snowplowing service, or digital caretaker. The service layer also includes information security and data management services.

The cognitive layer's meaning is to provide an environment to understand visualized and analyzed information. The information considered is beneficial for decision-makers who build and maintain smart buildings. On the cognitive layer, information gathered is being analyzed to form a contextual understanding of information for a decision-maker (Fig. 1).

4 Smart Buildings and Services

There are several types of smart buildings, such as smart homes, smart airports, smart hospitals, smart factories. For example, Alam et al. [3] define smart home as a home that has sensors and appliances, which communicate with each other and the smart meter that continuously pushes and receives information to and

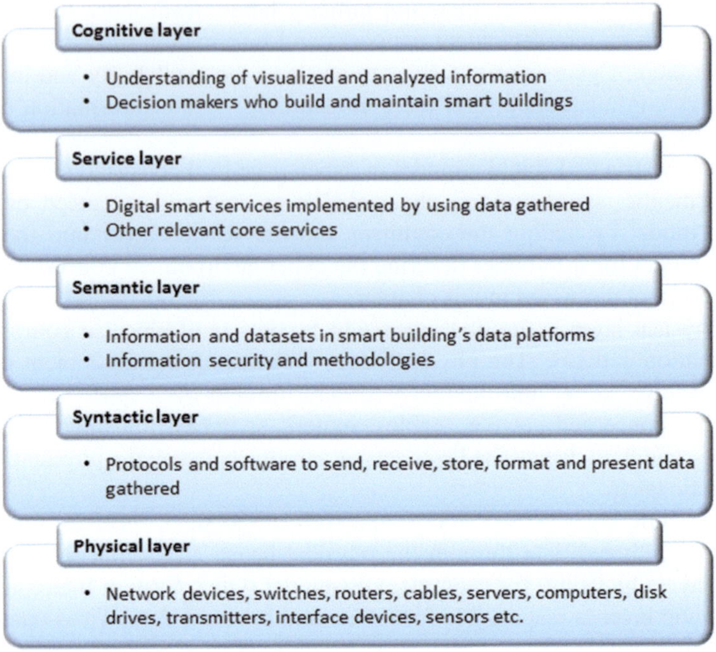

Fig. 1 Cyberspace Five Layer Model based on Libicki's model. (Modified by authors)

from the smart grid. The Smart grid forms from the union of information and communication technologies with the traditional power grid [43]. This information transfer is intended to minimize power consumption.

Authors define smart buildings loosely as buildings that have devices for energy consumption optimization of the structure while using sensors to gather data of the building conditions and actuators to maintain building conditions at an acceptable level for all inhabitants using some guiding method that could be perceived intelligent, such as AI. Devices and applications can be the Internet of Things (IoT) devices and operate under many different protocols, such as BTLE, PaaS & IaaS, ZigBee, SFP. These buildings are meant to protect both inhabitants and IoT devices against elements of nature. Smart buildings can include a multitude of sub-systems, such as smart homes, and altering energy sources and energy source combinations since some of them might be energy producers and consumers simultaneously. For example, Nagpal et al. [56] suggested a concept of cooperative energy consumption optimization for use with buildings. They showed that building automation systems (BAS) could be used together for a cluster of buildings with the results leading to up to 15% reduction of energy consumption. On a similar note, Wang et al. [80] trained an ensemble model for dynamic short-term cooling load forecasting of a building.

Smart buildings are also cyber-physical (CP) systems that combine both the physical aspects of the building and cyber (virtual) aspects of the cloud-based solution. Physical attributes include building, sensors, and actuators [49]. Building functions as the frame of the system, providing a place to integrate sensors and actuators while protecting these devices against weather conditions and manipulation. Sensors gather data from surroundings, which include the building itself. Actuators are devices that given a command they produce an action, which alters device settings and causes some change in the structure eventually. In the CP system, cyber refers to making decisions in the cloud. It can use knowledge of previous measurements and calculation results as well as the most current measurements from the sensors. It calculates new commands for the actuators.

Smart buildings and their sub-systems, such as structural health monitoring (SHM) systems and IoT -based devices, should be made to follow similar guidelines as CP systems since this could improve the gathering and utilization of data. According to Abate et al. [1], high reliability, availability, maintainability, and safety –standards are necessary. Jiang [45], although having a focus on smart factories, suggests an 8C architecture or guideline for CP systems as an improvement for design known as 5C architecture, those 8 Cs being connection, conversion, cyber, cognition, configuration, coalition, customer, and content. The last three Cs are providing more room for mass production and customization.

For collecting data, Legatiuk and Smarsly [49] recommend that these kinds of systems should be recorded mathematically. Sensors of the buildings or target structures ought to be modeled mathematically together with corresponding measured data. These should then be further formulated to cover also sensor groups and groups of sensor groups. This approach comes with the benefit of being general but also mathematically precise. On the downside, this method does not provide suitable information for every use case. However, Wang and Srinivasan [79], note that not all AI approaches need a high level of structural information about the building. Abate et al. [1] suggest using fault management trees (FMT) with smart building maintenance since they are dynamic event trees easing the decomposing of fault modes of the system.

Unused data has little value. Services that utilize IoT, wearable devices, portions of Big Data, and AI -based systems to ease the handling of the above-mentioned Big Data to provide continuous, traceable, and preemptive services for customers can be called smart services. These services are often novel.

On the one hand, with smart buildings and smart homes, it is well advised to consider the physical aspects of possible services. For example, when considering smart services for smart building energy usage control, Byun and Park [14] brought forth three main issues with smart services at the time: services were centralized which can lead to performance issues, fixed rule-based control is not necessarily capable of handling complex situations, and, lastly, physical parts have different lifetimes and sensor nodes tend to die out. Their proposed solution was a self-adapting intelligent system that had been designed to have distributed devices, which could alter their functioning based on measurements from the building, environment, and users. With a decentralized adaptive control system, it was possible to handle also the problem of dying batteries. On the other hand, transforming data into information can be vital for services. For example, Dao et al. [23] proposed a system based on cloud computing called EvIM that comprised both gathering data and using it for different services. The proposal made it possible to unite CP systems together with users and alleviated some of the rule-based rigidness mentioned above while providing real-time event handling.

Smart buildings bring forth yet another challenge, and that is the preservation of privacy since there are different kinds of intelligent buildings, and some might be controlled together. Therefore, smart services should be such that they do not expose user(s) or user data to third parties without user consent. Also, according to Sta [73], systems should be made capable of handling imperfect information when dealing with Big Data. Digital twin should be used with smart services, as well, since according to Qi et al. [58] digital twin is versatile and combines both physical aspects and virtual aspects with connections between them. Utilization of digital twins could lead to a simulation of various situations and eventually to better smart services. Lim et al. [53] remind that for service to succeed it needs to bring value to the system.

5 Cloud Services and Data Platforms

Cloud computing can be defined in various ways in the literature. It can be referred to as a way to store and access data over the Internet instead of one's computer's storage media. Through cloud computing, a user with a pay-as-you-go pricing business model can rent computing power, database storage, applications, and other relevant IT resources. According to Karthikeyan and Thangavel [46], Cloud computing can be thought of as a computing paradigm providing dynamically scalable infrastructure for application, data, and file storage. A large pool of various systems is connected in public and private networks to form the basis of the paradigm.

The concept of cloud computing can be traced back to the 1950s, the era of mainframe computers, which were accessible via thin terminal clients. The development towards nowadays cloud computing started in the 1980s with cluster computing. Cluster computing was followed by grid computing, focusing on solving significant problems with parallel computing. Grid computing lead to utility computing in the 1990s offering computing resources (clusters) as virtual platforms for computing with a metered service. Clusters are usually distributed locally using the same hardware and operating system, which provides the possibility to use them as a supercomputer by using the pay-per-use approach. In 2001 software as a service (SaaS) concept was introduced focusing on network-based subscription of applications. Figure 2 illustrates commonly agreed SPI (SaaS, PaaS, IaaS) framework of three primary services provided through the cloud.

Public cloud service is the most widely used service delivery model in cloud computing currently available. Public clouds can be owned, operated, and managed by third parties, such as government institutions, businesses, academic institutions, or a combination of them [16]. Public clouds are highly scalable, they provide large capacity, and shared resources require minimal IT investments and decrease

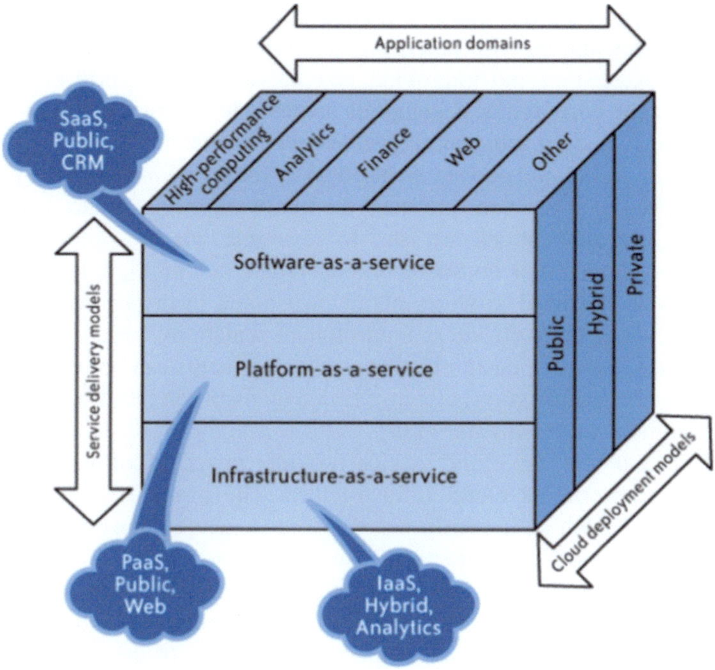

Fig. 2 SPI service model [55]

operating costs in the long run by using a pay-as-you-use –model [77]. All customers of public cloud providers share the same pool of security protections without a possibility to affect it. Major public providers in the market are Amazon Web Services (AWS), Microsoft, and Google.

Unlike public cloud services, private cloud services are intended for a single enterprise. Private clouds provide better controls and data security, which public cloud services are lacking. The private cloud divides into two categories, as follows: (1). On-Premise (internal) Private Cloud, and (2). Externally (External) Hosted Private Cloud. Internal clouds provide standardized process and protection, but size and scalability are limited and operated within one's own data center. Internal cloud fits for applications requiring control and configuring capabilities of the infrastructure and security. External clouds are externally operated with a cloud provider, which produces an exclusive cloud environment ensuring a high level of privacy. External clouds fit for companies, which require a highly secure cloud service not sharing of physical resources [46].

Hybrid clouds combine private and public cloud services. Hybrid cloud services increase flexibility as hybrid cloud providers can use third-party cloud provider services in full or partially depending on the need. The hybrid cloud enhances the capabilities of a private cloud, providing a possibility to use public cloud services when e.g., the computing power of a private cloud is not enough [46]. To eliminate security risks, an enterprise can use private cloud services to host sensitive and critical workloads and use 3rd party, public cloud provider services to host less-critical tasks i.e. testing and improving new services. The hybrid cloud reduces initial investment costs when developing services by using a pay-per-go model without a need to make a substantial investment beforehand.

In SPI SaaS model software is licensed on a subscription basis or by pay-per-use model. The cloud provider provides the hardware infrastructure and software applications, and applications are run, for example, via web portals. A single instance of the service runs on the cloud concerned, and multiple users can access it. Customers do not need to worry about investment in infrastructure, licensing, and maintenance of the software or environment scalability issues, as they are the provider's tasks. Security, customization, and components can be issues on SaaS – layer as customers cannot control them. Service on SaaS can be CRM, email, virtual desktop, communication, or games [35].

In SPI PaaS model, the cloud service provider provides software and product development tools on its cloud infrastructure. The provider's task is to offer system resources such as network, server, storage, operating systems, databases, development tools, and other relevant resources to customers. The customer can design, implement, and deploy his/her applications into the cloud service and run them there. The client must keep his/her deployed software updated to confirm security. The disadvantage of PaaS is the mandatory use of the service provider's API. Service on PaaS can be e.g., execution runtime, database, web server, and development tools [35].

In SPI IaaS model the cloud service provider controls and provides the infrastructure required to run customer's developed and deployed applications. IaaS layer offers storage and computing capabilities as a service. IaaS model also provides flexibility in the means of security as customers can also affect it. The customer needs to make sure the software deployed is up to date, configured, and appropriately integrated. Service on IaaS –layer can be virtual machines, servers, storage, load balancer, or network [35].

There is a need to introduce an additional Data-as-a-Service (DaaS) service layer into SPI–framework. DaaS provides a new architecture model in which, for example, private clouds can be located inside a public cloud service. DaaS is a service, which provides means and capabilities to transform raw stored data into meaningful assets e.g., smart service development and/or analysis from various data sources, such as databases, data warehouses, data lakes, filesystems, applications, data science platforms, applications, and BI tools. DaaS provides functions such as collection, integration, enrichment, curation, contextualization, aggregating, and analysis of the data [65].

Instead of copying or moving all the data from data sources into a data warehouse or a monolithic data lake, DaaS services can be implemented between them to gather the data required. Data lakes and data warehouses (DW) are centralized storage repositories that can store a significant amount of data. Repositories are different as data lakes can store data in native/raw (both structured and unstructured) format and DW handles only structured and cleansed data. Both repositories can be used as a source for DaaS and in conjunction to complement each other. DaaS can decrease redundancy and cut costs by placing relevant data into one location, providing data usage and modification for many users through one convenient service. Regardless of data location, structure, and size, DaaS enables users to examine, classify, and analyze the data. Users can use analytics tools they favor the most, such as Python, QlikSense, and R.

One way to use DaaS service is through Amazon (AWS) or Azure cloud services, which offer Data-as-a-Service functionalities in conjunction with open source Dremio DaaS platform solution. Cloud platforms generally provide various kinds of solutions and services for computing, security, AI, and storing, managing, and analyzing the data. Amazon object-oriented Simple Storage Service (S3) or Azure cloud service can be connected to Dremio DaaS service to discover and explore, curate, share, and analyze the data. The Dremio service includes data catalog, which provides a way to find and explore real and virtual datasets, which are automatically updated when new data source is added and when datasets evolve. Dremio also supports SQL syntax for advanced transformations, learning about the data and various kinds of transformations recommendations. Dremio can be deployed on-premises or in a public cloud service [27].

6 Cloud Service Security

6.1 Situational Awareness in the Cyber World

Digitalization is taking significant steps ahead continuously. Due to digital transformation advances, organizations are accelerating the migration of data to the cloud services creating an enormous increase in attack surface and numerous amounts of novel types of risks for organizations to manage. At the same time, cyberattackers are using more sophisticated attack methods to penetrate an organization's defenses that more and more located on the cloud service. Generally, organizations react after the cyber incident has already occurred. To prepare for cyber-attacks in advance, organizations should assess their cyber risk profile beforehand, fix current problems and proactively manage the defense. Situational awareness is the key to surviving in the cyber world.

Situational awareness is a crucial asset for an organization, as without it, organizations cannot build functioning cybersecurity resilience. Organizations would need to clarify what are potential threats, what kind of harm could they provide, and what do they mean to the organization. Organizations may feel familiar with attacks they confronted in the past years, but they may still lack the ability to deal with current more sophisticated, advanced, and emerging attack methods. Updating situational awareness concerning these kinds of emerging threats should be in high priority.

Threat, vulnerability, risk, and asset form an intertwined entity in the cyber world [50, pp. 3–29.]. According to Threat Analysis Group [75], the asset can mean people, tangible or intangible valued property and information, such as databases, software code, and information system records. The asset is a resource that has to be protected. The threat is a hazardous cyber event that can exploit the vulnerability, accidentally or intentionally to obtain, damage, or destroy an asset. Vulnerability is a weakness or gap in the security of the system that can be exploited by threats to get unauthorized access to an asset. Vulnerabilities can be divided into human actions, processes, or technologies according to where they exist. Risk is the potential of the expected damage, loss or destruction of an asset, and it can be seen as the intersection of assets, threats, and vulnerabilities. It can be assessed from the viewpoint of its economic consequences or loss of loss at face value [50, pp. 3–29].

According to ENISA [29] Threat Landscape Report 2018, an attack vector is a path or means by which a threat agent, for example, hacker or cracker, can gain access to a computer or network server, abuse weaknesses to achieve a specific outcome. Attack vectors include viruses, e-mail attachments, WWW-pages, chat rooms, and deception. Cybercriminals continuously seek new attack vectors they can utilize in attacking e.g., cloud service infrastructure. There exist various attack vectors, which threaten cloud services, but some of the common ones are AI/machine learning-based attacks, DoS/DDoS attacks, insider threats, IoT attacks.

6.2 Utilizing Artificial Intelligence in Cyber-Attacks

Artificial intelligence can be used when executing targeted attacks. An attacker can teach and utilize AI algorithms to recognize persons who are the most suitable target victims and provide them with malware. A perpetrator can also use AI to gain information from the target security solution through the perpetrator's reconnaissance actions on the target network. Attacks towards IoT devices are substantially growing, and due to underestimation of the situation, IoT devices generally lack necessary security measures and use relatively weak default device credentials opening a way to malware penetration. An attacker targeting IoT devices could use AI, for example, to generate credentials, find new vulnerabilities, learn the standard processes and behavior, distribute algorithms across all the nodes of a botnet for collective learning [47].

Artificial intelligence can be taught to find a way for new vulnerabilities by fuzzing, in which an attacker provides the algorithm with invalid, unexpected, or random input data. AI can also be a powerful technology to find the most effective way to attack. An attacker can abstract and combine attack techniques to identify the most effective ways of attacking. In the case of detection by the defender, the attacker needs to rerun the algorithm to follow a new learning path. Artificial intelligence can be utilized by the perpetrator in protecting himself by detecting intruders and suspicious nodes in their networks. The perpetrator can utilize artificial intelligence to spread disinformation, generate phishing emails and high-quality spam, and choose the best target, misuse a defender's AI model solution as a black box. He might use the same configuration in identifying what kind of traffic can pass through the defenses, and so on [47].

A perpetrator can also utilize adversarial examples when attacking machine learning models used, e.g., in cloud services, such as convolutional (CNN) neural networks or deep neural networks (DNNs), which can be used in implementing smart building services. Adversarial examples can be malicious inputs to DNNs providing erroneous model outputs while appearing to be unmodified in human eyes. This incident knocks out the classifier [61]. Adversarial input attacks are a threat to CNNs as instead of generalizing well and learning high-level representation (less prone to noise), they easily learn superficial dataset regularity [13]. Defending against adversarial attacks is difficult because the theoretical model of adversarial example crafting process is hard to construct. In theory, machine-learning models would be needed to defend against them to produce the right outputs for every possible input. In practice, ML models may only work on a relatively small number of potential data available that they face; models may block one type of an attack, but leave vulnerabilities open for the perpetrator to exploit [37].

6.3 Common Attack Vectors

6.3.1 DoS– and DDoS –Attacks

A Denial-of-Service (DoS) is a malicious attempt in which the perpetrator tries to disrupt data traffic to targeted service with limited bandwidth, machine or network resource by overloading the resource with a flood of traffic intending to make the service low or make it temporarily or entirely unusable [34]. The DoS attack can be described as a traffic jam hitting ordinary traffic on a highway by preventing its normal flow. A similar situation can happen when an overwhelming amount of people are in the midst of booking for concert tickets or buying discounted products at the same time when discounts are announced.

There are various DoS attacks, but popular ones are buffer overflow, ICMP flood (ping of death) and TCP SYN flood attacks. In a buffer overflow, attack the perpetrator sends more traffic than the target system can handle. In an ICMP attack, the attacker sends a huge amount of spoofed large-sized ICMP echo requests to target host enforcing each computer on the target network to ping instead of just the target computer attempting to switch it offline or keep it busy. TCP SYN flood uses a three-way handshake trying to make a connection with an invalid return address without completing the handshake [28].

A Distributed-Denial-of-Service (DDos) is similar to DoS attack, but it takes advantage of several compromised computers when carrying out an attack. Exploited 'cluster' of machines may consist of ordinary hacked computers or IoT 'bots' or 'zombies' devices and commanded by an attacker. Using bots provides means for attackers to use various IP addresses from different areas of the world at the same time, making it more complicated for service providers to defend themselves from incoming attacks as blocking one IP address will not make much of a difference [34]. Detecting the location of an attack is a challenging task as the attack system can be randomly distributed. Distributed DDoS attacks are used for the reason it is challenging to be detected; it is also efficient and cheap to execute due to hacked zombie computers. Usually, DDoS attacks focus to do damage to a victim for personal reasons, gain material benefits or popularity. Through DDoS attacks cloud services could get jammed, and the data become inaccessible.

6.3.2 IoT Based Attacks

IoT sensors can gather data from the real world by measuring the physical quantity and converting it into a signal and eventually data for the digital domain. These days, there are estimated to exist more than 50 billion sensors connected via IoT. According to HP security research [42], 80% had privacy concerns and bad passwords, 70% had lacked encryption, 60% had vulnerabilities in UI and insecure updates. According to Gartner, there are more than 6 billion relatively vulnerable IoT devices on the Globe. Therefore, IoT devices provide an excellent attack surface for cybercriminals targeting cloud services. A vast amount of IoT devices still use default login credentials, which makes penetrating the defenses an easy task. DDoS attacks also pose a significant risk on IoT devices, which became true in 2016 in the form of Mirai botnet malware reaching up to 1 Tbps of traffic through hundreds of thousands of compromised IoT devices [81]. In IoT poisoning attacks, utilizing adversaries can cause remarkable risks on sensors when manipulating the training data by altering the sensor's measurements. Poisoning attacks greatly decrease performance causing misclassification or other kind of bad behavior. Through poisoning attacks, backdoors and neural Trojans are sneaked in. To prevent this kind of incident, collecting poisonous data and training an arbitrary supervised learning model could work as a defense strategy [6].

IoT devices' vulnerabilities can be divided into system hardware or system software-based vulnerabilities. System hardware is vulnerable to exposure since the devices are often left unattended. Through exposure, the attacker might steal the device, extract sensitive data, modify the device's programming, or replace the original device with a malicious one [59]. In this case, the data in the cloud service might get tampered, and therefore the reliability is decreased.

IoT devices' software vulnerabilities are linked to application software, control software, and operating systems. Through system software, the attackers can execute, for example, access attacks or privacy attacks. In an access attack, an unauthorized person gains access to networks or devices in which they have no permission to enter [2]. Attack can also affect the cloud services, to which the devices are connected. According to Abomhara and Køien [2], the access attack can be targeted at the physical machine or to IP-connected devices.

Privacy attacks are directed, for example, to data mining, cyber espionage, and tracking. IoT devices' information can be highly sensitive since the collected data can be, in some cases linked to one's home or workplace. Through IoT device hacking an attacker might be able to tell when there are people inside the building, what are they doing at the moment, and so forth.

6.3.3 Insider Threats

Insider threats are substantial and increasing problem causing significant risk to organizations. Bonderud [8] claims that one in the four attacks start inside corporate networks. Insider threat is a current or former employee, contractor, or business partner who has or had authorized access to an organization's network, system, or data, and intentionally exceeded or misused access which negatively affected to the confidentiality, integrity, or availability of the organization's information system [20]. Cloud service can be targeted by an insider threat conducted by, for example, rogue administrator, an employee utilizing cloud weaknesses for unauthorized access, or an insider who uses cloud service resources to execute attacks against an organization's IT infrastructure. Motivations for conducting attacks can be e.g., financial aspect, theft of sensitive information, intellectual property, or fraud.

According to Ca Technologies report (2018), accidental or unintentional insider threat causes the most considerable risk (51%) to the organization, and malicious or deliberate risk is the second largest risk (47%). Regular employees pose the most significant security threat of 56% to the organization, privileged IT users/admins 55%, contractors/service providers, or temporary workers 42%. The most vulnerable data is confidential business information such as financials, customer data, or employee data. Cybercriminals are highly interested in the organization's databases (50%), fileservers (46%), cloud applications (39%) and cloud infrastructure (36%). According to the report, up to 90% of companies surveyed, felt vulnerable to insider threats.

Cloud services can be targeted by an insider threat conducted by e.g., rogue administrator, an employee utilizing cloud weaknesses for unauthorized access or an insider who uses cloud service resources to execute attacks against an organization's IT infrastructure. Motivations for conducting attacks can be, for example, financial aspect, theft of sensitive information, intellectual property, or fraud. Shaw et al. [74] identified a coherent cluster of risk factors characteristic of a vulnerable subgroup of critical information technology insiders. The factors that reduce inhibitions against potentially damaging acts are negative personal and social experiences, reduced loyalty towards the organization, personal and professional frustration, and ethical "flexibility," feeling of entitlement, anger, and lack of empathy. Also, stressors like family problems, substance abuse, disappointments at work, and threatened layoffs may trigger insider attacks [74].

7 Countering Cloud Cyberattacks

7.1 Encrypting the Data

Cloud services are becoming more and more popular due to the organization's interest in deploying applications and store their data into cloud service platforms. Cloud services are also gaining attention among smart building administrators. Cloud services provide many kinds of benefits, but one of the biggest worries is confidentiality. Organizations have to be sure that the cloud service provider has stored their data securely, and proper encryption methods have been used. Cryptographic algorithms provide a means to secure the data concerned, but they also limit the functionality of the cloud storage. According to Gupta et al. [40] two main categories of encryptions, symmetric (e.g., DES, AES, 3-DES, RC6, IDEA, Blowfish) and asymmetric (e.g., RSA, ECC, Elgamal), are being used in cryptography to achieve confidentiality, integrity, availability, and authentication. While using symmetric algorithms, encryption and decryption use the same algorithm and the same key to encipher and decipher the message. Symmetrical algorithms are useful to ensure confidentiality, integrity, and availability, but not authenticity. Asymmetric algorithms use two keys, one is a private key, which only recipient knows and the other is a public key, which everyone knows. Both of the keys can be used to encrypt and decrypt the message. Asymmetric algorithms provide better key sharing than symmetric algorithms, but they are slower than symmetric algorithms. Asymmetric encryption is slower than symmetric one due to the longer key lengths used and complexity of the encryption algorithms used. Conventionally known cloud service providers, e.g., Google cloud service platform, use AES128 and AES256, Amazon AWS AES128, AES192, and AES256 symmetrical algorithms and asymmetric RSA and Elliptic Curve Cryptography algorithms.

Encryption keys should be kept on a separate server on a storage block. Especially sensitive data needs to be encrypted after it is collected or created and uploaded to the cloud service data storage or an organization's private cloud service after the encryption process. The process mentioned may also bring out issues as if the data has been uploaded to the cloud service is encrypted and then later downloaded onto another media, it does not already have the decryption key providing useless encrypted data [12]. Using homomorphic encryption could circumvent this issue by allowing data to be sent the cloud service to be analyzed without having to decrypt it first. Using homomorphic encryption provides only users needed to be able to analyze the data leaving cloud service providers no chance to know what kind of information is contained on the data [54].

7.2 IoT Based Attacks

IoT devices typically are low powered, they have low storage, low computing resources, and they have been massively deployed and connected to each other. Partly due to a lack of resources, they are vulnerable to many kinds of cyber threats. Encryption provides an effective countermeasure, and nowadays, encryption is becoming a more and more crucial part of IoT sensor devices in various environments that formerly did not require it.

Ordinary cryptography methods, such as AES encryption and SHA-hashing, RSA signing is widely used in systems, which have enough processing power and memory. They are not fit for IoT sensor networks providing considerably less capability. Elliptic curve cryptography has been successfully applied on sensor nodes though. Therefore, lightweight cryptography methods are being developed and standardized to provide suitable means for IoT sensors with fewer resources. An adversary attack poses a real threat to an IoT sensor and sensor nodes by eavesdropping and modifying the data. Hence, be able to provide a secure routing protocol to ensure authentication, availability, and integrity is vital. Handful of lightweight cryptography protocols and primitives has been standardized as the ISO/ICE 29121 standard and primitives have been included in IPSec and TLS.

According to Buchanan, Li, and Asif [10], the disadvantage of lightweight cryptography is less secured than conventional ones due to limited resources on sensors. Lightweight cryptography implementations are usually bound to use short key sizes, which increase the risk for key-related attacks. Sometimes read-only (masking) technology is used to permanently burn keys into IoT device chips to decrease key space consumed. When considering lightweight cryptography, IoT device clock, memory, storing internal and key states should be evaluated.

Using proper authentication and data encryption alone is not enough for ensuring data security. According to Chang [18] adversary attacks can be injected into sensor nodes through compromised nodes. Intrusion detection systems (IDS) can be used to monitor suspicious and anomalous patterns of activity, which are different compared to ordinary and expected behavior. It is widely assumed that an intruder has significantly different behavioral patterns than legitimate users usually have in the network. Rule-based IDSs can be used to detect known patterns of intrusions, and anomaly-based IDSs can be used to detect new or unknown intrusions. Anomaly-based IDSs provide notably higher false alarm rates compared to rule-based IDSs.

Focusing on proper authentication and encryption and using intrusion detection systems can secure IoT devices. To prevent any incidents, collecting poisonous data, and training an arbitrary supervised learning model could work as a defense strategy [6]. IoT devices can be secured by focusing on confidentiality, integrity, authentication, accountability, auditing, and privacy [2]. Overall, the benefits of IoT devices are exceeding the downsides.

7.3 AI Based Tools in Countering Cloud Cyberattacks

7.3.1 Insider Attacks

Existing data protection techniques can be effective against insider attacks if implemented carefully and in the right way. Current technologies to prevent insider threats are Data Loss Prevention (DLP), encryption, identity, and access management solutions. In detecting active insider threats, organizations can utilize, for example, intrusion detection and prevention (IDS) services, log management, Security Information and Event Management System (SIEM) platforms, User Activity Monitoring (UAM), Privileged Access Management (PAM), DLP.

The monitoring of sensitive assets can be utilized in order to prevent and restrict insider threats that organizations are facing. According to Ca Technologies report [15], 78% of organizations inventory and monitor all or most of their key assets, and more than 93% of them monitor access to sensitive data. Due to the increase in insider threat volume, organizations have begun to utilize User Behavior Analytics (UBA) tools and solutions to detect, classify, and alert anomalous behavior. Finding insiders who cause the highest risk is a crucial part of threat prevention. Organizations can monitor their behavior and work patterns, such as hostility towards colleagues, missing work, an excessive amount of work outside ordinary working hours, declined performance. In addition to UBA monitoring, comprehensive data access, movement analysis, and security analytics can be utilized.

Various solutions can be used to tackle insider threat issues, such as Darktrace Vectra Cognito. Darktrace uses the Enterprise Immune System technology (EIS) utilizing machine learning algorithms and mathematical principles to detect anomalies. EIS can adapt and automatically learn user, device, or an information network behavior to identify behaviors reflecting threats, such as an insider threat. Darktrace uses mathematical approaches, such as Bayesian estimation to produce behavioral models for individual people and devices they use to detect unusual behavior and reveal possible insider attack. Darktrace [26] Vectra Cognito works in a bit similar way as Darktrace, and it continuously learns from an organization's network activity. Cognito uses data science, supervised and unsupervised machine learning, and behavioral analytics to reveal attack behaviors and attacks such as an insider attack. Vectra Cognito can monitor and detect suspicious access to critical assets, policy violations related to, for example, cloud service usage, or another means of moving data [78].

To lower the risk of insider attacks towards cloud services, an organization should avoid management errors. According to Shaw et al. [74], organizations should understand the personality and motivation of the at-risk employee. Clear and standardized rules about the use of company information systems should be created. Also, the consequences of misuse should be made clear, and rule violations should be enforced.

7.3.2 DoS/DDoS Attacks

DoS and DDoS attacks are ones of the most frequent, causing significant damage, and they impact cloud service performance. These kinds of attacks can be tricky to detect and block as the attack traffic can be easily tangled with legitimate traffic causing it to be challenging to trace. Especially application layer (Layer 7) DoS attacks can be hard to detect as the traffic appears to be like regular traffic with complete Transmission Control Protocol (TCP) connections and following protocol rules. Therefore, these attacks can target applications, which bypass the firewall [5].

Often security experts who deal with these kinds of issues are busy, so additional means to deal with these attacks are needed. There exist various tools to treat the problem, such as the PatternEx AI2 platform, that can predict incoming cyber-attacks, such as DoS or DDoS. AI2 uses three different unsupervised machine-learning methods and clusters data into patterns showing the top abnormal events to security analysts for further analysis to confirm attacks are real attacks. In the following phase, the platform builds a supervised model for the next set of data, which enables further active learning. This process will eventually improve the attack detection rate of the algorithms requiring less security analyst time. Currently, AI2 is able to detect up to 85% of attacks while false positives are reduced by factor 5 (Conner-Simons [19]).

Classical DDoS defense tools take advantage of rate-limiting and manual signature creation in mitigating cyber-attacks. Rate limiting tends to produce a significant amount of false positives while providing effective means in mitigating attacks. Manual signatures created can be then utilized to prevent or decrease the amount of false-positive results. Identifying the attack traffic is time-consuming as it requires human security analysts to analyze the attack vector, and it can be only done when the attack is already started. Hence, time to mitigation increases resulting in ineffective defense strategy [63].

Radware Defense Pro offers means to prevent, protect and mitigate DDoS and IoT botnet attacks, such as fast-moving, high volume, encrypted or short-duration attacks, and IoT attacks, such as Mirai, Pulse, Burst, DNS, TLS/SSL, PDoS and Ransom Denial-of-Service (RDoS). Defense Pro provides behavioral mitigation capabilities to circumvent the manual signature creation and rate-limiting problems. It uses automatic machine-learning algorithms to create signatures and adapt defenses in changing attack-vector environment. Defense Pro can learn real-time behavior of legitimate traffic and to quickly detect an attack when rate and rate-invariant parameters indicate an anomaly compared to legitimate traffic. Defense Pro offers negative and positive protection models and rate limiting, ensuring zero time to mitigation with scarce human cybersecurity professional intervention [64].

Reblaze offers DoS/DDoS protection solution that provides defense from DDoS botnet assaults until single malformed-packet DoS attempts. The Reblaze solution protection mechanism if effective against various forms of DoS/DDoS attack vectors, such as amplification and reflection attacks, application-layer vulnerabilities, malicious inputs, protocol exploits, volumetric flooding, resource depletion, and exhaustion. Reblaze provides DoS/DDoS protection towards attacks on ISO/OSI layers 3 (network), 4 (transport), and 7 (application) and blocks attacks in the cloud service. Full protection from DDoS attacks is not common to many so-called "DDoS solutions," but layers 3 and 4 are protected more comprehensively. Reblaze can run natively on Google Cloud Platform and integrate with Cloud Armor service. It augments Cloud Armor's capabilities and uses Machine Learning in self-learning and adapting when there will be changes in the cyber threat environment. The learning process is automated and constantly adapting. It provides pattern recognition and behavioral analysis to detect early-stage attacks generating a small amount of false-positive results [68].

7.4 Utilizing AI and ML Based Methods in Combating Cyberattacks

It is challenging to protect against insider threats, DoS/DDoS attacks, and adversarial attacks. Due to the increased amount of Big Data, AI, and ML methods are needed to combat these threats. Insider threats are challenging for AI, since not necessarily all malicious influence on the user can be prevented. According to Le et al. [48], insecure habits of the user can be used as adversarial examples for AI-based IDS. Therefore, human experts shouldn't be allowed to decide what data they label when using supervised learning on AI models. Gavai et al. [32] compared supervised and unsupervised ML models while investigating the usage of employees' social and web usage data, such as email frequency and machine access patterns, as possible features for the detection and prevention of insider threats. They found their unsupervised model to exceed their supervised model by a few percent. Zhang et al. [82] studied a way to classify possible insider threats based on user behavior logs. They used long-short term memory NN (LSTM), which is used typically for sequences, in order to find anomalies from role-based user log data.

DoS/DDoS attacks are dangerous attacks because they can be hard to detect in the early stages of the attack, malicious packages can hide between legitimate traffic, attacks can inconvenience the target server, and the attacker can hide among zombie computers, which might be IoT devices. AI and ML techniques are needed for the automated detection of DOS/DDOS attacks. According to Diro and Chilamkurti [25], the interconnectivity of smart cities is a potentially tempting playground for attackers. Rangaraju et al. [66] list in their article several ML techniques, such as Naïve Bayes (NB), Support Vector Machines (SVM) and genetic algorithms, that are used for detection and prevention of cyberattacks. NB as well as SVM are techniques based on probability while the genetic algorithm is an umbrella term for algorithms that are inspired by evolutionary theorem. Rathore and Park [67] introduced their fog-computing framework against distributed attacks that used an extreme learning machine (ELM) for faster generalization. Instead, Han et al. [41] proposed a defensive framework that focused on the detection of DDOS attacks both on data plane and on control plane while collaboratively distributing attack load on multiple defense applications. The NN model that the authors used was a stacked combination of autoencoder and softmax-classifier, which could detect a multitude of DDOS-attack types.

Adversarial attacks can be defended against with AI in some cases. Both CNN's and DNN's are commonly known to be susceptible to this cyber-attack type. Since no system is perfect, it is best to assume that all AI systems are vulnerable to adversarial attacks.

Adversarial training means modifying legitimate inputs to make AI classifier to learn to be more robust [76]. In other words, the use of various counts of modified inputs used together with unmodified inputs in the training stage helps NNs to compact against false (modified) inputs by broadening their "understanding," where understanding refers to what the inputs would mean to a human. The creation of modified inputs is typically done using two NNs of which the first one tries to produce falsified inputs, and the second one attempts to classify inputs as true or false. Many ways to alter input data exists. Ganin et al. [31] suggested training classifiers with either labeled or unlabeled training data, which come from a different distribution than the intended data but have the same features. This is also known as domain adaptation, hence the name domain-adversarial neural networks (DANN). However, with this method, features need to exist in both domains and remain the same. Samangouei et al. [69] proposed a new structure to protect NN used for classification called Defense-GAN. It finds similar input as given from its database and uses that as input for the classifying NN model. This has the benefit of protecting against adversarial attacks geared towards the classifier; however, it seems likely that the AI system would have to have a comprehensive and specific input domain to function properly as part of the CP system.

Defensive distillation has also been used in attempts to train NN models to resist adversarial attacks. Goldblum et al. [36] presented adversarial robust distillation (ARD) which can help smaller NNs to lean robustness of a bigger model. According to Papernot and McDaniel [60], defensive distillation is done by first training an NN model, where the last layer is a softmax-layer, with a labeled dataset. Then this model predicts new probability values from the training set. Using original input as input and output from the first NN, a second NN model (has softmax-layer) can be taught. However, there is a trick; a term called "temperature" in the softmax-layers. If the temperature is greater than 1, the probabilities get distributed more uniformly, meaning that for each class these probabilities are very nearly the same, when the temperature goes to infinity. If the temperature is 1, softmax-function will output probabilities closer to one that corresponds with most likely class labels. Both NNs are to be trained with the same temperature values that are greater than one, but after training of the second NN model is done, its temperature is set to one. According to a short article written by Papernot et al. [61], defensive distillation can work against adversarial attacks. However, according to Carlini and Wagner [17], defensive distillation does not necessarily work against carefully constructed adversarial attacks.

Adversarial noise removal refers to techniques that can help reduce the effect of input noise typical to adversarial attacks. Gu and Rigazio [38] tried both adding extra noise to image inputs and removing noise with the usage of autoencoders. They found that autoencoders work well for noise reduction but using them together with the original AI model leaves the compound model still vulnerable to even smaller adversarial noise. According to Liang et al. [51], when inputs are images, varying de-noising techniques should be used based on the image space, since over and under de-noising should be avoided. Possible input transformation techniques can be bit-depth reduction, compression, total variance minimization, image quilting [39], scalar quantization, and smoothing spatial filtering [51]. Guo et al. [39] managed to defend against 90% of black-box attacks, while Liang et al. [51] managed to get a high 94% F1-score.

Sometimes if one model does not work, its performance may be possible to increase with an ensemble. Ensembles typically refer to a NN that takes outputs of other NNs or ML models as its input and calculates new output for the entire system. The beauty of an ensemble is that it can produce more acceptable results compared to a single ML -model. For example, Jia et al. [44] devised an ensemble classifier model for the detection of DDOS attacks. This model consisted of Bagging, k-Nearest Neighbors (k-NN), and Random Forest, which were each trained and tested with cross-validation. Jia et al. [44] reported that their model reached similar classification results as the Random Forest method on its beating Bagging and K-NN with a significantly higher true negative score. Other kinds of ensembles exist. Sengupta [71] proposed a model that uses several NN models as defenders against adversarial attacks. This would make attacking a black-box model challenging, since defenders could alternate and therefore obscure decision boundaries. Tramèr et al. [76] used an ensemble of different adversarial attack models to train a defensive NN. They showed that, in some cases, learned robustness from some attack could be transferred and it can be used similarly against other attacks.

8 Conclusion

Artificial Intelligence in protecting smart building's cloud service infrastructure represents a potential research area as the importance of cybersecurity in cloud services is growing. This chapter presented a general overview of cyberspace, artificial intelligence, cloud services, smart buildings, and typical attack vectors a perpetrator can utilize when attacking towards cloud services.

Cloud services are becoming even more popular these days due to the organization's interest in deploying applications and store data into cloud services. Cloud services can provide many benefits, but they also pose risks in security issues. Organizations need to be sure that robust encryption methods are used in storing and transferring data. In the smart home context, data can be gathered through IoT sensors, which are commonly known as vulnerable towards cyberattacks. Ordinary cryptographic encryption methods cannot always be used due to lack of processing power, storage space, and computing resources of IoT sensors. Lightweight cryptography methods can be applied, but they are less secure than conventional methods. Even proper authentication and data encryption alone is not enough for ensuring data security allowing perpetrators to use e.g., adversarial attacks when attacking the cloud service. In countering cyber-attacks, proper AI- and ML-models can be utilized.

Various solutions, which can be utilized in countering cyberattacks towards common attack vectors, exist. The solutions presented in this paper are Darktrace Vectra Cognito, PatternEx AI2 platform, Radware Defence Pro, and Reblaze. Vectra Cognito utilizing ML algorithms and mathematical principles in detecting anomalies can be used in tackling e.g., with insider threat issues. The solution produces behavioral models for individual people and devices they use to detect unusual behavior and reveal possible insider attack. PatternEx AI2 platform can be used in predicting incoming cyberattacks, such as DoS and DDoS. AI2 utilizes unsupervised ML methods to present abnormal events to security specialists for further analysis to confirm attacks are real and produce a supervised model for further active learning. AI2 is estimated to reach 85% of detection accuracy. Radware Defence Pro offer means to detect and mitigate DDoS and IoT botnet attacks using ML algorithms to create signatures in real-time and adapt defenses. Defense Pro can separate anomalous attack traffic from legitimate one and to provide protection models ensuring zero time to mitigation. Reblaze's solution is to provide protection against DDoS botnet assaults until single malformed-packet DoS attempts. Reblaze's specialty is to provide more comprehensive protection than competing solutions against incoming DoS/DDoS attacks enabling ISO/OSI layers 3 (network) and 4 (transport) protection blocking incoming attacks towards cloud service. The solution integrates natively on public cloud service providers, such as Google Cloud Platform utilizing ML algorithms in self-learning and adapting under the continuous change in the cyber threat environment. The solution is able to detect early-stage attacks by using behavioral analysis and pattern recognition generating a small amount of false-positive results.

The results indicate that artificial intelligence can be used to prevent cyberattacks with some reservations. The architecture of chosen defensive AI model, defensive plan of the CP system, and how the model has been trained, determines how well artificial intelligence can combat attack vectors. Architecture and defensive plan of the CP system can help alleviate attacks on the CP system and the AI model, while under DDOS attack, for example, the system might start new defensive programs to mitigate load caused by the attack. When training models, utilizing different data manipulation schemes, such as adversarial training and defensive distillation, is important but not guaranteed to work perfectly. Classifiers trained with adversarial examples are more robust than classifiers trained with regular data, and this robustness can be transferred to other models. It was asserted that ensemble models could improve artificial intelligence performance, and it was found that ensemble training could do that as well. Training of models remains to be time-consuming.

Smart homes, smart building maintenance, and cloud services will get more popular over the years. Artificial intelligence is a huge part of that change. Artificial intelligence in protecting cloud services will gain more popularity in the future since current trends indicate that the number of cyber-attacks is increasing. For safety, various cyber threats towards cloud services should be researched thoroughly, and the adversarial attacks require further research.

References

1. Abate A, Budde CE, Cauchi N, Hoque KA, Stoelinga M (2018) Assessment of maintenance policies for smart buildings: application of formal methods to fault maintenance trees. In: European conference of the prognostics and health management society, vol 2018
2. Abomhara M, Køien GM (2014) Security and privacy in the Internet of Things: current status and open issues. In: Privacy and security in mobile systems (PRISMS), 2014 international conference on. IEEE, pp 1–8
3. Alam MR, Reaz MBI, Ali MAM (2012) A review of smart homes—past present and future", IEEE Trans Syst Man Cybern C Appl Rev.), vol. 42, no. 6, pp. 1190–1203
4. appliedAI. https://appliedai.com/use-cases/1. Accessed 5 Aug 2019
5. Ballal S, Prasad LS, Rajappa M, Khader A (2018) Bumper to bumper: detecting and mitigating DoS and DDoS attacks on the cloud. SecurityIntelligence. https://securityintelligence.com/bumper-to-bumper-detecting-and-mitigating-dos-and-ddos-attacks-on-the-cloud-part-1. Accessed 16 Aug 2019
6. Baracaldo N, Chen B, Ludwig H, Safavi A (2018) Detecting poisoning attacks on machine learning in IoT environments. IEEE international congress on internet of things (ICIOT), San Francisco, CA, USA
7. Bayuk JL, Healey J, Rohmeyer P, Sachs MH, Schmidt J, Weiss J (2012) Cyber security policy guidebook, 1st edn. Wiley, USA
8. Bonderud D (2018) Breaking bad behavior: can AI combat insider threats? Security Intelligence. https://securityintelligence.com/breaking-bad-behavior-can-ai-combat-insider-threats. Accessed 9 Aug 2019
9. BS ISO/IEC 27002 (2013) Information technology – security techniques – code of practice for information security management. The British Standards Institution. BSI Standards Limited, Switzerland
10. Buchanan WJ, Li S, Asif R (2018) Lightweight cryptography methods. J Cyber Secur Technol 1(3–4):187–201
11. Burnap P, Williams ML (2015) Cyber hate speech on twitter: an application of machine classification and statistical modeling for policy and decision making. https://doi.org/10.1002/poi3.85 . Accessed 20 Aug 2019
12. Business.com (2018) Cloud encryption: using data encryption in the cloud. https://www.business.com/articles/cloud-data-encryption. Accessed 4 Aug 2019
13. Busztein E (2018) Attacks against machine learning – an overview. https://elie.net/blog/ai/attacks-against-machine-learning-an-overview. Accessed 3 Aug 2019
14. Byun J, Park S (2011) Development of a self-adapting intelligent system for building energy saving and context-aware smart services. IEEE Trans Consum Electron 57(1):90–98
15. Ca technologies (2018) Insider threat report. Cybersecurity insiders. https://www.ca.com/content/dam/ca/us/files/ebook/insider-threat-report.pdf. Accessed 12 Aug 2019
16. Castro-Leon E, Harmon R (2016) Cloud as a service: understanding the service innovation ecosystem. Apress, Berkeley
17. Carlini N, Wagner D (2016) Defensive Distillation is Not Robust to Adversarial Examples. ArXiv: 1607.04311v1 [cs.CR] 14 Jul 2016
18. Chang Z. Wireless and internet of things (IoT) security. Department of Mathematical Information Technology. University of Jyväskylä, Finland. users.jyu.fi/~timoh/TIES327/Wireless.pdf. Accessed 9 Aug 2019

19. Conner-Simons (2016) System predicts 85 percent of cyber-attacks using input from human experts. http://news.mit.edu/2016/ai-system-predicts-85-percent-cyber-attacks-using-input-human-experts-0418. Accessed 3 Aug 2019
20. Costa D (2017) CERT definition of 'insider threat'. Software engineering institute, Cargenie Mellon, University. https://insights.sei.cmu.edu/insider-threat/2017/03/cert-definition-of-insider-threat%2D%2D-updated.html. Accessed 9 Aug 2019
21. Craigen D, Diakun-Thibault N, Purse R (2014) Security in cyberspace. Targeting nations, infrastructures, individuals. Bloomsbury publishing, New York
22. Cyberspace policy review. Assuring a trusted and resilient information and communications Infrastructure. https://www.energy.gov/sites/prod/files/cioprod/documents/Cyberspace_Policy_Review_final.pdf
23. Dao M-S, Pongpaichet S, Jalali L, Kim K, Jain R, Zettsu K (2014) A real-time complex event discovery platform for cyper-physical-social systems. ICMR 2014, April 1–4, Glasgow, UK
24. Deshpande N (2009) Artificial intelligence. Technical Publications. University of Pune, India
25. Diro AA, Chilamkurti N (2018) Distributed attack detection scheme using deep learning approach for Internet of Things. Future Gener Comput Syst 82:761–768
26. Darktrace (2018) Darktrace enterprise – detects and classifies cyber-threats across your entire enterprise. Darktrace. https://www.darktrace.com/en/products. Accessed 12 Aug 2019
27. Dremio (2019) Enabling Data-as-a-Service for AWS and R
28. Elleithy K, Blagovic D, Cheng W, Sideleau P (2006) Denial of service attack techniques: analysis, implementation and comparison. J Syst Cybern Inform 3:66–71
29. ENISA (2018) ENISA threat landscape report 2018 – 15 top cyberthreats and trends. European Union agency for network and information security
30. EUR-Lex (2013) Access to European Union law. Joint communication of the European parliament, the council, the European economic and social committee and the committee of the regions. Cyber Security strategy of the European Union: an open, safe and secure cyberspace. Document number 52013JC0001
31. Ganin Y, Ustinova E, Ajakan H, Germain P, Larochelle H, Laviolette F, Marchand M, Lempitsky V (2016) Domain-adversarial training of neural networks. J Mach Learn Res 17(2016):1–35
32. Gavai G, Sricharan K, Gunning D, Hanley J, Singhal M, Rolleston R (2015) Supervised and unsupervised methods to detect insider threat from enterprise social and online activity data. JoWUA 6(4). https://doi.org/10.22667/JOWUA.2015.12.31.047
33. Ghahramani Z (2015) Probabilistic machine learning and artificial intelligence. Nature 521:452–459
34. Gillespie A (2016) Cybercrime – Key issues and debates. Routledge, New York
35. Goel N, Sharma T (2014) Cloud computing – SPI framework, deployment models, challenges. International journal of emerging technology and advanced engineering. International conference on advanced deployments in engineering and technology, India
36. Goldblum M, Fowl L, Feizi S, Goldstein T (2019) Adversarially robust distillation. ArXiv:1905.09747v1 [cs.LG] 23 May 2019
37. Goodfellow I, Papernot N, Huang S, Duan R, Abbeel P, Clark J (2017) Attacking machine learning with adversarial examples. OpenAI. https://openai.com/blog/adversarial-example-research. Accessed 3 Aug 2019
38. Gu S, Rigazio L (2015) Towards Deep Neural Network Architectures Robust to Adversarial Examples. ArXiv:1412.5068v4 [cs.LG] 9 Apr 2015
39. Guo C, Rana M, Cissé M, van der Maaten L (2018) Countering adversarial images using input transformations. ArXiv:1711.00117v3 [cs.CV] 25 Jan 2018
40. Gupta D, Ghakraborty PS, Rajput P (2015) Cloud security using encryption techniques. International journal of advances research in computer science and software engineering, 5(2), SRM University, India
41. Han B, Yang X, Sun Z, Huang J, Su J (2018) OverWatch: a cross-plane DDOS attack defense framework with collaborative intelligence in SDN. Hindawi Secur Commun Netw 2018. https://doi.org/10.1155/2018/9649643

42. HP security research (2014) Internet of things research study. http://d-russia.ru/wp-content/uploads/2015/10/4AA5-4759ENW.pdf. Accessed 7 Aug 2019
43. Iyer G, Agrawal P (2010) Smart power grids. In: 42nd Southeastern Symposium on System Theory (SSST), IEEE (2010), pp 152–155
44. Jia B, Huang X, Liu R, Ma Y (2017) A DDOS attack detection method Based on hybrid heterogenous multiclassifier ensemble learning. Hindawi J Electr Comput Eng 2017. https://doi.org/10.1155/2017/4975343
45. Jiang J-R (2018) An improved cyber-physical systems architecture for Industry 4.0 smart factories. Adv Mech Eng 10(6):1–15
46. Karthikeyan P, Thangavel M (2018) Applications of security, mobile, analytic and cloud (SMAC) technologies for effective information processing and management, A volume in the advances in computer and electrical engineering (ACEE) book series. IGI Global, Hershey
47. Kubovič O, Košinár P, Jánošík J (2018) Can artificial intelligence power future malware? ESET white paper
48. Le DC, Khanchi S, Zincir-Heywood AN, Heywood MI (2018) Benchmarking evolutionary computation approaches to insider threat detection. Association for Computing Machinery. https://doi.org/10.1145/3205455.3205612
49. Legatiuk D, Smarsly K (2018) An abstract approach towards modeling intelligent structural system. 9th EWSHM, UK. CC-BY-NC license 4.0
50. Lehto M (2015) Phenomena in the cyber world. Cyber security: analytics, technology and automation. Springer, Berlin
51. Liang B, Li H, Su M, Li X, Shi W, Wang X (2019) Detecting adversarial image examples in deep neural networks with adaptive noise reduction. ArXiv:1705.08378v5 [cs.CR] 9 Jan 2019
52. Libicki MC (2007) Conquest in cyberspace – national security and information warfare. Cambridge University press, New York
53. Lim C, Kim K-H, Kim M-J, Heo J-Y, Kim K-J, Maglio PP (2018) From data to value: A nine-factor framework for data -based value creation in information-intensive services. Int J Inf Manag 39(2018):121–135
54. Machmeier C, Kunzke F (2019) How safeguarding sensitive data could lead to smarter AI. Sap News Center. https://news.sap.com/2019/01/homomorphic-encryption-safeguarding-sensitive-data-smarter-ai. Accessed 4 Aug 2019
55. Mather T, Kamaraswamy S, Latif S (2009) Cloud security and privacy – an enterprise perspective on risks and compliances. O'Reilly Media Inc., USA
56. Nagpal H, Basu B, Staino A (2018) Economic model predictive control of building energy systems in cooperative optimization framework. ICC, January 4–6, 2018, IIT Kanpur, India
57. Nicholson C. Skymind. https://skymind.ai/wiki/ai-vs-machine-learning-vs-deep-learning. Accessed 5 Aug 2019
58. Qi Q, Tao F, Zuo Y, Zhao D (2018) Digital twin service towards smart manufacturing. Procedia CIRP 72(2018):237–242
59. Padmavathi & Shanmugapriya et al (2009) A survey of attacks, security mechanisms and challenges in wireless sensor networks, arXiv preprint arXiv:0909.0576
60. Papernot N, McDaniel P (2016) On the effectiveness of defensive distillation. ArXiv:1607.05113v1 [cs.CR] 18 Jul 2016
61. Papernot N, McDaniel P, Goodfellow I, Jha S, Celic Z B, Swami A (2017) Practical black-box attacks against machine learning. In: Proceedings of the 2017 ACM Asia conference on computer and communications security, Abu Dhabi, UAE
62. Potluri S, Diedrich C (2016) Accelerated deep neural networks for enhanced intrusion detection system. 2016 IEEE 21st ETFA. https://doi.org/10.1109/ETFA.2016.7733515
63. Radware (2018) Machine-learning automation to ensure zero time to mitigation. https://www.radware.com/pleaseregister.aspx?returnurl=732862c3-5149-4806-b060-ba20d2bca6eb. Accessed 16 Aug 2019
64. Radware (2019) https://www.radware.com/products/defensepro/. Accessed 4 Aug 2019

65. Randal L (2016) What is data as a service? The 3 key dimensions. BDQ big data quarterly. http://www.dbta.com/BigDataQuarterly/Articles/What-is-Data-as-a-Service-The-3-Key-Dimensions-114568.aspx
66. Rangaraju NK, Sriramoju SB, Sarma S (2018) A study on machine learning techniques towards the detection of distributed denial of service attacks. Int J Pure Appl Math 120(6):7407–7423
67. Rathore S, Park JH (2018) Semi-supervised learning based distributed attack detection framework for IoT. Appl Soft Comput 72:79–89
68. Reblaze (2019) Comprehensive DDoS protection DoS/DDoS datasheet – web application & API security. https://www.reblaze.com/wp-content/uploads/2019/05 /Reblaze-DDoS-Datasheet.pdf. Accessed 22 Aug 2019
69. Samangouei P, Kabkab M, Chellappa R (2018). Defense-gan: Protecting classifiers against adversarial attacks using generative models. ArXiv:1805.06605
70. Schmidt M, Åhlund C (2018) Smart buildings as Cyber-Physical Systems: data-driven predictive control strategies for energy efficiency. Renew Sust Energ Rev 90:742–756. https://doi.org/10.1016/j.rser.2018.04.013
71. Sengupta S (2017) Moving target defense: a symbiotic framework for AI & security. In: Proceedings of the 16th international conference on autonomous agents and multiagent systems
72. Shahapure NH, Jayarekha P (2015) Replication: a technique for scalability in cloud computing. Int J Compute Appl (0975–8887) 122(5):13–18
73. Sta HB (2017) Quality and the efficiency of data in "Smart-Cities". Futur Gener Comput Syst 0167-739X 74(2017):409–416
74. Shaw E, Ruby K, Post J (1998) The insider threat to information systems. Secur Aware Bull 2(98):1–10
75. Threat Analysis Group (Tag) (2010) Threat, vulnerability, risk – commonly mixed up terms. https://www.threatanalysis.com/2010/05/03/threat-vulnerability-risk-commonly-mixed-up-terms. Accessed 31 Aug 2019
76. Tramèr F, Kurakin A, Papernot N, Goodfellow I, Boneh D, McDaniel P (2018) Ensemble adversarial training: attacks and defenses. ICLR 2018
77. Usman SH, Bawazir MA, Kabir AM (2014) Cloud computing: a strategy to improve the economy of Islamic societies. Int J Comput Trends Technol (IJCTT) 9(7):387–392
78. Vectra cognito (2019) Vectra security that thinks. Artificial intelligence powered automated threat hunting and network self-defense. https://www.beotech.rs/wp-content/uploads/2019/02/Vectra-Cognito-DataSheet.pdf. Accessed 12 Aug 2019
79. Wang Z, Srinivasan RS (2017) A review of artificial intelligence based building energy use prediction: contrasting the capabilities of single and ensemble prediction models. Renew Sust Energ Rev 75(3027):796–808
80. Wang L, Lee EW, Yuen RK (2018) Novel dynamic forecasting model for building cooling loads combining an artificial neural network and an ensemble approach. Appl Energy 228:1740–1753
81. Wani SY (2018) Internet of things (IoT) security and vulnerability. Research proposal. https://doi.org/10.13140/RG.2.2.29633.40801
82. Zhang D, Zheng Y, Wen Y, Xu Y, Wang J, Yu Y, Meng D. (2018). Role-based log analysis applying deep learning for insider threat detection. Assoc Comput Mach SecArch'18. https://doi.org/10.1145/3267494.3267495
83. Zhao D, Zhong M, Zhang X, Su X (2016) Energy consumption predicting model of VRV (Variable refrigerant volume) system in office buildings based on data mining. Energy 2016(102):287–297

Part IV
Smart Societies and Data Exploitation

Smart Distributed Ledger Technologies in Industry 4.0: Challenges and Opportunities in Supply Chain Management

Gregory Epiphaniou, Mirko Bottarelli, Haider Al-Khateeb, Nikolaos Th. Ersotelos, John Kanyaru, and Vinita Nahar

Abstract The rise of new digital economies and data-driven supply-chains seeks to revolutionalise the ways information is transferred, processed and analysed across different industry segments in the value-creation. This data-driven manufacturing revolution promises to increase productivity, democratise data sharing capabilities and foster industrial growth in scales never seen before. The traditional transactional models are to be re-visited, and distributed data storage architectures are to be re-designed to accommodate for optimised data flows across different organisation units. Data is increasingly becoming a strategic business resource that through innovation in existing sharing and processing approaches can decompose business bottlenecks in existing production lines and processes and disrupt traditional supply-chain models. This work seeks to articulate a state-of-the-art review of the application and impact of ML techniques and distributed Ledger technologies to further disrupt supply-chain capabilities with regards to data accuracy and completeness.

Keywords Blockchain · Supply chain management · Machine learning · Artificial intelligence

1 Introduction

Industry 4.0 promises to revolutionalise smart factories and production lines that employ entirely new approaches to production and addressing customer requirements (See Fig. 1). The data-driven value creation processes currently emerging can

G. Epiphaniou (✉) · M. Bottarelli · H. Al-Khateeb · N. T. Ersotelos · J. Kanyaru · V. Nahar
Schools of Mathematics & Computer Science, Wolverhampton Cyber Research Institute (WCRI), University of Wolverhampton, Wolverhampton, UK
e-mail: g.epiphaniou@wlv.ac.uk; m.bottarelli2@wlv.ac.uk; h.al-khateeb@wlv.ac.uk; n.ersotelos@wlv.ac.uk; j.kanyaru@wlv.ac.uk; v.nahar@wlv.ac.uk

H. Jahankhani et al. (eds.), *Cyber Defence in the Age of AI, Smart Societies and Augmented Humanity*, Advanced Sciences and Technologies for Security Applications, https://doi.org/10.1007/978-3-030-35746-7_15

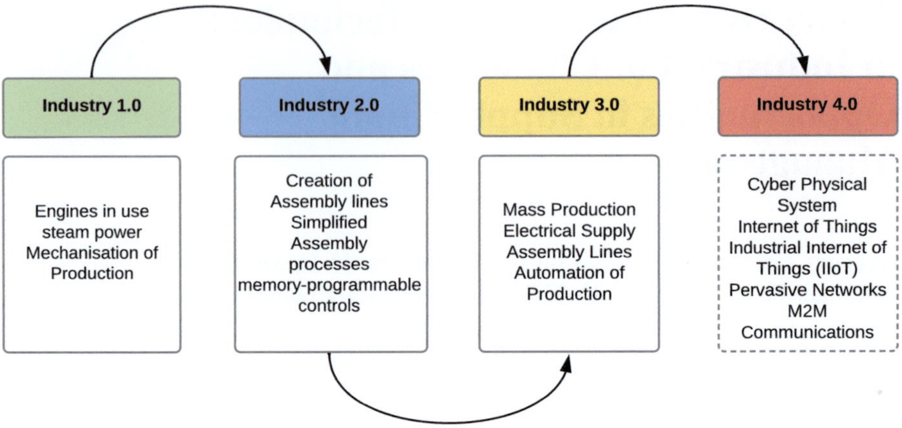

Fig. 1 The evolution to Industry 4.0

potentially enable rapid changes to production and logistics and flexibly respond to challenges identified in traditional supply-chain management (SCM) processes. Industry 4.0 also promises to change the way business models are constructed and manifested and facilitate optimised decision-making processes enabling more opportunities to small-scale businesses. Intelligent operations driven by strong data analytics can solve challenges around energy efficiency, resource optimisation and increased productivity while striking a better work-life balance for all entities within a supply chain (SC). A SC is often described as a group of interconnected business value creation and execution entities that offer a dedicated set of services and products to their end-users and customers [5]. These entities deploy automated storage management and logistics solutions to move value, products and services that cascade down the supply chain with different business owners and models.

Different stakeholders within the SC are also dependent to these solutions for identifying business process bottlenecks and optimisation processes in terms of delivery, material convention into products and distribution to multiple different intermediaries ranging from wholesale retailers to Internet-based companies [32]. Both mainstream and downstream operations have to increase their resilience to disruption and its adverse effects on corporate performance. Existing pre-disruption management processes are influenced by data processing and management aspects and associated technologies in this space. There is a systematic effort to orchestrate these processes in supply-chain (SC) ICT in a way that supply-chain redesign innovations can minimise disruption recovery times and optimise both the security and resilience of these entities [31]. Recently, elements such as the integrity of transactions executed and the immutability aspects required having in terms of audit trails for legal and regulatory purposes. Also, the increased necessity to maintain a certain degree of traceability and visibility mainly when processing large amounts of data in (SCM) has driven research efforts into the development and adaptation of new technologies such as blockchain in this domain.

Information sharing plays a crucial role in SCM with its flow between suppliers and providers being disrupted by the unification of digital technologies. The fourth industrial revolution places information and data processing capabilities at its core as one of the fundamental enablers to optimise workflows and productivity [55]. That is expected to have several implications to IT SCM systems due to the heterogeneity of sensors, communication technologies and production systems. Industry 4.0, which was first defined as the fourth industrial revolution in 2011 by Professor Siegfried Russwurm, Chief Technology Officer at Siemens AG [36], was divided by Herman [15] and Szozda [46], into four categories: (i) data and network connections, such as Big Data, IoT, RFID and Cloud Virtualization Storage, (ii) Cyber-physical Systems (CPS) describing the unification of digital with real workflows – digitisation and automation of work based on artificial intelligence and machine learning, (iii) human-machine interactions – touch GUI interfaces, portable devices and emerging technologies headsets, and (iv) automated machine production – new technological production tools, such as 3D printers and advanced robotics (see Fig. 2).

Referring specifically to the CPS, these tend to monitor natural processes, create a virtual copy of the physical world, make decentralised decisions and integrate digitised end-to-end processes, all of which impact supply chains by posing new challenges; creating flexible business models; reducing production costs; decreasing energy needs, combating overproduction; reducing waste, especially in product development phases; protecting natural resources; improving the environment of existing manufacturing plants [51]; driving more modular and adaptable automation by decentralising production; controlling themselves autonomously [35]; promoting business growth – for instance, in the automotive industry, once Industry 4.0 is fully implemented, productivity is expected to increases by 20%, [24]. Typically, SCM ties a broad range of activities to deliver product flows in a cost-effective manner, including the conversion of raw materials to products. SCM acts as an enabler for business to orchestrate activities between SC stakeholders (e.g. customers, suppliers, distributor) in order to minimise production and distribution costs while meeting market demands and customers' needs. This optimisation of processes often entails the integration of data management and processing solutions for the production and distribution of inventory systems and producing consumable intelligence on demand planning and financial capital [59]. The effective use of this software can provide a competitive advantage by the identification of opportunities or components within the SC that can increase productivity by removing business process bottlenecks [49]. The information produced is beneficial not only for the coordination and supply chain relationship management but equally, identifying the necessary steps to configure SC components from all its participating entities and measure the SC base with all associated costs. Data processing can also enable the clear identification of the extent of vertical integration and inform the decision for outsourcing activities [37].

The SC configuration is a strategic decision that can influence to a certain degree the operational SC coordination and monitoring of material and services flows to enable customer service forecasting. The precise identification of these

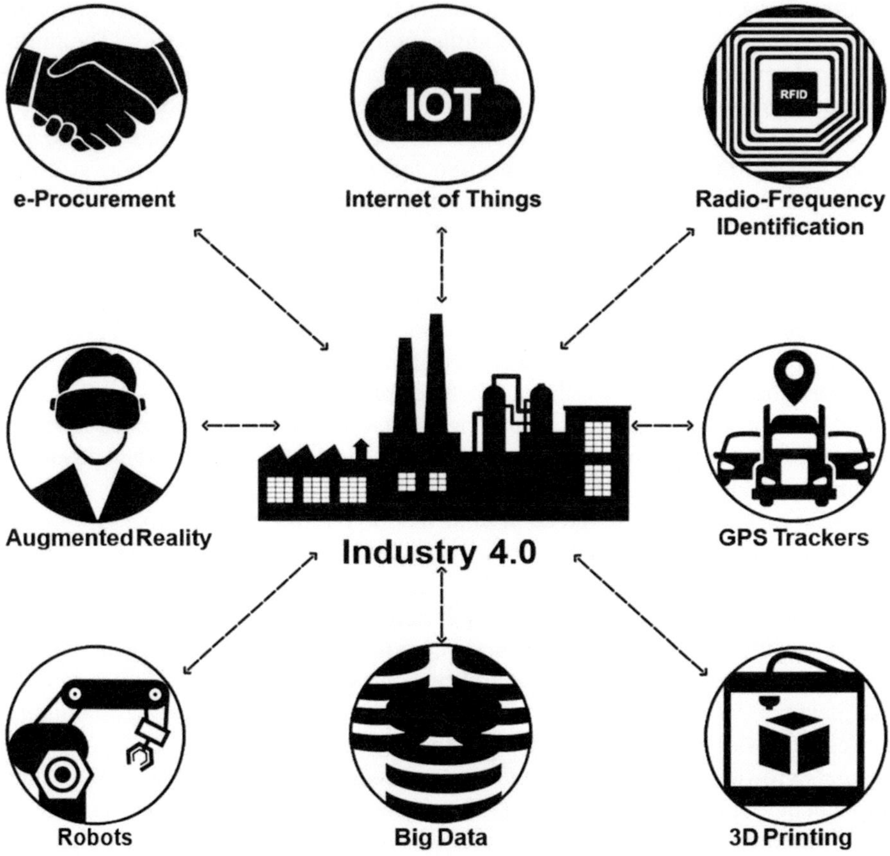

Fig. 2 Emerging technologies provide the enablers of the Industry 4.0 ecosystem

flows can decapsulate the relationships and exchanges between SC entities that can determine value, costs and long-term vision. The sales cycle is another component that can be affected by data processing capability. The sales cycle is often part of the final product/service offered regardless of the SC entities involved in the production stages [40]. It is generally acknowledged that SCM is an integral part of any business with potentials to disrupt sales, improve financial standing and improve customer service while reducing the operational costs. The adaptation of innovating technologies that process data and record transactions in real-time have proved their potentials to control SC expenditures, optimise asset management reduce transportation costs and increase the speed of products flows to end-users and customers.

The remainder of this chapter is structured as follows: Sect. 2 introduces the Blockchain technology and current and emerging works to its integration in traditional SCM approaches and models. In Sect. 3, we discuss BC-enabled distributed data storage and processing architectures for supply-chain management while Sect. 4 presents machine learning techniques in supply chain optimisation. Section 5 discusses emerging technologies and their impact on SCM. Finally, Sect. 6 concludes this work.

2 Blockchain Technology and Supply-Chains

To better understand the adaptation of BC technology in SCM, it is crucial to discuss both domains in terms of the benefits and limitations they present from their potential integration. In this section, we describe traditional SCM and the areas within which BC technology might have a transformational contribution in optimising specific processes and steps. The process of transformation from raw material to products and services undertakes a systematic process consisting of multiple components such as planning, developing, delivering and returning. The process is which the products to be developed might influence or influenced by demand is the initial stage in the SC process with a focus on profit maximisation [28]. The adaptation of BC technology in this step can offer enhanced data transparency and potentially reduce both delays and disputes while avoiding products and services stuck in the SC pipeline(s). Data transparency enabling companies to serve information more adequately and provide actionable intelligence to both manufacturers and suppliers most quickly and effectively. The adaptation of Blockchain technology can also offer a certain degree of scalability with decentralisation and data processing at its core. Since all the transaction records are saved and decentralised across multiple nodes in a distributed fashion, the data transparency and immutability aspects are always assured. This process offers a certain level of security since every block is linked to the previous one and records saved cannot be edited, erased or deleted (See Fig. 3). This feature seems to be extremely attractive in SCM to achieve specific targets such as a vast reduction in inventory management costs and quick identification of issues. The BC technology also promises to solve the information asymmetry within SCM systems due to the heterogeneous sources of information and increase SC's agility in contact market variability and changes. Due to the information asymmetry, SCM systems can also present certain anomalies in their planning algorithms often yielding inaccurate results with regards to the status and progress of specific processes within the SC [30].

2.1 Preliminaries on BC Technology in Traditional SCM

Blockchain was initially developed for financial transactions [11], however, this game-changer technology has taken hand-to-hand in various industries including healthcare, shipping, manufacturing, supply chain management, logistics and various other areas [12, 17, 53] because of its effectiveness and efficiency in recording,

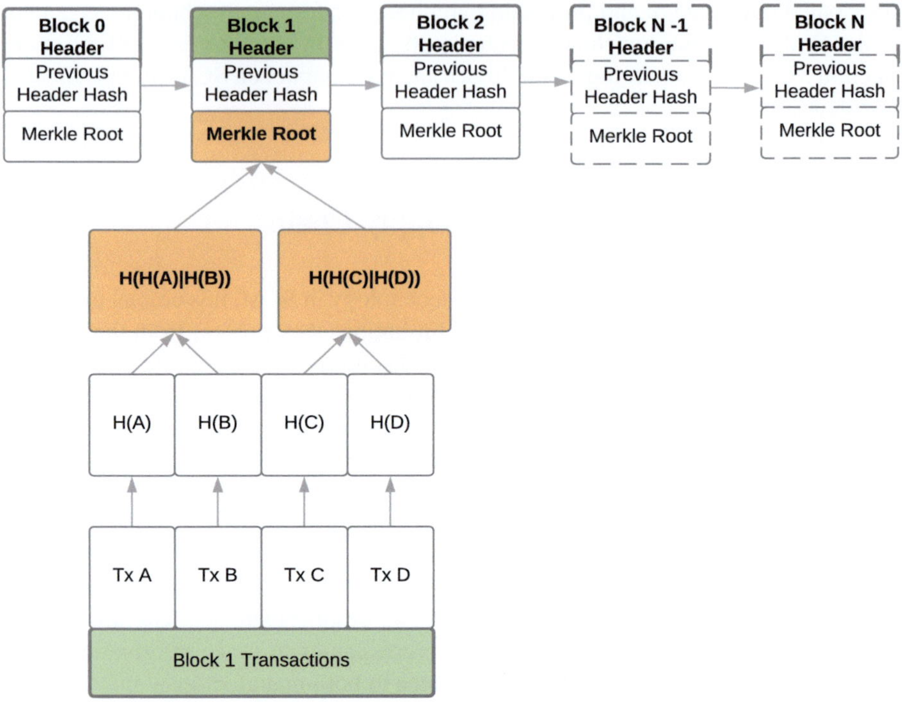

Fig. 3 Merkle Tree hash-based data structure

tracking and verifying assets in a temper-proof and minimal cost environment. Walmart implemented BC to improve food safety after the Spinach outbreak occurred in 2006, which caused illness and deaths. During the outbreak, it took several days to track down the source of the contaminated food item. Authors in [20] have implemented a BC-enabled food traceability system that promises to reduce drastically the overall food tracking time.

Fundamentally, BC can be seen as a distributed ledger, which is a collection of blocks, where each block stores the chronological order of all the transactions or activities in the system with a timestamp and digital signature information. A BC is built using (one-way) cryptography, which is a mathematical function that returns an alphanumeric string of fixed length for a given input (transactions) [52]. Each block has its own hash and for a new block, hash of the previous block along with its own transaction is used as an input. Each block refers to the hash of the previous block, thus forming a chain of blocks, known as BC. For a new block to be added into a BC, all nodes need to agree on the transactions and the order in which they have occurred using consensus otherwise new block invalidated and can't be added to the BC. Consensus ensures that each node has an exact same copy and avoid possible occurrence of fraudulent transaction entry. This means that for a given input, cryptographic hash function such as SHA256 should return the same output value. Therefore, for a new transaction to be added each block should produce a same hash. If the generated hash values are different, it indicates possible fraudulent transaction. A digital signature is used to verify that the transaction is originated by a specific sender (signed with private key) and not by an adversary.

When a new node (a typical supply chain member) enters a system, the node can register itself via a registrar that provides credentials and a unique identity to node. After successful registration, a public and private cryptographic key pair gets generated for the new node where the public key is used to recognise the node within the system and the private key is used to authenticate the node while interacting with the system. Through this member nodes can digitally process products within the system. For example, in a supply chain, the node which receives a product can only add the new data into the product profile using private key. Whereas, when the node passes this product to the next node, both the nodes need to sign a digital contract to authenticate the exchange [48].

The BC technology can influence key metrics in SCM such as transportation cost modelling and optimise existing warehouse efficiency models to decompose sources of information better and build a strong relationship with suppliers. By incorporating the different ways that the adaptation of distributed ledgers can create value, better auditing can be a toot for the supply chain manager with regards to all scheduling activities related to manufacturing, testing, packaging and preparation for delivery [27]. This can hence improve levels of production output and overall productivity from all differences within the SC while optimising specific processes within the logistics. Specific perceptual economic models based on BC have been developed in the public domain as a means to better articulate specific layers of value creation recording and actualisation [34]. The exponential increase in distributed applications (aka smart contracts) has revolutionised the SC optimisation tools by optimally allocating capital resources and business growth processes.

Eximchain project is a clear manifestation of BC technology in supply chain finance optimisation, taking under consideration the complexity of implementing supply chain finance (SCF) within and across the global supply chain limits. The authors offer the platform that implements smart contracts on SCF solutions with full visibility into the supply chain cash flow [19]. The solution seeks to address risk and operational efficiency better while providing investment opportunity in traditional corporate financing schemes. Digital supply integration dictates quicker access to customer demand needs and activity tracking and visibility in the whole supply chain. This can only be achieved through the seemingness integration and mapping of data from various sources that often increases processing costs and diffuse the actual value created during the data processing stages [27].

The strategic and operational information exchange place the foundations for competitive advantage and enhance communications between all SC entities. However, the security and integrity of that information is often a by-product of the data processing and sharing capabilities within and across the SC. The electronic regulation of such activities promises to minimise governance costs and enable more information to be processed at a scale never seen before. The automation of information flow processing can eliminate the necessity for manual interventions and reduce both investment costs and remove barriers to completion and further innovations. This is particularly important, especially in Industry 4.0, given the fact that all entities within manufacturing are often vertically integrated and information integration can create new value elements both in production systems and service delivery [39].

Any process that dictates product or service delivery from and to distribution centres requires a strict and meticulous SC strategy. Data-driven analytics can facilitate quicker and more accurate long-term projections of customer demand and adjust existing SC models and business models as appropriate. The accuracy and immutability of this data and transparent use of it can drive the decision-making on what enters the production and distribution lines and when by pushing the predictability boundaries and capabilities of the energy SC further [23]. For example, stockpiling raw materials in cases that prices are expected to increase is a collective action for manufacturers in alignment with the core SCM requirements such as the demand-supply forces and customers' needs.

BC can be disruptive to distribution and retailing production up to procurement processes within the SCM. The decomposition of information flows can underpin innovations to financial and IT and better identify RoI by actions such as products' traceability in a decentralised way. There are several approaches in the public domain on how this can be achieved leveraging the tamper-proof attributes of BC in food and drugs supply chains, where a certain degree of accountability can be assured for both auditing and compliance purposes [2, 29]. Also, being able to reduce operational costs while optimising customer services is a key performance indicator for SCM that often amalgamates management processes, core IT operations, suppliers' network and retail.

BC can help in tracking this flow of goods and services and act as an enabler to optimise logistics from tracking quantities and delivery/production times to counterfeit detection. The global supply-chain currently faces a significant issue to distinguish counterfeit parts from legitimate ones as the technology used by organised criminal groups (OCGs) has improved significantly. A cloned version of hardware has been found in the automotive industry, networking equipment and critical electronic systems posing a significant threat to security and safety. Cloning spans all levels of electronics with relatively straightforward techniques and processes, especially for equipment, which is high on demand [43, 47]. The counterfeiting types range from illegal manipulation of the legitimate circuit to fake circuits in circulation. There are cases that the integrated circuits (IC) has been replicated from old PCBs and re-branded to forged specifications and labelling to illegal contracts for fabrication or IC failed quality checks.

Frictionless transactions in the cyber-connected single market across the globe, demand common solutions, technology and standards on how to integrate business processes effectively pushing the boundaries of technological collaboration. BC promises to remove the necessity of trusted third parties to execute process and data integration on behalf of SC entities and minimise the number of intermediaries required. The integration of BC technology into these processes entails specific considerations around technology, service development challenges, transactions' volume and B2B integration models (e.g. manual, Cloud). These considerations will not only dictate the type of the BC deployment (private, public, federated) per case but equally establish a smooth transition from manual transactions to digitised information across all entities within the SC. Certain anonymisation aspects offered by the BC integration can de-associate electronically identified parties from a specific transaction. This can offer a certain level of protection against adversaries targeting sensitive or PII as part of these transactions in existing SCM models [8].

Different parts of the SC co-exist in different locations where there is a free trade agreement or low tariff between countries, in an attempt to reduce production costs. Because of that reason, SC has instead fragmented a phenomenon that can lead to significant losses in both financial and marketing terms. The integration of BC technology promises to optimise SC logistics while inform better decision-making for SC managers. It promises to expand and increase visibility for all downstream processes in SCM and adapt rapidly to any changes or issues related to customer needs, production, distribution and marketing. Most importantly, BC can offer a uniform way about the characteristics of any SC and widely deploy these across the sector. These will help to audit and record all changes at any moment in the SC, offering a high degree of flexibility across all entities that need to share information both vertically and horizontally.

3 Distributed Storage Architectures for Supply-Chain Management

In its purest form, the blockchain can be defined as a complete distributed platform for computation and information, much like to a decentralised database, which is maintained by multiple authoritative domains. Similarly, data can be stored in the blockchain in the form of transactions, sealed in crypto-protected blocks which are available to all blockchain peers, providing tamper-evidence, decentralisation and transparency, where the consensus method impersonates the unique impartial trustee. While these characteristics are appealing and advantageous in specific application fields, there are aspects which are too expensive to replace traditional approaches, such as verification and storage [45]. In this chapter, we focus on the latter, drawn by the interest in the continuous increase of digital assets and data gathered throughout the supply chain, for example, from IoT devices.

3.1 Storage Architectures

Traditionally, data is shared in centralised environments (Fig. 4a) where information is managed by a single party, referred to as the coordinator who has the duty to direct data flow like an orchestra conductor. In this architecture the latter represents

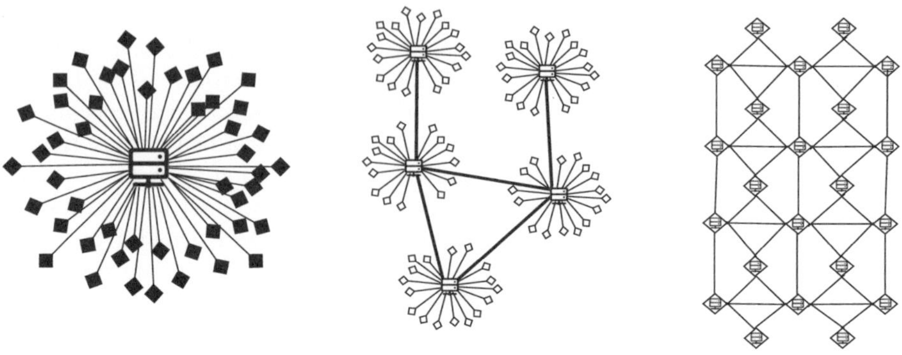

(a) (b) (c)

Fig. 4 Centralised (**a**), decentralised (**b**) and distributed (**c**) storage architecture

the single point of failure: if data is stored on-premises, an internal and expensive backup solution must be deployed to make the system failsafe, however, in case of a crash, you still need to wait for the recovery process to complete before restoring the standard processing capabilities. On the other hand, if data is sent to the cloud, backup/restore actions are automatically done by the service provider, but the risk is not annihilated, instead transferred to a possible lack of connectivity or reduced bandwidth. Moreover, as someone has full access to the central storage, possible changes made to data could eventually increase the risk of tampering, even in the presence of an audit system [45].

The decentralised system (Fig. 4b) arises as an evolution of the previous approach, where few coordinators are inter-connected among them as well as with the corresponding consumer nodes. In this particular architecture, even if multiple coordinator nodes fail, orphan nodes can potentially connect to the remaining available coordinators, thus maintaining a sufficient level of service. The single point of failure is now the network itself because it compensates for failed nodes by dynamically altering consumers-coordinator connections. This approach manifests two significant vulnerabilities: first, even with redundancy of coordinators, in the case consumer nodes are disconnected or the network itself gets partitioned, there is no way to recover such situations. The second weak spot is more subtle and regards how coordinators are elected/chosen: this choice usually reflects the dominant role of a member, as for example in the case of a company that becomes the hub through which all other members communicate, reducing the number of links needed. However, besides trivial supply-chains, it is hard to recognise such a central role, because a strong partnership does not connect the parties between them and hence their communication is time-limited, as for the duration of a single contract. In such cases, it is common to lean on an external entity, a specialised company which provides a platform for the exchange of business documents, representing another possible point of failure.

To overcome these problems, distributed architectural (Fig. 4c) patterns emerged in the last years where all nodes have a protagonist role in the network, by both processing and sharing information, while maintaining the ability to work, even if isolated from the rest (for an appropriate time duration). Unfortunately, everything comes to a price and this resilience is counterbalanced by its increased complexity during implementation, which is needed to guarantee the convergence of information and data originated in different parties that above all, do not trust each other. Blockchain comes to rescue matters, since it is the most popular representant of distributed ledger technologies (DLT), which provides a shared and ordered list of records stored in a distributed network of nodes, each of which is able to access and verify all the information, along with adding new transactions [16].

3.2 Storing Data in the Blockchain

Some blockchain implementations offer the possibility to append data to a transaction, affording a declared cost whose purpose is to repay the resource used in the creation process. In the case of Ethereum, this cost is referred to as "gas", which is depleted during the smart contract execution based on various rules. The Ethereum Virtual Machine (EVM) assigns a gas-cost to each executable operation, or OPCODE, according to its intrinsic complexity. For example, the addition of two numbers requires three units of gas whilst multiplying them requires five units of gas. The sum of the gas needed by all operations in a smart contract, multiplied by the current gas-price, gives you the total cost of creating a transaction. Nonetheless, things are even more complicated when someone wants to store data: according to the documentation [54] storage-related opcodes are extraordinarily expensive, for example saving 32 bytes requires 20.000 units of gas, which at the current gas-price is about 0.2 dollars or, equivalently, 5 million dollars per-GB [41]. In other blockchain implementations, it is even worse because the latter have not been designed to store any payloads besides regular finance transactions. In Bitcoin, for example, the block size is limited to 1 MB, which is not enough for a supply-chain scenario where physical goods may be associated with many various digital assets.

The core question is why the price of storing data on-chain is so high. Intuitively, that information will become an integral part of the system itself and will be replicated in every full node, being an unavoidable download by anyone that wants to contribute to the blockchain, soon making the entire system overstuffed and bloated [1, 56]. To better understand this fact, it is useful to consider traditional cloud storage: when a customer opts for cloud services, he agrees to pay a monthly fee for a fixed amount of space which is available to him for as long as he keeps paying. Otherwise, he risks that existing files would be cancelled. In other words, the provider has to hold a customer's data only during the paid time, a practice not applicable to blockchains. In the latter, the customer must pay upfront not just for the first month, but for all years to follow!

In current literature, there are several proposals to pass these limits without mixing blockchain technology with a traditional storage approach. An immediate solution would be to alter the block size, as for example in bringing it up to 2MB or even better to dynamically adapt it with time. However, this method also slows down the synchronisation process among nodes [22]. Even worse, updating the protocol rules implies a fork, a permanent change which can be backwards compatible (soft fork) or a risky complete divergence from the past (hard fork). Other proposals stem from the use of sidechains, which are separated blockchains attached to the parent one through a two-way peg, allowing a continuous and bidirectional flow of assets. An appealing idea could be to spawn a new sidechain for managing the data associated to an item or a group: in this way, only interested/permissioned peers have to download such sub-chain, while at the same time all of them are always able to verify the main channel. In other words, sidechains are capable of storing their own data and processing their own transactions, leaving the main-chain lighter, smaller and faster, thus providing a bit of breath to the scalability problems of the blockchain technology. Nonetheless, they also introduce new categories of risks, which must be considered and addressed. These risks include an augmented vulnerability to sophisticated attacks such as 51%-attack and, furthermore, the case when a sidechain becomes unreliable.

3.3 Blockchain and Databases

Comparing blockchains to databases, they suffer for limited scalability, slow transaction rate (i.e. Bitcoin takes 10 mins to accept a new transaction), low write throughput and minimal querying capabilities. All these characteristics do not make the "pure" smart ledger an appropriate ground to store (possibly big) supply-chain data. Traditional databases, on the other hand, possess a defined structure which translates into fast querying abilities and cheaper storage; however, they are vulnerable to data modifications and delicate synchronisation operations. A viable solution is to mix blockchain technologies with traditional databases, either by adding blockchain features to an existing database (Fig. 5a) or by creating a new system architecture where those entities cooperate, while preserving their independence (Fig. 5b).

BigchainDB [10] and HBasechainDB [42] are platforms that arise from distributed databases, MongoDB and Apache HBase respectively, by providing immutability and decentralised control on top of them, while inheriting their storage capability and high availability at the same time. The interesting contribution of BigchainDB is to postpone the validation after the block has been appended to the network by initiating a voting process among the nodes, a technique referred to as blockchain pipelining, whose aim is to increase transaction throughput. Mystiko [7] takes leverage of the Apache Cassandra distributed database as its storage engine, adding full-text search, while reducing network overhead with sharding-based data replication.

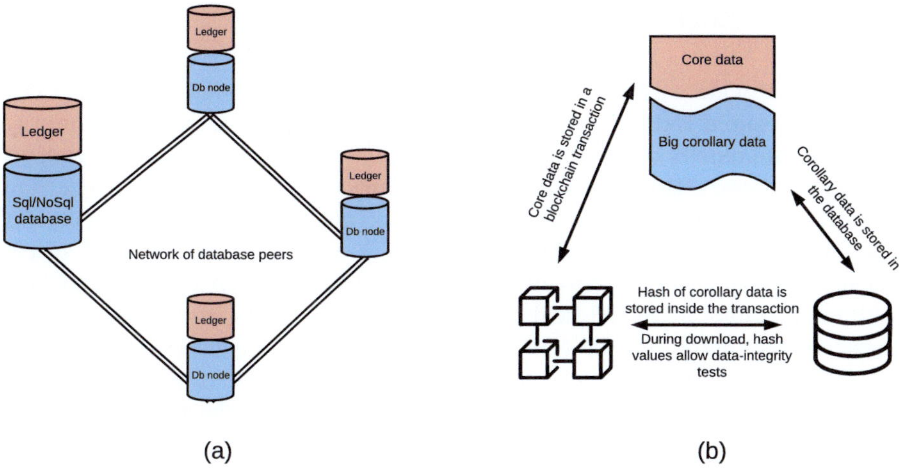

Fig. 5 Cooperation between blockchain and databases: (**a**) blockchain on top of distributed databases and (**b**) independent entities connected through a cryptographic anchor

As stated above, instead of using an already prepared "blockchain-powered database", a more versatile approach could be to connect a blockchain implementation to an external database of our choice, simultaneously storing small-core data in the former, and big-corollary data in the latter. Figure 5b shows the de-facto standard protocol. The first step in the latter is to identify which data should be stored outside the chain, followed by information being fed to a hashing function, for example SHA256 [38], in order to extract a fixed-length value which is its unique numerical representation. According to the avalanche effect of cryptographic hash functions, any small change in the input generates a significant difference in the result (footnote: for simplicity we do not consider the improbable case of collisions) hence, this value can be used as an authenticity signature and saved together with the core-data in the ledger. In this way, the properties of tamper-resistance and traceability are transferred to the external database, because a party can always check data integrity by re-computing the hash value and verify if it matches with the one immutably stored on-chain.

To evaluate the performance of this approach, in [14] authors stored personnel information (50 kb) in the blockchain for an increasing number of people (from 2.000 to 20.000) and compared these results with the same data which this time appeared as stored 20% (10 kb) on-chain and the remaining 80% (40 kb) to a central database. Tests showed that the time needed to add and to query data increased with its size, hence storing a large part externally is beneficial as it reduced the impact of consensus and nodes synchronisation, further staving off-blockchain inherent scalability problems. That implementation was based on Hyperledger Fabric [4], a permissioned blockchain system, highly modular and extensible which comes with different consensus methods (Kafka, RBFT, Sumeragi, PoET) and a smart contract execution layer named Chaincode.

3.4 Blockchain and File Systems

Databases provide a fast querying interface due to the structured way in which records are stored and this makes them the de-facto first choice when the priority is to generate reports or to feed data to a multidimensional OLAP cube to support business processes. However, when data is unstructured, for example when dealing with a large number of files such as audios or videos, storing such documents into the database could possibly be the wrong choice. In fact, even if recent database management systems offer the possibility to incorporate files together with tables, either as BLOB fields or separated streams, this has a non-negligible impact on backup/restore plans and timings, further increasing the costs of database administration. One alternative could be to store such documents on the filesystem, linking them back to the main blockchain application through the generation of file-anchors.

Considering the distributed nature of the blockchain, it is necessary to find a suitable filesystem to cooperate with and InterPlanetary File System (IPFS) [9] has been chosen in various system architectures. IPFS is a peer-to-peer distributed file system which arises from ideas successfully explored by previous peer-to-peer networks, as distributed hash tables (DSHT), BitTorrent file-sharing, Git source versioning system and self-certified filesystems (SFS). In IPFS, every node is associated to an ID which is the hash of a public-key generated by itself. This node can communicate with hundreds of other peers all over the internet using a network stack which can add reliability, connectivity, integrity and authenticity even if the underlying transport protocol does not provide them. Files are efficiently stored and distributed through a routing protocol based on distributed hash tables and a novel BitSwap file-sharing protocol which extends the original BitTorrent by allowing the simultaneous request of multiple blocks from different torrents. Furthermore, IPFS uses a Merkle Directed Acyclic Graph (DAG), borrowed from Git, to decompose a file in sub-parts which are connected by cryptographic hashes. This way, the data is addressed by the hash of its content, it is made tamper-resistant and finally, sub-blocks with the exact same content are possibly shared with other files, reducing the data duplication. All these characteristics make IPFS an ideal candidate to be used with blockchain. Similar to IPFS, SWARM is another distributed storage platform with a strong connection with Ethereum ecosystem [50] and a slightly different management of peers and an underlying protocol.

The standard process of uploading a document can be summarised as follows: first, the user sends the document through a frontend, usually web-based and the file is saved in the distributed filesystem obtaining its address as the result, i.e. the hash of the content. At this point, this hash value is merged with other user metadata (keywords, user public key, etc.) and passed as arguments for the execution of a smart contract whose duty is to encode and seal the transaction in the blockchain.

The symmetric action is to download this file by usually invoking another smart contract to search for files with specific keywords and then retrieving its payload from the storage. At this stage, the smart contract or the frontend also verifies that the information has not been modified by re-computing the hash value. This simplistic process is generally extended with permission management which permits us to define who has rights to access the file and in which way (read or write). Furthermore, as the entry points of all actions coincide to the execution of corresponding smart contracts, it is straightforward to keep an immutable log, or audit, which can become an interesting starting point for process optimisation opportunities.

In [58] researchers started from the consideration that transactional data occupies the large major part of blockchain blocks, hence it is beneficial to remove those transactions and to store them in IPFS. Consequently, miners can compress newly generated blocks by packing all IPFS hashes in a Merkle tree, together with the previous block hash. The protocol has been proposed for Bitcoin, but it is also applicable to private permissioned blockchains, thus suitable for supply-chain scenarios. In [44] IPFS is used together with Ethereum blockchain to store crop images during all stages before harvesting, enhancing this way the traceability in an agricultural supply chain. In [33] a IPFS-blockchain system is extended with smart contracts to act as a document versioning system, automating the actions of typical proposers and approvers in the process.

Since saving a large amount of information on the blockchain is not economically feasible, the road to be pursued seems to place corollary data alongside a traditional storage structure like an external database. In some cases, it has been decided to elevate existing distributed database nodes to be the blockchain peer, providing them with the necessary missing features. In other cases, for greater versatility, the entities are kept independent, by implementing a collaboration interface between them. The primary motivation for selecting a database as a blockchain storage companion is given by the possibility to define a schema, thus increasing the expressive power of queries. Unfortunately, this also increases the corresponding database administration costs which can be avoided or minimised by interfacing the blockchain with a filesystem. The IPFS InterPlanetary File System seems to adapt perfectly to this role, as it is composed of a distributed network of nodes and it relies on graphs of cryptographic links. Additionally, a file stored in IPFS is accessible through its content-based address which is finally stored in the blockchain inside the transaction payload, simultaneously guaranteeing resistance to modification and deduplication.

4 Machine Learning Techniques in Supply-Chain Optimisation

A key foundation of every organisation is the production of goods or services, and the delivery of these goods and services to the customer. The process steps involved this production (of goods and services) and their delivery often involves people, other organizations, resources and information all of which constitute what is in short called supply-chain [21]. Supply-chains adoption of technology means that collaborating agents and stakeholders can be seen as nodes in a graph or a network with complex dependencies. There is therefore a huge potential in the use of machine learning techniques in either predictive analytics (e.g., to cut delivery costs or reduce rejection rates) or to enhance the security of transactions as well as increase user satisfaction by using natural language processing and opinion mining [18, 21]. The implication is that parties involved in supply-chains can leverage the emerging computing applications and techniques such as Internet of Things (IoT), blockchain, and machine learning. The key driving motivation for leveraging these developments is to optimise supply-chains for the benefit of producers and consumers as well as the involved organisations in the network of activities.

The role and rise of cloud computing is a phenomenon that has impacted supply-chains over the years as organisations have showed an upward trajectory regarding the uptake of cloud-based infrastructure [24]. The need to reach new market sectors, and to scale production whilst minimising costs, accessibility by multiple parties are some of the reasons organisations have increasingly taken up cloud deployment [3]. The use of such technology as the cloud has many benefits besides boosting production and easing access to market. For instance, data generated and stored in the cloud can be used to provide insights regarding potential for new products, or requisite improvements on current ones based on customer feedback [13]. On the other hand, the emergence of blockchain technology adds enormous benefits with regards to security and verifiability of transactions. Organisations can be confident about the data because it's indelibly recorded on an immutable ledger. Additionally, organisations can use other blockchain capabilities, especially, secure transactions to extend business processes beyond organisational boundaries. It is for instance of huge potential to organisations can use decentralised applications encoded in 'smart contracts' that operate autonomously [57]. Another computing development that has gained stability over the past decade is the branch of artificial intelligence called machine learning. Machine learning is the application of mathematical methods (especially statistics and probabilities) in building systems that make decisions based on what they learn from data [25, 26]. So far, machine learning has found industrial and real-world application in areas such as machine vision and image recognition where deep learning algorithms such as convolutional neural networks have been very successful in this area. Another robust application of machine learning is in predictive analytic areas such as financial markets, as well spam email filtering. This article explores the potential for use of some machine learning algorithms in optimising the supply-chain.

4.1 Elements of Supply-Chain and Associated Challenges

A key characteristic of supply-chains, whether digitally-led or traditional is the constituent elements of it. A supply-chain comprises of a network of facilities that include production of goods (or services), means of transportation to the customer, value addition and manufacturing, storage, and distribution of materials and products. Practically, the main discrete elements of a supply-chain include producers or suppliers, manufacturers, distributors and customers. In this article, we argue that, a supply-chain in which the interplay of these elements produces optimal customer satisfaction and production of best quality at affordable cost is deemed optimal. Often, the question would then be what would be the measure or the metrics of the interplay between these elements to provide this level of a supply-chain. The key to answering this question is the ability of all the parties involved to learn from experience. Using data generated in the process can enable this learning and help mitigate delays, pilferage and to predict customer's mind-set change about products or processes. The mature field of predictive analytics and deep learning auger well for this type of question whereby volumes of data is available and is continually being generated, but has not been harnessed in ways that would feedback to the supply-chain process. Collaboration among parties in the supply chain is important in order to achieve accuracy in logistics and forecasts. Demand and supply forecasting are one of the key concerns in supply chain optimisation. It is likely that if an increase in forecasting accuracy is achieved, it might result in lower costs because of reduced production cost, and increased customer satisfaction that will result from an increase in on-time deliveries. In the whole, the main challenges involve reduction of time costs to market, prediction of customer behavior based on prior feedback, sentiment analysis, and also cross-selling.

4.2 Prediction as a Means of Improving Supply-Chain

A successful supply-chain is marked by rigorous planning. Since substantial knowledge is built into planning cycles, it is possible for predictive data mining to be applied to both select the best planning models as well as to predict outcomes. Additionally, best performing organisations can use this learned knowledge in advancing new product lines or gaining competitive advantage. There is also a potential for relevant sub-models to be decomposed from organisational structure as means of sharing logistical knowledge across different production plants within a production network. The key to application of data mining here is the use historical enterprise data to feed predictive models that support more informed decisions. It is important for organisations to tease out tacit inefficiencies to enhance production cost cuts as well as lower customer default risks. One of the key challenges faced in supply chains is the optimising logistics. Datasets gathered within logistics are quite varied, and of course machine learning algorithms can be applied to find patterns and therefore learn from such data [1]. For instance, courier companies such as FedEx and DHL have reported reduction in their logistics costs by mining patterns in data (tracking and tracing) which is obtained using IoT sensors [6].

4.3 *Optimising Supply-Chain with Machine Learning*

In this section, consideration is made of a selection of machine learning techniques most suited for optimising a technology led supply-chain. Machine learning constitutes the activity of writing computer programs that learn from data. The preceding sections indicate that the supply-chain is a heavily data driven process, and is there a strong candidate for application of machine learning.

4.3.1 Support Vector Machines

A Support Vector Machines (SVM) is a supervised machine learning algorithm which plots a training dataset into a higher-dimensional space so that it can be linearly separated. The goal of the SVM is to find the optimal hyperplane where categories of data are clearly defined. The hyperplane is the mathematical term for the boundary between the different categories of data points mapped by the algorithm. Once trained much of the training data is redundant, only a sub set of the data is used to find the boundary of the categories. These data points are called "Support Vectors", they represent vectors filled with variables and they support the creation of a boundary. When the algorithm is provided with new data after training, these new data points are classified by which category boundary they fall in to [19]. An important metric with SVMs is its margin; the margin represents the distance between two categories of data points, as such it can be said that the SVM algorithm essentially iterates over a data set to find the hyperplane which provides the largest geometric distance between category boundaries – more simply the clearest separation of categories [19]. A Support Vector Machine is therefore well suited in using categorical identifying optimal processes that produce best outcomes for production and customer satisfaction.

4.3.2 Neural Networks

Artificial neural networks are the primary building blocks of deep learning [4, 39]. Neural networks are at the core of image classification, speech recognition and sentiment analysis [4]. There are a wide variety of artificial neural networks, but the most relevant for supply chains are the feed-forward error back-propagation type neural nets. The mechanics of feed-forward neural nets is as follows. A layer's neuron produces an output signal which is then propagated to all the neurons of the next layer. The flow of neural activation is unidirectional, from one layer to another. The feed-forward error back-propagation neural nets have a minimum of two layers, since it needs an input and output layer in the least. It is of course possible to increase the networks computing power by adding more layers between the input and the output layers. With careful design and modelling, artificial neural networks can be<?pag ?>to provide an excellent level of prediction for supply-chains for the benefit of minimising risk and the modelling of customer behavior. This is a supervised type of training, in that the desired outputs are provided to the neural network during the course of training along with the inputs. That is, established and robust datasets can be used to train the neural network so that the future outputs based on unseen data are highly reliable.

4.3.3 Convolutional Neural Networks

Convolutional Neural Networks (CNN) work the same way as a foundational artificial neural network (ANN) in that a CNN receives input into various neurons and creates an output based on weight and biases [4]. The weights and biases are trained by exposing the neural network to a dataset (training set) that one would like the network to analyse [4]. Convolutional Neural Networks differ in that they mimic neurons in the visual cortex. Thus, they accept three-dimensional input and output (height, width, depth). This is useful in situations where the dataset is highly textual and that may require analytics and plotting of words in high-dimensional space (Support Vector Machines do this too). This allows for words to be clustered based on meaning and algorithm can then 'learn' the meaning of words by association. Clearly, this potentially applicable technique in supply chains where datasets are contributed to by multiple stakeholders in different formats. High-dimensional space can be seen as adding dimensions to words, for example, a customer can be seen to have dimensions such as location, item ordered, and selected delivery date among other dimensions. With a number of customers plotted into high- dimensional space clusters can be formed and insights derived regarding the best distribution/delivery methods.

4.3.4 Recurrent Neural Networks

Recurrent Neural Networks (RNNs) work by using inbuilt loops to enable information to persist. The importance of this persistence is to enable the network to make decisions about the future based on prior seen and stored information. Due to their loops and information persistence, RNNs have been shown to predict outcomes with excellent accuracy. For instance, RNNs have been used in financial markets, especially in stock price prediction. Another domain in which they have been successfully used is informing managers regarding fluctuations in inventory. Within the supply-chains, there is a likely central role of RNNs. For instance, RNNS can be used in predicting, and therefore helping anticipate when a product might be in high or low demand. This then positions producers and suppliers well to be able to determine when to place a product or service in the market. RNNs are also well suited to working with textual data, alongside natural language processing tasks. RNNS are well placed to with textual objects such as such as sentences, and documents making them extremely useful for natural language processing systems and well suited to analyzing customer sentiment. In general, RNNs allow output signals of some of their neurons to flow back and serve as inputs for the neurons of the same layer or those of the previous layers. This recursive behavior enables RNNs to become powerful tools for many complex problems that involve multiple stakeholders, a scenario so common within supply chains.

5 Emerging Technologies for Supply-Chain Management

For any business to be profitable it must have an efficient and seamless supply chain, since its customers expect the correct product to be supplied, effectively tracked and delivered swiftly to a convenient location. Later, they will expect to receive efficient after-sales support. Thus, both manufacturer and retailer expect to benefit from a supply chain that has been designed to meet their customers' needs at the lowest possible cost. The Industry 4.0 revolution means that many companies will use emerging technologies, such as robotics, smart warehousing, 3D printing, big data, the internet of things and artificial intelligence, to create global networks of machines situated in smart factories, automatically and autonomously exchanging information, and controlling each other. Thus, e-technologies will affect every component and every step in the supply chain, while the supply chain manager can be free to oversee and intervene all goods and materials at any time. Therefore, by utilizing machine learning capabilities and predictive analytics, businesses can ensure they meet demand while minimizing costs. This section discusses both the advantages and disadvantages of these emerging technologies, which should be taken into consideration by any company intending to use them in SCM.

5.1 Internet of Things in Supply-Chain Management

The Internet of Things (IoT), as described by Whitmore et al. [6], in 2014, enhances the supply chains by embedding everyday objects with technology (e.g. sensors, RFID, mobile network receivers or fixed internet connectors) and capabilities to process data, identify networks and connect to the Internet in order to communicate with devices or services over the Internet. The IoT can improve the procurement transactions, the efficiency of the production operations, can provide real-time products' tracking and tracing, can automatically make decisions for optimizing the retail industry and improve the customer services (Fig. 6). Eventually, more physical objects with embedded software to handle data, will connect to the internet than people [11]. Constantinides et al. [19] stated that IoT offers ubiquity, intelligence and autonomy, since by using GPS trackers (UCOT's sensors), goods and data transfers can be immediately located, losses or delays can be spotted early, thereby simplifying supply planning and reducing warehousing needs to a minimum. Thus, to facilitate supply chain integration, physical and digital worlds will be linked.

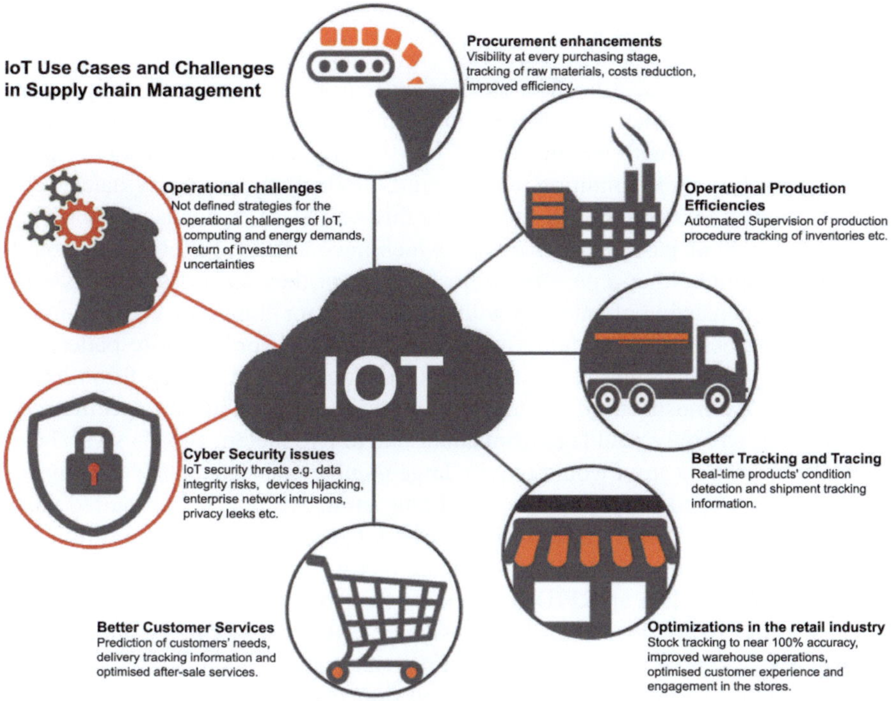

Fig. 6 IoT use cases and challenges in supply chain management

Although, Taliaferro et al. [24] state that sensors are expected to revolutionalise the retail sector, they also believe that what firms must do to improve supply chain performance by using IoT is unclear. Many companies do not have defined strategies for the operational challenges of IoT, hence they don't know how it can be used at supply chain level since much of its information will be stored, ubiquitously, in the cloud; neither do they know what kind of information should be stored or for how long. Furthermore, Santiago [39] believe that IoT could increase computing and energy demands, while Fernandez-Gago et al. [1] believe that a lack of security could impact on data integrity. Replacing the power feed of remote IoTs, building the appropriate (e.g. Big Data) platform to best serve the current and future supply chain operational requirements are additional challenges that need to be appropriately addressed. The need for highly customized solutions or an increase in interoperability and re-utilization, means that it may be more challenging to integrate IoT into the supply chain and make meaningful connections between physical objects' transmitted data and operational processes, compared to other market use cases e.g. metering [1, 24]. Finally, Dutton [2] believe that IoT deployment may not always offset its expense.

5.2 RFID

A Radio-frequency identification (RFID) system consists of microchips (tags) attached to objects which convert them into 'smart manufacturing objects' (SMOs) since they are fitted with an antenna for transmitting, through radio waves, information regarding their condition, temperature, production and expiry dates, serial number, current location, origin, etc. [3]. In this way SCM is enhanced. RFID tags are either active or passive. Active tags, which have a limited lifetime of three to five years, require either a powered infrastructure, or they use integrated batteries. Passive tags, consisting of antennas with semi-conductor chips attached, can fit into an adhesive label, have indefinite operational lives and don't require batteries or maintenance. The antenna in the tag alerts the circuit when energy flows into it. Passive tags are mostly used for tracking an inventory located in a specific room since readers can only read tags from about one to five meters away – so scaling a system to track a location would require large numbers of readers. For the physical location of a tagged object within a building, an active reader is required, since it can detect a tag further than 100 feet away. Consequently, in some instances, active systems can cover over 10,000 square feet using only one reader and a few reference points. When an RFID solution is selected by a company, both business and technology requirements must be considered before the type of tag is chosen [3]. All tags have essential components in common: an antenna, an integrated circuit and a printed circuit board. RFID is not without drawbacks since its uses are limited. Mandeep et al. [4] point out that the barriers to RFID are high implementation costs (infrastructure and storage), low investment returns, lack of global standards and regulations, and faulty tag manufacture. A more technical issue is a potential collision problem, when many chips in the same field are being read [5]. From a supply management perspective, RFID provides efficiency improvements, human error reductions, and manual processes elimination [7].

5.3 Big Data

According to Awwad et al. [8], Big Data is defined as high-volume, high-velocity and/or high-variety information assets, where volume is the amount of data retrieved from websites, portals and online, velocity is the speed at which information is created, and variety is structured and unstructured information generated by people or machines. The International Data Corporation expects digital data to increase from 2.8 trillion gigabytes in 2012 to 40 trillion by 2020, hence data sets generated by traditional software will become unmanageable [9]. Awwad et al. 2018 [8], have indicated the possible benefits of big data in its potentials to revolutionalise the SCM, customer demand predictions, product development, supply decisions, and distribution optimization. By using predictive big data techniques, firms like Amazon can foresee shoppers' product needs by recommending the right product to the ideal individual at the right time and send it to their nearest depot; thereby optimizing storage, flow, and availability [10]. Additionally, IoT and or RFID analytics, by using GPS tracking, which includes traffic and weather data, can optimize delivery routes, thereby avoiding delivery delays. Big data analytics can also help reduce both supply costs and time spent on suppliers' selection by providing access to supplier information, thus minimizing risks. Also, manufacturers can conduct suppliers' performance appraisals more simply than by traditional, static marketing benchmarks. However, big data, when used in supply chains, face numerous challenges, including (i) bad quality data, (ii) collecting and analysing data, which is time-consuming, (iii) skill deficits, including poor domain knowledge, poor analytic skills, and an inability to interpret data accurately, and (iv) a lack of shared data due to privacy and security laws. All of these challenges could impact negatively on SCM.

5.4 E-Procurement

One technological advance used in SCM is e-procurement (aka supplier exchange), which involves e-data transfers supporting operational, tactical and strategic e-purchases and sales. It first came into usage after the internet establishment in 1990s. However, it was first invented in 1960s and until the mid90s it was under the name electronic data interchange (EDI). The e-procurement is consisted by Internet tools and platforms that replace traditional procurement systems by identifying new direct and indirect suppliers of materials and services, including tendering, where the purchaser invites bids from potential suppliers, which, by agreeing a price, leads to either ordering or auctioning. E-procurement, firstly, gives purchasers the advantage of selecting the right item, from the right supplier at the right time and at the right price. Secondly, it allows suppliers to learn more about the purchaser's customers' needs than they would from within a normal supply chain. Hence, they can offer real-time information, together with an automatic service, which sends them materials when stock levels reach a low point. Such advantages take SCM to new levels. However, disadvantages of e-procurement are that, should a product fail to match the one described, or the product quality is poor, it will result in money being blocked (either during product replacement or refunding process) and time wasted. Moreover, other emerging technologies can also help the e-procurement process. 'Big Data' could further enhance the efficiency of e-procurement. IoT can give visibility at every purchasing stage. The 3D printing could become an alternative solution to produce low volume items, spare parts and prototypes instead of buying them from overseas suppliers [11].

5.5 Robotics and 3D Printing

Technological trends such as robotics, self-driven trucks and drones can further reduce the supply chain production and delivery cost by offering solutions for both warehousing and transportation. There prime advantage being that they can work twenty-four-hour days; however, they cannot make independent decisions. Some innovative companies have developed coordinated systems whereby robots do repetitive, routine jobs while humans focus on more complex tasks. A recent innovation, however, is a co-robot (cobot) that can interact with humans to perform routine tasks, such as selecting, loading and unloading warehouse items. According to the International Labor Organization [11] about 56% of Southeast Asian employees are at risk of displacement by robots over the next twenty years. Similarly, Clark [21] estimates that, across the economy, almost twenty-five million jobs will be lost worldwide to automation in the next ten years. However, he also expects new technology can create fifteen million jobs. Another technology which is expected to avail SCM is 3D printing, especially in managing 'high mix/low volume' product supplies, since it can be used to print spare parts, or aftermarket products. Thus, it contributes to inventory reduction and cost savings.

This section has considered supply-chains as business artefacts that are technology enabled. The article provides an overview of technology areas in which organisations find themselves in, especially areas that impact supply chains. These areas include blockchain as well as machine learning. The article is keen on machine learning techniques most attuned to enhancing supply chains. We argue that deep learning techniques, especially recurrent neural networks and artificial neural networks (and CNNs) are most suited to the large supply-chain optimisation.

6 Conclusion

The blockchain technology has now surpassed the limits of the financial industry and it is hence proposed as a milestone for disruptive technology winning support in sectors which are vital to companies, especially in the supply chain. In this field, the primary characteristics of the blockchain, such as immutability, transparency and decentralization, become necessary tools for the continuous optimization of processes and above all, they are qualities on which the consumer will increasingly rely on during his choices. However, there are some intrinsic limits to the scalability of the system and transactions rate, mainly due to the append-only nature of the data structure. Although these limits can be tolerated when transactions move only intangible assets (money), they become a major impediment when a high number of files, images and documents are associated to a physical good and especially during the product's transformation stages, during its journey from the manufacturer to the consumer.

References

1. Ahram T, Sargolzaei A, Sargolzaei S, Daniels J, Amaba B (2017) Blockchain technology innovations. In 2017 IEEE technology & engineering management conference (TEMSCON). IEEE, pp 137–141
2. Aich S, Chakraborty S, Sain M, Lee H, Kim H (2019) A review on benefits of IoT integrated blockchain based supply chain management implementations across different sectors with case study. In: 2019 21st international conference on advanced communication technology (ICACT), pp 138–141
3. Aivazidou E, Antoniou A, Arvanitopoulos-Darginis K, Toka A (2012) Using cloud computing in supply chain management: third-party logistics on the cloud
4. Androulaki E, Barger A, Bortnikov V, Muralidharan S, Cachin C, Christidis K, De Caro, A., Enyeart D, Murthy C, Ferris C, Laventman G, Manevich Y, Nguyen B, Sethi M, Singh G, Smith K, Sorniotti A, Stathakopoulou C, Vukolić M, Cocco SW, Yellick J (2018) Hyperledger fabric: a distributed operating system for permissioned blockchains. In: Proceedings of the 13th EuroSys conference, EuroSys 2018
5. Annarelli A, Battistella C, Nonino F (2019) How to trigger the strategic advantage of product service systems. Springer International Publishing, Cham, pp 95–141
6. Artificial intelligence in logistics; a collaborative report by DHL and IBM on implications and use cases for the logistics industry. https://www.businesswire.com/news/home/20180416006323/en/Artificial-Intelligence-Thrive-Logistics-DHL-IBM. Accessed 30 Aug 2019
7. Bandara E, Ng WK, De Zoysa K, Fernando N, Tharaka S, Maurakirinathan P, Jayasuriya N (2019) Mystiko – blockchain meets big data. In: Proceedings – 2018 IEEE international conference on big data, big data 2018, pp 3024–3032
8. Bartoletti M, Lande S, Pompianu L, Bracciali A (2017) A general framework for blockchain analytics. In: Proceedings of the 1st workshop on scalable and resilient infrastructures for distributed ledgers, SERIAL'17. ACM, New York, pp 7:1–7:6
9. Benet J (2014) IPFS – content addressed, versioned, P2P file system
10. BigchainDB GmbH (2018) BigchainDB: the blockchain database. BigchainDB. The blockchain database
11. Bitcoin: a peer-to-peer electronic cash system. https://s3.amazonaws.com/academia.edu.documents/54517945/Bitcoin_paper_Original_2.pdf?response-content-disposition=inline%3B%20filename%3DBitcoin_A_Peer-to-Peer_Electronic_Cash_S.pdf&X-Amz-Algorithm=AWS4-HMAC-SHA256&X-Amz-Credential=AKIAIWOWYYGZ2Y53UL3A%2F20190827%2Fus-east-1%2Fs3%2Faws4_request&X-Amz-Date=20190827T093650Z&X-Amz-Expires=3600&X-Amz-SignedHeaders=host&X-Amz-Signature=108d75c81898d73297135781e2d233f5c3521caf8e42e0bb58ab61928c7994a4. Accessed 30 Aug 2019
12. BOKREŠA ĐURI? (2017) Organisational metamodel for large-scale multi-agent systems: first steps towards modelling organisation dynamics. ADCAIJ: Adv Distrib Comput Artif Intell J 6:3
13. Carbonneau R, Laframboise K, Vahidov R (2008) Application of machine learning techniques for supply chain demand forecasting. Eur J Oper Res 184(3):1140–1154
14. Chen J, Lv Z, Song H (2019) Design of personnel big data management system based on blockchain. Futur Gener Comput Syst 101:1122–1129
15. Dawid H, Decker R, Hermann T, Jahnke H, Klat W, König R, Stummer C (2017) Management science in the era of smart consumer products: challenges and research perspectives. CEJOR 25(1):203–230
16. Dinh TTA, Liu R, Zhang M, Chen G, Ooi BC, Wang J (2018) Untangling blockchain: a data processing view of blockchain systems. IEEE Trans Knowl Data Eng 30(7):1366–1385
17. Epiphaniou G, Daly H, Al-Khateeb H (2019) Blockchain and healthcare. Springer International Publishing, Cham, pp 1–29

18. Eskandarpour M, Dejax P, Miemczyk J, Péton O (2015) Sustainable supply chain network design: an optimization-oriented review. Omega 54:11–32
19. Eximchain: supply chain finance solutions on a secured public, permissioned blockchain hybrid. https://eximchain.com/Whitepaper-Eximchain.pdf. Accessed 30 Aug 2019
20. Galvez JF, Mejuto J, Simal-Gandara J (2018) Future challenges on the use of blockchain for food traceability analysis. TrAC Trends Anal Chem 107:222–232
21. Garcia D, You F (2015) Supply chain design and optimization: challenges and opportunities. Comput Chem Eng 81:153–170
22. Gobel J, Krzesinski A (2017) Increased block size and Bitcoin blockchain dynamics. In: 2017 27th international telecommunication networks and applications conference (ITNAC). IEEE, pp 1–6
23. Imbault F, Swiatek M, de Beaufort R, Plana R (2017) The green blockchain: managing decentralized energy production and consumption. In: 2017 IEEE international conference on environment and electrical engineering and 2017 IEEE industrial and commercial power systems Europe (EEEIC/I CPS Europe), June 2017, pp 1–5
24. Taliaferro A, Guenette C-A, Agarwal A, Pochon M (2016) Industry 4.0 and distribution centers. Transforming distribution operations through innovation. https://www2.deloitte.com/us/en/insights/focus/industry-4-0/warehousing-distributed-center-operations.html?id=us:2sm:3li:dup3294:awa:dup:MMDDYY:4ir:author. Accessed 12 Sept 2016
25. Kim Y (2014) Convolutional neural networks for sentence classification. CoRR abs/1408.5882
26. Knoll D, Prüglmeier M, Reinhart G (2016) Predicting future inbound logistics processes using machine learning. Procedia CIRP 52:145–150. The Sixth International Conference on Changeable, Agile, Reconfigurable and Virtual Production (CARV2016)
27. Korpela K, Hallikas J, Dahlberg T (2017) Digital supply chain transformation toward blockchain integration
28. Long TB, Looijen A, Blok V (2018) Critical success factors for the transition to business models for sustainability in the food and beverage industry in the Netherlands. J Clean Prod 175:82–95
29. Lu Q, Xu X (2017) Adaptable blockchain-based systems: a case study for product traceability. IEEE Softw 34(6):21–27
30. Nakasumi M (2017) Information sharing for supply chain management based on block chain technology. In: 2017 IEEE 19th conference on business informatics (CBI), vol 01, pp 140–149
31. Namdar J, Li X, Sawhney R, Pradhan N (2018) Supply chain resilience for single and multiple sourcing in the presence of disruption risks. Int J Prod Res 56(6):2339–2360
32. Neubert G, Ouzrout Y, Bouras A (2018) Collaboration and integration through information technologies in supply chains. CoRR abs/1811.01688
33. Nizamuddin N, Salah K, Ajmal Azad M, Arshad J, Rehman M (2019) Decentralized document version control using ethereum blockchain and IPFS. Comput Electr Eng 76:183–197
34. Pazaitis A, Filippi PD, Kostakis V (2017) Blockchain and value systems in the sharing economy: the illustrative case of backfeed. Technol Forecast Soc Chang 125:105–115
35. Pereira A, Romero F (2017) A review of the meanings and the implications of the industry 4.0 concept. Procedia Manuf 13:1206–1214. Manufacturing engineering society international conference 2017, MESIC 2017, 28–30 June 2017, Vigo (Pontevedra)
36. Pfohl H-C, Yahsi B, Kurnaz T (2015) The impact of industry 4.0 on the supply chain. In: Innovations and strategies for logistics and supply chains, Jan 2015, epubli
37. Porter ME, Kramer MR (2019) Creating shared value. Springer Netherlands, Dordrecht, pp 323–346
38. Rachmawati D, Tarigan JT, Ginting ABC (2018) A comparative study of Message Digest 5(MD5) and SHA256 algorithm. J Phys Conf Ser 978:012116
39. Recommendations for implementing the strategic initiative industrie 4.0. https://www.din.de/blob/76902/e8cac883f42bf28536e7e8165993f1fd/recommendations-for-implementing-industry-4-0-data.pdf. Accessed 30 Aug 2019
40. Rokonuzzaman M The integration of extended supply chain with sales and operation planning: a conceptual framework. Logistics 2(2):8

41. Ryan D (2017) Calculating costs in ethereum contracts. https://hackernoon.com/ether-purchase-power-df40a38c5a2f. Accessed 30 Aug 2019
42. Sahoo MS, Baruah PK (2018) HBasechainDB – a scalable blockchain framework on Hadoop ecosystem. In: Lecture notes in computer science (including subseries lecture notes in artificial intelligence and lecture notes in bioinformatics), pp 18–29
43. Sahoo SR, Kumar KS, Mahapatra KK (2015) A novel ROPUF for hardware security. In: 2015 19th international symposium on VLSI design and test, pp 1–2
44. Salah K, Nizamuddin N, Jayaraman R, Omar M (2019) Blockchain-based soybean traceability in agricultural supply chain. IEEE Access 7:73295–73305
45. Sokolova M, Matwin S (2016) Personal privacy protection in time of big data. In: Challenges in computational statistics and data mining. Springer, Cham, pp 365–380
46. Szozda N (2017) Industry 4.0 and its impact on the functioning of supply chains. Logforum 13(4):2
47. Tehranipoor MM, Guin U, Forte D (2015) Counterfeit integrated circuits: detection and avoidance. Springer Publishing Company, Incorporated, Cham
48. Tian F (2017) A supply chain traceability system for food safety based on HACCP, blockchain AMP; internet of things. In: 2017 International conference on service systems and service management, pp 1–6
49. Trojanowska J, Kolinski A, Galusik D, Varela MLR, Machado J (2018) A methodology of improvement of manufacturing productivity through increasing operational efficiency of the production process. In: Hamrol A, Ciszak O, Legutko S, Jurczyk M (eds) Advances in manufacturing. Springer International Publishing, Cham, pp 23–32
50. Trón V, Fischer A, Nagy DA, Felföldi Z, Johnson N (2016) Swap, swear and swindle: incentive system for swarm. Ethereum Orange Paper
51. Waibel M, Steenkamp L, Moloko N, Oosthuizen G (2017) Investigating the effects of smart production systems on sustainability elements. Procedia Manuf 8:731–737; In: 14th global conference on sustainable manufacturing, GCSM, 3–5 Oct 2016, Stellenbosch
52. Walshe M, Epiphaniou G, Al-Khateeb H, Hammoudeh M, Katos V, Dehghantanha A (2019) Non-interactive zero knowledge proofs for the authentication of iot devices in reduced connectivity environments. Ad Hoc Netw 95:101988
53. Wang S, Qu X (2019) Blockchain applications in shipping, transportation, logistics, and supply chain. In: Qu X, Zhen L, Howlett RJ, Jain LC (eds) Smart transportation systems 2019. Springer, Singapore, pp 225–231
54. Wood G (2014) Ethereum: a secure decentralised generalised transaction ledger. In: Ethereum project yellow paper, pp 1–32
55. Xu LD, Xu EL, Li L (2018) Industry 4.0: state of the art and future trends. Int J Prod Res 56(8):2941–2962
56. Yue L, Junqin H, Shengzhi Q, Ruijin W (2017) Big data model of security sharing based on blockchain. In: 2017 3rd international conference on big data computing and communications (BIGCOM). IEEE, pp 117–121
57. Zheng Z, Xie S, Dai H, Chen X, Wang H (2017) An overview of blockchain technology: architecture, consensus, and future trends. In: 2017 IEEE international congress on big data (BigData congress), pp 557–564
58. Zheng Q, Li Y, Chen P, Dong X (2019) An innovative IPFS-based storage model for blockchain. In: Proceedings – 2018 IEEE/WIC/ACM international conference on web intelligence, WI 2018, Institute of Electrical and Electronics Engineers Inc., pp 704–708
59. Zhou C, Cao Q (2018) Design and implementation of intelligent manufacturing project management system based on bill of material. Cluster computing

Combating Domestic Abuse inflicted in Smart Societies

Joe Mayhew and Hamid Jahankhani

Abstract This chapter centres on the emergence of technology in cases of Domestic Abuse using two adjunct parts; (a) how digital coercive control using smart home devices is now an attack vector for abusers and (b) it sets out to answer if the UK Domestic Abuse bill is adequate to support victims of technology facilitated abuse.

Recent reports in the media have identified cases of Domestic Abuse where attackers are using smart home devices to exert coercive control over their partners or former partners.

This research importantly highlights a lack of awareness of technology facilitated Domestic Abuse by victims, support workers and law enforcement. This has resulted in the development of a new proposed framework titled SHADA Compliance – a Smart Home Anti Domestic Abuse framework. The research concludes that the Domestic Abuse bill does not adequately support the growing threat of technology in cases of Domestic Abuse. A list of recommendations for future study is included that could further the field of research for Domestic Abuse charities, law enforcement and also increase public awareness.

Keywords Domestic abuse · IOT · IOMT smart societies · Threat vector · Smart homes · Domestic abuse bill · DAPN · Social media

1 Introduction

Have you ever considered where certain sayings originate from? Take for example, 'the rule of thumb', used widely since the seventeenth century as an informal term

J. Mayhew
Ernst & Young LLP, London, UK
e-mail: Joe.Mayhew@uk.ey.com

H. Jahankhani (✉)
Northumbria University, London, UK
e-mail: Hamid.jahankhani@northumbria.ac.uk

H. Jahankhani et al. (eds.), *Cyber Defence in the Age of AI, Smart Societies and Augmented Humanity*, Advanced Sciences and Technologies for Security Applications, https://doi.org/10.1007/978-3-030-35746-7_16

of measurement [17]. However, this fairly innocently sounding phrase apparently had roots in domestic abuse. This was partly due to a seventeenth century English Judge Sir Francis Buller, satirised by James Gillray a famous Caricaturist of the period [11], Buller had a reputation for bestowing harsh punishments and had apparently made it law that a man could beat his wife with a rod no wider than the husband's thumb. Martin continues to state that the true origin remains unknown, likely because it could have been a throwaway comment or if anything was ever written down, it has likely been destroyed or hasn't been found yet. It's first recorded association with domestic violence was in the 1970s where it was first criticised by feminists [17]. On the face of it, there appears to be more articles trying to debunk the myth surrounding the origins of the phrase "rule of thumb" where a brief search on Google finds over 11 million results.

The most interesting aspect of this is: where did the association with the saying originate, was it just a public relations message invented by the feminists of the 1970's? [4, 23].

Whilst domestic abuse is not a new subject, it has transcended through the years. Although a vestigial patriarchal belief, where men were the head of the family and were encouraged to chastise their wives albeit without losing their temper, this was not always the case. Anna Clark, a professor of history at the University of Minnesota discusses domestic violence in a historical sense [3]. In Clark's paper, she cites work produced by Sarah Pomeroy [22], specifically a book she published titled "*The Murder of Regilla*" which documents the murder of *Appia Annia Regilla Atilia Caucidia Tertulla*, born around A.D. 125 and subsequent trial of her husband following the accusation of him killing his wife over a trivial matter.

During the 1950s–1980s, Domestic Abuse was satirised in comic strips. The twentieth century cartoonist Reg Smythe, creator of Andy Capp, styled his character on an English northern wife beater [30].

Modern understanding and acceptance of domestic abuse has changed over the years, the following timeline from the Guardian [27], provides a brief overview of changes in English law since the late nineteenth century.

A timeline of Domestic Violence legislation across England and Wales can be found in Table 1 below:

Based on this timeline developed by the Guardian and prior to 2014, the previous 2000 years have focussed solely on the physical aspect of abuse in intimate partner relationships. Aside from the obvious fact that electronic computing has only been around since the later part of the twentieth century, recent abuse has taken a more intangible path. Exerting abuse through non-physical means, such as mental, verbal or coercive control. This not a new thing, although the medium to inflict this abuse has moved to a more modern, digital attack vector such as through social media, tracking apps or reading personal emails and text messages.

Even more recently, advanced methods for exerting control has been highlighted through smart home devices, also referred to as IoT devices – Internet of Things. IoT is primarily a marketing term used for everyday devices connected to the internet so that they can be configured, controlled and/or viewed through a web client (e.g. Internet Explorer, Edge, Chrome or Safari) or application, typically through a

Table 1 Timeline of domestic violence legislation

1860 – Law of Coverture	At the point of marriage, a husband became legally responsible for the actions of both his wife and children. This meant he was entitled to use physical or verbal abuse to control their behaviour
1870 – Married Woman's Property Act	Before 1870, when a woman married, her property automatically became her husband's. After this act, any money she earned or inherited while married stayed hers
1895 – Curfew on wife beating	This city of London byelaw made hitting your wife between the hours of 10 pm and 7 am illegal – because the noise was keeping people awake
1923 – Matrimonial Causes Act	This act marked a big change in divorce law. Before, a wife had to prove her husband had been unfaithful and show evidence of other faults. After 1923, adultery could be a sole reason for divorce for women as well as men.
1956 – Sexual Offences Act	This was the first-time rape was defined under specific criteria, such as incest, sex with a girl under 16, no consent, use of drugs, anal sex and impersonation
1961 – Contraceptive Pill	Contraception was made available on the NHS irrespective of marital status or husband's permission
1971 – The first safe house	The charity Refuge opens the first safe house in Chiswick, west London, for women and children fleeing domestic abuse
1975 – First select committee	A government select committee on violence in marriage is created and recommends a minimum of one family place in a refuge per 10,000 people
1976 – Domestic Violence and Matrimonial Proceedings Act	This was the first legislation dedicated to combating domestic violence. It gave survivors new rights by offering civil protection orders (injunctions) for those at risk of abuse
1977 – Housing Act (Homeless Persons) 1977	Women and children at risk of violence were acknowledged as homeless. This meant they gained the right to state-funded temporary accommodation
1988 – Housing Act 1988	Rents became deregulated making it harder for survivors of domestic violence who tried to escape and find private accommodation. Though some argue it helped by developing more supported housing and refuge services
1989 – Children Act 1989	The law improved levels of child protection and parental responsibility but mostly ignored domestic violence
1991 – Criminal Rape Act	Before 1991 it was a husband's legal right to rape his wife – marriage implied consent for sexual intercourse. This was the first time a woman had legal protection from marital rape
1996 – Family Law Act 1996	Important changes to this law gave law enforcement automatic powers of arrest where violence had been used or threatened
2003 – Inter-ministerial group on domestic violence is established	This group received crucial evidence on the scale of domestic violence and use of refuges. Women's Aid (a charity dedicated to ending domestic violence) played a significant role in providing testimony
2004 – Domestic Violence, Crime and Victims Act	This made common assault an arrestable offence. This meant that law enforcement could arrest a suspect immediately, rather than leaving them with someone vulnerable while they applied for a warrant

(continued)

Table 1 (continued)

2005 – Home Office publishes Domestic Violence: A national report	This report was seen as government's first public commitment to taking responsibility for tackling the issue through policy
2010 – Government strategy is set out to end violence against women and girls	The strategy developed a 2011 plan which included financial commitments to support rape crisis centres and specialist training for health workers in the treatment of survivors
2014 – Clare's Law	A law is implemented across England and Wales giving people the right to ask law enforcement about a partner's history of domestic abuse

smart phone or tablet (another IoT device). Examples of IoT devices include digital assistants such as Amazon's Alexa, Google's Assist, Apple's Siri or Microsoft's Cortana; heating controls such as Hive, Nest, or Honeywell; doorbells or locking such as those from Ring or Yale and even home audio systems plus many more examples of IoT (fridges, cookers, toasters, coffee machines).

It should be noted that there is an argument that the recent abusive actions, digital coercive control, could be considered advanced, yet they could also be considered a simpler abuse method without consequences to the abuser. Why the argument? Consider the following, the attacks are performed using highly technologically advanced devices that connect to home networks, across the internet and finally to mobile devices (typically). In between the home automation device and the abuser is a vast amount of silicone, electronics and miles of cabling transmitting and receiving millions of binary ones and zeros orchestrated by millions of lines of software code (culminative). A highly complex mesh of electronics, with the ability to deliver an abuser's threat payload, inflicting misery on their targets. Conversely, with all of this technology in place, designed, manufactured and maintained by nearly a million IT professionals in the UK alone [26], attackers don't need to know how to wire a network, how to setup servers, how to code software. It's already been done, the delivery medium has already been created for other reasons than for someone to use the technology to inflict abuse towards a victim. The attack could be from any location where there is connectivity to the Internet. Providing there is access it can be instigated without geographic boundary, without the fear of being caught, or without seeing the misery inflicted and therefore without feeling there are consequences for their actions.

It is in this respect, that it is a simpler method for abusers to cause harm, as technology has advanced to create convenience for actions that previously took a moderate degree of thought and physical presence.

Work conducted by the Home Office on domestic abuse and the economic and social costs as a direct result state that in England and Wales in 2016/2017, this cost was c.£66billion [14]. This figure comprises costs to health services, policing, victim services and lost output among many other components. Based on details from the Office for National Statistics, the individual cost per victim equates to £34,015, however this is an average and depending on the exact crime, costs can be significantly more according to the Home Office.

Finally, the Home Office states the figure is under-estimated as it does not fully reflect the physical harm that victims sustain.

2 Smart Societies and Increase of Threat Vector in Domestic Abuse

A recent Ernst & Young (EY) study on the consumer attitudes towards smart home devices states that in 2018 22% of household respondents now have a digital home assistant compared to 11% in 2017 [10]. The report highlights an increase in the adoption of smart home devices. EY states that whilst 22% of households currently own a digital assistant, 41% replying said they will own one within five years. Smart heating was 2nd favourite, with 12% of respondents owning one, and 41% saying they will own one within 5 years (Fig. 1).

Whether this is due to advances in the technical features, accessibility for a wider group of users or buyers just want to have the latest technology, the EY report shows adoption of smart home devices is increasing.

Comparably, in July 2018, the New York Times reported on an increase in the rise of smart devices as a threat vector in cases of domestic abuse in the US [1, 13]. Citing front-liners that take calls or support emergency shelters, these volunteers

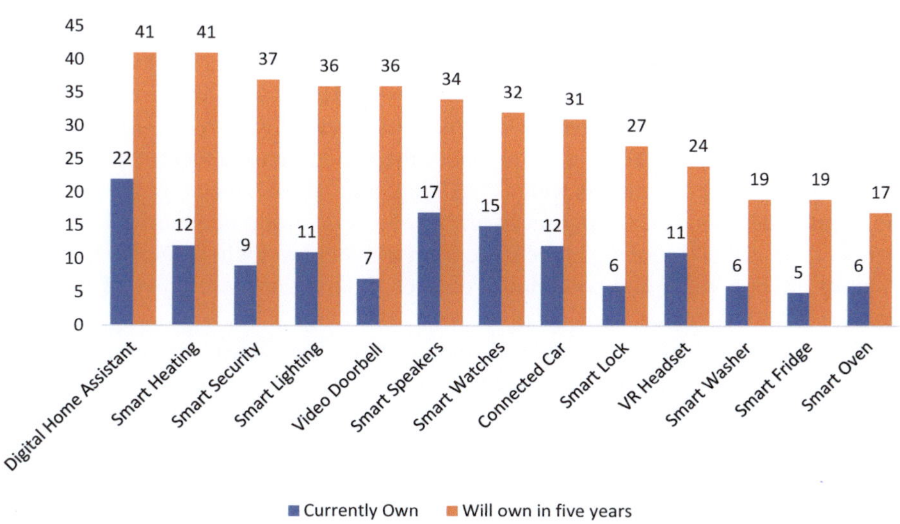

Fig. 1 Current and future take-up of smart home products by category – % Households [10]

recalled situations where victims had changed a setting on their heating system to have it mysteriously change again. Another, where her music system started blasting music for no reason. There are a number of examples, all of which point along similar lines where the attacker wanted to excerpt some level of control over their victims. This article was written by Nellie Bowles, a journalist that writes for the New York Times in San Francisco and has a history of writing technology related articles, covering socioeconomic topics, apps and items related to child online safety as well as this article on domestic abuse.

Reviewing this information in isolation uncovers very little, other than smart home adoption is likely to increase, and there have been occurrences of Domestic Abuse where technology has played an integral method to exert some level of coercive control in a digital manner. However, when combining both pieces of information, it could be hypothesised that an increase in smart home devices could lead to an increase in Domestic Abuse where technology is used, due to the simpler accessibility and easier abuse methods.

Digital Coercive Control or DCC, is a term conceived by Bridget A. Harris and Delanie Woodlock in their paper: "*Digital Coercive Control: Insights from two landmark domestic violence studies*" [13] in the British Journal of Criminology. It examines the use of technology in digital coercive control. Harris and Woodlock go on to frame Digital Coercive Control or DCC, to reference the method (digital) for intent (coercive) to cause harm through action (control). In the view of this author, this provides a frame for referring to domestic abuse through the use of technology as is it not specific to the type of technology utilised although the intent and action is broadly enough categorised to meaningfully reference.

Harris and Woodlock debate terminology and specifically the use of DCC vs previous terms to describe it such as "*technology facilitated stalking*" and "*technology facilitated abuse*". They refer to work by [31] as cited in Harris and Woodlock [13] where use of these terms to describe the mechanism for abuse confuses the actual problem into a technological one rather than a social issue. Vera-Gray further highlights other terminology coined by Henry and Powell 2015 cited in Vera-Gray [31] referring to "*technology facilitated sexual violence and harassment (TFSV)*". However, by being specific about the action being undertaken – 'sexual violence' and 'harassment', as well as 'stalking' and 'abuse', it could be argued that this could make for a confusing set of terms for a wide group of actions. This could detract from actually looking at the issue from a higher level where Digital Coercive Control offers a wider and potentially 'cleaner' framing of these types of criminal activity.

The work by Harris and Woodlock refers to DCC as 'spaceless' inferring it is not grounded in a place and it occupies no space, likely due to DCC performed across the Internet, although this specific paper then contradicts itself as Harris and Woodlock propose that geography increases DCC in rural and remote areas in their paper.

Two works [13, 33] discuss how rural environments create separation for abuse victims, separation from support services and separation from friends and family. Rurality is also framed within the sense of traditional meanings of non-urban environment isolation. However, the term of rurality could also be applied when abusers disassociate their victims from their support networks (family, friends and public services) as part of the cycle of abuse (Walker 1979 as cited in [12]) creating loneliness for the victim with a dependency on the abuser – *psycholigical rurality*.

In Delanie Woodlock's 2017 paper on "The Abuse of Technology in Domestic Violence and Stalking" [33], the main focus is an "emerging trend" in domestic violence where technology is used as a medium for intimate-partner stalking and abuse. Woodlock states that following a survey of 152 domestic abuse advocates and 49 victims, the use of phones, tablets, computers and social media is commonplace in cases of abuse. An important concept is the "omnipresence" of the abuser, isolating, humiliating and punishing victims, creating, as previously mentioned, a sense of psychological rurality.

Woodlock's introduction refers to the SmartSafe study conducted by the Domestic Violence Resource Centre Victoria (DVRCV) based in Collingwood, Australia which was a research project looking into the problem of technology facilitated stalking and abuse in Australia. The paper states that one in every five women above 15 years of age report they have been stalked according to statistics collected by DVRCV in 2006. It is worth noting that more recent statistics could provide higheraccuracy reflecting the current issues of technology facilitated stalking.

Research has primarily been conducted in Australia although Woodlock states that abusers of stalking are mainly male even internationally (Kuehner et al. 2012; Logan and Walker 2009; Strand and McEwan 2011 as cited in [33]). Research into this field is still limited [33], the purpose of the research was to understand if the use of technology was a growing method for abuse and if mobile technologies were used. Woodlock discssues E. Starks paper on policing coercive control (Stark 2007 as cited in [33]), specifically referring to intimate partner stalking as a form of coercive control.

At the time of publishing, the Woodlock paper only touchs on coercive control and the research seems to focus primarily on male perpetrators. Further to this, there are a number of references to older papers and research, signalling that there is an opportunity to perform more up to date research.

Interestingly, and even though Woodlock's paper was only published in 2017, Technology is providing the opportunity for abusers to control their partners and as far back as 2009, 25% of stalking victims in the U.S. stated they have been stalked via technology (Baum et al. 2009 as cited in [33]).

The SmartSafe Study that Woodlock references was an initiative of DVRCV, according to Woodlock this study was one of the first domestic violence organisations in the world to have online resources for victims. DVRCV provides websites, videos, blogs and online quizes. The study was conducted in 2012 and targetted both workers in domestic abuse refuges and victims/survivors of domestic abuse. This study was two online surveys, interviews and focus groups.

The use of social media as an attack vector becomes more apparent, and although there is no mention of smart home devices, there is prolifent use of mobile technologies to abuse victims in the results of the study where 78% of victim respondants state "*Used text messages, phone, and so on to call her names, harass her, or "put her down*" [33].

Whilst Woodlock's review of the SmartSafe study foccusses on GPS, phones, texts and social media in abuse, there is a single mention to smart home devices where a respondant in the worker survey stated:

> *it was discovered that her ex-partner had installed covert cameras both in the home and at the front gate that he had linked to his computer* [33].

In conclusion of [33] paper on the SmartSafe study, the study, whilst useful, it doesn't delve into a sufficiently broad enough technical landscape due to the growing use of smart devices in the modern home. Technology has evolved, with newer categories of device now available in the home and likely more ways in which an abuser can inflict misery onto a victim. Another key point is that the SmartSafe study is not only gender biased, but also biased on LGBTQ (Lesbian, Gay, Bisexual, Transgender & Queer), with no male respondents and only 9% of victim respondents identifying as bisexual.

The Bridget A. Harris and Delanie Woodlock paper from 2018 titled Digital Coercive Control: Insights from two landmark domestic violence studies begins to frame a more up to date view on technology in domestic abuse [13]. Spatiality becomes a key component of the topics discussed in this paper, where Harris and Woodlock consider the rural aspect of domestic violence. This is a key factor in this author's previously mentioned aspect of Digital Coercive Control (DCC) where abusers create a sense of psychological rurality for victims.

The Harris and Woodlock paper covers Woodlock's work on Smartsafe and Harris' work on Landscapes and rurality affects DCC (George and Harris 2014 as cited in [13]). The focus of the study was in the experiences of women in non-urban areas of domestic violence and by the works own admission, it was unexpected to see such a high level of DCC. Harris and Woodlock state that knowledge of DCC and initiatives to combat it are fragmented. Furthermore, they confirm at the time of writing, available literature centres on 'electronic dating violence', surveillance using social media as well as online misogyny.

The general theme throughout the majority of research is a gender bias towards female victims and male perpetrators. Harris and Woodlock highlight that at the time of writing, little had been done to examine differences between male or female abusers v.s. male or female victims where there could be a potentially larger issue within the LGBTQ community that is masked.

An interesting, although potentially worrying aspect of domestic abuse is that DCC is regarded as an issue that is less serious than other forms of abuse. Possibly due to the lack of knowledge by support workers or law enforcement as to the psychological abuse sustained by victims, but also due to the emerging aspect of DCC and that it is still a framing in its infancy due to the subtlety of that DCC can play sometimes and how it could be misconstrued as 'horse-play' [13]. Add this, to the speed with which technology is built, marketed and replaced in a constant cycle and it's no wonder it's a difficult area to track.

It would be easier to attribute the increase of domestic abuse as a result of the introduction of more technology into the home, where controlling behaviour over women increases with it's introduction. However, technology isn't the cause of domestic abuse, specifically DCC, but enables it in an expedited manner due to its 'spacelessness' [13] and the pervasive access through mobile technology.

Some of the research that Harris & Woodlock refers to, draws information from research undertaken in the Global North, where there was a startling level of digital coercive control. Harris & Woodlock stated that 22–93% of participants in multiple studies experienced some for of 'cyber aggression' (Picard 2007; Melander 2010; Burke et al. 2011; Bennett et al. 2011; Zweig et al. 2013; Dick et al. 2014; Leisring and Giumetti 2014; Borrajo et al. 2015; Barter et al. 2017; Ybarra et al. 2017 as cited in [13]). Harris & Woodlock further state that active members within the sphere of research, hypothesise youth and teenage subjects could view DCC behaviour as normal, although this has the potential to be flawed (George and Harris 2014 as cited in [13]).

In conclusion of the [13] paper, it identifies the spacelessness (without boundary) of DCC. Whilst not the originators of DCC, the paper discusses the further framing of DCC although lacks a wider identification of DCC such as the scope of DCC including other attack mediums outside of social media, texts, phoning etc. By their own admission, Harris & Woodlock do highlight the lack of research into DCC due to it being a new field of inquiry [13] and proposes further investigation.

Leading on to the draft UK Domestic Abuse bill, there is a consultation response document on the UK Government website titled "Transforming the Response to Domestic Abuse, Consultation Response and Draft Bill"[1] [5].

The consultation document is split into four sections with a further five Annexes. The document holds the key responses to the consultation that the Government conducted in 2018 with sections 1 to 4 focussed on the responses to key concerns. Annex A discusses nine projects that are funded through a Government Children's fund. Annex B discusses a list of offences taking extra-territorial jurisdiction to satisfy Government obligations under Article 44 of the Istanbul Convention [4]. Annex C documents a list of commitments. Annex D documents the draft bill in its entirety. Finally, Annex E provides a set of explanatory notes for the draft Domestic Abuse bill.

The Domestic Abuse bill is split into five parts –

- Part 1: Definition of "Domestic Abuse"
- Part 2: The Domestic Abuse Commissioner
- Part 3: Powers for Dealing with Domestic Abuse
- Part 4: Protection for Victims and Witnesses in Court
- Part 5: Miscellaneous and General

[1] https://assets.publishing.service.gov.uk/government/uploads/system/uploads/attachment_data/file/772202/CCS1218158068-Web_Accessible.pdf

The draft bill, understandably uses a high degree of legal language to define the issue it aims to address. Part 1, the definition of Domestic Abuse, is brief. Much briefer than expected, and could be interpreted as a disappointing attempt to frame a serious issue. Part 1, covering 46 lines of text, outlines two key areas: the definition of domestic abuse and the definition of personally connected.

The definition of abuse, outlines behaviour by a person ("A"), over the age of 16 towards another person ("B" – also over the age of 16) that they have a personal connection with.

It defines what abusive behaviour constitutes, whether it is physical or sexual, if it is violent or threatening, whether it is controlling or coercive, if it includes economic abuse or psychological, emotional or 'other'. Economic abuse, it states as having an adverse effect on B's ability to buy or own property, be in charge of their own money or buy goods and services of their own accord. And finally, abuse through another person (such as a victim's child).

The subject of this chapter is centred on the links between Domestic Abuse and smart devices in the home. These devices used to excerpt coercive control, to reduce the ability of the victim to think and act independently, to conduct themselves knowing that someone is watching, judging or controlling their normal activity – remotely and without regard for the psychological impact the abuser is having. This is clearly an unacceptable aspect to an intimate relationship where intimate relationships are built on trust, they require both parties to commit a sense of implicit trust in the other.

Woman's Aid, a non-profit organisation devoted to supporting victims of domestic abuse in the UK, describes coercive control as a method for controlling a person in order to make them dependent upon their attacker, through isolation, exploitation and deprivation of the victim's independence in their normal day to day activities [32].

Woman's Aid provides a non-exhaustive set of examples of coercive control:

- *"Isolating you from friends and family*
- *Depriving you of basic needs, such as food*
- *Monitoring your time*
- *Monitoring you via online communication tools or spyware*
- *Taking control over aspects of your everyday life, such as where you can go, who you can see, what you can wear and when you can sleep*
- *Depriving you access to support services, such as medical services*
- *Repeatedly putting you down, such as saying you're worthless*
- *Humiliating, degrading or dehumanising you*
- *Controlling your finances*
- *Making threats or intimidating you"* [32]

This useful, although basic list, provides some examples including ones that highlight control via digital technology such as monitoring via online communications tools or spyware. The Domestic Abuse bill only lists types of abuse, not a method or example of that type of abuse.

The list was written some time ago (c.2015 looking at the copyright on the page) and doesn't take into account newer technologies used in domestic abuse such as Internet connected doorbells or home heating controls. It highlights the need for a more up to date awareness campaigns.

The CPS (Crown Prosecution Service), admits that signs of coercive control may not be immediately obvious to victims [6]. Dating of this information has provided difficult, although an initial assessment dates it around 2015 or onwards (due to a 2015 case being referenced). The CPS states that domestic abuse takes many forms, typically person to person. However, their advice recognises the use of technology and specifically refers to mobile communications, internet communications (such as email and social media) but also mentions "other web-enabled methods" [6]. There is little more information other than this, but at least there is some mention of it. As this space (and the law) matures, there could be further examples provided along with existing, more well-known methods such as email, texts and social media.

The response document states a "*once-in-a-generation opportunity to transform the response*" specifically talking about Domestic Abuse ([5], Forward by Home Secretary and Justic Secretary). Although, Section 1 of the response states that "*Domestic Abuse is Complex*" and further states "*it is important that it is first properly recognised and understood*" ([5], pg. 5). The reason for highlighting this is that Part 2 of the Domestic Abuse bill outlines the role and responsibilities of the Domestic Abuse Commissioner. This section contains approximately seven times more content than the actual definition of domestic abuse itself Clearly the UK Crown Prosecution Service provides guidelines on Domestic Abuse [6] and it could be left intentionally vague so as not to restrictively define the issue and therefore could leave room for interpretation that could be both positive or negative depending on the framing, case situation and nature of the abuse. Even the Crown Prosecution Service guidance is out of date from modern attack methods and doesn't refer to Internet connected smart home devices.

Part 3 of the Domestic Abuse bill outlines powers for dealing with domestic abuse including the application for and giving of Domestic Abuse Protection Notices (DAPN) and Domestic Abuse Protection Orders (DAPO). The section briefly outlines conditions which must be met in order for a DAPN to be given. This responsibility is for a senior law enforcement officer, when they have reasonable grounds to believe that a person has been abusive towards another person and when both those people personally connected ([5], point 18).

For reference, the difference between a DAPN and a DAPO is that a DAPN is issued by a senior law enforcement officer as interim protection where the senior law enforcement officer suspects and has reasonable grounds to believe the recipient of the DAPN will continue to be abusive to the victim. It is a temporary protective notice whilst a magistrate is engaged to issue an order ([5], point 21). A DAPO application can be sought by a person who requires protection, chief officer of police a designate of the Secretary of State or a person with the leave of the court. Applications are made to a family court and issued by a magistrate.

In the issuance of a DAPN, there are provisions that can be made including prohibiting the abuser from entering the premises, requiring them to leave the premises and prohibiting the abuser from coming within a specified distance (e.g. restraining order). There is no provision for access to the premises digitally where the abuser could access home CCTV remotely over the Internet ([5], point 19). As an example, an abuser could be served with a DAPO, be required to leave the family home and remain a prescribed distance from their victim, although if the home was fitted with smart home devices, such as Internet connected heating controls, then the abuser can still exert coercive control remotely by changing the heating controls at a time of extreme temperature, such as switching the heating off during the winter. An additional reference to restricting access through digital means, for example email, text, social media and Internet connected smart devices in the home could reduce the risk of an attacker continuing to abuse their victim by to exert coercive control even though they are out of the home.

The Domestic Abuse bill provides a guideline following a breach of a DAPN ([5], point 18). It states, that should a constable have "*reasonable grounds*" to believe someone is in breach of a DAPN, then that person may be arrested. However, given the example earlier of an attacker continuing to exert control, even though they have left the home by using smart home technologies, how could this be proven? What evidence would need to be collected and without deep technical or deep network capture expertise or potentially features built into a smart device for reporting, this could be almost impossible to prove. It would be difficult for law enforcement or the courts to accept that an abuser had breached the DAPN without clear evidence. The Domestic Abuse bill includes a provision for making a DAPO without issuing a notice such as restraint against visiting or entering the premises, although it still remains possible for an attacker to abuse their victim remotely across the Internet even though they are denied access to the home ([5], point 30).

The Domestic Abuse bill further discusses the provisions that can be made by a DAPO, up to date references to technology facilitated abuse is not mentioned therefore there is still a gap. However, in both the DAPN and DAPO provisions, it is specified that the person who the DAPN/DAPO is for (the abuser), that they must not exclude the person it is protecting (the victim), from the premises. This could be the case when an Internet connected doorlock is at the property ([5], point 30).

Obtaining an arrest warrant for breach of an order could be difficult without the gathering of evidence from smart home devices. It could speculated that there is room to manoeuvre as the bill makes provisions for applications for breach substantiated by oath, or when a judge has reasonable grounds ([5], point 36).

There are a number of notification requirements included as part of an order, including the requirement that an abuser must notify the police if they change their address. A possible solution in the evidence gathering process, could be that, the abuser has to surrender their mobile phone for forensic analysis on a regular basis. Conversely, additional powers and training may be needed by the police to execute this ([5], point 37).

Due to the pervasive access to the Internet in most parts of the World, the Domestic Abuse bill could prove difficult to enforce when coercive control is exerted through digital methods. There is coverage for British Nationals and those under the protection of the United Kingdom, however this is primarily for offences that occur in the UK. Provisions do not appear to cater for offences that are perpretated overseas but received by a victim in the UK, especially if the overseas country does not recognise that offence ([5], point 55).

A literary review evidence can be found in Table 2.

The knowledge of Digital Coercive Control appears fragmented, in part to the lack of relevant research and wider public understanding of the issues. The research undertaken to date is valuable to begin framing the area of Digital Coercive Control, although it has primarily focussed on technology facilitated stalking [33]. Woodlock references social media and some mobile technology (texts/SMS) used in these examples.

This may be due to the rapid advancements in technology and support functions across law enforcement or social services unable to stay abreast of those advances and new methods for abuse through their regular training or knowledge sharing initiatives.

As the likelihood for smart home technology adoption increases [10], so could the increase of incidents of Domestic Abuse involving technology in Digital Coercive Control, especially as the EY report cited that 41% of the respondents indicated that they will be purchasing smart heating controls within the next 5 years.

The Domestic Abuse bill appears limited in its definition of coercive control and has little regard for more recent digital coercive control in the bill provisions, especially when a Domestic Abuse Protection Notice or Order is required to cease an abusers contact with their victim. This is especially relevant when law enforcement and the courts required evidence of a breach of those notices/orders.

Another important factor is that the reviewed research performed date may be biased in favour of heterosexual women and gives little, although in most cases no, reference to male or LGBTQ victims of Domestic Abuse.

In all of this, there is the emergence of two key questions such as, how can this bill reduce the impact of the spacelessness caused by digitial coercive control? Meaning, how does the bill limit or remove the ability of an abuser to create an environment where their victim feels helpless, where a victim of Digital Coercive Control has no control over their own privacy or have the security of support during a time of extreme anxiety, stress and hopelessness?

Secondly, how can the Domestic Abuse bill be enforced in the modern age when so little is understood about Digital Coercive Control, how DCC could be performed and how it can be proven?

This latter question is extremely important, in the performance of obtaining evidence to support Domestic Abuse Protection Notices and Orders. How can this evidence be gathered and presented to law enforcement or the courts when there is supsicion that a notice or order is in breach?

Table 2 Literary review evidence table

Author/date	Theoretical & conceptual framework	Research question	Methodology	Analysis/results	Conclusion	Implications for future research	Implications for future practice	Source reference
Who wrote it and when	What the title/area of study was	Details/abstract from the paper/research	Qualitive or quantitative	What was uncovered				References in Chapter 6
Bridget A. Harris and Delanie Woodlock 2018	Digital Coercive Control: Insights from two landmark domestic violence studies	*"This paper examines the use of digital technologies by domestic violence perpetrators, which we believe constitutes 'digital coercive control'. We draw on two Australian research projects and emerging research to provide definitional, conceptual and theoretical frames for harmful and invasive behaviours enacted through technology. Additionally, we highlight how such abuse intersects with other forms of violence but has unique and distinct features, including spacelessness"*	Previous papers, incorporating data obtained during interviews with victims (46) and sector professionals.	Fragmented knowledge sharing between support facilities, focus has been on technology-facilitated-abuse and technology-facilitated-stalking predominantly within the realm of electronic dating violence	Continues on previous research and highlights the lack of academic and industry research on DCC providing the opportunity to delve deeper in the topic.	Should look at a wider pool of technologies used for DCC	Basis to show previous research	[13]

Delanie Wood-lock 2017	The Abuse of Technology in Domestic Violence and Stalking	We focus on an emerging trend in the context of domestic violence—the use of technology to facilitate stalking and other forms of abuse. Surveys with 152 domestic violence advocates and 46 victims show that technology—including phones, tablets, computers, and social networking websites—is commonly used in intimate partner stalking. Technology was used to create a sense of the perpetrator's omnipresence, and to isolate, punish, and humiliate domestic violence victims. Perpetrators also threatened to share sexualized content online to humiliate victims. Technology facilitated stalking needs to be treated as a serious offense, and effective practice, policy, and legal responses must be developed	Qualitive and quantitative data including the SmartSafe study and the Landscapes study	The research intimates a gender bias in its results, although there could be a number of reasons why this is. Early research in the sphere of the topic as focus is predominantly on social media stalking and abuse	Relevant as conclusions could be drawn from earlier research and how the issues with DCC are maturing.	Should look at gender and LGBTQ bias and overcome this with further research	Basis to show previous research	[33]

(continued)

Table 2 (continued)

Author/date	Theoretical & conceptual framework	Research question	Methodology	Analysis/results	Conclusion	Implications for future research	Implications for future practice	Source reference
Who wrote it and when	What the title/area of study was	Details/abstract from the paper/research	Qualitive or quantitative	What was uncovered				References in Chapter 6
Isabel Lopez-Neira, Trupti Patel, Simon Parkin, George Danezis, and Leonie Tanczer, 2019	'Internet of Things': how abuse is getting smarter	"*From home thermostats you can control from your car, to home assistants ready to organise your diary at a spoken word, technology is playing a more central role in our daily lives. However, while networked home devices provide many advantages, they also offer abusers an abundance of opportunities to control, harass and stalk their victims. The Gender and Internet of Things project at University College London has been investigating how these devices are being misused, and what support survivors and services need to navigate these emerging risks*"	Qualitive	Discusses the use of IoT/smart home devices in cases of domestic abuse.	The conclusion of the paper was more training is required for staff and technology advances.	Provides a set of recommendations for further study including future-proofing legislation, risk-based approach to reduce tech abuse. However, there didn't appear to be much grounding or research to substantiate	Ideal thought-provoking piece that would support an argument for future research.	[16, 28]

HM Government Secretary of State for the Home Department 2019	Transforming the Response to Domestic Abuse: Consultation Response and Draft Bill	*"In February 2017, the Prime Minister announced plans for work to transform the way we think about and tackle domestic abuse, leading to the introduction of a new Domestic Abuse Bill. The commitment to introduce this bill was re-affirmed in the Queen's Speech at the opening of Parliament in June 2017"*	Qualitative	Outlines the plan for the Domestic Abuse bill, it includes the bill itself and the responses by the public on the draft and what the government plan to do	Better funding, improvements to the bill, appointment of a Domestic Abuse minister	Reference for legislation.	Reference for health practitioners, support staff and law enforcement	[5]
HM Government Home Office 2015	Controlling or Coercive Behaviour in an Intimate or Family Relationship Statutory Guidance Framework	*"This guidance provides information on:* *identifying domestic violence, domestic abuse and controlling or coercive behaviour;* *circumstances in which the new offence might apply;* *the types of evidence for the offence;* *the defence"*	Qualitative	Provides guidance on controlling or coercive behaviour	Guidance is outdated.	Adoption of more up to date coercive control methods should be investigate and included in an update	Reference for law enforcement and the courts.	[15]

3 Is the UK Domestic Abuse Bill Adequate to Combat Domestic Abuse Inflicted Through IoT in the Home?

To answer the question a varied and sensitive approach to both understanding the issues currently faced in the UK and the 'as-is' state of support and awareness provided by UK charities, the media, law enforcement and government needs to be followed in a form of a mixed methodology

Previous searches yielded some high-value research centred on digital coercive control; technology facilitated abuse the aspects of how rurality (predominantly in rural surroundings as opposed to urban). Academic papers such as these, in addition to news articles, statistics from the Office for National Statistics, Home Office guidance and the Domestic Abuse bill, all form quantitative methods in this research.

This approach should provide a balanced view of what the available information is and also what is happening both at the 'front line' from carers by questioning a group of them as well as reviewing previous studies. In order to maintain a low risk to ethics, the collection of primary data from abuse victims is a high risk and extremely sensitive given the nature of the data collected. The research proposed to collect data via a questionnaire from abuse victims using an intermediary such as a charity working with vulnerable adults and victims of domestic abuse. This was to ensure that the data received could be anonymised resulting in low risk to victims being identified which was the objective of conducting this primary data research via an intermediary. Although ethics was sought and agreed for this method, an amended plan was implemented to only seek to question Domestic Abuse support staff.

A UK charity was approached and have accepted to support this research aims, its methodology and offered to provide access to support staff through a questionnaire. The charity Hestia (http://www.hestia.org/), supports vulnerable adults and children in London and advocates locally and nationally on issues affecting them. Hestia also works closely with another UK charity, UK SAYS NO MORE (https://uksaysnomore.org/) which aims to end Domestic Abuse and Sexual Assault.

A questionnaire for Domestic Abuse support staff was created to identify if abuse through IoT was performed, it asked how victims identified it and what if anything was done about it. The questionnaire began with a set of diversity and inclusivity questions and demographics, followed by a set of questions to explore the field of study further.

In addition to the questionnaire, a Twitter poll was published to seek public opinion on technology facilitated abuse and the role of technology companies.

Whilst conducting the literary review, it is clear of the five W's that Dawson refers to, that this project is required to answer [8, 9].

In summary, these are:

- What: Is the UK Domestic Abuse bill adequate to combat domestic abuse inflicted through IoT in the home?

- Why: The hypotheses is that, there is insufficient awareness of digital coercive control related Domestic Abuse using smart home devices (IoT – Internet of Things devices). It is a relatively new abuse vector, one with little information or awareness and even less research.
- Who: Participants of the primary research will be Domestic Abuse support staff who are at the front line, providing support, advice and care for victims of Domestic Abuse.
- Where: The research questionnaire will be conducted using Survey Monkey and sent to c.100 Domestic Abuse support staff at the UK charity Hestia.
- When: Research was conducted during the summer of 2019.

Public opinion is important for such social, technological and legislative issues in examples of Domestic Abuse related to Digital Coercive Control. To gather data, two questions were created that were aimed to test whether public opinion was in favour of technology companies doing more to help and whether public opinion felt smart devices made domestic abuse easier.

Through Hestia, the UK charity 'UK Says No More' provided a platform via their own Twitter account to post the question to their 5441 followers.

The two specific questions were:

1. Do you think companies which build tech such as Alexa and Siri should build features to reduce the possibility of monitoring or stalking a partner?
2. Do you think smart devices make it easier to perpetrate abuse?

4 Smart Home Anti Domestic Abuse Framework, (SHADA)

Under advice from the charity, references to Digital Coercive Control were replaced with "technology facilitated abuse" as this terminology more closely matched the current awareness by support workers.

When asked if there was sufficient training to identify technology facilitated abuse, over 80% of the respondents answered 'No'. Comments received for this question centred around a lack of understanding by law enforcement and support staff and the respondents citing the lack of Tech Abuse training.

Of the respondents, 75% have met victims of Domestic Abuse. One comment received, referred to DASH which stands for The Domestic Abuse, Stalking & Honour Based Violence [7]. DASH is a risk identification, assessment and management model endorsed by law enforcement and charities supporting victims of Domestic Abuse [7]. On review of the 2009 DASH Risk Assessment process, there is no inclusion of coercive controlling behaviour in the questions other than financial dependency. Although the document appears to be dated 2009 (based on the copyright), which may explain why there is no reference to the updated criminal offence of coercive control.

When asked if cases of abuse using technology was on the increase, 100% responded 'yes'. Comments received mainly related to tracking-related applications including data apps and apps that were placed on a victim's device without their knowledge.

When asked if the victims had the knowledge to disable access to devices following a separation, over 90% responded 'No'. Respondents commented that Safety Planning was key to ensuring that victims had the knowledge to remove access. The responses mainly implied the restriction or removal of tracking applications and there was no mention of smart home devices. The comments also inferred that victims were unaware that technology was being used in the abuse and based on the responses, this would most likely be tracking applications on the victim's mobile phone.

It is clear that unfortunately children are also caught up in the different methods of Domestic Abuse related coercive control. The questionnaire asked if children had been gifted devices to be used in coercive control, over 90% responded 'Yes'. Interestingly, the comments received included an incident where a Father had bought a child a PlayStation that was used to spy on his former partner. Another incident cited was when a child was purchased a smart phone, the child uploaded pictures to social media and the father was able to track the Mother via these pictures. Whilst this could still have taken place if the Mother had purchased the phone, it does support the need for an improved quality of online hygiene performed by victims, where by default and therefore before providing smart devices, a list of settings should be checked to ensure that even some basic features are switched off and even educating young smart device users on the dangers of posting images that can be used to track and locate.

Within the comments for the same question, one piece of feedback that was received, simply stated 'Sanctuary Scheme'. Review of the Sanctuary Scheme guidance on the Gov.uk website provided no valuable guidance for securing technology, on page 75 of the guide, the advice was simply "*Change mobile phone and landline numbers*" [29].

The advice on room requirements may well still be valid, however the guidance on technology, in reality is years out of date having been written in 2010.

Finally, support staff were asked for the approximate number of male or female victims of Domestic Abuse they had worked with over the past three months where technology facilitated the abuse. Just under 90% of victims were Female and a little over 10% were Male.

The result from the Twitter poll via the @UKSAYSNOMORE Twitter profile with 5441 followers in September 2019 indicated the following:

Question 1 asked: "Do you think companies which build tech such as Alexa and Siri should build features to reduce the possibility of monitoring or stalking a partner?"

Of the 207 responses, 89% replied Yes, essentially that technology companies should introduce features that reduce the possibility of their devices being used in Digital Coercive Control and 11% replied No.

Question 2 asked: "Do you think smart devices make it easier to perpetrate abuse?"

The Twitter poll received 204 responses, 84% said Yes, they felt that smart devices made it easier for abusers of Domestic Abuse and 16% replied No.

The Twitter poll highlighted public opinion on both technology manufacturers role in reducing the opportunities for abusers to cause harm and if the public felt that smart devices were making it easier to abuse. The responses were reasonably clear with the public feeling strongly that technology companies should be doing more and that smart devices do make it easier.

Whilst the results are illuminating, it is worth noting that the pool of respondents could be considered low (approximately 400). It may not show a wholly accurate view but is useful to gauge a perspective on current public opinion.

5 2018 Domestic Abuse in England & Wales: Office for National Statistics

The Office for National Statistics (ONS), is "*the UK's largest independent producer of official statistics and the recognised statistical institute of the UK.*" [18]

Nationwide information and statistics for the UK can be found on this website, including information relating to Domestic Abuse which can provide insights into how Domestic Abuse in England and Wales is dealt with along with annualised crime data. The ONS website for Domestic Abuse [19–21] in the 2018 report states that there has been little change in the frequency of Domestic Abuse, although there has been an increase in the number of cases reported by law enforcement – through the crime survey. There are multiple sources of data in producing this information including criminal justice system and sector specific organisations providing services to victims.

The ONS site states that 599,549 domestic abuse-related crimes were committed in the year ended March 2018, this is up 23% from the previous year. Further, law enforcement made 225,714 arrests and conviction rates following prosecution was 76%, its highest level since 2010. By its admission, the ONS highlights that Domestic Abuse is frequently hidden, and therefore the statistics provided do not project a wholly accurate picture of Domestic Abuse in the UK [19–21].

Additionally, the ONS references the way in which law enforcement has improved the identification and recording of domestic abuse incidents as crimes.

There was a marginal difference in non-sexual Domestic Abuse between the statistics of 2018 versus 2017 with 5.6% and 5.5% of adults aged 16 to 59 respectively indicating a small increase.

The ONS report also cites a rise in Domestic Abuse related crimes recorded by law enforcement that now includes Coercive or Controlling behaviour. This is largely due to coercive or controlling behaviour becoming a specific criminal offence as of 29th December 2015 [19–21].

A total of 9053 coercive control offences were recorded in the year up to March 2018. The ONS report goes on to state that 960 offences of coercive and controlling behaviour resulted in a prosecution at a magistrate's court, three times more than the previous year where there were 309 offences in 2017 and just five in 2016 [19–21].

The ONS report cites that an increase is expected as law enforcement improves the identification and prosecution of these offences as well as public awareness increasing the ability to recognise coercive control as related to Domestic Abuse.

There was no clear distinction in the types of coercive control, nor the methods utilised by attackers. This information may be not be recorded by law enforcement or it may not be transferred into the ONS.

In considering the analysis of the research, three hypothetical solutions can be proposed. When combined, they provide a theoretical framework to support the reduction of Domestic Abuse, specifically in cases of digital coercive control.

The framework contents and image is original, however inspiration for the process was taken from guidance by the University of Southern California on how to create a theoretical framework when organising a social sciences research paper.

The following outlines a theoretical framework called 'SHADA Compliance', a Smart Home Anti Domestic Abuse framework created by the author Joe Mayhew. It is based on the analysis performed during this research.

The aim of this theoretical framework would focus of three core areas:

1. Legislative Amendments
   - Update the Domestic Abuse bill to include digital coercive control
   - Update CPS guidance that provides examples of digital coercive control
2. The role of Technology Developers
   - Define a functional standard to provide the ability to obtain evidence in the event a smart home product is used in cases of Domestic Abuse
   - Provide a mechanism for dual management of smart home devices to ensure that no single resident is in control
   - Use an RFC or ISO approach for SHADA Compliance that gives confidence to consumers
3. Awareness
   - Campaigns via TV advertisements, posters in social premises and social media aimed to raise public awareness of digital coercive control
   - Training for law enforcement on how recognise examples of digital coercive control and methods to collect evidence
   - Training for Domestic Abuse support staff to help victims identify examples of digital coercive control

The objectives of this framework would be to:

1. Ensure there is a legal policy is in place to prosecute abusers and protect victims of Domestic Abuse cases of digital coercive control
2. Ensure that technology is able to instil confidence to buyers of consumer electronics for the smart home and that it is difficult to be used in cases of

Domestic Abuse, potentially using a Kitemark quality standard as provided by the British Standards Institute [2]

3. Raise awareness for victims, law enforcement and Domestic Abuse support staff to be able to recognise digital coercive control, collect evidence and provide appropriate advice.

The challenges envisaged with this approach include:

- There is limited evidence that awareness campaigns work. Research into effective campaigns and ineffective campaigns would be advisable before working on raising awareness so that a campaign reaches its intended audience and has the desired outcomes.
- There could be technical challenges with adoption of some of the recommended features to aid evidence collection along with potential confidentiality issues.
- The pace of change of technology is at an unprecedented rate. Keeping abreast of technology advances in the home will prove challenging for all parties to ensure that awareness, advice and tools remain relevant.

The graphic below depicts an overview of the framework with the summarised points from the conclusion and recommendations (Fig. 2):

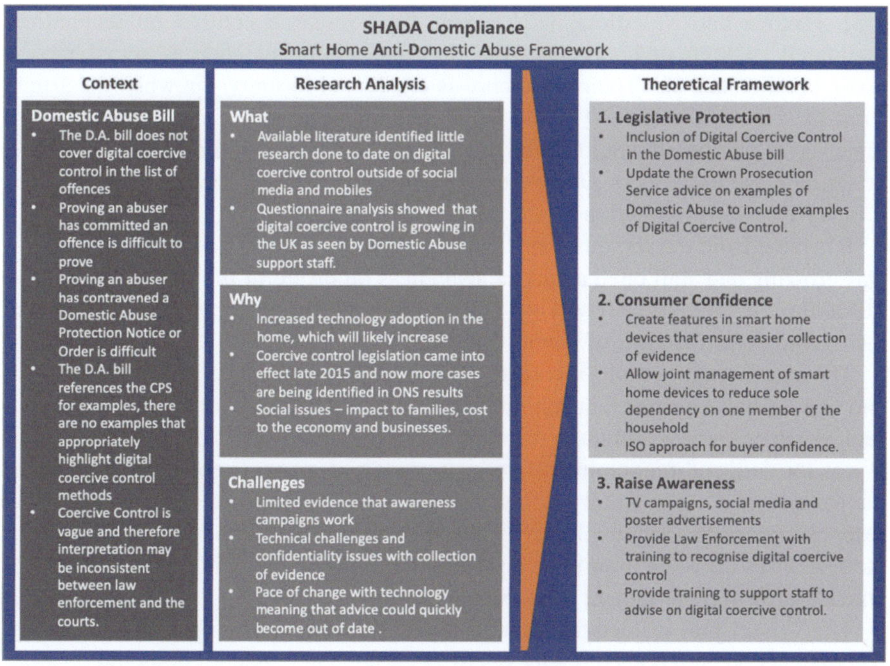

Fig. 2 Image of proposed SHADA framework, Source Joe Mayhew

6 Conclusions and Recommendations

The essence of this chapter is to understand if the UK Domestic Abuse bill adequately acknowledged, understood and incorporated technology into the fundamental core.

The Domestic Abuse bill [5] fell surprisingly short, by a long way. It is unclear if the bill intended to be as vague as it is when it discusses coercive control. It permits a varied range of interpretation of what is legal and that which is not and may result in law enforcement being confused and unable to support victims of Domestic Abuse at the level they require. It may be, that the interpretability is intentional, so as to provide a blanket cover potentially to ensure that the bill is non-specific and affords a mechanism to 'catch-all'.

The Domestic Abuse bill references the Crown Prosecution Service (CPS) and provides additional references [6] and examples for law enforcement and the courts. At the time of the research, the guidelines were drafted nearly four years ago. It is unclear if these are due to change any time soon as the guidelines outlines coercive control – restricting who their victims see, control their financial matters, spy on them, however there are no other examples such as smart heating systems or video doorbells.

The similar statement can be made for previous research, including by Woodlock [33], Harries and Woodlock [13] where their research centred on technology facilitated stalking or harassment that includes platforms such as social media, texting and emails etc. Their research was conducted recently, but when the article from the New York Times [1] and Ernst & Youngs Smart Home Survey [10] are reviewed in parallel. It would not be incomprehensible to hypothesise that there will be an increase in cases of Domestic Abuse where Digital Coercive Control is the primary case for prosecution by the courts.

It is clear from the responses that, whilst technology facilitated abuse was new, it was growing and staff cited a lack of awareness on behalf of the victims and a lack of training for support staff and law enforcement, as serious impediments to reduce the overall risk to victims of Domestic Abuse.

The questionnaire results included 100% of the respondents confirming that there had been instances where abusers are using technology as part of their abuse vector such as smart devices, although the majority of the time victims did not realise until it was too late and predominantly refer to mobile devices. It could be argued that the sample size is too small to draw any credible conclusions, however given the responses were all sought voluntarily and with no contact to the responders directly, the responses do begin to provide a picture of how technology is becoming an ideal attack platform by proficient abusers.

The Twitter poll yielded the strongest quantitative results with 89% and 84% of respondents on both polls replying yes to technology manufacturers should do more and yes smart device enable easier abuse methods respectively. The twitter poll, supports the technology manufacturer recommendation to incorporate features as per the SHADA framework.

The Office for National Statistics (ONS) was not detailed sufficiently to provide information on technology used in abuse such as digital coercive control or details of cases of primary or secondary offences (secondary to physical abuse as a primary offence for example). Coercive Control in intimate partner relationships only became unlawful at the end of 2015 and statistics show a three-fold increase from 2017 to 2018 (sub 1000 offences) and an $\times 60$ increase from the previous year (2016).

Offences that are registered with the ONS are categorised into sexual, non-sexual physical or emotional abuse. By its own admission, the ONS recognises that it does not capture the new offence of coercive control completely. The mechanism by which to do this must happen at source, this being information provided by law enforcement, information provided by Domestic Abuse support services and also the questionnaires that are provided to victims.

With the UK having the 2nd largest user penetration in consumer spending at 38.8% [25], and with 61.7% of consumer electronics users male and 38.3% female, men are more likely to buy smart home devices [24] and the addition of the findings by EY [10] on projected consumer spending over the next five years, plus the changes in legislation to now include coercive control, that a rise in Domestic Abuse related coercive control could be seen.

A public awareness campaign might help here but given previous campaigns, the right engagement and media channels will be imperative for a campaign's success and as suggested in the same blog, multiple methods need to be employed to get the messages across both digitally but also through other methods such as local support services.

References

1. Bowles, Nellie; NY Times (2018) Thermostats, locks and lights: digital tools of domestic abuse. Retrieved November 1, 2018, from https://www.nytimes.com/2018/06/23/technology/smart-home-devices-domestic-abuse.html
2. BSI (2019) BSI Kitemark™. Retrieved September 1, 2019, from https://www.bsigroup.com/en-GB/kitemark/product-testing/
3. Clark A (2011) Domestic violence, past and present. J Women's Hist 23(3):193–202
4. Conseil De L'Europe (2011) Council of Europe Convention on preventing and combating violence against women and domestic violence. Retrieved August 27, 2019, from https://rm.coe.int/168046031c
5. Crown Copyright (2019) Domestic abuse consultation response and draft bill. Retrieved from GOV.UK: https://www.gov.uk/government/publications/domestic-abuse-consultation-response-and-draft-bill
6. Crown Prosecution Service (2015) Domestic abuse guidelines for prosecutors. Retrieved July 20, 2019, from https://www.cps.gov.uk/legal-guidance/domestic-abuse-guidelines-prosecutors
7. DASH Risk Checklist (2019) Dash risk model. Retrieved August 27, 2019, from https://www.dashriskchecklist.co.uk/
8. Dawson C (2009a) Chapter 2 – How to decide upon a methodology. In: Introduction to research methods. How To Content, Oxford, pp 14–26

9. Dawson C (2009b) How to define your project. In: Introduction to research methods. How To Content, Oxford, pp 4–5
10. EY (2019) Taking new steps into the smart home. Retrieved May 27, 2019, from https://www.ey.com/Publication/vwLUAssets/EY-Taking-new-steps-into-the-smart-home/FILE/EY-Taking-new-steps-into-the-smart-home.pdf
11. Gillray J (1782) Sir Francis Buller, 1st Bt ('Judge Thumb'). National Portrait Gallery, London
12. Hammond C (2018) The narcissistic cycle of abuse. Retrieved July 6, 2019, from https://pro.psychcentral.com/exhausted-woman/2015/05/the-narcissistic-cycle-of-abuse/
13. Harris BA, Woodlock D (2018) Digital coercive control: insights from two landmark domestic violence studies. Br J Criminol azy052
14. Home Office – econ (2019) The economic and social costs of domestic abuse. Retrieved August 29, 2019, from https://assets.publishing.service.gov.uk/government/uploads/system/uploads/attachment_data/file/772180/horr107.pdf
15. Home Office (2015) Coercive or controlling behaviour – statutory guidance. Retrieved August 25, 2019, from https://assets.publishing.service.gov.uk/government/uploads/system/uploads/attachment_data/file/482528/Controlling_or_coercive_behaviour_-_statutory_guidance.pdf
16. Lopez-Neira I, Patel T, Parkin S, Danezis G, Tanczer L (2019) 'Internet of Things': how abuse is getting smarter. Safe – The Domestic Abuse Quarterly 63:22–26
17. Martin G (2019) The meaning and origin of the expression: rule of thumb. Retrieved May 27, 2019, from https://www.phrases.org.uk/meanings/rule-of-thumb.html
18. Office for National Statistics (2019) Retrieved August 27, 2019,from https://www.ons.gov.uk/
19. ONS (2018a) Domestic abuse in England and Wales – Appendix tables. Retrieved August 27, 2019, from https://www.ons.gov.uk/peoplepopulationandcommunity/crimeandjustice/datasets/domesticabuseinenglandandwalesappendixtables
20. ONS (2018b). Domestic abuse in England and Wales – data tool. Retrieved August 27, 2019, from https://www.ons.gov.uk/peoplepopulationandcommunity/crimeandjustice/datasets/domesticabuseinenglandandwalesdatatool
21. ONS (2018c) Domestic abuse in England and Wales: year ending March 2018. Retrieved August 27, 2019, from https://www.ons.gov.uk/peoplepopulationandcommunity/crimeandjustice/bulletins/domesticabuseinenglandandwales/yearendingmarch2018
22. Pomeroy SB (2007) The murder of regilla: a case of domestic violence in antiquity. Hardvard University Press, Cambridge, MA
23. Sommers CH (n.d.) The "rule of thumb for wife-beating" hoax. Retrieved July 1, 2019, from https://www.infotextmanuscripts.org/djetc/other_hoff_thumb.html
24. Statistica – age (2018) Consumer electronics. Retrieved August 29, 2019, from https://www.statista.com/outlook/251/100/consumer-electronics/worldwide#market-age
25. Statistica – pen (2019). Consumer electronics. Retrieved August 29, 2019, from https://www.statista.com/outlook/251/100/consumer-electronics/worldwide#market-globalRevenue
26. Statistica (2018) Number of employed information technology professionals in the United Kingdom (UK) in 2018, by occupation (1,000s). Retrieved June 22, 2019, from https://www.statista.com/statistics/778333/information-technology-professionals-employed-uk/
27. The Guardian Newspaper (2014) Domestic violence legislation in England and Wales: timeline. Retrieved May 27, 2019, from https://www.theguardian.com/society-professionals/ng-interactive/2014/nov/28/domestic-violence-legislation-timeline
28. UCL, University College London (2018) Gender and IoT. Retrieved December 28, 2018, from https://www.ucl.ac.uk/steapp/research/projects/digital-policy-lab/dpl-projects/gender-and-iot
29. UK Gov (2010) Sanctuary schemes for households at risk of domestic violence: guide for agencies. Retrieved August 27, 2019, from https://www.gov.uk/government/publications/sanctuary-schemes-for-households-at-risk-of-domestic-violence-guide-for-agencies
30. University of Kent – British Cartoon Archive (2016) Reg Smythe [Andy Capp]. Retrieved May 27, 2019, from https://www.cartoons.ac.uk/cartoonist-biographies/c-d/RegSmythe_AndyCapp.html

31. Vera-Gray F (2017) 'Talk about a cunt with too much idle time': trolling feminist research. Fem Rev 115(1):61–78
32. Women's aid (2015) What is coercive control? Retrieved July 19, 2019, from https://www.womensaid.org.uk/information-support/what-is-domestic-abuse/coercive-control/
33. Woodlock D (2017) The abuse of technology in domestic violence and stalking. Violence Against Women 23(5):584–602

Deep Convolutional Neural Networks for Forensic Age Estimation: A Review

Sultan Alkaabi, Salman Yussof, Haider Al-Khateeb, Gabriela Ahmadi-Assalemi, and Gregory Epiphaniou

Abstract Forensic age estimation is usually requested by courts, but applications can go beyond the legal requirement to enforce policies or offer age-sensitive services. Various biological features such as the face, bones, skeletal and dental structures can be utilised to estimate age. This article will cover how modern technology has developed to provide new methods and algorithms to digitalise this process for the medical community and beyond. The scientific study of Machine Learning (ML) have introduced statistical models without relying on explicit instructions, instead, these models rely on patterns and inference. Furthermore, the large-scale availability of relevant data (medical images) and computational power facilitated by the availability of powerful Graphics Processing Units (GPUs) and Cloud Computing services have accelerated this transformation in age estimation. Magnetic Resonant Imaging (MRI) and X-ray are examples of imaging techniques used to document bones and dental structures with attention to detail making them suitable for age estimation. We discuss how Convolutional Neural Network (CNN) can be used for this purpose and the advantage of using deep CNNs over traditional methods. The article also aims to evaluate various databases and algorithms used for age estimation using facial images and dental images.

Keywords Deep learning · CNN · Forensic investigation · Information fusion · Magnetic resonant imaging (MRI) · Dental X-ray

---

S. Alkaabi · S. Yussof
Institute of Informatics and Computing in Energy, Universiti Tenaga Nasional, Kajang, Malaysia

H. Al-Khateeb (✉) · G. Ahmadi-Assalemi · G. Epiphaniou
Wolverhampton Cyber Research Institute (WCRI), University of Wolverhampton, Wolverhampton, UK
e-mail: H.Al-Khateeb@wlv.ac.uk

H. Jahankhani et al. (eds.), *Cyber Defence in the Age of AI, Smart Societies and Augmented Humanity*, Advanced Sciences and Technologies for Security Applications, https://doi.org/10.1007/978-3-030-35746-7_17

1 Introduction

Forensic age estimation is one of the key research areas in the field of medical forensics. Although age estimation of unidentified cadavers or skeletal identification is a well- established forensic discipline, age estimation in living individuals is a relatively more recent area of applied research within forensic sciences that has attracted considerable attention [1]. The thriving integration of digital technologies into modern lives broadens the diversity and scope of forensic science and has created a need for new forensic science techniques including innovative computer vision and Machine Learning (ML) to support forensic investigations. Alongside the conventional forensic disciplines, Digital Forensics (DF) has developed as a branch of forensic science covering diverse digital technologies that can be exploited by criminals. Image-based evidence gained through sources like surveillance, monitoring or social media-driven intelligence that are commonly used by law enforcement in forensic investigations and by witnesses to describe suspects demonstrate the widening scope of forensic investigations. This creates specialised workload, generates backlog and requires highly specialised forensic practitioners [2, 3]. Therefore, more research is required to develop techniques and methods that are more efficient and automated thus reducing the backlog, workload and cost of the forensic investigation processes including the case studies when digital devices are involved as part of the crime scene or scope.

Soft biometric traits like age estimation, predicting a person's age using ancillary information from primary biometric traits like face, eye-iris, bones or dental structures, has attracted significant research in the past decade. Soft biometrics have a number of applications apart from medical forensics [1] including healthcare [4], age-related security control, human-computer interactions, law enforcement, surveillance and monitoring [5–7], socio-political related defence and security in border and immigration controls and to establish the age of illegal immigrants without valid proof-of-birth in adults or unaccompanied minors [8, 9], which is becoming an integral part of forensic practice [10]. Furthermore, without an accurate age estimation victims of child-trafficking, asylum seekers or illegal immigrants cannot receive the required instrumental support [11]. Due to the ease of online access, child sexual victimisation crimes are rising [12] with increased DF child exploitation investigations involving age estimation [13].

Apart from determining the age of cadavers or as part of the paleo-demographic analysis, the ability to estimate the age of living persons, which require accurate age estimation techniques, has become increasingly more important. In traditional approaches, most dental age estimation techniques like tooth emergence [14] or dental mineralisation [15] have limitations of age estimation beyond adolescence. Skeletal maturity with the development of X-ray was researched but due to the risks of exposure extensive X-ray based datasets were not produced. The development of highly detailed imaging techniques like ultrasound and Magnetic Resonant Imaging

(MRI), used to record dental and bone structures provide suitable opportunities for age determination of living persons [10].

Determining the age from image data is a highly complex task with numerous methods proposed by scientific research from measurement-driven analysis to the application of machine learning algorithms with constantly improving accuracy [16]. While a human face reflects significant amount of communicative information and facets about a person including gender, identity, ethnicity, expression and age, which humans have a capability to detect at a glance, there is a growing expectation that digital systems will have similar capabilities and recognition accuracy seamlessly [16–18]. Ancillary-related biological traits like the heterogeneity of the maturing process of human faces, bones, wrinkles, ethnicity or image-related traits including illumination, make-up or pose make age estimation challenging [19, 20]. Deep Learning (DL) methods result in higher accuracy compared to more traditional approaches like statistical [14], handcrafted methods that although require very small datasets, short training times and are computationally inexpensive their problem solving approach is modular relying on expert knowledge for complex feature extraction [21] or shallow learning which also requires feature extraction and classification [22]. Although DL methods require large-scale datasets, highly complex computational capability compared to the traditional approaches DL has automatic feature extraction with an end-to-end problem-solving approach that enables solving computer vision challenges [20, 23].

Furthermore, the large-scale availability of image dataset, the advantages of hardware, analysis techniques and parallel processing of High-Performance Computing (HPC) to deal with the computational requirement of image-based age estimation, although underexploited, are beneficial to the digital forensics' community and could reduce the computation time to expedite the processing and analysis of the DF investigation. Although traditionally GPU computing was considered difficult to utilise and targeted for very niche problem solving, the availability of multi-core CPU with GPU acceleration is increasingly more accessible and widely used in HPC enabling simpler programming models, better economies of scale and performance efficiency [2]. More precisely, recent research makes widespread use of deep Convolutional Neural Networks (CNN), automating and significantly increasing the age estimation accuracy. If applied, the use of CNN for automated age estimation could increase accuracy and reduce the human effort in forensic investigations.

This article addresses age estimation, introduces and discusses deep CNN in automated age estimation to support the medical community. The difference between the traditional approach and the deep learning approach for age estimation is discussed at length along with the reasons which made the deep learning approach more popular in recent years among researchers. A detailed comparison of deep CNN based methods for age estimation using different biological features is also covered including advantages and drawbacks of using dental MRI images for age estimation.

2 The Difference Between Traditional Approaches and Deep Learning for Age Estimation

We have found four distinctive approaches in the literature for estimating age from images. The first approach used statistical analysis of teeth and mandibular of child subjects [24]. proposed a method of age estimation based on the development of the seventh teeth from the left side of the mandible. And [25] proposes a method based on 14 stages of mineralization [25].

The second approach used handcrafted methods extracting features from the texture of the face, shape, the colour of the skin, appearance etc. [21] proposed a method for age estimation which can extract effective ageing pattern using a discriminant subspace learning algorithm. In [26], an automatic age estimation method based on ageing pattern representative subspace was proposed which mainly sorts face images by time order.

The third approach is related to shallow learning. It involves extracting features using local binary methods from the patches of the face and then classifying the extracted features using a classifier [27]. proposed Bio-inspired features which are widely used for age estimation, and [28] proposed the improvement base don using a scattering transform. This method added a filtering route to the biologically inspired future which improved the accuracy of age estimation [29]. proposed an orthogonal locality preserving projection technique (OLPP) which further increased the quality of features for age estimators. The second component in this method is a classifier or regressor. Classifiers can be a multi-layer Perceptron, k-nearest neighbours or Support Vector Machine (SVM). Polynomial regression [29] support vector regression and can be used as a regression method for age estimation. This method also requires some prior knowledge.

The fourth approach utilises deep learning algorithms to learn the hierarchical features automatically from images [30]. A detailed analysis of deep learning-based methods will be demonstrated in this article. These methods have the advantage of not requiring a feature selection process, instead, features are selected automatically according to the application.

The handcrafted and shallow learning approach requires a separate feature detection step, then these features are classified using a separate classifier. Whilst deep learning methods provides an end-to-end solution which removes the need of a separate classifier. However, the drawback of using deep learning can be manifested by the requirement of a big dataset and demand for a powerful processor. It has been observed that deep learning methods provide higher accuracy compared to other methods, but it is very difficult to interpret which features have been used to reach the conclusion with this higher level of accuracy.

Table 1 demonstrates differences between the traditional approaches namely shallow learning and hand-crafted feature learning methods, and deep learning methods.

Table 1 Comparison between deep learning and other traditional approaches

Comparison parameter	Deep learning	Shallow learning	Handcrafted methods
Data requirement	Large dataset	Small dataset	Very small dataset
Hardware requirement	CPU + GPU	CPU	Normal or embedded CPU
Feature extraction	Automatic	Handcrafted features + classification	Handcrafted features
Problem solving approach	End-to-end	Modular	Modular
Training time	Long	Short	Very short
Interpretability of features	Low	High	Very high

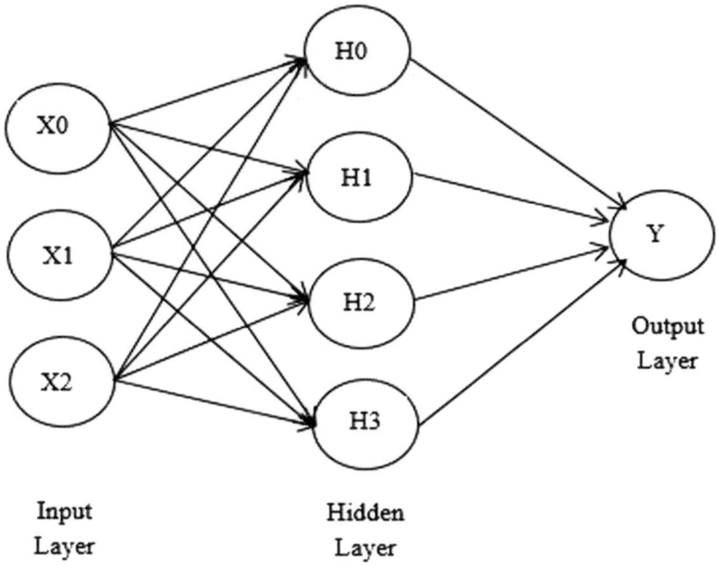

Fig. 1 Artificial neuron architecture

3 The Convolutional Neural Network (CNN)

The most widely used deep learning method for age estimation in literature is CNN. The basic type of neural network tries to mimic the behaviour of the human brain and is called Artificial Neural Network (ANN). The ANN architecture is a perceptron weighting a sum of inputs and applies a threshold activation function [31]. It contains multiple perceptrons connected with each other as shown in Fig. 1.

The ANN architecture in Fig. 1 contains an input layer with three neurons, an output layer with one neuron and a hidden layer with four neurons. The neurons in every layer are connected with each other so ANN is also known as a fully connected network. Each neuron performs the weighted sum of all the inputs and adds the bias term. This is a linear operation but most of the real word problems are non-linear.

Therefore, to make the network non-linear this sum is passed through an activation function. The output y for a neuron with k inputs can be represented as:

$$y = f\left(\sum_{i=0}^{k} X_i W_i\right) \quad (1)$$

The modern-day neural network contains many intermediate hidden layers so these networks are called Deep Neural Networks (DNN).

The number of weights between each layer can be calculated by multiplying neurons in a current layer by neurons in a previous layer. The number of weights will increase together with the number of neurons in the hidden layer. The number of hidden layers and number of neurons in each hidden layer is called hyperparameters which have to be chosen thoughtfully by the network designer according to the application.

The choice of activation function plays a very crucial role in determining the performance of the ANN. It will also determine how fast the network will converge while training and how much computational cost it requires. There are many activation functions used by network designers but Sigmoid, Tanh and ReLU are the most frequently used activation functions. The mathematical equations for these are given below.

$$\text{Sigmoid function}: f(y) = \frac{1}{1 + e^{-y}} \quad (2)$$

$$\text{Tanh function}: f(y) = \frac{e^{y} - e^{-y}}{e^{y} + e^{-y}} \quad (3)$$

$$\text{ReLU function}: f(y) = \max(0, y) \quad (4)$$

The Sigmoid function (3) is considered a smooth threshold function which is also differentiable. The output of a sigmoid function will be between 0 and 1. The issue with sigmoid function is that for a large value of activations it has a very small value of gradient so weights in initial layers will take a long time to update (also called the vanishing gradient problem). Tanh or hyperbolic tangent function as described in Eq. (4) is similar to sigmoid but it has an output in the range of -1 to 1. It will work better than sigmoid in most cases because it centres the data with zero Means. The vanishing gradient problem is also prevalent with the Tanh activation function. However, Rectified Linear Unit (ReLU) function as described in Eq. (5) can solve the problem of vanishing gradient. It is also easier to compute and the overall training of the network is relatively faster.

The final layer of ANN for a multiclass Image classification uses softmax activation function [32] described in Eq. (5) which is mainly an extension of the Sigmoid activation function. It gives the probability of each class by converting the vector to a range from 0 to 1.

$$\text{Softmax Activation function} : f(y) = \frac{e^{y_k}}{\sum_{k=1}^{k} e^{y_k}} \tag{5}$$

ANN uses weights and bias to store information related to the application. These weights and biases are updated during the training phase of the supervised learning approach by calculating the minima of a cost function. The cost function is an error function between the actual value and the predicted value and could be a Mean Square Error, Mean Absolute Error, Binary or sparse cross-entropy etc. The minima of the cost function can be found by using optimization algorithms like gradient descent, Adam, RMSProp etc.

There is a limit to using ANNs for computer vision tasks. The raw pixel values are used as input to the ANN. So for an image size of 1080×1080, there will be one million input neurons. Even if there is only one hidden layer with a small number of neurons, the network will have millions of trainable parameters which means a large dataset and a complex computational unit for training. The second drawback associated with using ANN for computer vision is that it does not take into account spatial neighbourhood information although it is essential for image processing.

These two drawbacks of ANN has led to the use of CNN in computer vision [33]. CNN uses convolution operation which takes into account the spatial neighbourhood information. It also uses the concept of parameter sharing which reduces the number of trainable parameters. It can do that because the same weights can be applied to find features from an entire image. A 3x3 Sobel filter can find edge features from an image of any size with only 9 weights.

The architecture of CNN for an age estimation problem is shown below in Fig. 2.

Figure 2 shows an input image (dental MRI scan) passing through a number of convolution and pooling layers. The convolutional layer tries to collect hierarchical features from the image. Then, the pooling layer is used to reduce the dimensions of the features map. The number of convolution operations in each layer along with the number of these layers should be chosen wisely by the network designer. The output is then converted to a single column vector by a Flattening layer. This single vector is given as an input feature vector to an ANN or a fully connected network for image classification.

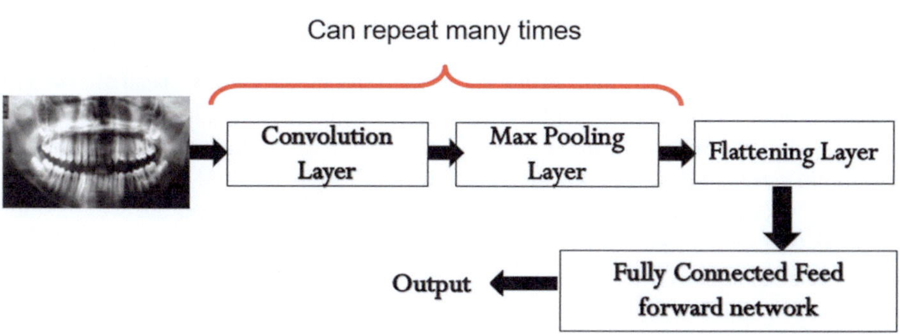

Fig. 2 CNN architecture

3.1 Techniques to Avoid Overfitting in CNN

When the network performs very well on the training data but poorly on the test data then it is called over-fitting. There are several techniques to avoid over-fitting. For instance, Regularization prevents the weights from getting too large. Batch normalization regularises the response after every convolution layer. Another technique is Dropout [34] where random neurons are dropped from the network during training, and the network will not be overly dependent on a single neuron.

3.2 Training of CNN

CNN stores information related to the application in the form of weights and bias and need to be trained for the given application. This can be done by showing the labelled training data to the CNN architecture, this approach is called Supervised Learning.

The weights and biases are initialized randomly with small values. Uniform Random distribution or Xavier initialization [35] is normally used to initialize the weights' value. When the labelled training image samples are given to the CNN architecture, it will calculate the prediction with a forward pass technique using the initialized weights. Then the error between the predicted output and the actual output will be calculated. Mean square error and Mean absolute error are two popular error function for regression problems. Binary cross-entropy is used for the binary classification problem, while the categorical or sparse cross-entropy is used as an error function for the multi-class classification problem.

The calculated error is backpropagated to update the weights using a gradient descent which is an optimization algorithm used to find the minimum of the error function. Other optimizers include Stochastic Gradient Descent, Adam, RMSProp and Adagrad.

There are different types of training methods depending on the number of times the weights are updated in a given timeframe. If weights are updates only once it is called full batch learning. The full batch learning method will take a long time to converge and it will require a large memory space to store images from the entire training set. The advantage of using full batch learning is that it will certainly converge to a global minimum. However, using a stochastic method as an alternative type of training updates the weights after every image, therefore, requires minimum memory and converges faster. It has the disadvantage of fluctuating around the minimum value. Moreover, an intermediate method is referred to as mini-batch learning where the training set is divided into several batches and the weights are updated after every batch of images.

4 Availability and Quality of Datasets for Age Estimation

The appropriateness and completeness of the training dataset can be the key factor to improve the accuracy of age estimation. CNN as a supervised algorithm requires a large number of labelled datasets for training. Datasets for age estimation should also contain a uniform distribution of images of all ages for accurate and inclusive detection. The widespread use of social networking sites has contributed to maintaining large scale facial datasets. Additionally, many open-source datasets designed specifically for age estimation have been created. Face and dental structure are the two most used biological features to estimate age in the literature. To investigate which of these have been successfully used in research studies, we have performed secondary data analysis of primary studies which we summaries in Tables 2 and 3.

Table 3 list all the datasets used in the literature to estimate age based on facial images.

The choice of datasets plays a very important role in getting an accurate result for a particular application. Therefore, a suitable database from the above tables can be chosen for estimating age using facial and dental images. In the next section, we compare between deep CNN methods trained using these databases for age estimation.

Table 2 Summary of dental datasets used for age estimation

Name	Number of subjects	Age range	Special note about the dataset
Southern Chinese Patient Dataset [36]	182	3–16 years	The dataset contained dental panoramic Tomograph (DPT) images from children and adults. The dataset contained the images in the range of 3 to 16 years. The selection of subjects was done from the archives of Prince Philip dental hospital, Hong Kong. The subjects were chosen randomly.
UK Caucasian Dataset [37]	5187	11–15 years	Aimed to develop a reference dataset for at the 13 year old threshold to support dental age assessment for Caucasian children.
French-Canadian Dataset [38]	274	2–21 years	This dataset is based on the dental maturity of French and Canadian population. This dataset overestimates the age by 6 months so you have to be very careful while choosing this dataset for a global population.
Darko Stern's collected MRI Dataset [39]	103	13–25 years	This custom dataset contains 103 3D MRI images of the hand, thorax and dental structure out of that 44 subjects were of minors.

Table 3 Summary of facial age estimation datasets

Database	Images	Age Range (Years)	Special Notes about Dataset
FG-NET [40]	1002	0–69	This dataset is widely is used for estimating age. It is not available for download from its official site but can be downloaded from other sources.
MORPH [41]	1724	27–68	This dataset is provided for age estimation in adults for academic distribution.
Yamaha gender and age (YGA) [21]	8000	0–93	The dataset contains five labelled frontal face images of the same person. The images have different facial expression and illumination.
WIT-DB [42]	5500	3–85	The WIT-DB dataset contains images with large illumination variation and a large age group. The number of images in a particular illumination condition is also unbalanced.
AI & R Asian [43]	34	22–61	This dataset contains images taken in the diverse scenarios like different poses, illumination, ages etc.
Burt's Caucasian face database [44]	147	20–62	This dataset is used to estimate age by combining visual features of colour and shape of facial components.
Lotus Hill research institute (LHI) database [45]	8000	9–89	This dataset contains images of Asians adults with a wide age range. It is also very large dataset which can be used for deep CNN models.
Human and object interaction processing (HOIP) [46]	306,600	15–64	The dataset is divided into ten age groups with each group containing images of 30 subjects. Each age group contain an equal distribution of male and female.
Iranian face database [47]	3600	2–85	The images in the dataset contain large variation in pose and expressions. Every subject has at least one image with the glass. The dataset contains images in the age group of 2–85 years with the majority of them are of subjects before 40 years. This dataset is appropriate for formative and middle age estimation.
Gallagher's web-collected database [48]	28,231	0–66	This database is designed for studying group photos so most of the images in the database are front-facing images with artificial poses. It is a large database which can be used to estimate age in a wide range.
Ni's web collected database [49]	219,892	1–80	This dataset is collected from the web search engines like Google specifically for age estimation in the wide age range. The size of the dataset makes it suitable to use this dataset in estimating an age for children, middle age and old age persons.

(continued)

Table 3 (continued)

Database	Images	Age Range (Years)	Special Notes about Dataset
Kyaw's Web-Collected Database [50]	963	3–73	This dataset is manually created for age estimation by finding out images from Microsoft search engine Bing. The images are aligned manually. The images are cropped to the patches of size 65 by 75.
Combination of LFW, and images from the web [51]	13,466	–	This dataset is collected by the biometric engineering research center. There is uniform illumination in all the images of the dataset along with no variation in facial expression. The uniform distribution of subjects is there in terms of gender and age group.
FDDB Dataset [52]	5171	–	This database contains face images taken in a wide range of difficulties that include occlusion, different poses, and different illumination. The images are taken in either colour or grayscale scenario.
Adience Benchmark [53]	2284	0–60	This dataset is prepared for the study of age and gender estimation from facial images. The dataset contains images with a different appearance, different lighting, noise etc. it intends to take in to account all the challenges of real-world imaging conditions.
Apparent age dataset [54]	4691	–	The images are taken in a real-time environment and have variation in pose, occlusion, lighting, illumination, background, ethnicity etc.
IMDB-WIKI Dataset [55]	524,230	–	This dataset contains the web crawled images of celebrities taken from IMDB and Wikipedia. This is the largest public dataset available of facial images which are widely used particularly for deep CNN applications.

5 Deep CNN Based Methods for Age Estimation

5.1 Deep CNN Based Methods for Age Estimation from Facial Images

Most CNN-based methods seem to utilise well-known architectures (e.g. AlexNet, GoogleNet, ResNet and VGGNet) pre-trained usually on the ImageNet [56] dataset. Very few methods try to develop a new CNN architecture from scratch. This approach is simpler and faster because it does not require fine-tuning. The second approach, however, fine-tunes the weights of well-known pre-trained CNN architectures on a new facial dataset. This approach is an end-to-end method which requires additional training on new facial datasets.

In [57], the CNN architecture consisted of three convolution layers and two pooling layers. It used a combination of CNN and Gabor filter for achieving higher accuracy. The study also showed that going wider instead of deeper with increased filter size can achieve a good result for age and gender classification. The proposed method does not use a complete end-to-end approach as it uses a Gabor filter to find features [58]. proposed a large-scale 22-layers deep CNN framework (AgeNet) for age estimation which used a combination of real value-based regression and label-distribution based classification to estimate the final age. It also proposed a learning method which can be really helpful in avoiding overfitting on a small dataset. However, this method required separate training for regression and classification models. Another study [55] proposed a system called Deep Expectation (DEX) using CNN. It used VGG-16 as a base architecture which is pre-trained on the ImageNet dataset and then fine-tuned the model on face images with age labels. The VGG-16 network used 16 trainable layers with a smaller filter size of 3x3 compared to larger filter sizes in earlier networks. The results showed improvement over direct age regression using CNN. The authors in [30] utilised pre-trained CNN architectures as well but only to perform feature extraction. They used Principal Component Analysis (PCA), Mutual Information and Statistical dependency techniques for dimensionality reduction and ANN for classification.

Age estimation via fusion of depthwise separable CNN was proposed in [59], this has reduced the number of parameters for training without sacrificing accuracy. Three state-of-the-art deep learning models Xception, Inception V3 and ResNet were modified to use depth wise convolution for enhancing the performance and lowering the computational requirement of the system. Empirical results based on four publically available datasets showed superior performance compared to other methods on those datasets when it comes to age estimation [60]. proposed a cluster CNN architecture which significantly reduces the preprocessing steps. The facial image is normalized to a standard size according to the distance between two eyes, This normalized image is fed to the cluster CNN architecture for prediction. The cluster is integrated into the CNN architecture which is capable of multimodal transformation. It is also differentiable so the parameters of it can be learnt using backpropagation. A ranking CNN architecture was proposed in [61]. It is a series combination of normal CNN architecture trained on ordinal age labels. The outputs from the individual CNN architectures are combines to predict the final age. This approach of estimating error seem to obtain better results compared to multi-class classification approach. The performance of the method was evaluated with the MORPH dataset and compared with other state-of-the-art methods.

CNN2ELM was proposed in [62] as a more complex design that incorporates CNN and Extreme Learning Machine (ELM). It consists of three CNN architectures Age-Net, Gender-Net and Race-Net to extract features related to age, gender and race from the image of the same person. The architectures are pre-trained on the ImageNet database. Then it uses ELM classifier for age grouping and ELM regressor for age estimation. The network is fine-tuned on IMDB WIKI dataset and it outperforms other architectures on well-known datasets. It does that because it uses decision fusing to achieve a robust decision. This approach finds more discriminative features from the image then combines the prediction on them to estimate age. However, the performance of the system was poor for a dataset with varied poses or turned/tilted faces.

A consistent limitation affecting all the above methods was the amount of labelled facial data available for age estimation. In response, [62] proposed a data augmentation technique to increase the size of the training data for age estimation. This has produced new training samples from existing images and can be accomplished by applying small transformation like translation, rotation, flipping to the images in the existing dataset. The proposed method also take in to account the intrinsic information about the human face while creating the augmented dataset. The MORPH [41] dataset was used with the same CNN model trained using original and augmented dataset seen a rise of 10% in F-score after utilising the augmented dataset.

Furthermore, [63] proposed a transfer learning-based method. They used VGG19 and VGG Face architecture to explore the performance of transfer learning in age estimation. Techniques such as input standardization, data augmentation and label distribution age encoding were employed to enhance the quality of training while transfer learning. Although the performance of the proposed system was good, it was performing poorly on minorities in the dataset such as old age people, females, people of Asian or African origin. The gender prediction was only based on the length of the hair. These flaws can be overcome by establishing a balanced dataset and changing the architecture or training technique. In [64], the authors proposed an age estimation system by combining CNN with the other popular deep learning architecture called Long Short Term Memory (LSTM). They called the system recurrent age estimation. CNN was used to find discriminative features from the facial images and LSTM were used for learning ageing patterns from a sequence of personalized features.

Further comparison of deep CNN based methods for age estimation is demonstrated in Table 4.

5.2 *Limitations When Using Facial Images for Age Estimation*

Relying on features extracted from the face to estimate age has many limitations. The human face matures in different ways at different ages. Bone growth and wrinkles will be different from one person to another. It is also observed in the literature that women are more likely to develop wrinkles in the perioral region than men [65]. Other challenges include changes in illumination, application of makeup on the face, different face poses and different backgrounds. Hence, the face alone is not always reliable for accurate age prediction.

Table 4 Comparison of deep CNN facial age estimation methods

Deep CNN Architecture	Dataset Used	Performance	Note
Wide CNN [57]	Adience Benchmark Dataset [53]	Age accuracy: 61.3% Gender accuracy: 88.9%	The paper solved the problem of age estimation as a classification problem with eight classes of different age groups so the accuracy is in percentages.
AgeNet [58]	Apparent age dataset provided by the ICCV2015 looking at people challenge	Mean normalized error = 0.2872 Mean absolute error = 3.3345	The paper used 2476 images for training, 1136 for validation and 1087 for testing. The performance of the network is measured in terms of mean normalized and mean absolute error.
Deep Expectation [55]	IMDB-WIKI Dataset [55]	Mean absolute error = 3.221 ε error = 0.278	The paper used mean absolute error and ε error for evaluation on IMDB-WIKI and the ChaLearn LAP dataset.
DSC- Xception [59]	IMDB-WIKI dataset	Mean absolute error = 6.2898	This network used the Xception module and depth wise separable convolution.
	MORPH II [41]	Mean absolute error = 3.25	
DSC- inception v3 [59]	IMDB-WIKI dataset	Mean absolute error = 6.3571	This network used inception v3 module and depth wise separable convolution.
	MORPH II	Mean absolute error = 3.32	
DSC- ResNet [59]	IMDB-WIKI dataset	Mean absolute error = 6.5099	This network used the ResNet architecture and depth wise separable convolution.
	MORPH II	Mean absolute error = 3.52	
DSC-Xception + Inception v3 + ResNet [59]	IMDB-WIKI dataset	Mean absolute error = 5.8865	This network used the fusion of Xception module, inception v3 module and Resnet along with depthwise separable convolution.
	MORPH II	Mean absolute error = 3.08	
VGG Face CNN + Dimensionality Reduction + ANN [30]	IMDB-WIKI dataset	Mean absolute error = 5.4	This technique used VGG face technique for feature extraction which was applied to various dimensionality reduction techniques and ANN for classification.
AlexNet CNN Dimensionality Reduction + ANN [30]	IMDB-WIKI dataset	Mean absolute error = 5.86	This technique used VGG face technique for feature extraction which was applied to various dimensionality reduction techniques and ANN for classification

(continued)

Table 4 (continued)

Deep CNN Architecture	Dataset Used	Performance	Note
Cluster CNN [60]	MORPH II	Mean absolute error = 2.71	The GoogleNet architecture trained on ImageNet database is used as a base network.
Ranking CNN [61]	MORPH	Mean absolute error = 2.96	The age estimation problem in the range of 16 to 66 years was considered. 43,490 samples were used for training and 10,872 were used for testing results. The results were carried out with five-fold cross-validation.
CNN2ELM [62]	MORPH	Mean absolute error = 2.61	The architecture is pre-trained on ImageNet database and then fine-tuned on IMDB WIKI and MORPH II database.
VGG19 and VGG Face Transfer Learning [63]	MORPH	Mean absolute error = 4.10	The VGG 19 architecture is pre-trained on ImageNet database. The MORPH II database is used to fine-tune the weights with 80% of the images are used for training and 20% of the images are used for testing.
Recurrent age estimation (RAE) [64]	MORPH	Mean absolute error = 1.32	Two public dataset MORPH and FG-net are used to evaluate the performance of the system. VGG-16 was used as a base CNN architecture.
	FG-net	Mean absolute error = 2.19	

5.3 *Deep CNN Based Methods for Age Estimation from Dental Images*

Teeth are among the more reliable features for estimating age especially until the age of 20. The various stages of teeth development can be utilised as features to estimate the age of a person but results are more accurate during the dentition development stage because the changes are very prominent and easy to observe. Sometimes the third molar is used for age estimation between 16 and 23 years old though this method is not so accurate. The tooth formation process is over after this age so it becomes very hard to estimate age. Instead, the 'wear' and 'age' regressive changes of hard and soft tissues in the teeth are analyzed to estimate the age for adults.

Examples of imaging techniques include the two-dimension intraoral and panoramic radiographs, 3-dimensional cone-beam computed tomography (CBCT) and Magnetic resonant Imaging (MRI). Many researchers are working using CNN architecture for various applications in dentistry but until recently very little work was directed at deep CNN for age estimation using dentistry.

In [66], the authors produced a method based on a modified Demirjian staging Technique that includes ten development stages. It used transfer learning on a pre-trained AlexNet CNN architecture and the ImageNet dataset. The analysis included 400 panoramic radiographic images and the results showed 10% improvement in classification accuracy. In another recent study [39] the proposal combined features from Dental and Skeletal MRI images. Age estimation is performed by fusing features of three different CNN architectures. Three CNN architectures are used to extract features from cropped wisdom teeth, hand and clavicle bones. Each CNN Architecture consists of three stages of two Convolution and one Max Pulling Layer followed by a fully connected layer. The data augmentation technique was used for the training making the results of the system more accurate and robust. However, it only contains 103 studied subjects so generalization of these results has to be done carefully. This method can be used to estimate the age range up to 25 years. In [67] the method aimed for chronological age estimation using panoramic dental X-ray images. The dataset was divided into three age groups of 2–11 years, 12–18 years and 19 years onwards respectively. The DenseNet-121 [68] architecture with channel-wise attention module was used for age estimation. The curriculum learning strategy was employed in which the network was first trained on images of subjects up to 11 years old and then it slowly included other subjects from 12–18 years and 19 years onwards. The method yielded promising results including the 19 years onwards age group with a giving mean absolute error of 4.398 years only.

Table 5 shows a comparison of deep CNN based methods for age estimation based on facial images.

5.4 Limitations When Using Dental Images for Age Estimation

The empirical findings in current literature are based on in-house datasets which prevent objective cross-method comparisons. Results from small datasets cannot be generalised while deep CNN requires a large amount of data for training, a problem that can be partially solved using data augmentation techniques. There is a need to create a large public dataset for dentistry which can be used for age estimation from dental images. It is also observed that most datasets in dentistry contain more images of children and less number of images for adults. Therefore, a more uniform distribution of images at all ages will be good to support further research in this area.

The size of dental images is relatively large, a problem usually solved by reducing image size before applying detection methods to cut computational cost. However, we could argue against this practice since important information can be lost. Nonetheless, current methods require manual intervention e.g. to fix the "region of interest". A process that can be automated as part of future work.

Table 5 Comparison of deep CNN dental age estimation methods

Deep CNN Architecture	Dataset	Accuracy	Note
AlexNet with Transfer Learning [64]	400 panoramic radiographic images	Mean accuracy = 0.51 Mean absolute difference = 0.6	The paper used the AlexNet CNN architecture trained on the ImageNet database as a base architecture which was fine-tuned on a custom dataset with 80% of images used for training and 20% of images used for testing. Five-fold cross-validation was used for training.
DCNN-MAJ-HAND [66]	103 3D MRI images of left hand, upper thorax and the jaw.	Classification accuracy = 90.3% Mean absolute error = 1.14 ± 0.96 years	The data consisted of images in the age ranges of 13–25 years. The data augmentation technique was used to increase the size of the dataset.
DenseNet-121 with channel-wise attention module [67]	Panoramic dental X-ray images of 9435 subjects.	Mean absolute error 2–11 years = 0.826 12–18 years = 1.229 19 years onwards = 4.398	The dataset contained an equal distribution of male and females with age range in 2–98 years. The size of the original image was 1024 × 2048 which was resized to 256 × 512 before giving it to CNN

Finally, developing age estimation methods is feasible up to 20 years of age, it is very challenging to develop a system to estimate age covering all age groups for several reasons. For example, there is a large variation in teeth conditions after puberty due to dietary habits and teeth management. Dental development is affected by various genetic, environmental, nutritional and endocrinal factors. Teeth eruption is also affected by a number of factors such as gender, ethnic origin, physical and sexual development. It is also observed that age estimation techniques developed for one population might not work for a population belonging to another ethnicity. As such, a typical error rate for adults using dental images is ±10 years which is a very large value. There is a need for researchers to minimize this error to as low as possible.

6 Conclusion

Age estimation plays a very important role in medical forensic as it provides confirmation of a most needed input based on biological features such as the face, bones, skeletal and dental structures. However, the application of age estimation goes beyond that to provide a form of authentication to computer systems. Think about a system's ability to offer personalised Human Computer Interaction (HCI) based on the user age group. Likewise, preventing unauthorized access to individuals as part of a proactive security and defence applications in connected cars [69], border control and more. Clearly, features from facial images or a live feed of the face will be more feasible to utilise for most of these applications.

The research in age estimation has seen a great amount of transformation in recent years following a surge in the use of deep learning algorithms for computer vision. This can be attributed to the availability of medical image datasets and an increase in computer processing power with GPUs or through Cloud Computing services. In this article, we have covered how deep CNN emerged and discussed several recent proposals to highlight the advantages and limitations associated with each approach. Performing secondary data analysis of deep CNN is inevitable to understand research gaps and opportunities.

References

1. Alkass K, Buchholz BA, Ohtani S, Yamamoto T, Druid H, Spalding KL (2010) Age estimation in forensic sciences, application of combined aspartic acid racemization and radiocarbon analysis. Mol Cell Probes 9(5):1022–1030. https://doi.org/10.1074/mcp.M900525-MCP200
2. Lillis D, Becker B, O'Sullivan T, Scanlon M (2016) Current challenges and future research areas for digital forensic investigation. arXiv preprint arXiv:1604.03850
3. Boddington R (2016) Practical digital forensics. Packt Publishing Ltd, Birmingham
4. Kim K, Choi Y, Hwang E (2009) Wrinkle feature-based skin age estimation scheme, pp 1222–1225. Published. https://doi.org/10.1109/ICME.2009.5202721
5. Guo G, Fu Y, Huang TS, Dyer CR (2008) Locally adjusted robust regression for human age estimation, pp 1–6. Published. https://doi.org/10.1109/WACV.2008.4544009
6. Han H, Otto C, Jain AK (2013) Age estimation from face images: human vs. machine performance, pp 1–8. Published. https://doi.org/10.1109/ICB.2013.6613022
7. Ahmadi-Assalemi G, Al-Khateeb HM, Epiphaniou G, Cosson J, Jahankhani H, Pillai P (2019) Federated blockchain-based tracking and liability attribution framework for employees and cyber-physical objects in a smart workplace, pp 1–9. Published. https://doi.org/10.1109/ICGS3.2019.8688297
8. Schmeling A, Garamendi PM, Prieto JL, Landa MI (2011) Forensic age estimation in unaccompanied minors and young living adults. In: Forensic medicine—from old problems to new challenges. InTech, Rijeka, pp 77–120. https://doi.org/10.5772/19261
9. Hjern A, Brendler-Lindqvist M, Norredam M (2012) Age assessment of young asylum seekers. Acta Paediatr 101(1):4–7. https://doi.org/10.1111/j.1651-2227.2011.02476.x
10. Schmeling A, Black S (2010) An introduction to the history of age estimation in the living. In: Age estimation in the living, pp 1–18. https://doi.org/10.1002/9780470669785.ch1
11. Sauer PJJ, Nicholson A, Neubauer D, Advocacy and Ethics Group of the European Academy of Paediatrics (2016) Age determination in asylum seekers: physicians should not be implicated. Eur J Pediatr 175(3):299–303. https://doi.org/10.1007/s00431-015-2628-z
12. Seigfried-Spellar KC (2012) Measuring the preference of image content for self-reported consumers of child pornography, pp 81–90. Published. https://doi.org/10.1007/978-3-642-39891-9_6
13. Gladyshev P, Marrington A, Baggili I (2015) Digital forensics and cyber crime. Springer, Berlin. https://doi.org/10.1007/978-3-642-35515-8
14. Demirjian A, Goldstein H, Tanner J (1973) A new system of dental age assessment. Hum Biol 45:211–227

15. Moorrees CF, Fanning EA, Hunt EE Jr (1963) Formation and resorption of three deciduous teeth in children. Am J Phys Anthropol 21(2):205–213. https://doi.org/10.1002/ajpa.1330210212
16. Anda F, Lillis D, Le-Khac N, Scanlon M (2018) Evaluating automated facial age estimation techniques for digital forensics, pp 129–139. Published. https://doi.org/10.1109/SPW.2018.00028
17. Sehrawat D, Gill NS (2018) Emerging trends and future computing technologies: a vision for smart environment. Int J Adv Res Comput Sci 9(2):839. https://doi.org/10.1109/TIFS.2014.2359646
18. Shejul AA, Kinage KS, Reddy BE (2017) Comprehensive review on facial based human age estimation, pp 3211–3216. Published. https://doi.org/10.1109/ICECDS.2017.8390049
19. C. f. D. C. a. P (2019) Chronic diseases: the leading causes of death and disability in the United States. 01/08/2019. https://www.cdc.gov/chronicdisease/resources/infographic/chronic-diseases.htm
20. Dantcheva A, Elia P, Ross A (2015) What else does your biometric data reveal? A survey on soft biometrics. IEEE Trans Inf Forensics Secur 11(3):441–467. https://doi.org/10.1109/TIFS.2015.2480381
21. Fu Y, Huang TS (2008) Human age estimation with regression on discriminative aging manifold. IEEE Trans Multimedia 10(4):578–584. https://doi.org/10.1109/TMM.2008.921847
22. Guo G, Mu G, Fu Y, Huang TS (2009) Human age estimation using bio-inspired features, pp 112–119. Published. https://doi.org/10.1109/CVPR.2009.5206681
23. Tian Q, Chen S (2015) Cumulative attribute relation regularization learning for human age estimation. Neurocomputing 165:456–467. https://doi.org/10.1016/j.neucom.2015.03.078
24. Demirjian A, Goldstein H, Tanner JM (1973) A new system of dental age assessment. Hum Biol 45(2):211–227
25. Moorrees CFA, Fanning EA, Hunt EE Jr (1963) Formation and resorption of three deciduous teeth in children. Am J Phys Anthropol 21(2):205–213. https://doi.org/10.1002/ajpa.1330210212
26. Wang J, Shang Y, Su G, Lin X (2006) Sim0075lation of aging effects in face images. In: Intelligent computing in signal processing and pattern recognition. Springer, Berlin, pp 517–527
27. Guo G, Guowang M, Fu Y, Huang TS (2009) Human age estimation using bio-inspired features, pp 112–119. Published. https://doi.org/10.1109/CVPR.2009.5206681
28. Chang K, Chen C (2015) A learning framework for age rank estimation based on face images with scattering transform. IEEE Trans Image Process 24(3):785–798. https://doi.org/10.1109/TIP.2014.2387379
29. Guo G, Fu Y, Dyer CR, Huang TS (2008) Image-based human age estimation by manifold learning and locally adjusted robust regression. IEEE Trans Image Process 17(7):1178–1188. https://doi.org/10.1109/TIP.2008.924280
30. Anand A, Labati RD, Genovese A, Muñoz E, Piuri V, Scotti F (2017) Age estimation based on face images and pre-trained convolutional neural networks, pp 1–7. Published. https://doi.org/10.1109/SSCI.2017.8285381
31. Rojas R (2013) Neural networks: a systematic introduction. Springer Science & Business Media, Berlin
32. Bishop CM (2006) Pattern recognition and machine learning. Springer, New York
33. Fukushima K (1988) Neocognitron: a hierarchical neural network capable of visual pattern recognition. Neural Netw 1(2):119–130. https://doi.org/10.1016/0893-6080(88)90014-7
34. Srivastava N, Hinton G, Krizhevsky A, Sutskever I, Salakhutdinov R (2014) Dropout: a simple way to prevent neural networks from overfitting. J Mach Learn Res 15(1):1929–1958
35. Kumar SK (2017) On weight initialization in deep neural networks. arXiv preprint arXiv:1704.08863
36. Jayaraman J, King N, Roberts G, Wong H (2011) Dental age assessment: are Demirjian's standards appropriate for southern Chinese children? J Forensic Odontostomatol 29(2):22

37. Chudasama PN, Roberts GJ, Lucas VS (2012) Dental age assessment (DAA): a study of a Caucasian population at the 13 year threshold. J Forensic Legal Med 19(1):22–28. https://doi.org/10.1016/j.jflm.2011.09.008
38. Jayaraman J, Wong HM, King NM, Roberts GJ (2013) The French–Canadian data set of Demirjian for dental age estimation: a systematic review and meta-analysis. J Forensic Legal Med 20(5):373–381. https://doi.org/10.1016/j.jflm.2013.03.015
39. Štern D, Kainz P, Payer C, Urschler M (2017) Multi-factorial age estimation from skeletal and dental MRI volumes, pp 61–69, Published
40. Panis G, Lanitis A, Tsapatsoulis N, Cootes TF (2016) Overview of research on facial ageing using the FG-NET ageing database. IET Biometrics 5., https://digital-library.theiet.org/content/journals/10.1049/iet-bmt.2014.0053
41. Ricanek K, Tesafaye T (2006) MORPH: a longitudinal image database of normal adult age-progression, pp 341–345. Published. https://doi.org/10.1109/FGR.2006.78
42. Ueki K, Hayashida T, Kobayashi T (2006) Subspace-based age-group classification using facial images under various lighting conditions, pp 6–48. Published. https://doi.org/10.1109/FGR.2006.102
43. Fu Y, Zheng N (2006) M-face: an appearance-based photorealistic model for multiple facial attributes rendering. IEEE Trans Circuits Syst Video Technol 16(7):830–842. https://doi.org/10.1109/TCSVT.2006.877398
44. Burt DM, Perrett David I (1995) Perception of age in adult Caucasian male faces: computer graphic manipulation of shape and colour information. Proc R Soc Lond Ser B Biol Sci 259(1355):137–143. https://doi.org/10.1098/rspb.1995.0021
45. Suo J, Wu T, Zhu S, Shan S, Chen X, Gao W (2008) Design sparse features for age estimation using hierarchical face model, pp 1–6. Published. https://doi.org/10.1109/AFGR.2008.4813314
46. Fu Y, Guo G, Huang TS (2010) Age synthesis and estimation via faces: a survey. IEEE Trans Pattern Anal Mach Intell 32(11):1955–1976. https://doi.org/10.1109/TPAMI.2010.36
47. Azam B, Melika Abbasian N, Mohammad Mahdi D (2007) Iranian face database with age, pose and expression, pp 50–55. Published. https://doi.org/10.1109/ICMV.2007.4469272
48. Gallagher AC, Chen T (2009) Understanding images of groups of people, pp 256–263. Published. https://doi.org/10.1109/CVPR.2009.5206828
49. Ni B, Song Z, Yan S (2009) Web image mining towards universal age estimator. In: Proceedings of the 17th ACM international conference on multimedia, Beijing, China, pp 85–94. https://doi.org/10.1145/1631272.1631287
50. Sai Phyo K, Wang J, Eam Khwang T (2013) Web image mining for facial age estimation, pp 1–5. Published. https://doi.org/10.1109/ICICS.2013.6782962
51. Sun Y, Wang X, Tang X (2013) Deep convolutional network cascade for facial point detection, pp 3476–3483. Published. https://doi.org/10.1109/CVPR.2013.446
52. Jain V, Learned-Miller E (2010) Fddb: A benchmark for face detection in unconstrained settings, UMass Amherst Technical Report. http://works.bepress.com/erik_learned_miller/55/
53. Levi G, Hassncer T (2015) Age and gender classification using convolutional neural networks, pp 34–42. Published. https://doi.org/10.1109/CVPRW.2015.7301352
54. Escalera S, Fabian J, Pardo P, Baró X, Gonzàlez J, Escalante HJ, Misevic D, Steiner U, Guyon I (2015) ChaLearn looking at people 2015: apparent age and cultural event recognition datasets and results, pp 243–251. Published. https://doi.org/10.1109/ICCVW.2015.40
55. Rothe R, Timofte R, Gool LV (2015) DEX: deep expectation of apparent age from a single image, pp 252–257. Published. https://doi.org/10.1109/ICCVW.2015.41
56. Berg A, Deng J, Fei-Fei L (2010) Large scale visual recognition challenge (ILSVRC), 2010, vol 3. URL http://www.image-net.org/challenges/LSVRC
57. Hosseini S, Lee SH, Kwon HJ, Koo HI, Cho NI (2018) Age and gender classification using wide convolutional neural network and Gabor filter, pp 1–3. Published. https://doi.org/10.1109/IWAIT.2018.8369721

58. Liu X, Li S, Kan M, Zhang J, Wu S, Liu W, Han H, Shan S, Chen X (2015) AgeNet: deeply learned regressor and classifier for robust apparent age estimation, pp 258–266. Published. https://doi.org/10.1109/ICCVW.2015.42
59. Liu K, Liu H, Chan PK, Liu T, Pei S (2018) Age estimation via fusion of depth-wise separable convolutional neural networks, pp 1–8. Published. https://doi.org/10.1109/WIFS.2018.8630776
60. Shang C, Ai H (2017) Cluster convolutional neural networks for facial age estimation, pp 1817–1821. Published. https://doi.org/10.1109/ICIP.2017.8296595
61. Chen S, Zhang C, Dong M, Le J, Rao M (2017) Using ranking-CNN for age estimation:742–751. Published. https://doi.org/10.1109/CVPR.2017.86
62. Duan M, Li K, Li K (2018) An ensemble CNN2ELM for age estimation. IEEE Trans Inf Forensics Secur 13(3):758–772. https://doi.org/10.1109/TIFS.2017.2766583
63. Smith P, Chen C (2018) Transfer learning with deep CNNs for gender recognition and age estimation, pp 2564–2571. Published. https://doi.org/10.1109/BigData.2018.8621891
64. Zhang H, Geng X, Zhang Y, Cheng F (2019) Recurrent age estimation. Pattern Recogn Lett 125:271–277. https://doi.org/10.1016/j.patrec.2019.05.002
65. Paes EC, Teepen HJLJM, Koop WA, Kon M (2009) Perioral wrinkles: histologic differences between men and women. Aesthet Surg J 29(6):467–472. https://doi.org/10.1016/j.asj.2009.08.018
66. De Tobel J, Radesh P, Vandermeulen D, Thevissen PW (2017) An automated technique to stage lower third molar development on panoramic radiographs for age estimation: a pilot study. J Forensic Odontostomatol 35(2):42–54
67. Kim J, Bae W, Jung KH, Song IS (2019) Development and validation of deep learning-based algorithms for the estimation of chronological age using panoramic dental x-ray images
68. Huang G, Liu Z, Maaten L v d, Weinberger KQ (2017) Densely connected convolutional networks, pp 2261–2269. Published. https://doi.org/10.1109/CVPR.2017.243
69. Al-Khateeb H, Epiphaniou G, Reviczky A, Karadimas P, Heidari H (2018) Proactive threat detection for connected cars using recursive Bayesian estimation. IEEE Sensors J 18(12):4822–4831. https://doi.org/10.1109/JSEN.2017.2782751

Secure Implementation of E-Governance: A Case Study About Estonia

Rodrigo Adeodato and Sina Pournouri

Abstract The purpose of this research is to identify how feasible it is to securely and effectively implement e-governance in a developed country. The project uses Estonia as a case study, and analyses the path the country chose to take in order to achieve its current state of e-governance. The country's answer to the existing risks embracing this transition included employing a distributed ledger technology and a proprietary solution acting as a data exchange layer, which promoted improvement on transparency and efficiency, and resulted on an increase on citizen's trust. The study establishes a direct relationship between e-government and digital security, and compares Estonia's level of preparedness in cyber security with other nations. This study also investigates the adoption of Cyber Situational Awareness program as an element of secure implementation of e-governance.

Examining Estonia's results attained over the past two decades, and comparing its e-governance and cyber security index rankings with other developed countries, it becomes clear that the digital transition has acted as a lever in successfully developing the nation and maintaining it at the forefront of cyber security internationally.

Keywords E-government · Cyber attack · Cyber situational awareness · Estonia · Cyber security · Governance · Blockchain · Distributed ledger technology

1 Introduction

This chapter aims to investigate secure implementation of e-governance in a developed country. The study describes the concept of e-governance, a brief history of its implementation worldwide, its benefits, challenges and the risks of digitising sensitive information. It attempts to establish how difficult it is to implement e-

R. Adeodato · S. Pournouri (✉)
Sheffield Hallam University, Sheffield, UK
e-mail: Rodrigo.GradowskiAdeodato@student.shu.ac.uk; s.pournouri@shu.ac.uk

H. Jahankhani et al. (eds.), *Cyber Defence in the Age of AI, Smart Societies and Augmented Humanity*, Advanced Sciences and Technologies for Security Applications, https://doi.org/10.1007/978-3-030-35746-7_18

governance in a secure and effective way. In regard to the Estonian successful case, what is the X-Road (or X-tee) technology used, and how does a secure distributed data exchange layer platform helped to enable the transition between styles of governance. This study also scrutinises how the distributed ledger technology has helped to achieve the level of trust required to achieve success on e-governance. Finally, it analyses how does the Estonian solution compare to other developed countries, and proposes guidelines for secure implementation of e-government.

It is essential to establish the difference between electronic governance and electronic government (e-government). E-governance is a wide-scope subject that defines the entire relationship between governments and the application of technology, involving a multitude of independent actors. It can be summarised as a collection of processes and procedures that have impact on a state's exercise of power, openness, participation, responsibility, effectiveness and consistency [35]. E-government, on the other hand, is a smaller part of the whole governance concept, developing online services for government, citizens, employees and business to bring all actors together, and allowing them to exchange information in a transparent, simplified, and secure way [35]. Simply put, e-governance is a natural evolution of the current manner of governing in order to match the information age in which we are living. It is useful as it offers better delivery of service, improved interactions with business/industries, citizen empowerment, easier management, cost reductions and convenience. In the same way computers changed in the last three decades the way we live and work, now ubiquitous digital information is pushing society for the next step, where data confidentiality, integrity and availability needs to be kept safe at all times and exchanged securely.

Although this study will encompass the broader e-governance concept to provide an overarching understanding of the security challenges, it will emphasise in the Estonian e-government implementation, as it provides more specific and in-depth approaches on how to tackle the security issues. Furthermore, there are several challenges for e-governance/e-government implementation, varying from technical, organisational, social to financial barriers, but this paper will try to focus on the security aspect of it.

E-government consists of the interactions between multiple elements, namely the government (G), their citizens (C), employees (E) and businesses (B) [63]. These interactions at various levels create a web of data exchanged and are typically displayed in the format *G2G* (abbreviation for *government* to *government,* for example). G2G refers to data exchange between government organisations, agencies or departments. It is crucial to establish coordination and cooperation between governments as well, by sharing databases, resources and tools. G2C (*government* to *citizens*), G2E (*government* to *employees*), G2B (*government* to *businesses*) and C2G (*citizen* to *government*) complete the list with the most common interactions.

G2C and C2G establish the relationship between government and citizens, and arguably the core of e-government. Populations' trust in the governance model stimulates acceptance and embracing technology in a large scale strengthens the entire governance system, which creates a more efficient, convenient, transparent, simplified, less bureaucratic and cost-effective environment [60].

Once a high level of commitment is achieved from these actors to start interacting and exchanging data digitally, the concern for privacy and security arise. From the security perspective, there are several identifiable challenges, from infrastructure requirements, need for qualified professionals, lack of regulations, difficulty in obtaining citizens' trust and adoption, arduousness in establishing the right cryptographic standards, certification and authentication methods, and regulatory issues amongst many others.

For the past two decades, multiple countries have come up with different solutions and have also embraced technology in dissimilar levels. Not every nation has the resources, the knowledge and suitable timing to transition and secure the way information flows. However, as old methods of governance start to become obsolete, slow and cost-inefficient, technology innovation presents itself as a possible solution to address these issues.

The Estonian answer for most of these problems was committing to embrace technology in a very fast pace, transitioning from the traditional methods of governance to high-level state of e-governance. In order to achieve it, the nation has adopted a distributed ledger technology (DLT), commonly referred to as Blockchain technology (even though the terms are not technically the same), which allowed data to be stored decentralised, securely, in an endlessly verifiable and unchallengeable form. Similar technology has been used for diverse purposes (such as cryptocurrencies) worldwide with fair confidence and great results. This chapter investigates implementation of this solution based on an analysis of the Estonian implementation over the years, and how does that compare to e-governance adoption elsewhere.

2 Background

Larsson & Grönlund [33] define e-governance as the use of technologies to achieve a better government. It aims to solve efficiency problems in the public sector by reducing costs and improving services to its citizens. Dawes [10] argues e-governance is a "dynamic socio-technical system encompassing interactions among societal trends, human elements, changing technology, information management, interaction and complexity, and the purpose and role of government". The increasing use of technology has been redesigning governing tools and solutions, offering a myriad of benefits but also presenting new challenges.

The e-governance benefits are numerous and clear. Limiting to the e-government aspect alone, it can help improving [41]:

- Efficiency – Mass processing tasks and public administration operations. Even further efficiency is achieved when governments start sharing data among themselves.
- Services – Seamless and customer-oriented online services provided to citizens, who do not need to understand the complexity infrastructure and relationships behind them.

- Reform – Necessity for public management modernisation and reform can be a trigger point for e-governance, which provides globalisation and addresses fiscal demands.
- Citizens' trust – Promotes transparency and accountability, preventing corruption and therefore, gaining citizens' trust in the system.

Citizens' trust is one of the main beneficial elements of establishing a high level of e-governance, but paradoxically it also plays an essential role in the foundation and initial e-governance adoption by citizens and the private sector. Establishing this preliminary trust can be fairly challenging depending on the country's current governance, size, economy and social behaviour. Golesca [20] defines the two more important determinants of trust in e-government are perceived organisational trustworthiness and trust in technology. Parent, Vandebeek, & Gemino [49] defend trust cannot be improved by increasing government online presence, as distrustful citizens will not change their behaviour regardless of the way they interact. The fact is, without citizen's participation and confidence, alongside relevant public and private partaking, governance would hardly change from its traditional ways.

In terms of worldwide adoption, the United Nations (UN) e-government survey [71] shows Denmark, Australia and South Korea amongst the top countries to implement e-government. Overall, European countries lead the chart, followed by American, Asian and then African nations, and the UN website [71] establishes a positive correlation between the country's income level and its e-government ranking. Even though Estonia did not make the top 10 positions in that year, its governmental digital presence is abundant and studied by many authors, such as the article from Solvak et al. [66], which investigates the country's massive citizens' acceptance to e-governance adoption. The article corroborates what the Estonian official webpage [14] states, establishing a linear growth on e-government diffusion and fast user reception. Alketbi, Nasir & Abu Talib [1] also mention countries that have greatly adopted it and highlighting the Estonian case, but questioning how secure the current identity storage system is, emphasising that it lacks proper infrastructure and data security.

Telecommunication infrastructure is indeed vital for a country switching to electronic governance, alongside education, as citizens need to be able to understand and exploit the services offered [5]. This requires state commitment and investment, which not only could lead to a better state of e-governance, but also would improve cyber security levels; digital data security plays a vital role when a government uses technology to store and exchange information.

The solution presented by Estonia (among many other nations) to achieve a high level of e-government while maintaining privacy and security was developing a system based on distributed ledger technology, often referred (or misnamed) to as *Blockchain*. This technology allows the administration to use a decentralised system, without a single point of failure [12]. Sekhar, Siddesh, Kalra, and Anand [65] cover what this technology means, its relationship with smart contracts, and how they help maintaining the chain, demonstrating its applicability for different domains and case studies. These authors defend the use of distributed ledger technology to store information, and propose a working model to replace the conventional methods of information storage, which are allegedly further prone to attacks.

Konashevych & Poblet [31] also mention security risks in Estonia's project, stating different organisations in the EU and UK have recently released reports on issues using distributed ledger technology. According to Konashevych & Poblet [31], these reports are much less specific about the design of Blockchain-based e-government systems and how to implement it in particular areas, stressing it might be premature due to the fact it is still in early stages and supports a public key infrastructure (PKI) solution instead. Hoberman [24], however, published an in-depth explanation supporting Blockchain usage, benefits and impact in government use, justifying it is mature enough for adoption. Santhana [62] recommends caution, due to the risks imposed by this or any other nascent technology, but admits it is a transformational model which should be embraced alongside a strong risk management strategy, governance and controls framework.

Deloitte [11] published an article considering multiple-source risks whilst using a Blockchain-like framework, taking into consideration aspects such as business continuity, strategic, reputational regulatory, operational, data confidentiality, key management, consensus protocol, legal and information security. It summarises the risks of implementing a distributed ledger solution and their potential mitigation steps, as in governance, policies/standards, management process, risk metrics and culture. Despite being widely adopted, some researchers are still sceptical about it. Most of the criticism about Blockchain security, though, is related to Bitcoin, one of the main cryptocurrencies which use this technology since 2009. Keenan [30] sustains Blockchain has a bad reputation due to possible attacks, such as the *51% problem* and debates the difference between Proof-of-Work (PoW), Proof-of-Stake (PoS) and Proof-of-Authority (PoA) protocols, and how they affect reliability in this case. These are relevant concerns to take into consideration, but they primarily pose a problem for distributed ledgers using public and permission-less infrastructure, as opposed to a private government-controlled permissioned distributed system, which usually plays a part on e-governance adoption, such as the Estonian case.

Distributed ledgers are often tangled with smart contracts, digital agreements that are automatically performed once a consensus on the result of an event is created on the platform [9]. Smart contracts help minimising the mistrust between parties as they are stored on the Blockchain and are autonomous, decentralised and auto sufficient. That creates opportunities for insurance companies, supply chain management, copyright protection, digital identity, financial data recording and more [39], by reducing intermediary services to facilitate a transaction. O'Hara [42], raises the concern that smart contracts rely exclusively on software, which could be compromised, leaving parties exposed to legal issues and financial loss. Gatteschi, Lamberti, Demartini, Pranteda, & Santamaria [19] establish that smart contracts increase the processing speed for transactions, reducing costs and mistakes and increasing trust and acceptance. Generally, most authors agree with the use of smart contracts, and defend the implementation of security steps to prevent misuse – which does not differ much from non-digital forms of contracts, it only excludes the human factor from the equation.

Governments worldwide have started using the distributed ledger platform to store data, and Estonia has become an example of successful use of combined technologies to achieve excellence, as pointed out by Stephany [67]. This author investigates how some factors helped achieving it, such as being small country, having a relatively young population, trustworthy institutions and necessity of technological renewal. The Estonian model has also been positively depicted by Vassil [73], Oxford Analytica [47] and Jaffe [27], but has been questioned by Drechsler [13], who argues Estonia has been being presented as a leading digital governance country, but states the title might be overhyped. Hartmann and Steup [23] present a more technical discussion about the Estonian e-governance backbone system called X-Road (or X-Tee), and how it strives towards an EU-wide expansion. Hartmann and Steup [23] also address general aspects of the design of secure transnational data exchange frameworks.

Regardless of using distributed ledger or not, the thrive of Estonia's economy for the past decade indicates the right choice was made. Hartmann and Steup [23] published a comparative analysis of existing e-governance systems within Europe based on well-defined security aspects. It explored how decisions made in the design may affect the security of the underlying network and its components. The range of online public services provided have increased over the past decades and compared to other European countries, Estonia still has the edge for the diversity of government online services offered [74] such as e-ID, qualified electronic signature, tax system, births registering, social security benefits, public certificates, residence and relocation, or setting up a new company. The higher the number of quality services provided, the better the country perform in governance indicators.

Detailed comparison indices for e-government keep being published by the United Nations [72] every other year with detailed methodology on how to measure development in electronic governance. The way nations accomplish the requirements to increase their index ranking somewhat diverges. Joseph and Advic [29] argue the Nordic countries reached a high level of development in this field through extensive public sector reforms. Young-Jin [74] defends South Korea achieved it via heavy stakeholder's engagement, political commitment and clear objectives. Ott, Hanson and Krenjova [46] published a document detailing the Australian solution, which follows the similar guidelines established by Estonia.

As government systems get progressively more digital, cyber security becomes a concern. Cyber security indices, directly related to e-government implementation levels are also available online, such as the National Cyber Security Index [40] and the International Telecommunications Union [26]. Each index uses its particular method to rank nations and show how prepared different countries are against cyber threats. Even though there is some consensus between what is important and what not, the path to achieve security in e-governance is not yet that well-defined.

Regardless of the level of success achieved in preparedness, the number of attacks keep rising. Perhaps due to the fact data is progressively made more available, the risk of having it breached has also increased (Estonian Information System [18]).

In an escalated worldwide scenario, and unrelatedly to the type of defence technology adopted, signs of cyber war have allegedly been present and pose a clear threat to security. While Rid [53] defended the reasons why digital war would never take place, more recently, Stockburger [68] analysed the denial-of-service attacks suffered by Estonia in 2007 and Georgia 2008 impacting core infrastructure, supposedly indicating they had been performed by Russia (which has never admitted it). It is challenging to pinpoint the source of cyber-attacks, and link such activities to a specific state government would have dire consequences which are in no one's interest. McGraw [37] states even though the risk of cyber was is over-hyped, the absence of proper security would inevitably lead to digital warfare. Making sensitive data available and moving services online seem to be a favourable idea, but the potential perils can dramatically reshape the relationship between countries.

Regardless of risks, the path to e-governance seems to be a logical and inevitable step forward for developed countries. It is the platform that will enable governments to use technology to embed good governance principles and achieve public policy goals through transparency and trust [7], and will likely be the foundation of developed governments for the next decades. The question is really how to get to that point efficiently, maintaining security and privacy whilst implementing it.

3 E-Governance, Cyber Security and the Estonian Case

3.1 About Estonia

Estonia is a northern European country with a total area of 45,336 km^2 [17], bordering Finland, Sweden, Latvia and Russia. Estonia got its independence from Russia recognised in 1920 at the end of World War, but only became a parliamentary democratic republic after its *de facto* independence in 1991, after the collapse of the Soviet Union. It has then joined the European Union and NATO in 2004, and the Eurozone in 2011.

In 2019, Estonia's population is estimated to be around 1.3 million [72] with a nominal GDP per capita of US$ 23,510 [25]. According to the World Bank [70], its economic growth was steady around 7% per year between 2000 and 2008, one of the fastest rising economies in the European Union (Fig. 1).

3.2 Switching to e-Government, e-ID and E-Residency

One of the reasons for this remarkable growth was arguably switching its civil services to a nearly full digital form of e-government, which has become worldwide a case study of a successful implementation of e-governance [69]. That change on the countries strategic direction allowed increasing efficiency, convenience, transparency, in a cost-effective manner, simplified, and less bureaucratic.

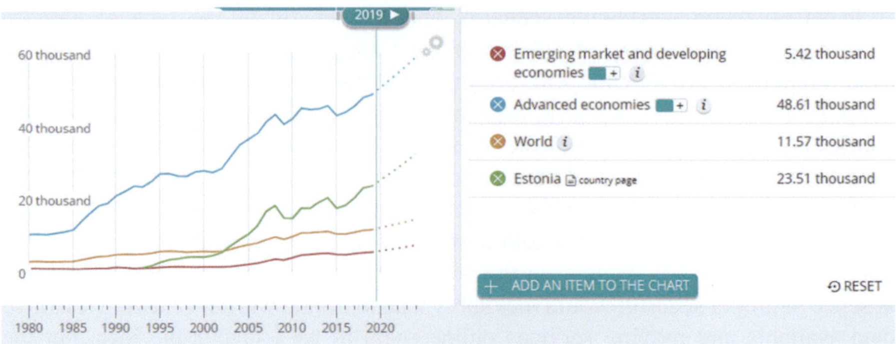

Fig. 1 Estonia's economic growth and projection [25]

Estonia heavily invested in this transition. Before the digital project started in 2002, around 70% of the population had never used the internet [17]. Limited access to computers, poor internet connections, absence of skills and motivation were factors that contributed to a low digital adoption. The successful history shows that Estonia managed to, in less than two decades, reduce the gap and embrace digital services as a society.

Conceivably the foundation for the Estonian *G2C* success was based on an electronic identification (e-ID) card, which has been launched in 2002 and became mandatory for all citizens [73]. Making a digital document compulsory [73] in that scenario was probably risky, but the government commitment to educate the population on the matter and private sector engagement helped the adoption. Still according to Vassil [73], in 2014 the e-ID was used more than 80 million times for authentication and 35 million times for digital transactions.

The e-ID facilitated developing a more connected society, including public and private services and making Tallinn, the nation's capital, an innovation hotspot. Estonia became such a reference to digital identification that other developed countries, such as Japan, looked up to Estonia's e-ID as a reference for their own digital identification programme.

The e-ID uses a rooted Certificate Authority (CA) chain of trust and its certificates use 2048-bit public key encryption. It features an electronic chip which can be read in nearly any card reader, whether it is on a computer or establishment/kiosk. The card is based on public key infrastructure (PKI) technology [36] and has two certificates; one for authentication and the second one for digital signatures. Each private key requires a different personal identification number (PIN), which is used for different purpose [73]. In a scenario where a user would access their online banking, they would be prompted for their authentication PIN, while if they would like to pay a bill, they would need the second code.

The e-ID became widely used for both public and private sectors, creating a convenient network for citizens to use. Even though it is still susceptible to vulnerabilities as any other piece of hardware, it has been considered consistently trustworthy by the majority of the population [69].

In 2014, Estonia expanded its e-government features to foreigners, creating an electronic residency programme (e-residency) that allowed entrepreneurs living elsewhere to start a business in Estonia. E-residents are able to remotely use Estonian public and private sector services, such as business registration, banking services, buying/selling properties, and goods and services trading [69].

E-residents are not technically Estonian citizens or permanent residents, and therefore have limited services available to them. For the government, e-residency allows economy to expand beyond their borders, generating business and developing the country. From the security point of view, e-residency is important because it strengthens the use of the e-ID to perform online business, which would be very difficult to achieve without a dependable platform.

According to Sullivan and Burger [69], Estonia's goal is to have ten million e-residents by 2025 – having a population of just 1.3 million itself. Since the programme began, the number of e-residents has been growing steady to about 50,000 people from 157 countries, creating 6000 new companies in the country [32]. Although that does not necessarily translate directly in taxes (as many e-residents are exempt from taxes), the e-residency programme does boost Estonia's businesses and integrate even more the use of the electronic ID and services.

Estonia publicised in November 2015 that was cooperating with Bitnation [69], an emerging initiative based on Blockchain technology intended to replace traditional governance systems with an open-source governance, also known as *Governance 2.0* [43]. Bitnation takes a step further on governance decentralisation, redefining how jurisdictions are determined, health, education, social security, security; all these aspects would abstract from a centre government, creating a digital information borderless world. Even though this seems to be far from reality for most people, Estonia's e-residency seems to be pointing to that direction.

3.3 The X-Road/X-Tee

Undoubtedly the most pressing concern regarding the security and privacy aspects is the technology itself, which supports secure communication between government and any other government, citizen, business or employee. The challenge is how to provide secure interoperability between different technologies ensuring a seamless operation, as if the entire process was being handled by a single system.

In order to address this issue, in 2001 Estonia developed the X-Road, a secure distributed *Data Exchange Layer* (DXL) platform that enabled internet-based data exchange between information systems. Its English name has changed in 2018 from X-Road to X-tee [22], but literature is still widely found referring to both terms and they are often used interchangeably. The X-Tee is not fundamentally a new creation, but a set of existing technologies harnessed in particular way within governmental environment, effectively becoming the backbone of the country's digital infrastructure.

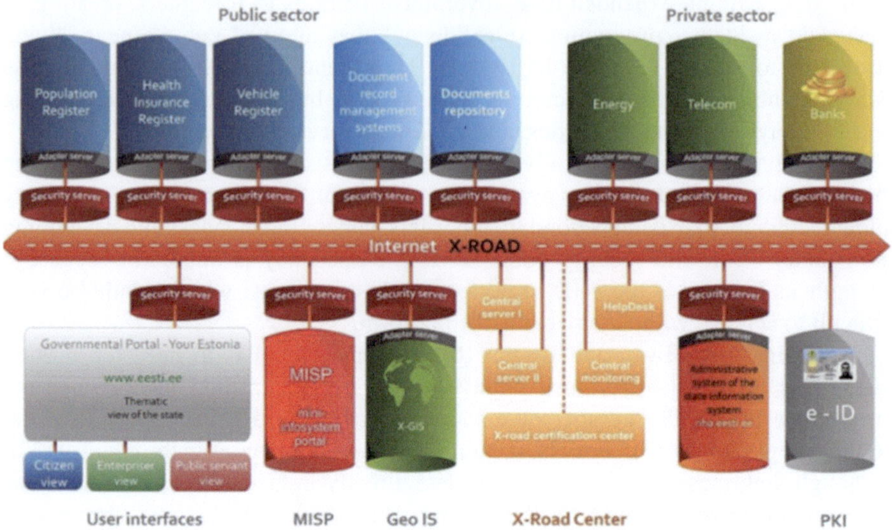

Fig. 2 A schematic of Estonia's X-Road data exchange [73]

Initially built to be accessible only by public sector units, it did not take long for the government to acknowledge how convenient would be to integrate its services with the private sector as well. The X-tee has also been implemented in several other countries, like Finland, Namibia, Iceland, Ukraine, Kyrgyzstan, and the Faroe Islands. Two X-tee ecosystems (two different governments) can publish and utilise services as if they were in the same environment, allowing cross-border data exchange, such as the one established between Estonia and Finland in 2018.

An important aspect of the X-tee is the fact that application development is already built into the platform, so new software that successfully joins the system will already have potential access to multiple government-sanctioned services. Amongst these services, a few ones are vital for security, such as authentication, multi-level authorisation, registry services, data entry and exchange that is encrypted and signed, high-level system for processing logs, query tracking and others [73] (Fig. 2).

Another feature of this platform is its decentralized nature, encouraging every joining organisation to share their data, which is particularly useful for queries involving multiple databases.

Operating principles of X-tee [22]:

- **Independence of platform and architecture:** The X-tee allows interoperability between information systems. It also should work over low bandwidth connections, accessible to all.
- **Multilateralism:** X-tee members can request access to multiple data services.

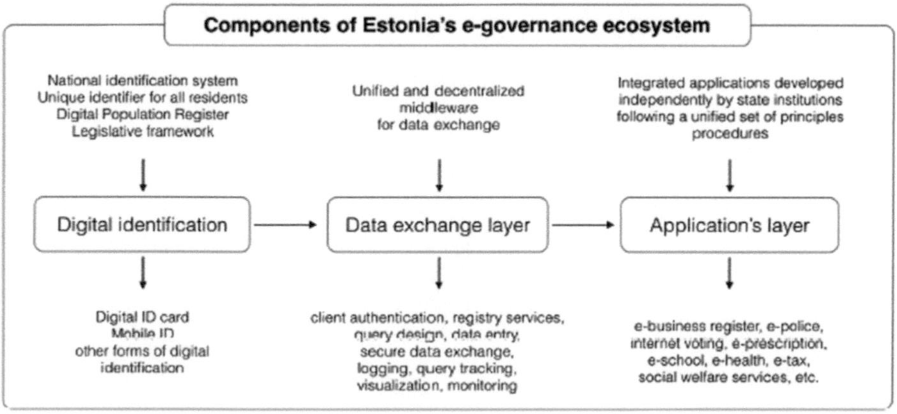

Fig. 3 Estonia's e-governance ecosystem [73]

- **Availability and standardisation:** International standards and protocols are used from development and management.
- **Security:** The X-tee claims to exchange data without compromising integrity, availability or confidentiality. All communications are encrypted and data owners can enforce additional security conditions if desired. Each connected point should be identifiable through a cryptographic certificate.

The X-tee supports complex queries to multiple databases. An organisation's membership provides access to any particular data provider that has been previously agreed upon, as long as the person initiating the request provides a valid identification.

The Estonian X-tee is maintained by the State Information System Authority (abbreviated as '*RIHA*' in Estonian), which is in charge of accepting new members, issuing certificates, defining the Code of Conduct and monitoring the usage patterns. Additionally, standard cryptographic services such as timestamping and certification can be provided by Trust Service Providers.

Multiple X-tee security servers are accountable for tasks such as encrypting the traffic, enumerating target databases, translating register names into IP addresses, preventing eavesdropping, and unauthorised change, loss and duplication. The servers are in fact Open Systems Interconnection (OSI) level 7 Application Gateways, responsible for redirecting any requests initiated by client users via X-tee Messaging Protocol, which need to formatted and encrypted prior to be forwarded to the X-tee pipeline. According to RIHA, access controls are well-established and members will not have access to databases if end-users do not have proper rights to its data (Fig. 3).

On the data providers' side, security also plays an important role. Data providers need to be properly registered with RIHA and have an Adapter Server, also known as "integration components". These servers are in charge of analysing requests from customers, configuring them into a standardised X-tee format and assigning them back to the X-tee – and eventually back to customers. Low budgeted organisations legally required to provide data can still be a part of the X-tee; utilising MISP (Mini Information Service Portal), a web gate that allows minor requests to be processed – usually ones that will not require access to confidential information.

Not restrained to its own e-government services, Estonia has taken the next step and connected its data exchange layer to Finland's (both use the X-Road technology) in 2016, for seamless cross-border data transfer. This interconnection represents multi-state trusted online services, with transactions supported by two sovereign nations. In a smaller scale but yet impressive move, Estonia has opened in 2018 its first e-residency collection centre in South Korea, which also allowed interoperability between countries via the X-Road platform to authenticate electronic residents across the globe.

3.4 Distributed Ledger Technology

The concept of data centralisation in a single large database is subject of debate for a long time. University of Cambridge's professor Ross John Anderson [3] has been arguing for years that large databases will always be targeted for security breaches. Commonly attributed to him and popularly known as *Anderson's rule* states that a database cannot be built with scale, functionality and security, because if a system is designed to be wide, to ease access, it becomes insecure; if that database is made watertight it becomes unbearable to use [3].

Aiming to maintain data integrity, the X-Tee uses a distributed ledger technology (DLT) solution. The DLT is a database spread across several nodes, often misnamed as *Blockchain*, which makes the event record reliable and consistent. Requests are digitally signed and get a timestamp that allows data to be secured and easily searched at any point, adding a non-repudiation quality to it. New information can only be added to the ledger if the nodes that support the data structure have mutually agreed to it. Its decentralisation reduces dependency on a central actor and makes it harder against data manipulation, and all nodes can track the history of transactions of the distributed ledger.

This technology employed in e-governance has many benefits, such as reduction on costs and complexity, sharing trusted processes, improving discoverability of audit trails and trusted recordkeeping [48]. The DLT has been used for the past decade in many applications, such as currency (Bitcoin is the most renowned example), financial services, real state, healthcare, and has now started to gain traction on government levels.

Blockchain is a subset of a distributed ledger. By chaining time-ordered events in distributed servers in forms of immutable blocks, it effectively delivers non-repudiation aspects to its data, creating a trusted digital foundation for information storage. Any new data added to chain is replicated throughout all other databases which can vouch for the information integrity and once added, data cannot be removed. Although Blockchain is a type of DLT, not every DLT is necessarily Blockchain.

Distributed ledgers will not necessarily replace every database. They are particularly more suitable for scenarios where stored items tend to often change their status; or when there are different systems and applications dispersed on a large network [38]. The same government might have multiple databases, designed for particular functionalities, which may or may not be decentralised.

DLT can be divided into two distinct classes: permission-less or permissioned chains. The first accept new nodes to participate in the network without vetting from an administrator. The latter, which is the case for the Estonian DLT, requires approval of a centralised unit, which reduces many risks known to the platform. Permissioned private DLTs have lower validation costs, shorter validation times, reduced risk of attacks and better privacy [19], creating an ideal scenario for a government-backed technology to store and exchange data.

It is worth mentioning there has been some debate over the exact nomenclature of the distributed ledger technology used by Estonia. Some of the official government websites state they use Blockchain as a solution to store data, but the Estonian system actually predates the *Bitcoin Blockchain* introduced in 2008. Jeffries [28] mentioned that Estonia's CIO, Siim Sikkut, would have confirmed in a Blockchain conference that the Estonian solution is a distributed ledger rather than Blockchain, even though the media generally uses these terms interchangeably. DLTs do not need proof of work and do not necessarily need data structured in blocks; the actual common factor between the X-Tee and Blockchain is that both use cryptographic hash functions for linking data items on a decentralised distributed ledger [16].

3.5 Smart Contracts

Gatteschi et al. [19] define smart contracts as small codes programmed to autonomously behave in a certain way whenever specific conditions are met. Smart contracts intend to facilitate, verify and enforce a contract through automation. The concept of smart contracts was introduced in 1994 by Nick Szabo, a legal scholar and cryptographer [59], but it started to get relevance when Blockchain gained popularity.

In a decentralised environment, elements in the network interact in a distributed manner, removing the need for a trusted third-party, and therefore, allowing smart contracts to thrive. That means cheaper, faster, reliable and more efficient transactions [19].

However, since the conditions to trigger its self-execution are based on information from the Blockchain, they depend on data input into the distributed ledger. That data-feeding services to the chain are commonly referred to as *oracles*, a piece of software or hardware that enables information to be sent from and to the Blockchain. Although oracles are vital for the execution of smart contracts, wrong information fed into the chain could generate an undesired automatic response. Another potential issue is the fact the code for smart contracts protocol cannot afford to be flawed, as possible vulnerabilities could be exploited and automatically accepted as legitimate.

Gatteschi et al. [19] introduced a more complex application of smart contracts and oracles, presenting the idea of Decentralised Autonomous Organisations (DAO), another computer code where smart contracts are connected together and function as a government mechanism. However, as with any new technology, in 2016 hackers have proved that a function in the DAO could be exploited and used it to steal *Ethereum* (another renowned cryptocurrency). The attack illustrates how adopting a new technology that automates services can become a problem. That also opens the path for discussion in future conflicts between regulations, whether it has been established by the law or coded in smart contracts.

Despite the risks, smart contracts are fundamental in the implementation of a Blockchain-like network, as they increase the processing speed for transactions, while at the same time reduce the costs and mistakes normally related to the manual processing of data. Transparency to investigate the contractual conditions of the smart contract is also an advantage, generating trust. The importance of discussing smart contracts in the context of e-governance is merely to take into consideration the emphasised threats and understand regulatory issues this solution may present.

3.6 Risks and Resilience

The potential solutions to secure a database are not limited to the distributed ledger, but its recent widespread popularity indicates a trend. As the DLT protocol has become part of many organisations' core infrastructure, it is vital to establish the possible risks associated with decentralised databases. The adoption and operation of the DLT is strongly dependent on the appropriate management of the risks related to it, and it is being viewed as the foundational technology for the future of risk management [62]. Amongst the potential risks and approaches to secure government information systems:

- **Strategic risk**
  Each government needs to assess whether it is the right time to adopt new technologies or if the administration should wait for it to mature, given the circumstances at where the country is and its technological strategy [11]. Moreover, using the X-tee, Estonia has created a full software and service development environment intertwined with its DLT, but not all countries would have this option. The usage of a Blockchain-like platform could create limitations on the services delivered by the government if this integration is not natively provided.
- **Infrastructure and business continuity**
  Not only DLT requires a fairly good infrastructure to maintain the network and nodes, but if it becomes the core of identity management and authentication, the service must work seamlessly and cannot afford to be interrupted at any time. In order to maintain availability, redundant infrastructure and a fast-response business continuity plan are paramount.

- **Scalability and performance**
  The DLT requires a strong infrastructure to be able to handle high-volume applications, and systems with high throughput rates or high transaction volumes tend to experience relative slowness in terms of system latency periods (Deutsche [12]). Compared to centralised databases, public DLTs are less efficient, more difficult to scale up and have less flexibility [45]. Even in private DLTS, depending on the consensus mechanism and technical configurations chosen, scalability could be affected if supporting infrastructure does not match the needs of the government.
- **Operational risk**
  Technical barriers and lack of qualified professionals pose a risk. Dealing with legacy systems, maintaining interoperability, speed, and scalability are some of the main challenges in this aspect. Policies and procedures must be adjusted and IT personnel ought to be trained accordingly to implement and support the system. Also, safe-keeping the credentials that allow users to input data into the DLT is paramount. Endpoints (where humans meet the Blockchain) represent a vulnerability, as credentials needed for that data input could be compromised affecting integrity of the entire chain.
- **Data integrity and availability**
  Integrity is a major concern when dealing with databases. The majority of the existing government systems is centralised and replicated to multiple database servers [15]. The problem with this centralised architecture is that it becomes a target for cyber-attacks which, and if successfully performed, could compromise its integrity. For instance, hackers could launch a denial-of-service (DoS) attack which could make the network unavailable or upload a malware that could impact integrity and confidentiality, corrupting or disclosing sensitive information. Therefore, a potential lack of synchronisation amongst databases would simply mean data is conflicting, and therefore becoming untrustworthy.

  By using a decentralised system, there is no single point of failure, boosting security of assets stored in the blocks. If one node is compromised by a malicious actor, other copies of the distributed ledger containing the original data would be used to restore the node [12].

  In the Estonian case, not only the government decided to use a decentralised system, but it has also implemented in 2017 the concept of data-embassy, a mixture of private and public redundancy cloud storage services elsewhere (the pilot project was set in Luxembourg), that allows the government to still run its services in possible major incidents [47].
- **Data confidentiality**
  Privacy is a critical topic in governance and digitising services must be accompanied by comprehensive policies [2]. This will ensure correct access controls will be used, guaranteeing that collected data will be used only for legitimate purposes.

  Blockchain networks have on its design some important privacy issues, as each node that processes transactions inherently has access to the data itself, although in an encrypted state. Even though the stored transactions are encrypted by default, metadata might be available to other network members, which could expose data on the activity of any public address in the Blockchain framework. This metadata, alongside statistical analysis could show information from the encrypted data, paving the way for pattern recognition [64].

- **Consensus protocol**
  Data is stored in the distributed ledger when there is consensus among participant nodes through a cryptographic protocol. Failure to achieve this accord would result in the abortion of the data exchange. This is the base for the still hypothetical *51% attack*, where a malicious actor could have control over the majority of the nodes in the network, resulting in the interference of the process of recording and manipulation of the outcome of the new blocks. In a permissioned system, such as the Estonian case, nodes are controlled by the government and this risk is mitigated, but maintaining control of the trusted network is fundamental at all times.
- **Standards and regulation**
  There is not a universally accepted standardised framework defined for distributed ledgers yet [8], and governments will use whichever technology seems fit. As long as the database remains within a centralised administration, this is manageable and it abides by the laws of that sovereign state. However, once DLTs interconnect and start sharing data, regulations need to be imposed, to specify enforcement for smart contracts and lawful standards. The legal liability largely still stays unclear for improper or malicious use of a smart contracts in the network.
- **Trust**
  Prior to implementing an electronic form of governance, a nation should consider its culture, metabolism and governance to understand population's trust in technology. Citizens' trust is a factor in new technologies' adoption rate, and chances of success in their application. Individuals that already trust their government tend to have it reinforced through electronic interaction provided by e-government systems. However, the opposite is also true; distrustful citizens will not increase their trust, regardless of how they interact [49]. It is up to the government to work on these trust levels, and gain private sector support prior to committing to switch governance to an electronic form of it.

  Gaining trust in the digital governance has suffered multiple setbacks in the past decade, and every time it happens, the privacy element is questioned and fear of hyper-surveillance strengthens. Edward Snowden's National Security Agency (NSA) leak in 2013, for example, illustrates how users can come to distrust the internet for communication, sharing and storage, as citizens became aware that the government could be spying on them at any moment. Society has become somewhat divisive about who to trust, demanding platforms that could reassure preservation of users' privacy and security.
- **Blockchain immutability and General Data Protection Regulation**
  The fact that in a Blockchain solution the written blocks cannot be easily altered makes it a strong candidate to maintain integrity in public databases. That immutability has been recently called into question in Europe, due to the

rule established in the General Data Protection Regulation (GDPR), known as *right to be forgotten* [50]. That regulation challenges the whole concept of the Blockchain, forcing it to be editable so that attend requests to modify, redact or delete information as requested. In order to address that risk, technical workarounds and advanced cryptographic techniques have been proposed to resolve this issue, but it still constitutes an obstacle to the adoption of Blockchain-like networks. This risk must be taken into consideration not only from the technological aspect, but from the legal one as well, as financial fines for privacy breaches are escalating sharply.

- **Cyber war**
  Governance of information systems that control major aspects of modern society has gradually been digitised. When public transportation, power grids, financial services, authentication/authorisation systems become online, the results of having a security breach could be devastating. In that scenario, the risk of having state or governmental-sanctioned actors intervening on other sovereign countries' digital infrastructure could constitute an act of war.

  Despite being considered over-hyped by some [37], the risk of cyber war has been built up over the past decades. One of the incidents that has made this clear was Stuxnet, uncovered in 2010, a highly sophisticated computer worm targeting Supervisory Control and Data Acquisition (SCADA) systems to damage the Iranian nuclear programme. Even though no country has admitted being responsible for its deployment, due to its sophistication level, it is believed to been designed by a state actor, allegedly often linked to a joint United States/Israel operation [37].

  While in one side there are signs of government action behind these cyber-attacks, on the other hand the world has seen multilateral intelligence agreements, such as the Five Eyes or the Club of Berne, to promote cooperation and exchange information amongst allied countries. Though some of these alliances precede the computer era, intelligence-gathering teams eventually evolved into a sophisticated cyber security network, suggesting powerful nations have been preparing themselves for digital warfare for a long time. The main risk is that insufficient security combined with ever-increasing online government information systems could lead to state-backed actions to destabilise enemy's cyber environments.

3.7 Framework Regulation

Besides understanding the risks, attention to regulatory compliance is also decisive for the establishment of e-governance. An unsuccessful attempt to foresee the fast growth of the technology can lead to governance failure [61], especially when dealing with citizens' sensitive data. By accurately anticipating legal problems due to technological innovations, the government can securely adjust its regulation to cope with novelty.

In the Estonian case, the e-government legal framework is robustly regulated and has adopted measures to provide privacy and security of citizens' personal data. Together, these legislations (such as the Personal Data Protection Act [56], Public Information Act [54], Population Register Act [55], Electronic Communication Act [57]) deliver institutional background for the country's e-governance, establishing a connection between technology and regulatory acts, paving the way for a functional digital government. These protocols define the processes by which citizens, business, and institutions may request and receive access to data stored and exchanged by government databases [73]. The Estonian government vouches for the digital signature, which is equivalent to a hand-written signature. This dramatically increases the level of responsibility on the security aspect of the electronic signatures, as Estonian authorities are compelled to accept digitally signed documents. At the same time, it increases population trust on the system, which becomes more reliant on the overarching digital governance, including the e-ID, the e-residency and the X-Tee.

3.8 E-Governance Implementation Worldwide

Governments worldwide are testing technologies such as distributed ledgers to implement functionalities such as electronic ID, money tracing, electronic voting, and records of all types (passports, criminal records, tax records, legal enforcement) [44]. Not all countries use DLT to support this transition, though, and the ones that do, not necessarily use it in every single database. There are many individual academic researches available presenting e-governance adoption rates based on a specific nation's requirements, potential, resources and appetite.

Though the concept of governance is broad and somewhat difficult to measure, the United Nations' e-government development index (EGDI) [71] tries to compare digital excellence amongst member states. This index includes infrastructure and educational levels that enable citizens to have access to online services, telecommunication connectivity and human capacity (Fig. 4).

Estonia's EGDI ranked as 16th (out of 193 UN members) in 2018, and the relevance of its digital information adoption cannot be understated. About 95% of the tax declarations were filed electronically in 2017, 99% of medical prescriptions were in digital format in 2018, and 30% of votes in elections are cast over the internet [66], among many other examples.

Solvak et al.'s [66] study about Estonia also showed that e-government adoption rate grows linearly with a high peak acceptance rate, concluding that public and private initiative are decisive to encourage the population to adopt it. The number of digital authentications and digital signatures shown on Fig. 5 [73] shows the slow uptake in the first 5 years, with a progressive rise on population acceptance over time.

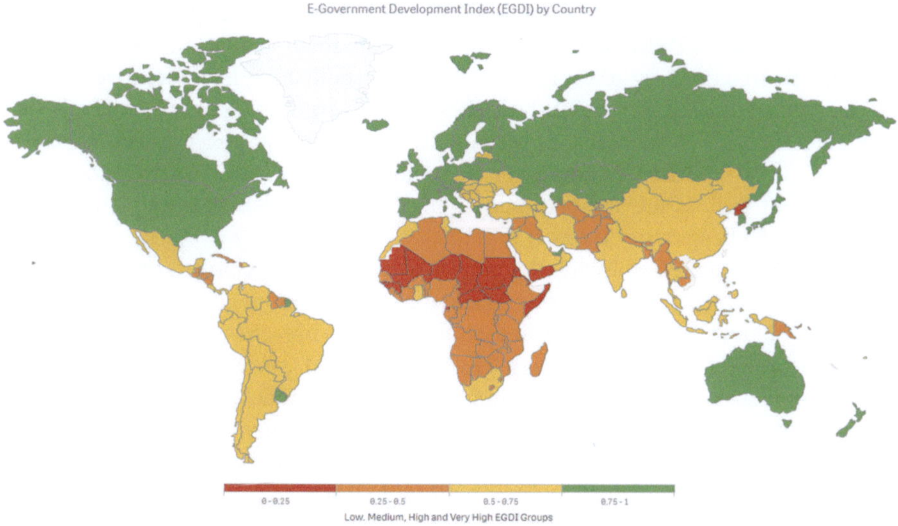

Fig. 4 E-government development index by country in 2018 [71]

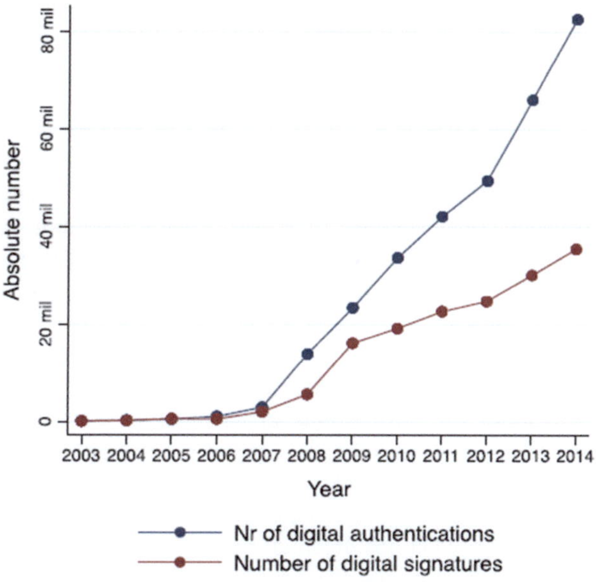

Fig. 5 Growth of digital authentications and signatures [73]

3.9 *Cyber Security Role in e-Governance*

Using technology to store and exchange citizens personal data require a highly developed cyber security state. Information security properties, namely confidentiality, integrity and availability, underpin services as authentication, authorisation, accountability and reliability, which combined, define the foundation of a secure digital governance. Pursuing cyber resilience is the main vector to address security risks whilst transitioning to digital governance; establishing how effective a country's cyber security is, generally shows how mature e-governance has reached.

Using indices to measure governments' commitment also raises awareness towards cyber security. Examples of these initiatives, with up-to-date countries' development indices are the National Cyber Security Index (NCSI) [40] and the ITU Global Cybersecurity Index (GCI) [26].

The NCSI was created by the Estonian Government to quantify a country's level of preparedness to prevent cyber threats and handle cyber incidents, such as denial-of-service attacks, data integrity and confidentiality breaches. Data is voluntarily provided by over 100 participating governments and evidences are uploaded to the NCSI website, which is validated, analysed and indexed according to their methodology. According to NCSI's index, Estonia currently ranks in second place in the charts; an impressive achievement given the history of the country. That ranking means cyber security is dealt with in a steady and organised method, and that investing in a secure platform has paid off throughout the years. Top countries include Czech Republic, Estonia, Spain, Lithuania and France (Fig. 6).

The GCI is maintained by the International Telecommunication Union (ITU), a United Nations specialised agency for information and communication technologies (ICTs). ITU was founded in 1865 to facilitate international connectivity, and provides works such as allocating global radio spectrum, satellite orbits, and developing technical standards to ensure networks interconnect (Fig. 7).

The GCI was created for the same reasons as the NCSI, promoting awareness in cyber security. The governmental aspects analysed by GCI are divided in the categories before being aggregated into a single score: legal measures, technical measures, organisational measures, capacity building and cooperation. The index is then, annually publicised and made available online so governments evaluate their level of commitment to stopping cybercrimes (Fig. 8).

The GCI also produces an ICT Development Index (IDI) since 2009, which combines 11 indicators into one benchmark score and it is used to compare ICT development between different countries. The publication presents a quantitative analysis of the information society and shows emerging trends, which makes it easier to understand the connection between e-government levels, IDI and cyber security (Fig. 9).

Not coincidently, Estonia once again stands out and ranks close to other major cyber security nations. In an effort to continually improve this and demonstrate government commitment, the country has founded the Estonian Information Security Association (EISA) in 2018, which role is to improve cross-sectorial cooperation among government, academia and the private sector (Fig. 10).

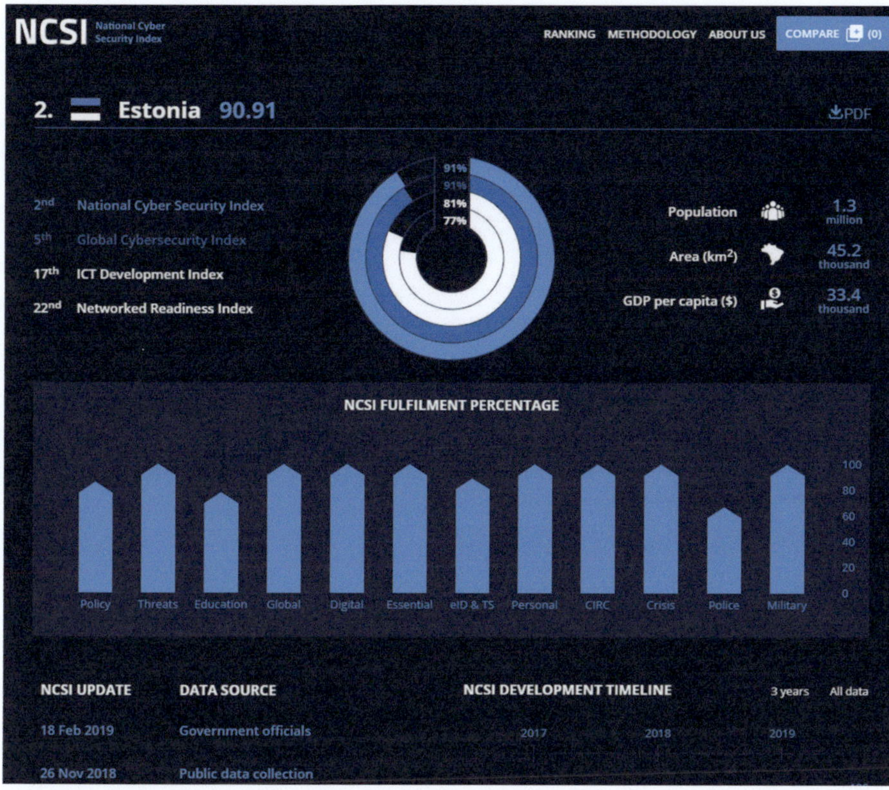

Fig. 6 Estonia's National Cyber Security Index

3.10 Estonian Security and Privacy Issues

The Estonian Information System Authority publishes yearly the Annual Cyber Security Assessment, which summarises breaches, malware campaigns, denial-of-service attacks, data leaks and vulnerabilities found throughout the year [18]. Even though the publication shows a sharp rise in cyber incidents in 2018 (nearly twice as previous years), the number of critical incidents were reduced (Fig. 11).

As this report indicates, the highest number of security incidents was related to breach of integrity, produced by malware campaigns aiming to create botnets through vulnerabilities in exploitable devices. Ransomware and phishing e-mails were also very present during that period of time and other minor attacks contributed to the total number of incidents. It is important to state that the increased user awareness and improved detecting capabilities might have also contributed to raising these numbers, which means not necessarily the number of attacks have increased, as the ability to identify the existing ones has been amplified as well. Perhaps the fact that Estonia has become a high-profile country in the cyber security field worldwide

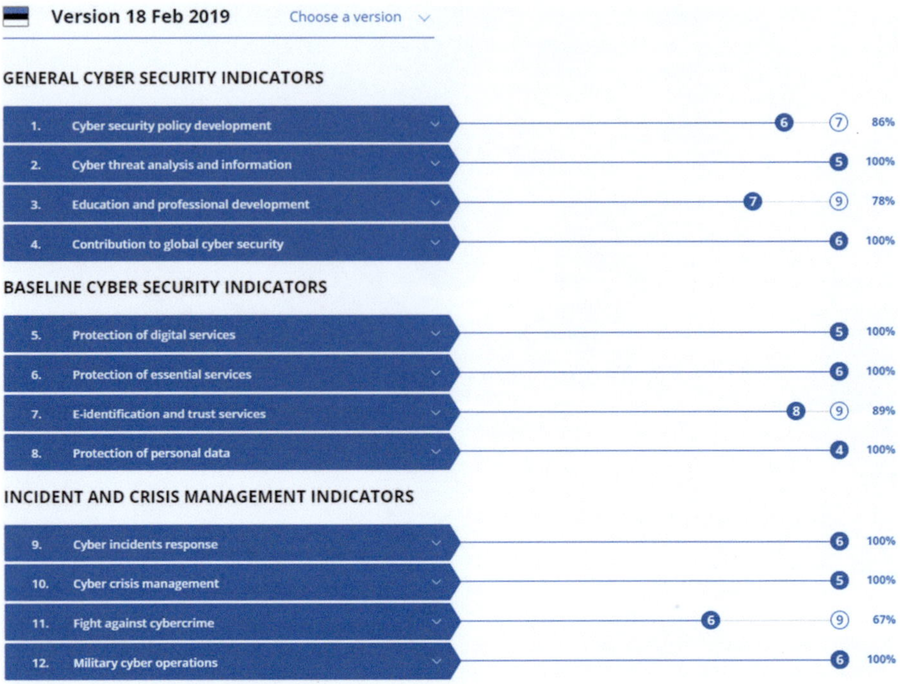

Fig. 7 Estonia's NCSI indicators

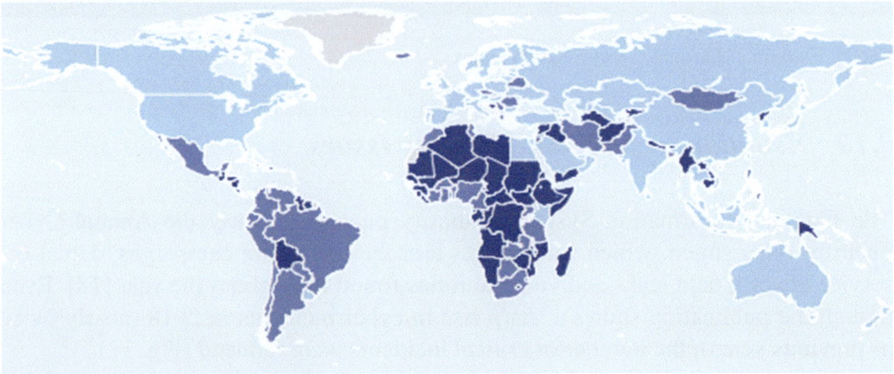

Fig. 8 Heat map showing geographical commitment [26]

might also have contributed with the surge on numbers of incidents, as hackers might be more motivated to test their skills against top secured environments (Fig. 12).

Rank	Member States	GCI Score	Legal	Technical	Organizational	Capacity building	Cooperation
1	United Kingdom	0.931	0.200	0.191	0.200	0.189	0.151
2	United States of America	0.926	0.200	0.184	0.200	0.191	0.151
3	France	0.918	0.200	0.193	0.200	0.186	0.139
4	Lithuania	0.908	0.200	0.168	0.200	0.185	0.155
5	Estonia	0.905	0.200	0.195	0.186	0.170	0.153
6	Singapore	0.898	0.200	0.186	0.192	0.195	0.125
7	Spain	0.896	0.200	0.180	0.200	0.168	0.148
8	Malaysia	0.893	0.179	0.196	0.200	0.198	0.120
9	Norway	0.892	0.191	0.196	0.177	0.185	0.143
10	Canada	0.892	0.195	0.189	0.200	0.172	0.137
11	Australia	0.890	0.200	0.174	0.200	0.176	0.139

Fig. 9 GCI most committed countries globally in 2018 [26]

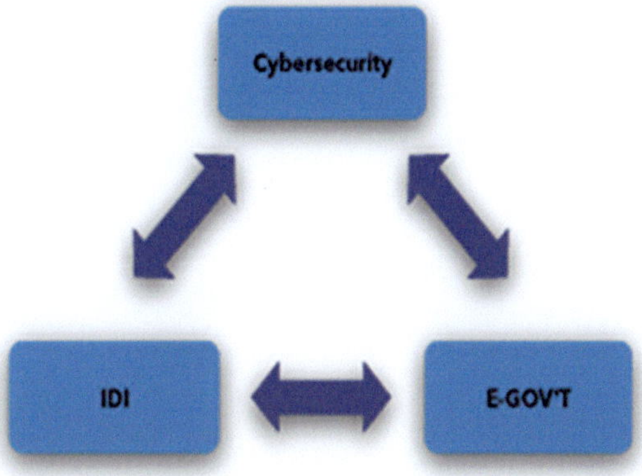

Fig. 10 Linking cybersecurity to development and e-governance [26]

Regardless of the reason for the rise on figures, the numbers below from the official Estonian Government annual assessment [17] help putting this number into perspective:

Still according to the same report, in 2018 there were no major international campaigns (such as Non-Petya and WannaCry malwares from the previous years) and the pillars of Estonia's digital environment (the X-tee and the ID card) sustained no further issues. The report allegedly credited it to the government

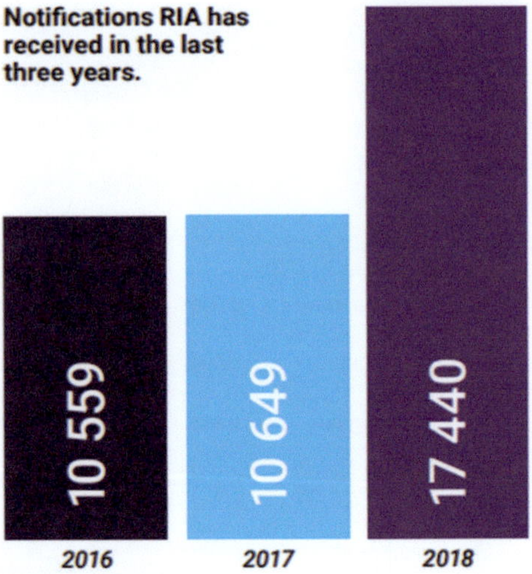

Fig. 11 Estonian security incidents 2016–2018 [18]

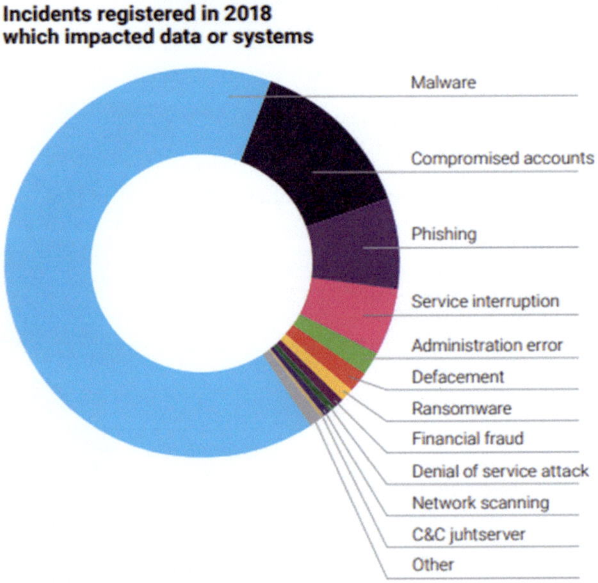

Fig. 12 Categories for top security incidents [18]

commitment, providing additional funding to the ICT sector and complying with legal and regulatory frameworks [18]. The Estonian Government understands that risk management is cost-efficient, therefore heavily investing in prevention and incident management has seemingly paid off. Furthermore, the government works tirelessly to raise security awareness and offer citizens and organisations solutions for authentication, digital signing and exchange of data [18].

Since the implementation of the X-tee, Estonia has sustained two particular incidents that deserve a more cautious examination; The first one, in 2007, a

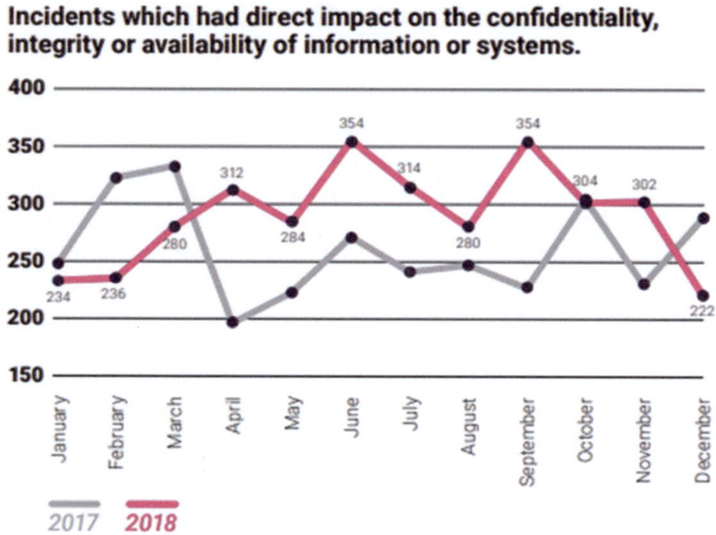

Fig. 13 Meaningful security incidents 2017–2018 [18]

distributed denial-of-service (DDoS) attack targeting the Estonian parliament, banks, ministries, newspapers and broadcasters. Even though the occurrence has disrupted major ICT infrastructure, shutting down servers all over Estonia, it did not compromise data storage. The nature of the decentralised database, backed up by the distributed ledger technology, makes it very hard for a successful data destructive hit. It is suspected that Russia launched this attack even though the country has been denied any wrongdoing [68]. The incident did coincide with the moment Estonia removed the *Soviet Bronze Soldier* monument from Tallinn, which sparked outrage from the Russian government. The episode was labelled by the media as *Cyber War* [68] and was highly coordinated and sophisticated. A similar event happened in Georgia in 2008, and inevitably makes society wonder how secure these nations are when core ICT infrastructure is interconnected with the public network (Fig. 13).

The second issue was observed in 2017 is related the e-ID cards, one of the foundations of the Estonian digital security. A new chip was introduced in October 2014, which had a private key-related vulnerability that could allow personal identification and digital signing without the physical card and relevant PIN codes. The vulnerability was only found in August 2017 and by then, over 760,000 vulnerable e-ID cards had been issued and required immediate replacement.

New vulnerabilities will always be found in innovative technologies. The way the government addresses the issue, however, dictates how the citizens' trust on e-government shifts – or not. Despite the seriousness and impact of those attacks, the Estonian Government used these to improve its defences and learn from mistakes made. Also, none of the perpetrated attacks affected information on its distributed ledger or apparently reduced trust on the country's security platform [18].

4 Proposed Guidelines Based on Estonian Success

Based on the secure implementation of E-governance in Estonia there is a need to facilitate countries to have a common platform both at national and international levels. By reducing the number of different and independent systems, interconnectivity can be streamlined, more easily manageable and cost-efficient. The recommendation would be creating an open-source framework, based on distributed ledger technology, with e-governance-friendly built-in software development tools. Software, network, apps should be built within the framework instead of having these secured independently.

Government commitment alongside strong private sector investment were essential to the Estonian case (and other successful digital governments) and are paramount for any nation pursuing a high level of governance. E-governance implementation relies on a delicate balance and commitment of all participants involved in this ecosystem.

As well as technology, citizen's awareness needs to be taken into consideration. The perception of administrative trustworthiness and confidence in technology are key elements on e-government and therefore, e-governance [20]. By being part of the transition, the population understands its role, embraces changes and becomes more mindful of how critical cyber security is.

Continuous improvement and investment in technology and incident response. It has become evident that the financial impact of attacks such as denial-of-service, malware, phishing among others justify the investment on up-to-date ICTs, training, policies, regulatory compliance and incident response team.

Sharing of experiences, problems and solutions on implementation and technical levels increase cyber resilience. Sharing cyber security knowledge is commonly perceived as an effective measure to help governments and organisations manage security incidents, reduce uncertainty and sustain a collaborative environment [34].

Employment of e-governance and cyber security indices to keep up with high standards and to be able to compare and understand what measures have led other countries to accomplish a higher level of preparedness.

Finally, understand e-governance takes time. The Estonian adoption took several years to become evidently efficient, after major adoption of public and private institutions. Therefore, other countries should not expect immediate results; most countries referred in this project accomplished quality e-governance after a decade of full commitment to implementing it.

4.1 Implementation of the Guidelines

The research and development of the guidelines intends to provide a path to follow, based on existing evidences of the case study and similar effective governments. The accomplishment of the implementation of these recommendations is subjected to the consideration of the existing barriers and the strategies to overcome these issues. The guidelines are not a systematically developed statement to assist a government to implement digital governance, but an overall compilation of the steps successful countries used to achieve it. There should be a coherent research programme for each government scenario, ensuring that the approach towards e-governance is appropriately tailored to each situation, and is cost-effective.

It is important to stress that e-government is not the answer to every nation's public services issues. Also, as context is vital to determine the steps to take, there is no singular strategy universally recommended.

5 Implementation of Cyber Situational Awareness in E-Government

Although secure implementation of e-government is a crucial factor, having a cyber defence system which can protect the components of e-government is highly needed. The concept of Cyber Situational Awareness (CSA) has come to the attention of cyber security experts as a way to protect against different and dynamic cyber threats [51]. According to Antonik [4] situational awareness is often referred to different elements in an environment which can contribute to understand current events and to predict future incidents. If an e-government wants to adopt a Cyber Situational Awareness program as its cyber defence system, 3 main elements can be protected, and they are as follows:

1. Financial sectors: Financial sectors such as banks and stock markets are always hit by different cyber-attacks. In order to mitigate the risk of cyber breaches in financial organization cyber situational awareness can be considered as general strategy. Over the past few years, cyber-attacks have become more complicated and sophisticated and damages caused by them can have significant negative influence on business and economy in larger scale. Security professionals can address these issues by having a program for improving CSA as a wider strategy for mitigation of cyber threats. By getting more details about these threats not only they can increase the level of security inside their institution, but also, they can feed their clients and customers with cyber security instruction. CSA in financial sector can improve knowledge of managers about ongoing security and help them to prioritize the security needs by identification of needs, vulnerabilities, and weaknesses. IT experts can use regular security reports including merging threats against financial sectors and by analysing them conclude a wider and more extensive CSA program to increase level of security in both institution side and client side. For instance in bank side by allocation of suitable and effective IT resources they can prevent or decrease the damage of cyber-attacks and in client side, also they can improve the knowledge of customers about cyber threats such as spear phishing and different malware in order to decrease the risk and the probability of cyber-attacks.

2. Health care Industry: health care institution such as hospitals and clinics also need CSA program as a security strategy. For instance, their database can contain sensitive and classified information about their customers and patients and that makes the more responsible and accountable to adopt effective CSA improvement program. CSA programs and tools can come together and help healthcare industries to improve the level of security and protect their patients against any sensitive information leakage. An extensive improving program applied to CSA should have following competences [6]:
   (a) Identification of vulnerabilities and weaknesses within the computer systems in health care institution.
   (b) Monitoring merging cyber threats
   (c) Classification of merging cyber threats based on their impact on the health care industry.
   (d) Updating security countermeasures regularly in order to be prepared for new and unknown cyber attacks
   (e) Holding educational program for staff in the health care industry about cyber security issues.
3. Critical Infrastructure: critical infrastructures are the most important and sensitive to governments and nations and they play a crucial role in e-governance. A cyber-attack to CI can damage heavily and lead to unprecedented loss to victims. US army highlights CSA in critical infrastructure therefore it has been concluded that CSA program should cover following capabilities in protecting critical infrastructure [21]:
   (a) Monitoring cyber activities: this process should be done continuously, and intelligence should be gathered in terms of cyber activities to address new issues timely.
   (b) Identification of vulnerabilities: this task should be continuously carried out by the CSA program in order to resolve and patch security bugs.
   (c) Profiling cyber adversary behaviour: an extensive CSA should use past intelligence about past cyber-attacks in order to profile cyber attackers for adopting effective security countermeasures against known enemies such as state sponsored cyber attackers [52].

6 Conclusion

This study investigated the information security aspect of e-governance implementation in developed countries, using Estonia as a study case. This paper established how this nation successfully transitioned to a nearly-fully digital society, shortly after becoming independent from the Soviet sphere of influence. This success involved employing its own security framework, backed up by public sector commitment and private sector participation. The X-Tee uses a distributed ledger to integrate the data layers and provide security to data exchange, and it has effectively been used by other countries as well – with a similar rate of success. It is difficult to categorically affirm that this would be a universal solution, but it certainly copes with some database security risks and provides practical data interoperability.

This study showed that Estonia decided to work close to its citizens in order to gain public trust, whether promoting security awareness or strongly incentivising the use of e-government tools. By improving the perception of the administrative trustworthiness, citizens' adoption rate has increased, which strengthened government integration with residents even further. A few major security breaches suffered by Estonia in the past decades have been covered in this study, which seemingly have not diminished the level of trust on the government's system. The Estonian government seems to treat the security issues with a high level of transparency, which triggers confidence from the population instead of damaging trust previously grown.

This study compared Estonia's current performance in e-governance and cyber security rankings with other nations, establishing a positive correlation between those two elements. This research was able to identify challenges and risks involved in governance transition, and understanding these difficulties allowed guidelines to be proposed for other countries to follow. However, even though this set of guidelines to mitigate risk has been recommended, each country needs to evaluate its own strategy, and adjust its electronic governance transition according to its requirements and resources.

6.1 Discussion and Evaluation of the Results

The clear results obtained by Estonia, though, would not necessarily be exactly the same, even if the equivalent methods were applied to other countries, in other circumstances. Nevertheless, the nation's approach seems to match e-governance and cyber security levels achieved by other developed countries; therefore, suggests to indicate the right path to follow.

By choosing a permissioned distributed ledger technology, the government keeps a controlled, decentralised database which is more reliable and encourages citizens' trust. The DLT shifts the trust factor of the data storage from a human archetypal to an algorithm [11]. In this manner, the DLT has the potential to improve e-governance by disintermediating processes, improving efficiency and transparency, which directly translates to more cost-efficient, faster and auditable transactions. The benefits are undisputable, but so are the inherent risks. Technology, in that sense, is not correcting problems with the old model, but replacing those with newer risks that should be addressed by an effective risk management programme and controls framework.

6.2 Future Work

Even though there has been a gradual increase on the number of research performed about e-governance for over a decade, the research on this field is still relatively scarce, and mostly done by American and European universities [58]. There is demand by governments for solutions to be implemented but no standards or frameworks globally defined.

Due to the fact that e-governance and cyber security and reasonably novel concepts, there is room for additional research to be conducted alongside governments, main providers of relevant primary data. Further analysis could be conducted and compared worldwide to determine a cause-and-consequence relationship whilst adopting new technologies, such as the distributed ledger or a data exchange layer, both analysed in this study.

Also, this research focused exclusively in developed countries. It is understandable that developing nations might have different complications ensuring security whilst implementing e-governance. Additional research could try to cross-reference recommendations on projects and identify which principles could be employed on both cases as a guideline for e-governance adoption.

References

1. Alketbi A, Nasir Q, Abu Talib M (2018) Blockchain for government services -use cases. In: Security benefits and challenges. Learning and technology conference. IEEE, Piscataway, pp 112–119
2. Alshehri M, Drew S (2011) E-government principles: implementation, advantages and challenges. Int J Electron Bus 9(3):255–270
3. Anderson R (2008) Security engineering: a guide to building dependable distributed systems. Wiley Publishing, Inc., Indianapolis
4. Antonik J (2007, October) Decision management. In: MILCOM 2007-IEEE military communications conference. IEEE, pp 1–5
5. Ayanso A, Chatterjee D, Cho D (2011) E-government readiness index: a methodology and analysis. Gov Inf Q 28(4):522–532
6. Beavers J, Pournouri S (2019) Recent cyber attacks and vulnerabilities in medical devices and healthcare institutions. In: Blockchain and clinical trial. Springer, Cham, pp 249–267
7. Bhattacharya S, Goswami J (2011) Study of E-governance: the attractive way to reach the citizens. In: 2nd national conference – computing, communication and sensor network
8. Bizarro P, Mankowski R, Mankowski H (2018) Blockchain technology: benefits, risks and the future. Intern Audit 33(4):12–16
9. Cong L (2018) Navigating the next wave of blockchain innovation: smart contracts. MIT Sloan management review
10. Dawes S (2009) Governance in the digital age: a research and action framework for an uncertain future. Gov Inf Q 26:257–264
11. Deloitte (2017) Blockchain risk management – risk functions need to play an active role in shaping blockchain strategy. Fonte: https://www2.deloitte.com/content/dam/Deloitte/us/Documents/financial-services/us-fsi-blockchain-risk-management.pdf
12. Deutsche Bundesbank (2017) Distributed ledger technologies in payments and securities settlement: potential and risks. Monthly Report of the Deutsche Bundesbank, 35–49

13. Drechsler W (2018) E-Estonia as the β-version. JeDEM – eJ eDemocr Open Govern 10:1–22
14. E-estonia (2018) E-governance. Fonte: https://e-estonia.com/solutions/e-governance/
15. Elisa N, Yang L, Chao F, Cao Y (2018) A framework of blockchain-based secure and privacy-preserving E-government system. Wirel Netw:1–11
16. Estonia Government (2018) X-Road not to be confused with blockchain. Fonte: https://e-estonia.com/why-x-road-is-not-blockchain/
17. Estonian Government (2019) Overview. Estonia, Fonte. https://estonia.ee/overview/
18. Estonian Information System Authority (2019) Annual cyber security assessment 2019. Fonte: https://www.ria.ee/sites/default/files/content-editors/kuberturve/ktt_aastaraport_eng_web.pdf
19. Gatteschi V, Lamberti F, Demartini C, Pranteda C, Santamaria V (2018) Blockchain and smart contracts for insurance: is the technology mature enough? Future Internet 10(2):20
20. Golesca S (2009) Understanding Trust in e-Government. Economics of engineering decisions
21. Gould J (2015) US Army seeks leap-ahead cyber defense tech. Retrieved September 20, 2019, from Defense News website: https://www.defensenews.com/2015/07/01/us-army-seeks-leap-ahead-cyber-defense-tech/
22. Government of Estonia (2018) Data exchange layer X-tee. Information System Authority, Fonte. https://www.ria.ee/en/state-information-system/x-tee.html
23. Hartmann K, Steup C (2015) On the security of international data exchange services for e-governance systems. Datenschutz und Datensicherheit – DuD 39:472–476
24. Hoberman S (2018) How Blockchain changes the rules of the game. Technics Publications, Basking Ridge
25. International Monetary Fund (2019) Fonte: IMF DataMapper: https://www.imf.org/external/datamapper/NGDPDPC@WEO/OEMDC/ADVEC/WEOWORLD/EST
26. International Telecommunications Union (2019) Fonte: Global Cybersecurity Index: https://www.itu.int/en/ITU-D/Cybersecurity/Pages/global-cybersecurity-index.aspx
27. Jaffe E (2016) How Estonia became a global model for e-government. Medium, Fonte. https://medium.com/sidewalk-talk/how-estonia-became-a-global-model-for-e-government-c12e5002d818
28. Jeffries A (2018) Blockchain is meaningless. The Verge, Fonte. https://www.theverge.com/2018/3/7/17091766/blockchain-bitcoin-ethereum-cryptocurrency-meaning
29. Joseph S, Advic A (2016) Where do the Nordic Nations' strategies take e-government? Electron J E-Gov 14(1):3–17
30. Keenan TP (2017) Alice in blockchains: surprising security pitfalls in PoW and PoS blockchain systems. In: 15th annual conference on privacy, security and trust (PST). IEEE, Calgary
31. Konashevych O, Poblet M (2018) Is blockchain hashing an effective method for electronic governance? In: 31st international conference on legal knowledge and information systems (JURIX 2018). Groningen
32. Korjus K (2018) E-residency is 4 years old so here's 4 surprising facts about the programme. Medium, Fonte. https://medium.com/e-residency-blog/e-residency-is-4-years-old-so-heres-4-surprising-facts-about-the-programme-c3a9d64c988d
33. Larsson H, Grönlund Å (2014) Future-oriented e-Governance: the sustainability concept in eGov research, and ways forward. Gov Inf Q 31(1):137–149
34. Luiijf E, Kernkamp A (2015) Sharing cyber security information. Global Conference on CyberSpace, Fonte. https://publications.tno.nl/publication/34616508/oLyfG9/luiijf-2015-sharing.pdf
35. Mahajan N (2015) E-governance: its role, importance and challenges. Int J Curr Innov Res 1(10):237–243
36. Martens T (2010) Electronic identity management in Estonia between market and state governance. Identity Inf Soc 3(1):213–233
37. McGraw G (2013) Cyber war is inevitable (unless we build security in). J Strateg Stud 36(1):109–119
38. McLean S, Deane-Johns S (2016) Demystifying blockchain and distributed ledger technology – hype or Hero? Comput Law Rev Int 17(4):97–102

39. Mehrotra S (2018) Why are smart contracts so important. Fonte: https://medium.com/acycliclabs/why-are-smart-contracts-so-important-81883d93a0cc
40. NCSI (2019) Fonte: national cyber security index: https://ncsi.ega.ee
41. OECD (2003) The e-government imperative: main findings. Policy Brief, Fonte. http://unpan1.un.org/intradoc/groups/public/documents/APCITY/UNPAN015120.pdf
42. O'Hara K (2017) Smart contracts – dumb idea. IEEE Internet Comput 21(2):97–101
43. Ojo A, Estevez E, Janowski T (2010) Semantic interoperability architecture for governance 2.0. Inf Polity 15(1/2):105
44. Olnes S, Ubacht J, Janssen M (2017) Blockchain in government: benefits and implications of distributed ledger technology for information sharing. Gov Inf Q
45. Ølnes S, Ubacht J, Marijn J (2017) Blockchain in government: benefits and implications of distributed ledger technology for information sharing. Gov Inf Q 34(3):355–364
46. Ott A, Hanson F, Krenjova J (2018). Introducing integrated e-government in Australia. Fonte: https://www.acs.org.au/content/dam/acs/acs-publications/E-Gov%20Report.pdf
47. Oxford Analytica (2016) Estonia: E-governance model may be unique. ProQuest, Fonte. https://search-proquest-com.hallam.idm.oclc.org/docview/1831820370
48. Palfreyman J (2015) Blockchain for government? IBM, Fonte. https://www.ibm.com/blogs/insights-on-business/government/blockchain-for-government
49. Parent M, Vandebeek C, Gemino A (2005) Building citizen trust through E-government. Gov Inf Q 22(4):720–736
50. Politou E, Casino F, Alepis E, Patsakis C (2019) Blockchain mutability: challenges and proposed solutions. Fonte: https://arxiv.org/pdf/1907.07099.pdf
51. Pournouri S, Akhgar B (2015) Improving cyber situational awareness through data mining and predictive analytic techniques. In: International conference on global security, safety, and sustainability. Springer, Cham, pp 21–34
52. Pournouri S, Zargari S, Akhgar B (2018) Predicting the cyber attackers; a comparison of different classification techniques. In: Cyber criminology. Springer, Cham, pp 169–181
53. Rid T (2011) Cyber war will not take place. J Strateg Stud 35(1):5–32
54. Riigi Teataja (2000) Public Information Act. Fonte: https://www.riigiteataja.ee/en/eli/ee/Riigikogu/act/529032019012/consolide
55. Riigi Teataja (2017) Population Register Act. Fonte: https://www.riigiteataja.ee/en/eli/ee/Riigikogu/act/522032019005/consolide
56. Riigi T (2018) Personal data protection act. Riigi Teataja, Fonte. https://www.riigiteataja.ee/en/eli/ee/Riigikogu/act/523012019001/consolide
57. Riigi Teataja (2019) Electronic communications act. Fonte: https://www.riigiteataja.ee/en/eli/ee/Riigikogu/act/520032019015/consolide
58. Rodriguez M, Alcaide L, Lopez A (2010) Trends of e-government research: contextualization and research opportunities. Int J Digit Account Res 10:87–111
59. Rozario A, Vasarhelyi M (2018) Auditing with smart contracts. Int J Digit Account Res 18:1–27
60. Sadashivam T (2010) A new paradigm in governance: is it true for e-governance? J Knowl Econ 1(4):303–317
61. Sadikin M, Purwanto S (2018) The implementation of E-learning system governance to deal with user need, institution objective, and regulation compliance. Telkomnika 16(3):1332–1344
62. Santhana P (2016) Risks posed by blockchain-based business models. Deloitte, Fonte. https://www2.deloitte.com/us/en/pages/risk/articles/blockchain-security-risks.html
63. Satyabrata D, Subhendu K (2016) E-governance paradigm using cloud infrastructure: benefits and challenges. Proc Comput Sci 85:843–855
64. Schou-Zibell L, Phair N (2018) How secure is blockchain? World Economic Forum, Fonte. https://www.weforum.org/agenda/2018/04/how-secure-is-blockchain/
65. Sekhar S, Siddesh G, Kalra S, Anand S (2019) A study of use cases for smart contracts using Blockchain technology. Int J Inf Sys Soc Change (IJISSC) 10(2):15–34

66. Solvak M, Unt T, Rozgonjuk D, Võrk A, Veskimäea M, Vassil K (2019) E-governance diffusion: population level e-service adoption rates and usage patterns. Telematics Inform 36:39–54
67. Stephany F (2018). It is not only size that matters: How unique is the Estonian e-governance success story? Agenda Austria, Working Papers 15
68. Stockburger P (2016) Known unknowns: state cyber operations, cyber warfare, and the jus ad bellum. Am Univ Int Law Rev 31(4):545–591
69. Sullivan C, Burger E (2017) E-residency and blockchain. Comput Law Secur Rev: Int J Technol Law Pract 33(4):470–481
70. The World Bank (2018) GDP growth. Fonte: https://data.worldbank.org/indicator/NY.GDP.MKTP.KD.ZG?locations=EE
71. United Nations (2018) UN E-government survey. UN E-Government Knowledgebase, Fonte. https://publicadministration.un.org/egovkb/en-us/Reports/UN-E-Government-Survey-2018
72. United Nations (2019) Estonia. Fonte: http://data.un.org/en/iso/ee.html
73. Vassil K (2015) Estonian e-government: foundation, applications, outcomes. Digital Dividends
74. Young-Jin S (2011) E002Dgovernment and universal administrative information service in South Korea. Fonte: http://unpan1.un.org/intradoc/groups/public/documents/UNGC/UNPAN043625.pdf
75. Zefferer T (2015) E-government services in Europe – a comparison of seven countries. Fonte: https://www.vodafone-institut.de/wp-content/uploads/2015/09/VFI_eGovServices_EN.pdf

Insider Threat

James Bore

Abstract This chapter discusses the threat arising from within the organisation, whether from negligence, malice, or exploitation by an external party. The trusted insider is one of the greatest challenges facing organisations today. The analysis considers the balance to be struck between allowing insiders access and privileges to show trust and increase productivity, and securing that access at the cost of good will and with an increased risk of workarounds being found, placing vulnerabilities at the heart of an organisation's policies and processes. The tactics of social engineering and exploitation of human psychology to compromise or completely bypass technical and procedural security measures are considered, along with the effectiveness of training and difficulties of raising cultural awareness of security on a long term basis in a rapidly changing technological landscape.

Keywords Authentication · Autonomous devices · Impersonation · Insider threat · Smart devices · Trust · Social engineering

1 Insider Threat

A common misconception is that cyber security incidents, particularly data breaches, are the result of sophisticated attackers who have found a way through a company's defences. Even where there is good awareness of the threat that insiders can pose, an assumption is often made that it is the malicious insiders who are at fault and seek to deliberately cause harm to the organisation. Fortunately these misconceptions are quick to disprove. Data gathered by the Information Commissioner's Office (ICO) clearly shows that the overwhelming majority of data breaches occur due to mistakes made by trusted parties, whether they are

J. Bore (✉)
Independent Researcher, London, UK
e-mail: james@coffeefueled.org

H. Jahankhani et al. (eds.), *Cyber Defence in the Age of AI, Smart Societies and Augmented Humanity*, Advanced Sciences and Technologies for Security Applications, https://doi.org/10.1007/978-3-030-35746-7_19

configuration errors in systems, messages and files sent to the wrong party, or simply misplacing paperwork [15].

The ICO data discussed above only provides detail on incidents which have been both detected and reported. Given the common desire for individuals to cover up or deny mistakes, combined with the under reporting of cyber security incidents suggested by [16], it is likely that there are significantly more incidents caused by insiders which are never reported even within the organisations affected.

As the potential for interconnectivity and greater communication options opens up, so too do the opportunities for simple mistakes or omissions to cause incidents. Greater data storage, more access to sensitive information and systems, and the increase in automation with reduced manual oversight means that a simple mistake can quickly magnify from a minor incident to one with severe impacts on nations, or even internationally. Recent cloud outages caused by administrator configuration mistakes illustrate this well, a Border Gateway Protocol (BGP) misconfiguration by a Verizon customer, along with a failure of filtering by Verizon, affected several thousand networks and tens of thousands of individual IP addresses. Packet losses caused damaging issues for several major providers, including an estimated failure of 15% of traffic to the service provider Cloudflare along with impacts to Amazon and Linode. All the damage was due to a relatively simple configuration error made purely through negligence.

All of this is without considering the potential of genuinely malicious insiders, or of external attackers seeking to exploit trusted parties to use as a proxy into organisations or systems. While this is much less prevalent in terms of causing incidents, the damage from a deliberate, targeted attack by either a malicious insider or one who is being manipulated by an external attacker is significantly greater. An e-mail sent to the wrong recipient is much less likely to reveal sensitive information to a person equipped and motivated to make use of it. With a targeted attack, the attacker knows exactly which information they are after, or which system they want to target, and will have prepared to exploit it effectively beforehand.

Even where negligence on the part of a trusted party does not directly cause an incident, it can contribute greatly to opportunities for malicious attackers. Both the well-publicised British Airways and Equifax breaches were due to errors made by trusted parties. In the case of Equifax a simple oversight in vulnerability management led to the compromise of records covering nearly half of the United States population, along with millions of records related to inhabitants of the United Kingdom and Canada. While this is generally seen as a breach by an outside attacker, the role of the insider in enabling the attack cannot be overlooked.

When we add new technologies under development to this picture, we can clearly see a path where the insider threat will continue into the future. The development of technologies such as Deepfakes [22] potentially allows attackers to exploit a trusted insider's position without any direct interaction with the victim. With the increase of remote working and newer methods of communication such as holographic communications the potential for attacks enabled by simulation and impersonation of trusted insiders becomes clear. As the Internet of Things (IoT) expands there are also questions about autonomous and smart devices which require human oversight

– an illustrative example is autonomous cars, where even the definition of the insider becomes challenging.

For discussion purposes we will define the insider as anyone or anything fulfilling a trusted role. This allows the inclusion of impersonators who make use of presumed insider status in order to exploit an organisation, those who cause or enable incidents through negligence, and implicitly trusted devices within systems. The insider threat can then be defined as any use of a permission arising from a trusted role which goes against the intended purposes.

While there are certain basic common defences for insider threats arising from negligence, malice, and impersonation each have certain attributes that demand a more focused examination for effective mitigation. There are also, as always, aspects of the insider threat which fall into grey areas such as shadow Information Technology (IT)[1] which does not fall neatly into either the negligence or malice category, but would perhaps be better thought of as misapplied good intentions leading to increased risk.

1.1 Preventing Negligence

Negligence is the most common and insidious form of insider threat, difficult to predict and nearly impossible to defend against comletely. As the poet, Alexander Pope, stated "to err is human". There is no evidence that this is going to change in the future, and indeed as the technological landscape expands our opportunities to get things wrong it is much more likely to increase. Given that, at a fundamental level, all insider threat can be reduced to the idea that people use a system or process for a purpose or in a way that is not intended, the methods we can use to treat negligence are universally applicable across any form of insider threat.

The simplest applicable defence, regardless of technology level, person, or role, is the principle of least privilege or principle of least authority. Comprehensively enforced, the least privilege principle prevents individuals or devices from being able to do anything with a system that they are not intended to be able to do. Unfortunately this is not a perfect control to prevent insider threat, whether malicious or negligent, as it is abuse of the permissions that they are given that causes the threat to manifest. However, placing strict limits on the functions available at least makes the scope of the threat definable for a particular role, and allows other appropriate controls to be put in place.

One of the major challenges of this approach is that overly strict controls, without effective training and communication, can enhance rather than reduce insider threat as individuals find alternative paths to complete work when systems are seen as

[1]The use of solutions or systems by individuals or groups within an organisation outside the oversight of a governing IT or cyber security function, more common and often seen as inevitable within large organisations.

awkward to use. Whether the controls are policy and procedure based, or enforced through technology, this will apply – causing procedures to be ignored and the rise of shadow IT. Worse yet, the deployment of shadow IT outside the oversight of departments responsible for governing security is sometimes seen as a positive by the senior leadership of an organisation for knowledge sharing and collaboration [20], as the rules and standards put in place by security teams are still occasionally seen as overly strict, and unjustified.

Even where the principle of least privilege has been thoroughly applied, assuring that no single entity has any more rights than required to fulfil their role, there is a second requirement that is necessary. Ensuring that no single person can perfect potentially abusive actions requires separation of duties between different roles – a typical example is that the person who makes a payment should require authorisation by another role. In a small organisation, the principle of least privilege may not prevent a person from authorising their own payments, and so we must apply separation of duties.

Separation of duties entails ensuring that no one person or other entity has permissions to both enact and authorise any potentially harmful action. Even in small organisations where persons may need to both authorise and enact payments this can be implemented through the mechanism of ensuring that no person can authorise their own payments. Despite being simple in concept, there are many difficulties in implication and often enforcement will become a case of monitoring for breaches rather than a technological control preventing the actions.

Either malice, or willful negligence, can cause separation of duties to be circumvented by persons working together in undesirable way. Preventing collusion is more challenging than preventing simple abuse of privileges, and any organisation must make a decision on how they wish to strike this balance given their risk appetite and the potential for abuse. Again we return to the issue of trust, as an organisation may decide to invest higher trust in its employees, allowing for greater and more efficient productivity with the increased risk of negligent or malicious activity causing greater harm.

Even where collusion may be guarded against through layers of independent approvals and independent audit, one similar area of insider threat is where conflicts of interest arise. A senior employee with a vested interest in a particular supplier may have the authority to direct purchases their way, which can cause not only issues with fraud and corruption but also weaken security controls since full due diligence and effective evaluation of services and products may be bypassed. Particularly for senior decision makers guarding against conflicts of interest is essential to reducing insider threat.

Without effective communication and cultural change, security controls applied to individuals may not only be ineffective (as they are worked around), they may actively encourage the insider threat to arise. The balance of encouraging insiders to feel trusted and responsible rather than restricted, not hindering productivity, and yet still installing effective security controls whether through organisational or technological means is challenging. As it depends almost entirely on human factors, each organisation must consider carefully the approach appropriate both to their level of accepted risk and the culture which exists. Where there is a greater sense of responsibility among individuals, greater controls can be placed without significantly increasing the chance of them either being worked around or motivating insiders to become disgruntled and act against the organisation.

Implementing the form of cultural change needed to reduce insider threat is a challenge, depending largely on the specific organisation involved as well as the wider surrounding culture or cultures. This is especially difficult for multinational organisations, as different national cultures have a significant impact on the perception of cyber security risk and the degree of responsibility individuals feel towards their employer. It is clear that the simple approach of a general cyber security awareness campaign is often ineffective [3] and a more targeted approach is required. Any such approach must take into account the different audiences within an organisation, aiming to foster feelings of ownership as well as understanding of the security issues.

The simplest way to defend against negligence as an insider threat is to apply automation wherever possible. The debate about the positives and negatives of automation aside from in terms of security is still ongoing and heated, and there is no need ot delve into it here. In terms of security and especially the insider threat, however, automation is a way to prevent simple human error in a process from causing damage. As a bonus, automation also makes the malicious insider threat, whether through an active insider or manipulation by an outsider, significantly more challenging. Fewer human entry points into a system allows for more thorough defense of those that remain.

In essence there are five broad types of overlapping human-enabled insider threat that can apply to an organisation.

Willfully ignorant There are those who are willfully ignorant, refusing to take part in training or awareness programs and thus become ready vectors for a malicious attacker, or for the negligence and collusion possibilities previously discussed. An effective audit of training and awareness programs will highlight these over the longer term, and an organisation must act to address them. This class overlaps heavily with the next in that those who have fallen foul to phishing attacks previously, and refuse training, are more likely to fall foul to repeated attacks.

Negligent As mentioned, negligence is the most common form of insider threat as well as the most challenging to resolve. While it is possible to carry out training and awareness, it is a simple fact that even those who normally exhibit highly secure behaviours can and will make mistakes. Even when reviews and approvals are put in place, it is possible that highly trained reviewers will miss things, meaning negligence can never be fully resolved as a threat. Negligence is also the most common vector for external attacks to succeed, through targeted social engineering attacks, system misconfigurations, or other mechanisms.

Collusion While this area does overlap somewhat with willful ignorance and negligence, it tends to appear more in combination with malice. Collusion is most commonly the domain of fraud and similar activities, though theft of intellectual property can also occur. Collusion may also not purely involve the insider, but may be an active deliberate cooperation with an outside actor.

Resentment A previously trustworthy employee who feels mistreated can be the most damaging threat to an organisation. One of the most dramatic examples was heard in [24], where a network administrator, Terry Childs, after a dispute about having lied during his application process, was found to have seized sole control over the FibreWAN networking infrastructure for the City and County of San Francisco, and refused to disclose the passwords required for access. While the passwords were recovered in a short time, between the 9th to 21st of July the city government had no control over the FibreWAN infrastructure and, due to the mode that the configurations were stored, were unable to risk any power disruptions which may have affected the network.

Persistent insiders While a resentful insider is likely to cause the greatest single incident of damage to an organisation, a persistent insider is a much greater long term threat. Usually they are employees seeking to use company resources for a supplemental income, and often have the access and authority to prevent discovery for a significant amount of time. It is only as the damages they cause increase over time that detection becomes more likely.

1.2 Prediction and Detection

Sanzgiri and Dasgupta [28] provides a useful taxonomy to look at the methods available to both predict and detect insider threat activities. The methods range from technological controls such as the well established Role Based Access Control (RBAC) model and deployment of decoy or honeypot[2] files, to predictive methods based on psychological and behavioural factors drawn from monitoring of employees. The addition of threat modeling techniques [14] allows for prioritisation in deploying these detection methods, as well as giving guidance on where and which controls to prevent incidents should be deployed.

Behavioural analysis, in particular anomaly detection, is currently a popular way to detect potential insider threats. Network traffic, application activities, use of permissions, working hours, and various other factors can be and are considered as inputs to a model applicable to a particular role. The baseline is normally built up over several months before alerting is switched on, meaning that with a sufficient sample size it is unlikely that consistent anomalous behaviour on the part of one individual will affect the model for a role. Of course, there are significant ethical concerns with these models and, while they are effective at detecting deviancies from normal practices, there are multiple explanations available. It is important that

[2]Files or information designed to appeal to a malicious party, often with dummy sensitive information. In more advanced implementations the files may have dynamically generated watermarking such that individuals will be linked to unique files, making investigation a simple process.

any such behavioural analysis is not only carried out with careful consideration of the ethical issues involved with observing employees, but also that any alerts which arise are handled with consideration and care since mishandling of an anomaly arising due to causes unrelated to insider threat are more likely to make such a threat manifest than prevent it.

To give examples of how behavioural analysis can be applied, and the associated difficulties, we will look at working patterns. If we take the example of an employee whose working hours are normally between 9am and 5pm, with some variance for travel disruptions or staying late to finish projects, a baseline can quickly be established. If there are then consistent anomalies outside of that baseline, let us say working until 8pm several days in a row when very few others are in the office, gives an indication that something has changed. Various causes could be put forwards, maybe a particularly complex project the employee is working on, a change in financial circumstances meaning overtime pay is required, a personal circumstance that means they want to stay longer at the office, or possibly something more sinister such as collecting sensitive information for sale or distribution. This is where the ethical concerns, and the need to treat the situation carefully, come into play. With any of the causes which are personal, trying to restrict access to information or starting any overt investigation could easily be perceived as persecution by the employee causing an otherwise trustworthy employee to become less so. Of course, the other causes could also be a trigger for a less malicious insider threat to manifest, so not doing anything to mitigate the situation is also not an option.

All of the behavioural analysis methods for detection converge in the concept of the digital twin, and it seems likely that existing technologies for insider threat detection will begin to move in this direction – if they have not already done so. Providing behaviour profiles for individuals in the form of a digital twin[3] which can be subjected to much more thorough analysis than the individual otherwise would does have significant potential for improving the prediction of insider threat, though the ethical issues in such deep profiling by employers, including the incorporation of information from social media platforms and various other sources, are difficult to consider. At the least it is likely that digital twin-based predictive technologies will be reserved, initially, for highly sensitive organisations where the requirement for security outweighs the right to privacy for individuals. This already occurs, to some degree, with the highest levels of security screening.

Authenticity analysis with machine learning algorithms assisting human participants will become increasingly vital to insider threat detection and prevention as emerging technologies, such as holographic, Augmented Reality (AR),[4] and

---

[3]A synthetic behavioural model of a person or other entity used to predict behaviour, whether predicting drug interactions in medical research or purchasing patterns in online shopping.

[4]Augmented Reality is use the use of technology to enhance the real world using technologies such as digital overlays to provide contextual information, or insert artificial digitally generated 3D objects into a real-time landscape.

Augmented Human (AH)[5] communications for collaboration become well established. Given the existence of deceptive technologies which will be discussed in the next section, and the pace of development, it will become essential to make use of detective technologies in any sensitive remote communications, whether video, audio, or holographic, to assist human participants in guaging authenticity [6]. This is already occurring at a basic level with e-mail communications, with many systems now raising notifications that a recipient is external, or that an e-mail is suspicious. Given the profitability of deception-based cybercrime exploiting insider threat it is not likely that bad actors will ignore the potential to use new technologies to support their attacks, and continue to manipulate insiders [25].

1.3 The Outside Insider

Insider threat does not always relate directly to human employees. The key requirement is that an attacker somehow gains access as a trusted individual, and there are many examples of how this can occur without ever directly engaging with an organisation. One of the more recently named techniques which illustrates this effectively is warshipping [13]. Named as a portmanteau of shipping and wardriving, which involves individuals travelling around searching for wireless network signals in order to gain access, warshipping exploits the delivery network of companies. Even where a company may take great precautions to prevent their wireless network from being publicly accessible, warshipping bypasses all of the physical protections and provides an attacker with an evil access point within the organisation's premises.

Conceptually warshipping is a very simple attack: the attacker will build a small, battery powered computer with both wireless and mobile connectivity options. The mobile connectivity is used to remotely control the device, to then compromise any wireless networks detected when it has reached its destination. Location is monitored using mobile connectivity to determine which cell the device resides in, or GPS. Once it has arrived the device will continue working for days or weeks depending on the battery available. Known exploits against wireless networks are used to gain access, allowing the attacker to carry out Man-in-the-Middle (MitM) attacks or other compromises.

Naturally warshipping, while recently named, is not a new technique though its prevalence has not been studied in depth. Of course, often there is not even a need to go to the lengths of paying for shipping, as often an outsider can pose as a visitor to an organisation and gain access to public areas. This gives easy access to any organisational wireless networks in the building, and in some cases direct network access where port management is not correctly implemented. Even where port management is properly managed an outsider who can gain access to secure

[5]Using wearable or implanted technology to enhance the capabilities of a human.

areas, whether by pretexting as an employee, visitor, courier, supplier, or another role, can make use of a wide variety of tools to manifest an insider threat with only limited contact with employees. A well known attack which exploits the insider, with minimal risk to an attacker, is simply dropping USB drives in an area near an organisation's premises, ideally marked with some interesting and relevant label to exploit the curiousity of employees. Once inserted the device carries a payload of malware which allows the attacker to act in the role of the manipulated insider on the networks and systems.

For better-resourced attackers rather than travelling in person, or paying for shipping, drones may be used to the same effect. Some of these are autonomous, compromising wireless networks and building networks of insider bots on the fly [26].

Smart devices often used with voice activated smart home systems, or with smart phone control, such as lights, kettles, aquarium control systems, pet feeders, and similar convenient items. These are often provided with trusted access to home and office networks, and the lack of any comprehensive security standard governing their design and manufacture makes them ideal entry or manipulation points for an attacker. Since all of these devices will hold, at least, the access keys to a network the insider threat is clear. Of course, the threat profile of some devices is significantly higher than others, especially when safety features are implemented in manipulable code rather than mechanical. It is not a great stretch to think of a kettle with temperature regulation governed by a thermostat controlled in software, to allow for custom temperature adjustments for the perfect cup of tea. Unfortunately such a device, without a mechanically-implemented safety cut-out to prevent overheating could easily be the cause of a fire, moving the insider threat from a threat against information to one in the physical realm.

Plenty of high-profile examples covering why these devices should not be trusted, and should at best be isolated to dedicated, quarantined networks are available. One of the most popular tales is of a casino, which will remain unnamed, which serves as an example of insider threat both through a smart device, and through the actions of a customer or visitor as an insider. Originally discovered by Darktrace, the threat manifested when a visitor to the casino connected to the fish tank control system. The control system was connected to the internet in order to regulate temperature and feed the fish when needed, but was also connected to the casino's internal network. Once on the network the attacker's located the high-roller database, and exfiltrated it through the thermostat to cloud storage [30]. The attack on the casino also highlights the importance of applying least privilege principles to devices as well as individuals: there is no need for a thermostat to be able to transmit data to cloud storage, or even run a fully functional computing system. The scenario could easily have been prevented by limiting the device to only accepting and sending temperature data, with anything else being dropped automatically.

For a more literal view of the insider threat, implantation of medical devices has saved many lives. As technology advanced, developments have meant that controlling, adjusting, or updating medical devices no longer requires invasive surgery as it would before, reducing the immediate risk to patients. Unfortunately this does raise other risks instead, as demonstrated by [27]. Butts and Rios discovered vulnerabilities in various medical devices, with the most dramatic being the control system for a pacemaker. As the AH becomes more common, whether with life-saving medical devices, enhancements, or simply convenience devices such as implanted chips it is vital that this area of insider threat, sitting inside the target's own body, receive the attention it deserves. The ease and simplicity of other attack vectors, along with the morality involved in compromising a medical device, may help to keep the threat reduced, but it is not hard to envision a form of disturbing ransomware deliberately targeting devices if they are not adequately protection.

Until the ongoing rise in smart devices is matched by increased security awareness and understanding in manufacturers, this is an area of insider threat that will continue to grow.

Combining two of these vectors, the SkyNET botmaster system and similar networks can be used in order to find and compromise smart IoT devices in addition to wireless networks. The first widely popularised system making use of this strategy was developed by the security firm Praetorian in 2015, and is continually developed to this day. While their demonstration does not attempt to compromise discovered smart devices, it provides an effective picture of the attack surface for those whose attempts are more malicious and was designed to highlight this potential. A similar service, without the use of scouting drones, is provided by the search engine Shodan and searches for smart devices directly attached or routed to the public internet. All of these devices are a potential vector for insider threat to manifest, as they are almost always considered trusted participants in their host networks.

1.3.1 Insecurity by Design

Issues with the security of smart or other computing devices are not always due to failings in design. Deliberately introduced vulnerabilities, or backdoors, are a major concern for organisations from the scale of small businesses to national governments, and even individuals. Whether introduced for the convenience of developers and not removed after the development process is complete, or maliciously inserted into the design process by a hostile actor, the threat posed by backdoors in hardware or software is increasing.

Software backdoors can be treated after discovery in the same way as any other vulnerability, with patching or software updates. Unfortunately many backdoors have a hardware component that makes their removal range from challenging to ouright impossible. One of the earliest well-documented cases occured in 2008 with the leak of an Federal Bureau of Investigation (FBI) presentation detailing the investigation into counterfeit Cisco components which had been installed across the United States, including in sensitive military and government sites. While Cisco claimed the counterfeit equipment was for profit reasons rather than corporate or governmental espionage, the FBI considered the threat of backdoors being included severe enough to describe it as a 'critical infrastructure threat'. Worsening the situation were claims that Cisco's approved vendors had also purchased and resold counterfeit equipment.

The potential of a compromised trusted supply chain leading to the installation of malicious equipment in key systems is not taken lightly, and is not simple to resolve. Full inspection of every piece of equipment, down to circuit board and silicon wafer level, would be required to remove any trust requirements given the sheer complexity of modern systems. Simultaneously that level of deep inspection is not merely impractical, but impossible with current resources and tools. Trust is required for any such chain to function effectively, and as with individuals the more controls and checks are applied the more impact there will be on productivity. Indeed, as with employees it is entirely possible that excessive arduous checks will simply lead a supplier to decide to cut their losses and walk away. When this is a matter of specialist hardware, the receiving party may simply not have an alternative but to allow the checks to be bypassed.

The Department of Defense (DoD) takes the situation seriously [8], as do many other national governments and organisations. Unfortunately given the complexity of systems, as the report states, it is extremely challenging to very some of the claims made about supplier compromise, or malicious suppliers, and even more so to separate the claims and subsequent actions from other political or economic motivations. When claims of security risks from supposedly trusted sources have been shown to serve an ulterior motive, even our own sources of security intelligence can become an incidental or malicious insider threat. Take as an example a claim that a major equipment manufacturer is in some way compromised. If this claim is made by a significant and trusted authority, such as an intelligence agency, it is likely to cause equipment to be removed, supply agreements to be broken, and similar. If this claim was made not on the basis of genuine security concerns, but for national economic interests, then at best it is a waste of effort and funds, while at worst it could lead to further threats arising from the use of substandard systems.

1.3.2 Inside Voice

It is not always necessary for an insider threat to be overtly hostile as in the case of warshipping. More and more homes are now adding some automated capability through the use of voice activated or remotely controlled smart home devices. While there is no denying the convenience of these devices, they do have a substantial and often overloooked threat profile. The instances of children, or in one notable case a pet parrot [4], purchasing substantial amounts of food, treats, or other undesired items (at least undesirable to the person paying for the purchase) are well recorded.

In 2014 various owners of Microsoft's, at the time, new XBox One console began complaining of the devices activating in response to a new television advert which began with the words "XBox On". At the time those affected viewed the issue as a minor nuisance, with the greatest damage being from those who would find their XBox activation would switch the inputs of their televisions [17]. In April 2017 Burger King took things a step further, by using their television advert to deliberately query the Google Home smart device with the question "OK, Google, what is the Whopper burger?". The advert, and resulting irritation of viewers, led to a brief edit war over the Whopper wikipedia entry before Google stepped in to disable the voice query [31]. A less deliberate instance occurred in early 2018, when an advert for Purina cat food contained the words "Alexa, reorder Purina cat food". While that instance did not result in a confirmed order, it followed a spate of similar instances which did and caused vendors of smart home devices to add layers of confirmation before orders would complete, along with blocking their own adverts (and any others they were notified of) from triggering the voice recognition [1]. Given the sheer number of voice activated devices now carried and trusted by users, the potential for insider threat through voice recognition is obvious.

More concerning is the fact that much of the voice recognition processing for these devices is carried out on cloud-based systems, and it has come to light that to assist these machine learning models human contractors have been provided with audio recordings of conversations believed to be private in order to improve recognition success [7]. Any voice activated system should now be considered suspect and a potential insider threat, no matter how much convenience they add. Voice recordings and transcripts can be requested under various privacy legislation, and in previous cases a request has resulted in the wrong individual's data being sent, though it can be hoped that this was a one-off case and the lessons learned were quickly implemented [2]. The misinterpretation of a conversation as voice commands is not so easily addressed, and even has a higher potential for damage since it may result in private conversations being shared with friends and family [18].

1.3.3 When Security Is a Threat

Even where devices and infrastructure are secured, much of the time they are now secured by various third party managed services. Alternatively many companies, recognising that the expertise required to provide cyber security capabilities is a specialist area, will delegate responsibility for their security to third party providers who take on not only the administration and configuration, but the hosting of their tools.

This approach has definite upsides as economies of scale kick in, particularly with shared providers. The problem comes in when those shared services also raise potential threats themselves. A Managed Security Service Provider (MSSP), or even Managed Service Provider (MSP), can certainly provide security expertise which may not be available in house, but they simultaneously increase the threat landscape for their customers. By becoming a single trusted provider to multiple customers they make themselves a valuable target for attackers or any form of malicious insider, the trusted access to customer systems providing a single point of entry for multiple further targets. Given the size of many MSPs and MSSPs organisations their structure and infrastructure have significantly increased complexity, which opens up many more opportunities for anyone with both access and malice in mind.

Due to this companies need to consider carefully what access a third party provider should have to their systems, and again apply principles of least privilege and separation of duties. Due diligence to ensure that a provider is secure themselves is also essential. A threat actor targeting an MSP is likely to be much more sophisticated and better resourced than one targeting a single small to medium enterprise, and while security by obscurity is very negatively perceived in modern times, a lower profile against motivated and well-resourced attackers is definitely of benefit.

1.4 The Future Insider

As technology advances, our systems grow in complexity and become ever more interconnected. This interconnection between technologies and people, and the increasing rise of AH through wearable and companion devices blurring the line between users and participants in networks, means that the threat landscape is constantly growing at the same time as more and more participants and devices are becoming insiders. A city-wide public wireless network, as an example, means that every person within that city with any wireless device becomes a potential insider and target. To add to the landscape we must also consider non-human network participants such as more advanced, autonomous smart devices, and even the overarching machine learning programs that govern these systems as potentials for insider threat.

On top of these, the growth of AR and communications technologies to permit long distance and asynchronous communications, whether through telephony, email, or collaboration tools, means that the social network of insiders continues to expand alongside the technological underpinnings of society. When physical presence is no longer necessary, establishing the authenticity of a participant becomes exponentially more challenging. Even with technologies to allow for visual communications, emerging systems make it harder to establish whether the parties at the other end of a long distance connection should be treated as trusted insiders, or quarantined from a system. Finally, such long distance communications remove one of the most effective controls on insider threat – face to face communications allow for a much more effective assessment of an individual's behaviour and attitudes in most cases than a remote call, and many insider threats have been prevented by the simple expedient of colleagues in a workplace noticing that someone is behaving oddly, unusually angry, or seems tired and has been making mistakes due to personal circumstances.

1.4.1 Smart Transport and Smart Cities

Deserving of separate consideration to simple smart appliances, as almost the ur-example of a futuristic cyber security threat in both media and research, are autonomous vehicles [29]. A popular scene in modern films looking towards the future is a car chase (sometimes with flying cars) where between one to all of the vehicles involved are compromised by an attacker, acting as hostile agents in a trusted network. While such scenes once seemed outlandish, and are usually presented as a hostile outside attacker, the use of trusted, open networks to carry out this form of attack means we can consider it as an insider threat.

Another example, older even than the autonomous car compromise, is the attacker who targets a smart city's networked traffic management system to similar effect. Vehicles under the control of humans are attacked through causing traffic control systems to show contradictory signals. While again this is arguably an attack by a hostile outsider, the threat itself manifests through the misbehaviour of a trusted network system. Such an attack would be ineffective if the system were not trusted and relied upon by all participants.

At this point even loose definition of insider with which we began this chapter becomes a challenge. Is a car which is misled through bad radar information or control codes a threat to the driver? Is a driver through inattention and poorly-judged manual override the threat to the car and autonomous road systems? Is a pedestrian, or third party driver a potential insider threat to an autonomous road network in a smart city, or vice versa? In these scenarios, are one, both, or neither insider threats or external threats? There is no clear answer to these questions, and any answer given depends entirely upon individual perspective and the context of the discussion.

For simplicity we will say for now that in a smart city, autonomous travel system, or similar shared publicly accessible network, all users and authorised participants can be considered as insiders, and potentially give rise to insider threat. The open nature of such systems increases the number of insiders and simultaneously decreases the amount of implicit trust they can be safely granted, since unlike employees or contractors there are fewer options for instilling an effective security culture, and the greater variety of participants leads to a much higher likelihood of undetected disgruntled or even malicious insiders.

1.4.2 The Unpredictable

Big data, machine learning, and digital evolution for problem solving are popular approaches as our technological landscape becomes too large and complex for individual, or teams of, humans to comprehend in any realistic timescale. Machine learning is ever more sought after as an approach to analyse events in pursuit of anomalies that might indicate security incidents in large computing networks, in addition to being applied to much more diverse problems such as landing planes, planning road layouts, managing traffic, profiling customers for targeted advertising, monitoring human users for abnormal behaviour, carrying out medical analyses, and many other purposes. As these models and methods are developed and applied, the outputs are given a huge amount of innate trust, in some instances more even than the analysis that might instead be produced by a team of humans.

As machine learning models become trusted participants and sources of intelligence, we must be on the watch for insider threat. This is not because our current machine learning models are likely to become self-aware and take over the world, but because they are unpredictable and, in many cases, comprehensible only after mistakes are realised. In many cases these systems are tested in isolated, simulated worlds, set with a certain collection of goals before being deployed to any real-world function. A few of the narratives described by [19] indicate the dangers of implementing any system which has not been fully understood by its creators – when we are discovering exploitable vulnerabilities in light bulbs, there is a very valid cause for concern in implementing algorithms which are not comprehensible to their creators and the risks posed must be considered carefully.

It is inevitable that, as these systems are implemented, a proportion will have been set up with incomplete initial goals, leading to such situations as the algorithm to land a plane instead bringing it down with enough force to overload the sensors, fortunately in simulation rather than practice. While this is an extreme example, where consequences are less dramatic there is a lower chance of detection during development and testing, meaning that decisions will be taken based on incorrect analysis. With the complex environment we experience today it is inevitable that machine learning algorithms utilising big data collection will be implemented in order to solve challenges, and we must be ready to deal with this new vector for insider threat to manifest.

This does not account for people misusing the systems implemented using greater information to their own benefit. Multiplayer online gaming illustrates that participants in a system, in pursuit of competition, will learn to exploit flaws in the system to their own advantage and the detriment of others. Where such systems are implemented in a wider environment in order to govern rules of behaviour it is likely that a number with advanced knowledge of the system may do exactly the same, learning to feed carefully determined misinformation to the system in order to benefit themselves at the expense of others. Indeed, this happens currently in many systems of bureaucracy where fraud is a serious concern. The attempted impeachment in 2018 of the Supreme Court of Appeals of West Virginia serves to highlight the effectiveness of privileged actors abusing their positions, as well as the difficulties of detecting such behaviour and mitigating the damages after the fact.

1.4.3 Consequences of Mitigation

As with most insider threats, the most effective mitigations for these large scale environments with a broad population of insiders can be reduced down to two options. One is to reduce the amount of trust in the insider, applying greater and greater automation to these systems with fewer points of interaction and reduced control for participants. This reduces the potential attack surface for a disgruntled insider, but does have consequences in terms of human autonomy in such environment by reducing options and applying ever-stricter controls in pursuit of a secure environment. Alternative there is the option of attempting to inculcate a security-focused and highly responsible culture using the autonomous environments themselves. The measurement of trust and scaling of potential interactions based on rule-following is undoubtedly a security measure, but the human cost involved in terms of privacy loss, reduction of rights, and potential for abuse is not something to be dismissed lightly.

China has been implementing and developing one such cultural regulation system since 2014, and the ethical debate across the rest of the world is no less heated years later than when the system was first proposed [5]. Concerns can also be raised over the increased importance such a system places on digital twins over the behaviour of people when not observed by the system, and the potential for calculated attacks against the behavioural profile of participating citizens through a co-ordinated misinformation campaign are of serious concern. Considering the real life impact which can occur as a result of co-ordinated social media shaming or harassment campaigns already, the possibility for such attacks leading consequences enforced directly by a social credit score supported by governments is truly terrifying. All such incidents are a form of insider threat, as those targeted by and co-ordinating the campaign must be authorised participants in the system.

Equally the disruption that could be caused to a state relying on a governing social capital system, tied into all automated systems and with oversight on booking and other commercial systems, by a single bad actor with high level privileges is almost staggering. Whether against an individual, or against a large group, an administrator would potentially be capable of bringing an entire society almost to a halt by maliciously adjusting the privilege thresholds for citizens. In such a society, those who directly control the system and its design have the greatest potential to cause damage to the society as a whole.

1.4.4 Impostor

Currently spoofing and impersonation is largely limited to changing source email addresses, display names, or phone numbers. SMS spoofing in particular is becoming prevalent as banks and other service providers switch towards using One Time Passwords (OTPs)) sent via mobile networks. An attacker using a spoofed number can then impersonate the service provider, and often will ask for the OTP sent to be forwarded on. Similarly a phone call from a spoofed number can lead to the same result, circumventing the security normally provided by Multi-Factor Authentication (MFA). All spoofing and impersonation attack techniques can fall under insider threat, as they attempt to exploit the implicit trust placed in insiders.

The use of deceptive technologies such as Deepfakes [11] in order to spread misinformation is already well documented. As the technology develops and is combined with face tracking and recreation systems [33] the potential for use as a tool for malicious insider threats, or indeed carrying out an insider threat without requiring any participation on the part of an insider, becomes clear. Given the data and behavioural profiling that is considered a standard, it is entirely possible to conceive of a dedicated, motivated attacker constructing the profile of an insider based on personal data leakage in order to carry out an attack. While this may seem outlandish, it is already one of the most effective techniques in the social engineering arsenal, with impersonation being one of the most successful an profitable mechanisms to extract funds from organisations [21] with the use of nothing more sophisticated than changing e-mail display names.

Arguments have been made that an impostor is not truly an insider threat. To counter this it is clear they are acting as an insider and looking to exploit the implicit trust involved to their own ends. The only additional precautions to be taken against an impostor threat, beyond those that should be taken against an insider threat, are methods to verify identity and to put insiders on guard against potential impostors. Otherwise, all other controls placed for a true insider threat are equally effective against an impostor.

It is obvious that this particular insider threat is going to increase with time. Currently many uses are for entertainment, though there are already instances where Deepfake systems have been used to generake fake pornography of both celebrities and individuals, whether for blackmail, exploitation, or revenge purposes. Currently it is usually possible to establish the authenticity of such videos, but this technology is still in the earliest stages and the possibility of a world where we cannot reliably trust any video recording or communication as being authentic is a genuine concern. New authenticity mechanisms will be required, whether they will involve a resurgence in physical meetings or other security mechanisms such as assuring secure connections through the use of security passphrases which are provided by some secure communications applications.

Several sites already exist allowing reading aloud of any text in the voice of a particular individual, and faked videos of personalities such as Barack Obama have circulated since 2017 [10]. While it may take some time before this deceptive technology expands to include holographic communications through AR, or to insert false avatars into Virtual Reality (VR) or even classical video conferencing environments, the tools are all available with voice recognition providing a realtime input mechanism, facial tracking allowing for ever more realistic imitation of an artificial face in terms of emotion, lip-syncing, and general expression, and Deepfakes applied to create the artificial avatar and replace the impostor's voice.

We are likely heading towards a future where the current cyber security arms race between defenders and attackers will expand along yet another front, with authenticity of communications previously trusted implictly outside of highly sensitive environments become an ever-greater concern. While research is rapidly taking place into various methods to defeat deceptive technologies, many of the answers depend on particular artifacts which are a result of the immaturity of the technology and will be rapidly addressed as its usefulness as a tool of attack becomes apparent [22, 23, 32].

As an aside, it is interesting to note that these sophisticated attacks on authenticity do not apply only to information and communications, but have crossed into the physical world. Especially in the art world, where machine learning algorithms are often used to establish authenticity, the threat posed by algorithms which can produce 'replicas' which appear authentic to an artist's style is clear [12]. With more direct implications on general security, the potential for counterfeit documentation of all kinds can only cause trust to be lost in physical copies, requiring more dependence on computational authenticity mechanisms such as cryptographic signing authorities as are now provided by the Estonian government [9].

1.5 Overview

We have seen that the insider threat is a broad subject, with many different attack vectors ranging from simple innocent negligence to a well-resourced and motivated attacker posing as an insider. As our technological systems develop opportunities for these threats to manifest continue to increase and develop, providing new vectors for attack which leverage brand new technologies and quickly adopting new capabilities to serve their goals.

New communications channels will require new methods of working to ensure authenticity, where we have not achieved this even with well established channels such as email. Ever more complex systems give rise to more potential for damage to occur through both negligence and malice. The increase in technologies such as AR and particularly the rise of AH, along with smart devices and increasing dependencies on interconnected systems mean that threats from insiders, including participants in systems, can manifest in new and unpredictable ways.

Whether the insider threat is a true insider, an imposter, or the manipulation or exploitation of an insider to achieve an attacker's ends, certain founding principles can help to mitigate the threat. Building systems, where possible, or re-engineering systems to apply least privilege principles and enforce separation of duties will help to prevent insider threat from manifesting for an organisation, while instilling a culture of security awareness and responsibility where insiders can be trusted will reduce the possibilities for manipulated insiders to cause damage to an organisation.

Regardless of the measures taken to prevent insider threat from manifesting, the nature of people and the requirements for trust in order to achieve anything as any organisation means that the threat and its accompanying risks will always be present. Properly ascertaining and addressing these risks can help to minimise them, but they can never be completely eradicated from any organisation and constant monitoring is required to ensure that when they arise they can be dealt with in a timely and effective manner.

References

1. Ad That Fooled Amazon Device Cleared (2018) In: BBC news. Business. https://www.bbc.com/news/business-43044693 (visited on 27 Aug 2019)
2. Alexa User Accesses Stranger's Chats (2018) In: BBC news. Technology. https://www.bbc.com/news/technology-46637427 (visited on 27 Aug 2019)

3. Bada M, Sasse AM, Nurse JRC (2019) Cyber security awareness campaigns: why do they fail to change behaviour? arXiv: 1901.02672 [cs]. http://arxiv.org/abs/1901.02672 (visited on 12 Aug 2019)
4. BBC (2018) Naughty parrot keeps using Alexa to buy things online – CBBC newsround. In: Newsround. https://www.bbc.co.uk/newsround/46566019 (visited on 20 Aug 2019)
5. Chen Y, Cheung ASY (2017) The transparent self under big data profiling: privacy and chinese legislation on the social credit system. SSRN scholarly paper ID 2992537. Social Science Research Network, Rochester. https://papers.ssrn.com/abstract=2992537 (visited on 20 Aug 2019)
6. Chesney R, Citron DK (2018) Deep fakes: a looming challenge for privacy, democracy, and national security. SSRN scholarly paper ID 3213954. Social Science Research Network, Rochester. https://papers.ssrn.com/abstract=3213954 (visited on 02 Sept 2019)
7. Day M, Turner G, Drozdiak N (2019) Amazon workers are listening to what you tell Alexa. In: Bloomberg.com. https://www.bloomberg.com/news/articles/2019-04-10/is-anyone-listening-to-you-on-alexa-a-global-team-reviews-audio (visited on 27 Aug 2019)
8. Defense Science Board (2017) DSB task force on cyber supply chain. https://www.acq.osd.mil/dsb/reports/2010s/DSBCyberSupplyChainExecutiveSummary-Distribution_A.pdf (visited on 31 Aug 2019)
9. E-identity (2019) https://e-estonia.com/solutions/e-identity/ (visited on 02 Sept 2019)
10. Fake Obama Created Using AI Video Tool (2019) https://www.bbc.com/news/av/technology-40598465/fake-obama-created-using-ai-tool-to-make-phoney-speeches (visited on 02 Sept 2019)
11. Fikse TD (2018) Imagining deceptive deepfakes: an ethnographic exploration of fake videos. Master's thesis, p 58
12. Floridi L (2018) Artificial intelligence, deepfakes and a future of ectypes. Philos Technol 31(3):317–321. ISSN:2210-5441. https://doi.org/10.1007/s13347-018-0325-3 (visited on 02 Sept 2019)
13. Henderson C (2019) Package delivery! Cybercriminals at your doorstep. https://securityintelligence.com/posts/package-delivery-cybercriminals-at-your-doorstep/ (visited on 12 Aug 2019)
14. Homoliak I et al (2019) Insight into insiders and IT: a survey of insider threat taxonomies, analysis, modeling, and countermeasures. ACM Comput Surv 52(2):1–40. ISSN:03600300. https://doi.org/10.1145/3303771. http://dl.acm.org/citation.cfm?doid=3320149.3303771 (visited on 12 Aug 2019)
15. ICO (2018) ICO data security trends Q4 2017-18. https://ico.org.uk/media/action-weve-taken/reports/2014675/data-security-trends-pdf.pdf (visited on 29 July 2019)
16. ISACA (2019) State of cyber 2019. https://view.ceros.com/isaca/state-of-cyber-2019 (visited on 29 July 2019)
17. Kelion L (2014) Xbox one ad switches on consoles. In: BBC news. Technology. https://www.bbc.com/news/technology-27827545 (visited on 27 Aug 2019)
18. Lee D (2018) Amazon Alexa heard and sent private chat. In: BBC news. Technology. https://www.bbc.com/news/technology-44248122 (visited on 27 Aug 2019)
19. Lehman J et al (2018) The surprising creativity of digital evolution: a collection of anecdotes from the evolutionary computation and artificial life research communities. In: arXiv: 1803.03453 [cs]. http://arxiv.org/abs/1803.03453 (visited on 21 Aug 2019)
20. Mallmann GL, Gastaud Maçada AC, Oliveira M (2018) The influence of shadow IT usage on knowledge sharing: an exploratory study with IT users. Bus Inf Rev 35(1):17–28. ISSN:0266-3821. https://doi.org/10.1177/0266382118760143 (visited on 01 Aug 2019)
21. Mansfield-Devine S (2016) The imitation game: how business email compromise scams are robbing organisations. Comput Fraud Secur 2016(11):5–10. ISSN:1361-3723. https://doi.org/10.1016/S1361-3723(16)30089-6. http://www.sciencedirect.com/science/article/pii/S1361372316300896 (visited on 12 Aug 2019)

22. Maras M-H, Alexandrou A (2019) Determining authenticity of video evidence in the age of artificial intelligence and in the wake of deepfake videos. Int J Evid Proof 23(3):255–262. ISSN:1365-7127. https://doi.org/10.1177/1365712718807226 (visited on 22 July 2019)
23. Matern F, Riess C, Stamminger M (2019) Exploiting visual artifacts to 17 expose deepfakes and face manipulations. In: 2019 IEEE winter applications of computer vision workshops (WACVW), pp 83–92. https://doi.org/10.1109/WACVW.2019.00020
24. The PEOPLE, Plaintiff and Respondent, v. Terry CHILDS, Defendant and Appellant (2013) Court of Appeal, First District, Division 4, California
25. Ponemon Institute and Accenture (2017) 2017 cost of cybercrime study. https://www.accenture.com/t20170926t072837z_w_/us-en/_acnmedia/pdf-61/accenture-2017-costcybercrimestudy.pdf (visited on 12 Aug 2019)
26. Reed T, Geis J, Dietrich S (2011) SkyNET: a 3G-enabled mobile attack drone and stealth botmaster. In WOOT, pp 28–36
27. Rios B, Butts J (2018) Black Hat USA 2018. https://www.blackhat.com/us-18/briefings/schedule/#understanding-and-exploiting-implanted-medical-devices-11733 (visited on 27 Aug 2019)
28. Sanzgiri A, Dasgupta D (2016) Classification of insider threat detection techniques. In: Proceedings of the 11th annual cyber and information security research conference on – CISRC'16. The 11th annual cyber and information security research conference. ACM Press, Oak Ridge, pp 1–4. ISBN:978-1-4503-3752-6. https://doi.org/10.1145/2897795.2897799. http://dl.acm.org/citation.cfm?doid=2897795.2897799 (visited on 12 Aug 2019)
29. Thing VLL, Wu J (2016) Autonomous vehicle security: a taxonomy of attacks and defences. In: 2016 IEEE international conference on internet of things (iThings) and IEEE green computing and communications (GreenCom) and IEEE cyber, physical and social computing (CP-SCom) and IEEE smart data (SmartData), pp 164–170. https://doi.org/10.1109/iThings-GreenCom-CPSCom-SmartData.2016.52
30. Vamosi R (2018) Casino's aquarium leaks high rollers' personal data, 17 Apr. https://blogs.synopsys.com/from-silicon-to-software/2018/04/17/casinos-aquarium-leaks-high-rollers-personal-data/ (visited on 27 Aug 2019)
31. Wakefield J (2017) Burger king ad sabotaged on Wikipedia. In: BBC news. Technology. https://www.bbc.com/news/technology-39589013 (visited on 27 Aug 2019)
32. Yang X, Li Y, Lyu S (2019) Exposing deep fakes using inconsistent head poses. In: ICASSP 2019 – 2019 IEEE international conference on acoustics, speech and signal processing (ICASSP), pp 8261–8265. https://doi.org/10.1109/ICASSP.2019.8683164
33. Zollhöfer M et al (2018) State of the art on monocular 3D face reconstruction, tracking, and applications. Comput Graphics Forum 37(2):523–550. ISSN:1467-8659. https://onlinelibrary.wiley.com/doi/abs/10.1111/cgf.13382 (visited on 12 Aug 2019)

MIX
Papier aus verantwortungsvollen Quellen
Paper from responsible sources
FSC® C105338

If you have any concerns about our products,
you can contact us on
ProductSafety@springernature.com

In case Publisher is established outside the EU,
the EU authorized representative is:
Springer Nature Customer Service Center GmbH
Europaplatz 3, 69115 Heidelberg, Germany

Printed by Libri Plureos GmbH
in Hamburg, Germany